STUDENT SOLUTIONS MANUAL

VOLUME 2: CHAPTERS 21-44

SEARS & ZEMANSKY'S

UNIVERSITY PHYSICS

12TH EDITION

YOUNG AND FREEDMAN

A. LEWIS FORD

TEXAS A & M UNIVERSITY

PEARSON

Addison Wesley

San Francisco   Boston   New York
Cape Town   Hong Kong   London   Madrid   Mexico City
Montreal   Munich   Paris   Singapore   Sydney   Tokyo   Toronto

Publisher:	Jim Smith
Executive Editor:	Nancy Whilton
Editorial Manager:	Laura Kenney
Associate Editor:	Chandrika Madhavan
Executive Marketing Manager:	Scott Dustan
Managing Editor:	Corinne Benson
Production Supervisor:	Nancy Tabor
Production Service:	WestWords PMG
Illustrations:	Rolin Graphics
Cover Design:	Yvo Riezebos Design and Seventeenth Street Design
Cover and Text Printer:	Bind-Rite Graphics
Cover Image:	Jean-Philippe Arles/Reuters/Corbis

ISBN-13: 978-0-321-50038-0
ISBN-10: 0-321-50038-5

1 2 3 4 5 6 7 8 9 10— BR —10 09 08 07

www.aw-bc.com

Contents

PREFACE

This *Student Solutions Manual,* Volumes 2 and 3, contains detailed solutions for approximately one third of the Exercises and Problems in Chapters 21 through 44 of the Twelfth Edition of *University Physics* by Hugh Young and Roger Freedman. The Exercises and Problems included in this manual are selected solely from the odd-numbered Exercises and Problems in the text (for which the answers are tabulated in the back of the textbook). The Exercises and Problems included were not selected at random but rather were carefully chosen to include at least one representative example of each problem type. The remaining Exercises and Problems, for which solutions are not given here, constitute an ample set of problems for you to tackle on your own. In addition, there are the Challenge Problems in the text for which no solutions are given here.

This manual greatly expands the set of worked-out examples that accompanies the presentation of physics laws and concepts in the text. This manual was written to provide you with models to follow in working physics problems. The problems are worked out in the manner and style in which you should carry out your own problem solutions.

The author will gratefully receive comments as to style, points of physics, errors, or anything else relating to this manual. The Student Solutions Manual Volume 1 companion volume is also available from your college bookstore

A. Lewis Ford
Physics Department MS 4242
Texas A&M University
College Station, TX 77843
ford@physics.tamu.edu

21.1. **(a) IDENTIFY and SET UP:** Use the charge of one electron $(-1.602 \times 10^{-19} \text{ C})$ to find the number of electrons required to produce the net charge.

EXECUTE: The number of excess electrons needed to produce net charge q is

$$\frac{q}{-e} = \frac{-3.20 \times 10^{-9} \text{ C}}{-1.602 \times 10^{-19} \text{ C/electron}} = 2.00 \times 10^{10} \text{ electrons.}$$

(b) IDENTIFY and SET UP: Use the atomic mass of lead to find the number of lead atoms in $8.00 \times 10^{-3} \text{ kg}$ of lead. From this and the total number of excess electrons, find the number of excess electrons per lead atom.

EXECUTE: The atomic mass of lead is $207 \times 10^{-3} \text{ kg/mol,}$ so the number of moles in $8.00 \times 10^{-3} \text{ kg}$ is

$$n = \frac{m_{tot}}{M} = \frac{8.00 \times 10^{-3} \text{ kg}}{207 \times 10^{-3} \text{ kg/mol}} = 0.03865 \text{ mol.}$$ N_A (Avogadro's number) is the number of atoms in 1 mole, so the

number of lead atoms is $N = nN_A = (0.03865 \text{ mol})(6.022 \times 10^{23} \text{ atoms/mol}) = 2.328 \times 10^{22}$ atoms. The number of

excess electrons per lead atom is $\dfrac{2.00 \times 10^{10} \text{ electrons}}{2.328 \times 10^{22} \text{ atoms}} = 8.59 \times 10^{-13}.$

EVALUATE: Even this small net charge corresponds to a large number of excess electrons. But the number of atoms in the sphere is much larger still, so the number of excess electrons per lead atom is very small.

21.7. **IDENTIFY:** Apply Coulomb's law.

SET UP: Consider the force on one of the spheres.

(a) EXECUTE: $q_1 = q_2 = q$

$$F = \frac{1}{4\pi\epsilon_0} \frac{|q_1 q_2|}{r^2} = \frac{q^2}{4\pi\epsilon_0 r^2} \text{ so } q = r\sqrt{\frac{F}{(1/4\pi\epsilon_0)}} = 0.150 \text{ m} \sqrt{\frac{0.220 \text{ N}}{8.988 \times 10^9 \text{ N} \cdot \text{m}^2/\text{C}^2}} = 7.42 \times 10^{-7} \text{ C (on each)}$$

(b) $q_2 = 4q_1$

$$F = \frac{1}{4\pi\epsilon_0} \frac{|q_1 q_2|}{r^2} = \frac{4q_1^2}{4\pi\epsilon_0 r^2} \text{ so } q_1 = r\sqrt{\frac{F}{4(1/4\pi\epsilon_0)}} = \tfrac{1}{2} r \sqrt{\frac{F}{(1/4\pi\epsilon_0)}} = \tfrac{1}{2}(7.42 \times 10^{-7} \text{ C}) = 3.71 \times 10^{-7} \text{ C.}$$

And then $q_2 = 4q_1 = 1.48 \times 10^{-6} \text{ C.}$

EVALUATE: The force on one sphere is the same magnitude as the force on the other sphere, whether the sphere have equal charges or not.

21.9. **IDENTIFY:** Apply $F = ma$, with $F = k \dfrac{|q_1 q_2|}{r^2}$.

SET UP: $a = 25.0g = 245 \text{ m/s}^2$. An electron has charge $-e = -1.60 \times 10^{-19} \text{ C.}$

EXECUTE: $F = ma = (8.55 \times 10^{-3} \text{ kg})(245 \text{ m/s}^2) = 2.09 \text{ N}$. The spheres have equal charges q, so $F = k \dfrac{q^2}{r^2}$ and

$$|q| = r\sqrt{\frac{F}{k}} = (0.150 \text{ m})\sqrt{\frac{2.09 \text{ N}}{8.99 \times 10^9 \text{ N} \cdot \text{m}^2/\text{C}^2}} = 2.29 \times 10^{-6} \text{ C.} \quad N = \frac{|q|}{e} = \frac{2.29 \times 10^{-6} \text{ C}}{1.60 \times 10^{-19} \text{ C}} = 1.43 \times 10^{13} \text{ electrons.}$$ The

charges on the spheres have the same sign so the electrical force is repulsive and the spheres accelerate away from each other.

EVALUATE: As the spheres move apart the repulsive force they exert on each other decreases and their acceleration decreases.

21.13.   **IDENTIFY:**   Apply Coulomb's law. The two forces on q_3 must have equal magnitudes and opposite directions.

SET UP: Like charges repel and unlike charges attract.

EXECUTE: The force $\vec{F}_2$ that q_2 exerts on q_3 has magnitude $F_2 = k\dfrac{|q_2 q_3|}{r_2^2}$ and is in the $+x$ direction. $\vec{F}_1$ must be in the $-x$ direction, so q_1 must be positive. $F_1 = F_2$ gives $k\dfrac{|q_1||q_3|}{r_1^2} = k\dfrac{|q_2||q_3|}{r_2^2}$.

$$|q_1| = |q_2|\left(\frac{r_1}{r_2}\right)^2 = (3.00 \text{ nC})\left(\frac{2.00 \text{ cm}}{4.00 \text{ cm}}\right)^2 = 0.750 \text{ nC} .$$

EVALUATE: The result for the magnitude of q_1 doesn't depend on the magnitude of q_2 .

21.17.   **IDENTIFY** and **SET UP:**   Apply Coulomb's law to calculate the force exerted by q_2 and q_3 on q_1. Add these forces as vectors to get the net force. The target variable is the x-coordinate of q_3.

EXECUTE: $\vec{F}_2$ is in the x-direction.

$F_2 = k\dfrac{|q_1 q_2|}{r_{12}^2} = 3.37 \text{ N}$, so $F_{2x} = +3.37 \text{ N}$

$F_x = F_{2x} + F_{3x}$ and $F_x = -7.00 \text{ N}$

$F_{3x} = F_x - F_{2x} = -7.00 \text{ N} - 3.37 \text{ N} = -10.37 \text{ N}$

For F_{3x} to be negative, q_3 must be on the $-x$-axis.

$F_3 = k\dfrac{|q_1 q_3|}{x^2}$, so $|x| = \sqrt{\dfrac{k|q_1 q_3|}{F_3}} = 0.144 \text{ m}$, so $x = -0.144 \text{ m}$

EVALUATE: q_2 attracts q_1 in the $+x$-direction so q_3 must attract q_1 in the $-x$-direction, and q_3 is at negative x.

21.19.   **IDENTIFY:**   Apply Coulomb's law to calculate the force each of the two charges exerts on the third charge. Add these forces as vectors.

SET UP: The three charges are placed as shown in Figure 21.19a.

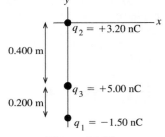

Figure 21.19a

EXECUTE: Like charges repel and unlike attract, so the free-body diagram for q_3 is as shown in Figure 21.19b.

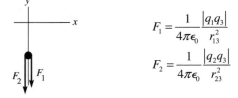

$$F_1 = \frac{1}{4\pi\epsilon_0}\frac{|q_1 q_3|}{r_{13}^2}$$

$$F_2 = \frac{1}{4\pi\epsilon_0}\frac{|q_2 q_3|}{r_{23}^2}$$

Figure 21.19b

$$F_1 = (8.988\times10^9 \text{ N}\cdot\text{m}^2/\text{C}^2)\frac{(1.50\times10^{-9} \text{ C})(5.00\times10^{-9} \text{ C})}{(0.200 \text{ m})^2} = 1.685\times10^{-6} \text{ N}$$

$$F_2 = (8.988\times10^9 \text{ N}\cdot\text{m}^2/\text{C}^2)\frac{(3.20\times10^{-9} \text{ C})(5.00\times10^{-9} \text{ C})}{(0.400 \text{ m})^2} = 8.988\times10^{-7} \text{ N}$$

The resultant force is $\vec{R} = \vec{F}_1 + \vec{F}_2$.

$R_x = 0.$

$R_y = F_1 + F_2 = 1.685\times10^{-6} \text{ N} + 8.988\times10^{-7} \text{ N} = 2.58\times10^{-6} \text{ N}.$

The resultant force has magnitude $2.58\times10^{-6} \text{ N}$ and is in the $-y$-direction.

EVALUATE: The force between q_1 and q_3 is attractive and the force between q_2 and q_3 is repulsive.

21.21. **IDENTIFY:**   Apply Coulomb's law to calculate each force on $-Q$.

SET UP: Let $\vec{F}_1$ be the force exerted by the charge at $y = a$ and let $\vec{F}_2$ be the force exerted by the charge at $y = -a$.

EXECUTE: **(a)** The two forces on $-Q$ are shown in Figure 21.21a. $\sin\theta = \dfrac{a}{(a^2 + x^2)^{1/2}}$ and $r = (a^2 + x^2)^{1/2}$ is the

distance between q and $-Q$ and between $-q$ and $-Q$.

(b) $F_x = F_{1x} + F_{2x} = 0$. $F_y = F_{1y} + F_{2y} = 2\dfrac{1}{4\pi\epsilon_0}\dfrac{qQ}{(a^2 + x^2)}\sin\theta = \dfrac{1}{4\pi\epsilon_0}\dfrac{2qQa}{(a^2 + x^2)^{3/2}}$.

(c) At $x = 0$, $F_y = \dfrac{1}{4\pi\epsilon_0}\dfrac{2qQ}{a^2}$, in the $+y$ direction.

(d) The graph of F_y versus x is given in Figure 21.21b.

EVALUATE: $F_x = 0$ for all values of x and $F_y > 0$ for all x.

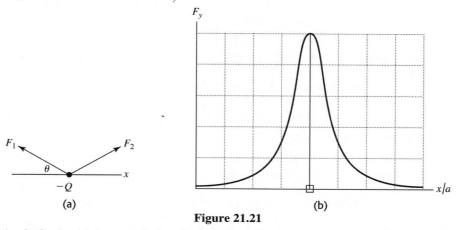

(a)   (b)

Figure 21.21

21.23. **IDENTIFY:**   Apply Coulomb's law to calculate the force exerted on one of the charges by each of the other three and then add these forces as vectors.

(a) SET UP: The charges are placed as shown in Figure 21.23a.

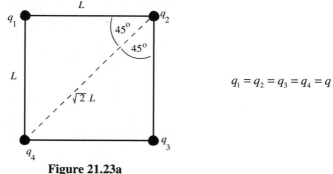

$q_1 = q_2 = q_3 = q_4 = q$

Figure 21.23a

Consider forces on q_4. The free-body diagram is given in Figure 21.23b. Take the y-axis to be parallel to the diagonal between q_2 and q_4 and let $+y$ be in the direction away from q_2. Then $\vec{F}_2$ is in the $+y$-direction.

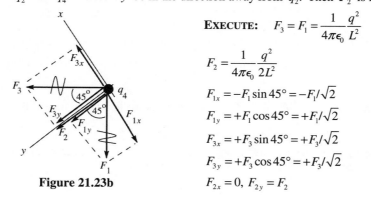

Figure 21.23b

EXECUTE: $F_3 = F_1 = \dfrac{1}{4\pi\epsilon_0}\dfrac{q^2}{L^2}$

$F_2 = \dfrac{1}{4\pi\epsilon_0}\dfrac{q^2}{2L^2}$

$F_{1x} = -F_1\sin 45° = -F_1/\sqrt{2}$

$F_{1y} = +F_1\cos 45° = +F_1/\sqrt{2}$

$F_{3x} = +F_3\sin 45° = +F_3/\sqrt{2}$

$F_{3y} = +F_3\cos 45° = +F_3/\sqrt{2}$

$F_{2x} = 0$, $F_{2y} = F_2$

(b) $R_x = F_{1x} + F_{2x} + F_{3x} = 0$

$R_y = F_{1y} + F_{2y} + F_{3y} = (2/\sqrt{2})\dfrac{1}{4\pi\epsilon_0}\dfrac{q^2}{L^2} + \dfrac{1}{4\pi\epsilon_0}\dfrac{q^2}{2L^2} = \dfrac{q^2}{8\pi\epsilon_0 L^2}(1+2\sqrt{2})$

$R = \dfrac{q^2}{8\pi\epsilon_0 L^2}(1+2\sqrt{2})$. Same for all four charges.

EVALUATE: In general the resultant force on one of the charges is directed away from the opposite corner. The forces are all repulsive since the charges are all the same. By symmetry the net force on one charge can have no component perpendicular to the diagonal of the square.

21.27.   **IDENTIFY:** The acceleration that stops the charge is produced by the force that the electric field exerts on it. Since the field and the acceleration are constant, we can use the standard kinematics formulas to find acceleration and time.

(a) SET UP: First use kinematics to find the proton's acceleration. $v_x = 0$ when it stops. Then find the electric field needed to cause this acceleration using the fact that $F = qE$.

EXECUTE: $v_x^2 = v_{0x}^2 + 2a_x(x - x_0)$. $0 = (4.50 \times 10^6 \text{ m/s})^2 + 2a(0.0320 \text{ m})$ and $a = 3.16 \times 10^{14} \text{ m/s}^2$. Now find the electric field, with $q = e$. $eE = ma$ and $E = ma/e = (1.67 \times 10^{-27} \text{ kg})(3.16 \times 10^{14} \text{ m/s}^2)/(1.60 \times 10^{-19} \text{ C}) = 3.30 \times 10^6 \text{ N/C}$, to the left.

(b) SET UP: Kinematics gives $v = v_0 + at$, and $v = 0$ when the electron stops, so $t = v_0/a$.

EXECUTE: $t = v_0/a = (4.50 \times 10^6 \text{ m/s})/(3.16 \times 10^{14} \text{ m/s}^2) = 1.42 \times 10^{-8} \text{ s} = 14.2 \text{ ns}$

(c) SET UP: In part (a) we saw that the electric field is proportional to m, so we can use the ratio of the electric fields. $E_e / E_p = m_e / m_p$ and $E_e = \left(m_e / m_p\right)E_p$.

EXECUTE: $E_e = [(9.11 \times 10^{-31} \text{ kg})/(1.67 \times 10^{-27} \text{ kg})](3.30 \times 10^6 \text{ N/C}) = 1.80 \times 10^3 \text{ N/C}$, to the right

EVALUATE: Even a modest electric field, such as the ones in this situation, can produce enormous accelerations for electrons and protons.

21.29.   **(a) IDENTIFY:** Eq. (21.4) relates the electric field, charge of the particle, and the force on the particle. If the particle is to remain stationary the net force on it must be zero.

SET UP: The free-body diagram for the particle is sketched in Figure 21.29. The weight is mg, downward. For the net force to be zero the force exerted by the electric field must be upward. The electric field is downward. Since the electric field and the electric force are in opposite directions the charge of the particle is negative.

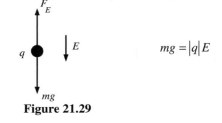

Figure 21.29

EXECUTE: $|q| = \dfrac{mg}{E} = \dfrac{(1.45 \times 10^{-3} \text{ kg})(9.80 \text{ m/s}^2)}{650 \text{ N/C}} = 2.19 \times 10^{-5} \text{ C}$ and $q = -21.9 \ \mu\text{C}$

(b) SET UP: The electrical force has magnitude $F_E = |q|E = eE$. The weight of a proton is $w = mg$. $F_E = w$ so $eE = mg$

EXECUTE: $E = \dfrac{mg}{e} = \dfrac{(1.673 \times 10^{-27} \text{ kg})(9.80 \text{ m/s}^2)}{1.602 \times 10^{-19} \text{ C}} = 1.02 \times 10^{-7} \text{ N/C}$.

This is a very small electric field.

EVALUATE: In both cases $|q|E = mg$ and $E = (m/|q|)g$. In part (b) the $m/|q|$ ratio is much smaller ($\sim 10^{-8}$) than in part (a) ($\sim 10^{-2}$) so E is much smaller in (b). For subatomic particles gravity can usually be ignored compared to electric forces.

21.31.   **IDENTIFY:** For a point charge, $E = k\dfrac{|q|}{r^2}$. The net field is the vector sum of the fields produced by each charge. A charge q in an electric field $\vec{E}$ experiences a force $\vec{F} = q\vec{E}$.

SET UP: The electric field of a negative charge is directed toward the charge. Point A is 0.100 m from q_2 and 0.150 m from q_1. Point B is 0.100 m from q_1 and 0.350 m from q_2.

EXECUTE: **(a)** The electric fields due to the charges at point A are shown in Figure 21.31a.

$$E_1 = k\frac{|q_1|}{r_{A1}^2} = (8.99 \times 10^9 \ \text{N} \cdot \text{m}^2/\text{C}^2)\frac{6.25 \times 10^{-9} \ \text{C}}{(0.150 \ \text{m})^2} = 2.50 \times 10^3 \ \text{N/C}$$

$$E_2 = k\frac{|q_2|}{r_{A2}^2} = (8.99 \times 10^9 \ \text{N} \cdot \text{m}^2/\text{C}^2)\frac{12.5 \times 10^{-9} \ \text{C}}{(0.100 \ \text{m})^2} = 1.124 \times 10^4 \ \text{N/C}$$

Since the two fields are in opposite directions, we subtract their magnitudes to find the net field.
$E = E_2 - E_1 = 8.74 \times 10^3 \ \text{N/C},$ to the right.

(b) The electric fields at points B are shown in Figure 21.31b.

$$E_1 = k\frac{|q_1|}{r_{B1}^2} = (8.99 \times 10^9 \ \text{N} \cdot \text{m}^2/\text{C}^2)\frac{6.25 \times 10^{-9} \ \text{C}}{(0.100 \ \text{m})^2} = 5.619 \times 10^3 \ \text{N/C}$$

$$E_2 = k\frac{|q_2|}{r_{B2}^2} = (8.99 \times 10^9 \ \text{N} \cdot \text{m}^2/\text{C}^2)\frac{12.5 \times 10^{-9} \ \text{C}}{(0.350 \ \text{m})^2} = 9.17 \times 10^2 \ \text{N/C}$$

Since the fields are in the same direction, we add their magnitudes to find the net field. $E = E_1 + E_2 = 6.54 \times 10^3 \ \text{N/C}$, to the right.

(c) At A, $E = 8.74 \times 10^3 \ \text{N/C}$, to the right. The force on a proton placed at this point would be
$F = qE = (1.60 \times 10^{-19} \ \text{C})(8.74 \times 10^3 \ \text{N/C}) = 1.40 \times 10^{-15} \ \text{N},$ to the right.

EVALUATE: A proton has positive charge so the force that an electric field exerts on it is in the same direction as the field.

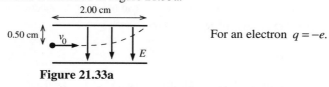

(a)   (b)

Figure 21.31

21.33.   **IDENTIFY:**   Eq. (21.3) gives the force on the particle in terms of its charge and the electric field between the plates. The force is constant and produces a constant acceleration. The motion is similar to projectile motion; use constant acceleration equations for the horizontal and vertical components of the motion.

(a) SET UP: The motion is sketched in Figure 21.33a.

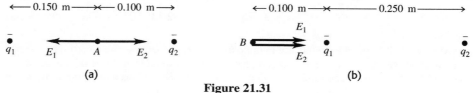

For an electron $q = -e$.

Figure 21.33a

$\vec{F} = q\vec{E}$ and q negative gives that $\vec{F}$ and $\vec{E}$ are in opposite directions, so $\vec{F}$ is upward. The free-body diagram for the electron is given in Figure 21.33b.

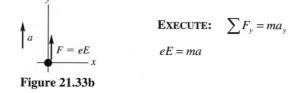

EXECUTE: $\sum F_y = ma_y$

$eE = ma$

Figure 21.33b

Solve the kinematics to find the acceleration of the electron: Just misses upper plate says that $x - x_0 = 2.00$ cm when $y - y_0 = +0.500$ cm.

x-component

$v_{0x} = v_0 = 1.60 \times 10^6$ m/s, $a_x = 0$, $x - x_0 = 0.0200$ m, $t = ?$

$x - x_0 = v_{0x}t + \frac{1}{2}a_x t^2$

$t = \dfrac{x - x_0}{v_{0x}} = \dfrac{0.0200 \ \text{m}}{1.60 \times 10^6 \ \text{m/s}} = 1.25 \times 10^{-8}$ s

In this same time t the electron travels 0.0050 m vertically:

<u>y-component</u>

$t = 1.25 \times 10^{-8}$ s, $v_{0y} = 0$, $y - y_0 = +0.0050$ m, $a_y = ?$

$y - y_0 = v_{0y}t + \frac{1}{2}a_y t^2$

$a_y = \dfrac{2(y - y_0)}{t^2} = \dfrac{2(0.0050 \text{ m})}{(1.25 \times 10^{-8} \text{ s})^2} = 6.40 \times 10^{13}$ m/s^2

(This analysis is very similar to that used in Chapter 3 for projectile motion, except that here the acceleration is upward rather than downward.) This acceleration must be produced by the electric-field force: $eE = ma$

$$E = \frac{ma}{e} = \frac{(9.109 \times 10^{-31} \text{ kg})(6.40 \times 10^{13} \text{ m/s}^2)}{1.602 \times 10^{-19} \text{ C}} = 364 \text{ N/C}$$

Note that the acceleration produced by the electric field is <u>much</u> larger than g, the acceleration produced by gravity, so it is perfectly ok to neglect the gravity force on the elctron in this problem.

(b) $a = \dfrac{eE}{m_p} = \dfrac{(1.602 \times 10^{-19} \text{ C})(364 \text{ N/C})}{1.673 \times 10^{-27} \text{ kg}} = 3.49 \times 10^{10}$ m/s^2

This is much less than the acceleration of the electron in part (a) so the vertical deflection is less and the proton won't hit the plates. The proton has the same initial speed, so the proton takes the same time $t = 1.25 \times 10^{-8}$ s to travel horizontally the length of the plates. The force on the proton is downward (in the same direction as $\vec{E}$, since q is positive), so the acceleration is downward and $a_y = -3.49 \times 10^{10}$ m/s^2.

$y - y_0 = v_{0y}t + \frac{1}{2}a_y t^2 = \frac{1}{2}(-3.49 \times 10^{10} \text{ m/s}^2)(1.25 \times 10^{-8} \text{ s})^2 = -2.73 \times 10^{-6}$ m. The displacement is 2.73×10^{-6} m, downward.

(c) EVALUATE: The displacements are in opposite directions because the electron has negative charge and the proton has positive charge. The electron and proton have the same magnitude of charge, so the force the electric field exerts has the same magnitude for each charge. But the proton has a mass larger by a factor of 1836 so its acceleration and its vertical displacement are smaller by this factor.

21.37.   IDENTIFY and **SET UP:** The electric force is given by Eq. (21.3). The gravitational force is $w_e = m_e g$. Compare these forces.

(a) EXECUTE: $w_e = (9.109 \times 10^{-31} \text{ kg})(9.80 \text{ m/s}^2) = 8.93 \times 10^{-30}$ N

In Examples 21.7 and 21.8, $E = 1.00 \times 10^4$ N/C, so the electric force on the electron has magnitude

$F_E = |q|E = eE = (1.602 \times 10^{-19} \text{ C})(1.00 \times 10^4 \text{ N/C}) = 1.602 \times 10^{-15}$ N.

$\dfrac{w_e}{F_E} = \dfrac{8.93 \times 10^{-30} \text{ N}}{1.602 \times 10^{-15} \text{ N}} = 5.57 \times 10^{-15}$

The gravitational force is much smaller than the electric force and can be neglected.

(b) $mg = |q|E$

$m = |q|E/g = (1.602 \times 10^{-19} \text{ C})(1.00 \times 10^4 \text{ N/C})/(9.80 \text{ m/s}^2) = 1.63 \times 10^{-16}$ kg

$\dfrac{m}{m_e} = \dfrac{1.63 \times 10^{-16} \text{ kg}}{9.109 \times 10^{-31} \text{ kg}} = 1.79 \times 10^{14}$; $m = 1.79 \times 10^{14} m_e$.

EVALUATE: m is much larger than m_e. We found in part (a) that if $m = m_e$ the gravitational force is much smaller than the electric force. $|q|$ is the same so the electric force remains the same. To get w large enough to equal F_E, the mass must be made much larger.

(c) The electric field in the region between the plates is uniform so the force it exerts on the charged object is independent of where between the plates the object is placed.

21.41.   IDENTIFY and **SET UP:** Use $\vec{E}$ in Eq. (21.3) to calculate $\vec{F}$, $\vec{F} = m\vec{a}$ to calculate $\vec{a}$, and a constant acceleration equation to calculate the final velocity. Let $+x$ be east.

(a) EXECUTE: $F_x = |q|E = (1.602 \times 10^{-19} \text{ C})(1.50 \text{ N/C}) = 2.403 \times 10^{-19}$ N

$a_x = F_x/m = (2.403 \times 10^{-19} \text{ N})/(9.109 \times 10^{-31} \text{ kg}) = +2.638 \times 10^{11}$ m/s^2

$v_{0x} = +4.50 \times 10^5$ m/s, $a_x = +2.638 \times 10^{11}$ m/s^2, $x - x_0 = 0.375$ m, $v_x = ?$

$v_x^2 = v_{0x}^2 + 2a_x(x - x_0)$ gives $v_x = 6.33 \times 10^5$ m/s

EVALUATE: $\vec{E}$ is west and q is negative, so $\vec{F}$ is east and the electron speeds up.

(b) **EXECUTE:** $F_x = -|q|E = -(1.602 \times 10^{-19} \text{ C})(1.50 \text{ N/C}) = -2.403 \times 10^{-19} \text{ N}$

$a_x = F_x / m = (-2.403 \times 10^{-19} \text{ N})/(1.673 \times 10^{-27} \text{ kg}) = -1.436 \times 10^8 \text{ m/s}^2$

$v_{0x} = +1.90 \times 10^4 \text{ m/s}, \; a_x = -1.436 \times 10^8 \text{ m/s}^2, \; x - x_0 = 0.375 \text{ m}, \; v_x = ?$

$v_x^2 = v_{0x}^2 + 2a_x(x - x_0)$ gives $v_x = 1.59 \times 10^4 \text{ m/s}$

EVALUATE: $q > 0$ so $\vec{F}$ is west and the proton slows down.

21.45.   **IDENTIFY:**   Eq.(21.7) gives the electric field of each point charge. Use the principle of superposition and add the electric field vectors. In part (b) use Eq.(21.3) to calculate the force, using the electric field calculated in part (a).

(a) **SET UP:**   The placement of charges is sketched in Figure 21.45a.

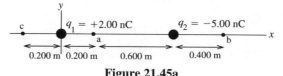

Figure 21.45a

The electric field of a point charge is directed away from the point charge if the charge is positive and toward the point charge if the charge is negative. The magnitude of the electric field is $E = \dfrac{1}{4\pi\epsilon_0}\dfrac{|q|}{r^2}$, where r is the distance between the point where the field is calculated and the point charge.

(i) At point a the fields $\vec{E}_1$ of q_1 and $\vec{E}_2$ of q_2 are directed as shown in Figure 21.45b.

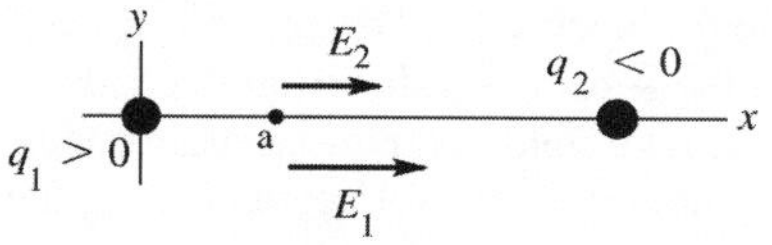

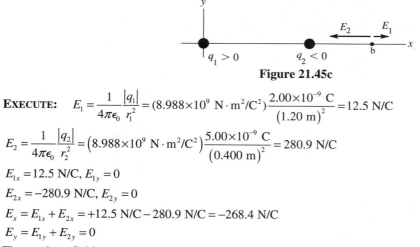

Figure 21.45b

EXECUTE: $E_1 = \dfrac{1}{4\pi\epsilon_0}\dfrac{|q_1|}{r_1^2} = (8.988 \times 10^9 \text{ N} \cdot \text{m}^2/\text{C}^2)\dfrac{2.00 \times 10^{-9} \text{ C}}{(0.200 \text{ m})^2} = 449.4 \text{ N/C}$

$E_2 = \dfrac{1}{4\pi\epsilon_0}\dfrac{|q_2|}{r_2^2} = (8.988 \times 10^9 \text{ N} \cdot \text{m}^2/\text{C}^2)\dfrac{5.00 \times 10^{-9} \text{ C}}{(0.600 \text{ m})^2} = 124.8 \text{ N/C}$

$E_{1x} = 449.4 \text{ N/C}, \; E_{1y} = 0$

$E_{2x} = 124.8 \text{ N/C}, \; E_{2y} = 0$

$E_x = E_{1x} + E_{2x} = +449.4 \text{ N/C} + 124.8 \text{ N/C} = +574.2 \text{ N/C}$

$E_y = E_{1y} + E_{2y} = 0$

The resultant field at point a has magnitude 574 N/C and is in the $+x$-direction.

(ii) **SET UP:**   At point b the fields $\vec{E}_1$ of q_1 and $\vec{E}_2$ of q_2 are directed as shown in Figure 21.45c.

Figure 21.45c

EXECUTE: $E_1 = \dfrac{1}{4\pi\epsilon_0}\dfrac{|q_1|}{r_1^2} = (8.988 \times 10^9 \text{ N} \cdot \text{m}^2/\text{C}^2)\dfrac{2.00 \times 10^{-9} \text{ C}}{(1.20 \text{ m})^2} = 12.5 \text{ N/C}$

$E_2 = \dfrac{1}{4\pi\epsilon_0}\dfrac{|q_2|}{r_2^2} = \left(8.988 \times 10^9 \text{ N} \cdot \text{m}^2/\text{C}^2\right)\dfrac{5.00 \times 10^{-9} \text{ C}}{(0.400 \text{ m})^2} = 280.9 \text{ N/C}$

$E_{1x} = 12.5 \text{ N/C}, \; E_{1y} = 0$

$E_{2x} = -280.9 \text{ N/C}, \; E_{2y} = 0$

$E_x = E_{1x} + E_{2x} = +12.5 \text{ N/C} - 280.9 \text{ N/C} = -268.4 \text{ N/C}$

$E_y = E_{1y} + E_{2y} = 0$

The resultant field at point b has magnitude 268 N/C and is in the $-x$-direction.

(iii) **SET UP:**   At point c the fields $\vec{E}_1$ of q_1 and $\vec{E}_2$ of q_2 are directed as shown in Figure 21.45d.

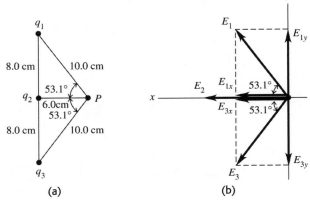

Figure 21.45d

EXECUTE: $E_1 = \dfrac{1}{4\pi\epsilon_0}\dfrac{|q_1|}{r_1^2} = (8.988\times10^9 \ \text{N}\cdot\text{m}^2/\text{C}^2)\dfrac{2.00\times10^{-9} \ \text{C}}{(0.200 \ \text{m})^2} = 449.4 \ \text{N/C}$

$E_2 = \dfrac{1}{4\pi\epsilon_0}\dfrac{|q_2|}{r_2^2} = \left(8.988\times10^9 \ \text{N}\cdot\text{m}^2/\text{C}^2\right)\dfrac{5.00\times10^{-9} \ \text{C}}{(1.00 \ \text{m})^2} = 44.9 \ \text{N/C}$

$E_{1x} = -449.4 \ \text{N/C}, \ E_{1y} = 0$

$E_{2x} = +44.9 \ \text{N/C}, \ E_{2y} = 0$

$E_x = E_{1x} + E_{2x} = -449.4 \ \text{N/C} \ + 44.9 \ \text{N/C} = -404.5 \ \text{N/C}$

$E_y = E_{1y} + E_{2y} = 0$

The resultant field at point b has magnitude 404 N/C and is in the $-x$-direction.

(b) SET UP: Since we have calculated $\vec{E}$ at each point the simplest way to get the force is to use $\vec{F} = -e\vec{E}$.

EXECUTE: (i) $F = (1.602\times10^{-19} \ \text{C})(574.2 \ \text{N/C}) = 9.20\times10^{-17} \ \text{N}, \ -x$-direction

(ii) $F = (1.602\times10^{-19} \ \text{C})(268.4 \ \text{N/C}) = 4.30\times10^{-17} \ \text{N}, \ +x$-direction

(iii) $F = (1.602\times10^{-19} \ \text{C})(404.5 \ \text{N/C}) = 6.48\times10^{-17} \ \text{N}, \ +x$-direction

EVALUATE: The general rule for electric field direction is away from positive charge and toward negative charge. Whether the field is in the $+x$- or $-x$-direction depends on where the field point is relative to the charge that produces the field. In part (a) the field magnitudes were added because the fields were in the same direction and in (b) and (c) the field magnitudes were subtracted because the two fields were in opposite directions. In part (b) we could have used Coulomb's law to find the forces on the electron due to the two charges and then added these force vectors, but using the resultant electric field is much easier.

21.47.   **IDENTIFY:** $E = k\dfrac{|q|}{r^2}$. The net field is the vector sum of the fields due to each charge.

SET UP: The electric field of a negative charge is directed toward the charge. Label the charges q_1, q_2 and q_3, as shown in Figure 21.47a. This figure also shows additional distances and angles. The electric fields at point P are shown in Figure 21.47b. This figure also shows the xy coordinates we will use and the x and y components of the fields $\vec{E}_1$, $\vec{E}_2$ and $\vec{E}_3$.

EXECUTE: $E_1 = E_3 = (8.99\times10^9 \ \text{N}\cdot\text{m}^2/\text{C}^2)\dfrac{5.00\times10^{-6} \ \text{C}}{(0.100 \ \text{m})^2} = 4.49\times10^6 \ \text{N/C}$

$E_2 = (8.99\times10^9 \ \text{N}\cdot\text{m}^2/\text{C}^2)\dfrac{2.00\times10^{-6} \ \text{C}}{(0.0600 \ \text{m})^2} = 4.99\times10^6 \ \text{N/C}$

$E_y = E_{1y} + E_{2y} + E_{3y} = 0$ and $E_x = E_{1x} + E_{2x} + E_{3x} = E_2 + 2E_1 \cos 53.1° = 1.04\times10^7 \ \text{N/C}$

$E = 1.04\times10^7 \ \text{N/C}$, toward the $-2.00 \ \mu\text{C}$ charge.

EVALUATE: The x-components of the fields of all three charges are in the same direction.

Figure 21.47

21.49.   **IDENTIFY:**   The electric field of a positive charge is directed radially outward from the charge and has magnitude $E = \dfrac{1}{4\pi\epsilon_0}\dfrac{|q|}{r^2}$. The resultant electric field is the vector sum of the fields of the individual charges.

SET UP: The placement of the charges is shown in Figure 21.49a.

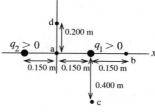

Figure 21.49a

EXECUTE: **(a)** The directions of the two fields are shown in Figure 21.49b.

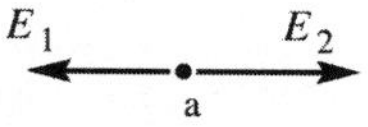

$E_1 \xleftarrow{\hspace{2cm}} \bullet \xrightarrow{\hspace{2cm}} E_2$
$\phantom{E_1 \xleftarrow{\hspace{2cm}}} a$

$E_1 = E_2 = \dfrac{1}{4\pi\epsilon_0}\dfrac{|q|}{r^2}$ with $r = 0.150$ m.

$E = E_2 - E_1 = 0$; $E_x = 0$, $E_y = 0$

Figure 21.49b

(b) The two fields have the directions shown in Figure 21.49c.

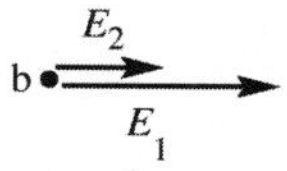

$b\,\bullet\,\xrightarrow{\;E_2\;}$
$\xrightarrow{\hspace{2cm}}$
E_1

$E = E_1 + E_2$, in the $+x$-direction

Figure 21.49c

$E_1 = \dfrac{1}{4\pi\epsilon_0}\dfrac{|q_1|}{r_1^2} = (8.988\times10^9\ \text{N}\cdot\text{m}^2/\text{C}^2)\dfrac{6.00\times10^{-9}\ \text{C}}{(0.150\ \text{m})^2} = 2396.8\ \text{N/C}$

$E_2 = \dfrac{1}{4\pi\epsilon_0}\dfrac{|q_2|}{r_2^2} = (8.988\times10^9\ \text{N}\cdot\text{m}^2/\text{C}^2)\dfrac{6.00\times10^{-9}\ \text{C}}{(0.450\ \text{m})^2} = 266.3\ \text{N/C}$

$E = E_1 + E_2 = 2396.8\ \text{N/C} + 266.3\ \text{N/C} = 2660\ \text{N/C}$; $E_x = +2260\ \text{N/C}$, $E_y = 0$

(c) The two fields have the directions shown in Figure 21.49d.

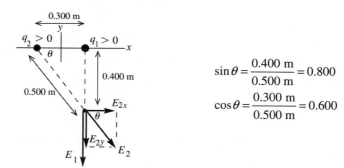

$\sin\theta = \dfrac{0.400\ \text{m}}{0.500\ \text{m}} = 0.800$

$\cos\theta = \dfrac{0.300\ \text{m}}{0.500\ \text{m}} = 0.600$

Figure 21.49d

$E_1 = \dfrac{1}{4\pi\epsilon_0}\dfrac{|q_1|}{r_1^2}$

$E_1 = (8.988\times10^9\ \text{N}\cdot\text{m}^2/\text{C}^2)\dfrac{6.00\times10^{-9}\ \text{C}}{(0.400\ \text{m})^2} = 337.1\ \text{N/C}$

$E_2 = \dfrac{1}{4\pi\epsilon_0}\dfrac{|q_2|}{r_2^2}$

$$E_2 = (8.988 \times 10^9 \text{ N} \cdot \text{m}^2/\text{C}^2) \frac{6.00 \times 10^{-9} \text{ C}}{(0.500 \text{ m})^2} = 215.7 \text{ N/C}$$

$E_{1x} = 0, \ E_{1y} = -E_1 = -337.1 \text{ N/C}$

$E_{2x} = +E_2 \cos\theta = +(215.7 \text{ N/C})(0.600) = +129.4 \text{ N/C}$

$E_{2y} = -E_2 \sin\theta = -(215.7 \text{ N/C})(0.800) = -172.6 \text{ N/C}$

$E_x = E_{1x} + E_{2x} = +129 \text{ N/C}$

$E_y = E_{1y} + E_{2y} = -337.1 \text{ N/C} \ -172.6 \text{ N/C} \ = -510 \text{ N/C}$

$E = \sqrt{E_x^2 + E_y^2} = \sqrt{(129 \text{ N/C})^2 + (-510 \text{ N/C})^2} = 526 \text{ N/C}$

$\vec{E}$ and its components are shown in Figure 21.49e.

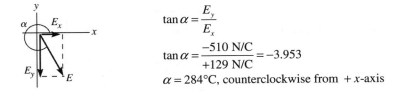

$$\tan\alpha = \frac{E_y}{E_x}$$

$$\tan\alpha = \frac{-510 \text{ N/C}}{+129 \text{ N/C}} = -3.953$$

$\alpha = 284°\text{C}$, counterclockwise from $+x$-axis

Figure 21.49e

(d) The two fields have the directions shown in Figure 21.49f.

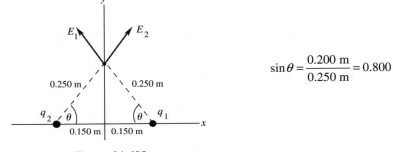

$$\sin\theta = \frac{0.200 \text{ m}}{0.250 \text{ m}} = 0.800$$

Figure 21.49f

The components of the two fields are shown in Figure 21.49g.

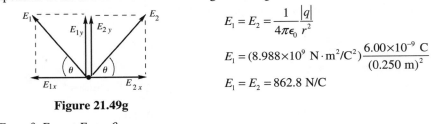

$$E_1 = E_2 = \frac{1}{4\pi\epsilon_0} \frac{|q|}{r^2}$$

$$E_1 = (8.988 \times 10^9 \text{ N} \cdot \text{m}^2/\text{C}^2) \frac{6.00 \times 10^{-9} \text{ C}}{(0.250 \text{ m})^2}$$

$E_1 = E_2 = 862.8 \text{ N/C}$

Figure 21.49g

$E_{1x} = -E_1 \cos\theta, \ E_{2x} = +E_2 \cos\theta$

$E_x = E_{1x} + E_{2x} = 0$

$E_{1y} = +E_1 \sin\theta, \ E_{2y} = +E_2 \sin\theta$

$E_y = E_{1y} + E_{2y} = 2E_{1y} = 2E_1 \sin\theta = 2(862.8 \text{ N/C})(0.800) = 1380 \text{ N/C}$

$E = 1380 \text{ N/C}$, in the $+y$-direction.

EVALUATE: Point a is symmetrically placed between identical charges, so symmetry tells us the electric field must be zero. Point b is to the right of both charges and both electric fields are in the $+x$-direction and the resultant field is in this direction. At point c both fields have a downward component and the field of q_2 has a component to the right, so the net $\vec{E}$ is in the 4th quadrant. At point d both fields have an upward component but by symmetry they have equal and opposite x-components so the net field is in the $+y$-direction. We can use this sort of reasoning to deduce the general direction of the net field before doing any calculations.

21.51.   **IDENTIFY:**   The resultant electric field is the vector sum of the field $\vec{E}_1$ of q_1 and $\vec{E}_2$ of q_2.

SET UP: The placement of the charges is shown in Figure 21.51a.

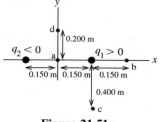

Figure 21.51a

EXECUTE: **(a)** The directions of the two fields are shown in Figure 21.51b.

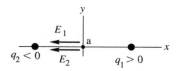

Figure 21.51b

$$E_1 = E_2 = \frac{1}{4\pi\epsilon_0}\frac{|q_1|}{r_1^2}$$

$$E_1 = (8.988\times10^9 \ \text{N}\cdot\text{m}^2/\text{C}^2)\frac{6.00\times10^{-9}\ \text{C}}{(0.150\ \text{m})^2}$$

$$E_1 = E_2 = 2397 \ \text{N/C}$$

$E_{1x} = -2397$ N/C, $E_{1y} = 0$ $E_{2x} = -2397$ N/C, $E_{2y} = 0$

$E_x = E_{1x} + E_{2x} = 2(-2397 \ \text{N/C}) = -4790 \ \text{N/C}$

$E_y = E_{1y} + E_{2y} = 0$

The resultant electric field at point a in the sketch has magnitude 4790 N/C and is in the $-x$-direction.

(b) The directions of the two fields are shown in Figure 21.51c.

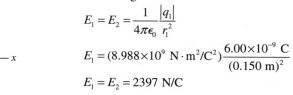

Figure 21.51c

$$E_1 = \frac{1}{4\pi\epsilon_0}\frac{|q_1|}{r_1^2} = (8.988\times10^9 \ \text{N}\cdot\text{m}^2/\text{C}^2)\frac{6.00\times10^{-9}\ \text{C}}{(0.150\ \text{m})^2} = 2397 \ \text{N/C}$$

$$E_2 = \frac{1}{4\pi\epsilon_0}\frac{|q_2|}{r_2^2} = (8.988\times10^9 \ \text{N}\cdot\text{m}^2/\text{C}^2)\frac{6.00\times10^{-9}\ \text{C}}{(0.450\ \text{m})^2} = 266 \ \text{N/C}$$

$E_{1x} = +2397$ N/C, $E_{1y} = 0$ $E_{2x} = -266$ N/C, $E_{2y} = 0$

$E_x = E_{1x} + E_{2x} = +2397 \ \text{N/C} - 266 \ \text{N/C} = +2130 \ \text{N/C}$

$E_y = E_{1y} + E_{2y} = 0$

The resultant electric field at point b in the sketch has magnitude 2130 N/C and is in the $+x$-direction.

(c) The placement of the charges is shown in Figure 21.51d.

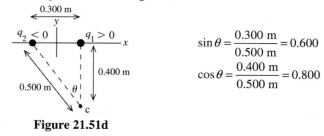

Figure 21.51d

$$\sin\theta = \frac{0.300\ \text{m}}{0.500\ \text{m}} = 0.600$$

$$\cos\theta = \frac{0.400\ \text{m}}{0.500\ \text{m}} = 0.800$$

The directions of the two fields are shown in Figure 21.51e.

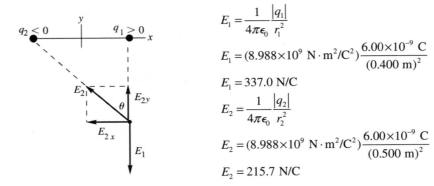

$$E_1 = \frac{1}{4\pi\epsilon_0} \frac{|q_1|}{r_1^2}$$

$$E_1 = (8.988\times10^9 \text{ N}\cdot\text{m}^2/\text{C}^2)\frac{6.00\times10^{-9} \text{ C}}{(0.400 \text{ m})^2}$$

$$E_1 = 337.0 \text{ N/C}$$

$$E_2 = \frac{1}{4\pi\epsilon_0} \frac{|q_2|}{r_2^2}$$

$$E_2 = (8.988\times10^9 \text{ N}\cdot\text{m}^2/\text{C}^2)\frac{6.00\times10^{-9} \text{ C}}{(0.500 \text{ m})^2}$$

$$E_2 = 215.7 \text{ N/C}$$

Figure 21.51e

$E_{1x} = 0, E_{1y} = -E_1 = -337.0 \text{ N/C}$

$E_{2x} = -E_2 \sin\theta = -(215.7 \text{ N/C})(0.600) = -129.4 \text{ N/C}$

$E_{2y} = +E_2 \cos\theta = +(215.7 \text{ N/C})(0.800) = +172.6 \text{ N/C}$

$E_x = E_{1x} + E_{2x} = -129 \text{ N/C}$

$E_y = E_{1y} + E_{2y} = -337.0 \text{ N/C} +172.6 \text{ N/C} = -164 \text{ N/C}$

$E = \sqrt{E_x^2 + E_y^2} = 209 \text{ N/C}$

The field $\vec{E}$ and its components are shown in Figure 21.51f.

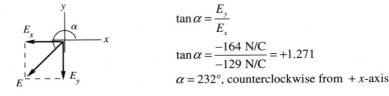

$$\tan\alpha = \frac{E_y}{E_x}$$

$$\tan\alpha = \frac{-164 \text{ N/C}}{-129 \text{ N/C}} = +1.271$$

$\alpha = 232°$, counterclockwise from $+x$-axis

Figure 21.51f

(d) The placement of the charges is shown in Figure 21.51g.

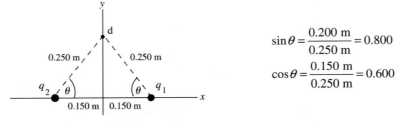

$$\sin\theta = \frac{0.200 \text{ m}}{0.250 \text{ m}} = 0.800$$

$$\cos\theta = \frac{0.150 \text{ m}}{0.250 \text{ m}} = 0.600$$

Figure 21.51g

The directions of the two fields are shown in Figure 21.51h.

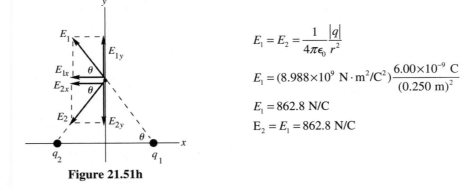

$$E_1 = E_2 = \frac{1}{4\pi\epsilon_0} \frac{|q|}{r^2}$$

$$E_1 = (8.988\times10^9 \text{ N}\cdot\text{m}^2/\text{C}^2)\frac{6.00\times10^{-9} \text{ C}}{(0.250 \text{ m})^2}$$

$$E_1 = 862.8 \text{ N/C}$$

$$E_2 = E_1 = 862.8 \text{ N/C}$$

Figure 21.51h

$E_{1x} = -E_1 \cos\theta,\ E_{2x} = -E_2 \cos\theta$

$E_x = E_{1x} + E_{2x} = -2(862.8 \text{ N/C})(0.600) = -1040 \text{ N/C}$

$E_{1y} = +E_1 \sin\theta,\ E_{2y} = -E_2 \sin\theta$

$E_y = E_{1y} + E_{2y} = 0$

$E = 1040 \text{ N/C}$, in the $-x$-direction.

EVALUATE: The electric field produced by a charge is toward a negative charge and away from a positive charge. As in Exercise 21.45, we can use this rule to deduce the direction of the resultant field at each point before doing any calculations.

21.57. **IDENTIFY:** By superposition we can add the electric fields from two parallel sheets of charge.

SET UP: The field due to each sheet of charge has magnitude $\sigma/2\epsilon_0$ and is directed toward a sheet of negative charge and away from a sheet of positive charge.

(a) The two fields are in opposite directions and $E = 0$.

(b) The two fields are in opposite directions and $E = 0$.

(c) The fields of both sheets are downward and $E = 2\dfrac{\sigma}{2\epsilon_0} = \dfrac{\sigma}{\epsilon_0}$, directed downward.

EVALUATE: The field produced by an infinite sheet of charge is uniform, independent of distance from the sheet.

21.61. **IDENTIFY:** Use symmetry to deduce the nature of the field lines.

(a) SET UP: The only distinguishable direction is toward the line or away from the line, so the electric field lines are perpendicular to the line of charge, as shown in Figure 21.61a.

Figure 21.61a

(b) EXECUTE and EVALUATE: The magnitude of the electric field is inversely proportional to the spacing of the field lines. Consider a circle of radius r with the line of charge passing through the center, as shown in Figure 21.61b.

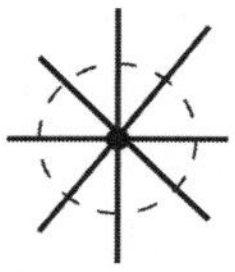

Figure 21.61b

The spacing of field lines is the same all around the circle, and in the direction perpendicular to the plane of the circle the lines are equally spaced, so E depends only on the distance r. The number of field lines passing out through the circle is independent of the radius of the circle, so the spacing of the field lines is proportional to the reciprocal of the circumference $2\pi r$ of the circle. Hence E is proportional to $1/r$.

21.63. **(a) IDENTIFY and SET UP:** Use Eq.(21.14) to relate the dipole moment to the charge magnitude and the separation d of the two charges. The direction is from the negative charge toward the positive charge.

EXECUTE: $p = qd = (4.5\times10^{-9}\text{ C})(3.1\times10^{-3}\text{ m}) = 1.4\times10^{-11}\text{ C}\cdot\text{m}$; The direction of $\vec{p}$ is from q_1 toward q_2.

(b) IDENTIFY and SET UP: Use Eq. (21.15) to relate the magnitudes of the torque and field.

EXECUTE: $\tau = pE\sin\phi$, with ϕ as defined in Figure 21.63, so

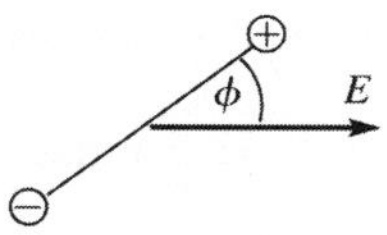

$E = \dfrac{\tau}{p\sin\phi}$

$E = \dfrac{7.2\times10^{-9}\text{ N}\cdot\text{m}}{(1.4\times10^{-11}\text{ C}\cdot\text{m})\sin 36.9°} = 860 \text{ N/C}$

Figure 21.63

EVALUATE: Eq.(21.15) gives the torque about an axis through the center of the dipole. But the forces on the two charges form a couple (Problem 11.53) and the torque is the same for any axis parallel to this one. The force on each charge is $|q|E$ and the maximum moment arm for an axis at the center is $d/2$, so the maximum torque is

$2(|q|E)(d/2) = 1.2\times10^{-8}\text{ N}\cdot\text{m}$. The torque for the orientation of the dipole in the problem is less than this maximum.

21.69.   **IDENTIFY:**   The torque on a dipole in an electric field is given by $\vec{\tau} = \vec{p} \times \vec{E}$.

SET UP: $\tau = pE\sin\phi$, where ϕ is the angle between the direction of $\vec{p}$ and the direction of $\vec{E}$.

EXECUTE: **(a)** The torque is zero when $\vec{p}$ is aligned either in the *same* direction as $\vec{E}$ or in the *opposite* direction, as shown in Figure 21.69a.

(b) The stable orientation is when $\vec{p}$ is aligned in the *same* direction as $\vec{E}$. In this case a small rotation of the dipole results in a torque directed so as to bring $\vec{p}$ back into alignment with $\vec{E}$. When $\vec{p}$ is directed opposite to $\vec{E}$, a small displacement results in a torque that takes $\vec{p}$ farther from alignment with $\vec{E}$.

(c) Field lines for E_{dipole} in the stable orientation are sketched in Figure 21.69b.

EVALUATE: The field of the dipole is directed from the + charge toward the − charge.

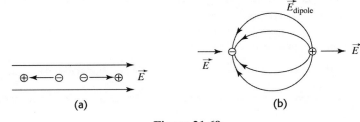

(a)   (b)

Figure 21.69

21.71.   **(a) IDENTIFY:**   Use Coulomb's law to calculate each force and then add them as vectors to obtain the net force. Torque is force times moment arm.

SET UP: The two forces on each charge in the dipole are shown in Figure 21.71a.

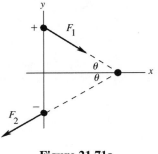

$\sin\theta = 1.50/2.00$ so $\theta = 48.6°$
Opposite charges attract and like charges repel.
$F_x = F_{1x} + F_{2x} = 0$

Figure 21.71a

EXECUTE: $F_1 = k\dfrac{|qq'|}{r^2} = k\dfrac{(5.00\times10^{-6}\ \text{C})(10.0\times10^{-6}\ \text{C})}{(0.0200\ \text{m})^2} = 1.124\times10^3\ \text{N}$

$F_{1y} = -F_1\sin\theta = -842.6\ \text{N}$

$F_{2y} = -842.6\ \text{N}$ so $F_y = F_{1y} + F_{2y} = -1680\ \text{N}$ (in the direction from the $+5.00$-μC charge toward the -5.00-μC charge).

EVALUATE: The *x*-components cancel and the *y*-components add.

(b) SET UP: Refer to Figure 21.71b.

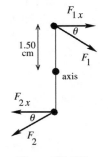

The *y*-components have zero moment arm and therefore zero torque.
F_{1x} and F_{2x} both produce clockwise torques.

Figure 21.71b

EXECUTE: $F_{1x} = F_1\cos\theta = 743.1\ \text{N}$

$\tau = 2(F_{1x})(0.0150\ \text{m}) = 22.3\ \text{N}\cdot\text{m}$, clockwise

EVALUATE: The electric field produced by the -10.00μC charge is not uniform so Eq. (21.15) does not apply.

21.73.   **(a) IDENTIFY:**   Use Coulomb's law to calculate the force exerted by each Q on q and add these forces as vectors to find the resultant force. Make the approximation $x \gg a$ and compare the net force to $F = -kx$ to deduce k and then $f = (1/2\pi)\sqrt{k/m}$.

SET UP: The placement of the charges is shown in Figure 21.73a.

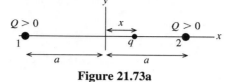

Figure 21.73a

EXECUTE: Find the net force on q.

Figure 21.73b

$$F_x = F_{1x} + F_{2x} \text{ and } F_{1x} = +F_1, \ F_{2x} = -F_2$$

$$F_1 = \frac{1}{4\pi\epsilon_0}\frac{qQ}{(a+x)^2}, \ F_2 = \frac{1}{4\pi\epsilon_0}\frac{qQ}{(a-x)^2}$$

$$F_x = F_1 - F_2 = \frac{qQ}{4\pi\epsilon_0}\left[\frac{1}{(a+x)^2} - \frac{1}{(a-x)^2}\right]$$

$$F_x = \frac{qQ}{4\pi\epsilon_0 a^2}\left[+\left(1+\frac{x}{a}\right)^{-2} - \left(1-\frac{x}{a}\right)^{-2}\right]$$

Since $x \ll a$ we can use the binomial expansion for $(1-x/a)^{-2}$ and $(1+x/a)^{-2}$ and keep only the first two terms: $(1+z)^n \approx 1 + nz$. For $(1-x/a)^{-2}$, $z = -x/a$ and $n = -2$ so $(1-x/a)^{-2} \approx 1 + 2x/a$. For $(1+x/a)^{-2}$, $z = +x/a$ and $n = -2$ so $(1+x/a)^{-2} \approx 1 - 2x/a$. Then $F \approx \frac{qQ}{4\pi\epsilon_0 a^2}\left[\left(1-\frac{2x}{a}\right) - \left(1+\frac{2x}{a}\right)\right] = -\left(\frac{qQ}{\pi\epsilon_0 a^3}\right)x$. For simple harmonic motion $F = -kx$ and the frequency of oscillation is $f = (1/2\pi)\sqrt{k/m}$. The net force here is of this form, with $k = qQ/\pi\epsilon_0 a^3$. Thus $f = \frac{1}{2\pi}\sqrt{\frac{qQ}{\pi\epsilon_0 m a^3}}$.

(b) The forces and their components are shown in Figure 21.73c.

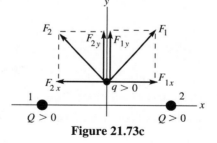

Figure 21.73c

The x-components of the forces exerted by the two charges cancel, the y-components add, and the net force is in the $+y$-direction when $y > 0$ and in the $-y$-direction when $y < 0$. The charge moves away from the origin on the y-axis and never returns.

EVALUATE: The directions of the forces and of the net force depend on where q is located relative to the other two charges. In part (a), $F = 0$ at $x = 0$ and when the charge q is displaced in the $+x$- or $-x$-direction the net force is a restoring force, directed to return q to $x = 0$. The charge oscillates back and forth, similar to a mass on a spring.

21.75.   **IDENTIFY:**   Use Coulomb's law for the force that one sphere exerts on the other and apply the 1st condition of equilibrium to one of the spheres.

(a) SET UP: The placement of the spheres is sketched in Figure 21.75a.

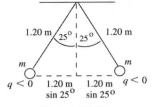

Figure 21.75a

The free-body diagrams for each sphere are given in Figure 21.75b.

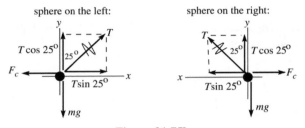

Figure 21.75b

F_c is the repulsive Coulomb force exerted by one sphere on the other.

(b) EXECUTE: From either force diagram in part (a): $\sum F_y = ma_y$

$T\cos 25.0° - mg = 0$ and $T = \dfrac{mg}{\cos 25.0°}$

$\sum F_x = ma_x$

$T\sin 25.0° - F_c = 0$ and $F_c = T\sin 25.0°$

Use the first equation to eliminate T in the second: $F_c = (mg/\cos 25.0°)(\sin 25.0°) = mg\tan 25.0°$

$$F_c = \frac{1}{4\pi\epsilon_0}\frac{|q_1 q_2|}{r^2} = \frac{1}{4\pi\epsilon_0}\frac{q^2}{r^2} = \frac{1}{4\pi\epsilon_0}\frac{q^2}{[2(1.20\text{ m})\sin 25.0°]^2}$$

Combine this with $F_c = mg\tan 25.0°$ and get $mg\tan 25.0° = \dfrac{1}{4\pi\epsilon_0}\dfrac{q^2}{[2(1.20\text{ m})\sin 25.0°]^2}$

$$q = (2.40\text{ m})\sin 25.0° \sqrt{\frac{mg\tan 25.0°}{(1/4\pi\epsilon_0)}}$$

$$q = (2.40\text{ m})\sin 25.0° \sqrt{\frac{(15.0\times 10^{-3}\text{ kg})(9.80\text{ m/s}^2)\tan 25.0°}{8.988\times 10^9\text{ N}\cdot\text{m}^2/\text{C}^2}} = 2.80\times 10^{-6}\text{ C}$$

(c) The separation between the two spheres is given by $2L\sin\theta$. $q = 2.80\,\mu C$ as found in part (b).

$F_c = (1/4\pi\epsilon_0)q^2/(2L\sin\theta)^2$ and $F_c = mg\tan\theta$. Thus $(1/4\pi\epsilon_0)q^2/(2L\sin\theta)^2 = mg\tan\theta$.

$$(\sin\theta)^2\tan\theta = \frac{1}{4\pi\epsilon_0}\frac{q^2}{4L^2 mg} = (8.988\times 10^9\text{ N}\cdot\text{m}^2/\text{C}^2)\frac{(2.80\times 10^{-6}\text{ C})^2}{4(0.600\text{ m})^2(15.0\times 10^{-3}\text{ kg})(9.80\text{ m/s}^2)} = 0.3328.$$

Solve this equation by trial and error. This will go quicker if we can make a good estimate of the value of θ that solves the equation. For θ small, $\tan\theta \approx \sin\theta$. With this approximation the equation becomes $\sin^3\theta = 0.3328$ and $\sin\theta = 0.6930$, so $\theta = 43.9°$. Now refine this guess:

θ	$\sin^2\theta\tan\theta$	
45.0°	0.5000	
40.0°	0.3467	
39.6°	0.3361	
39.5°	0.3335	
39.4°	0.3309	so $\theta = 39.5°$

EVALUATE: The expression in part (c) says $\theta \to 0$ as $L \to \infty$ and $\theta \to 90°$ as $L \to 0$. When L is decreased from the value in part (a), θ increases.

21.77. **IDENTIFY** and **SET UP:** Use Avogadro's number to find the number of Na^+ and Cl^- ions and the total positive and negative charge. Use Coulomb's law to calculate the electric force and $\vec{F} = m\vec{a}$ to calculate the acceleration.
(a) EXECUTE: The number of Na^+ ions in 0.100 mol of NaCl is $N = nN_A$. The charge of one ion is $+e$, so the total charge is $q_1 = nN_A e = (0.100 \text{ mol})(6.022 \times 10^{23} \text{ ions/mol})(1.602 \times 10^{-19} \text{ C/ion}) = 9.647 \times 10^3 \text{ C}$

There are the same number of Cl^- ions and each has charge $-e$, so $q_2 = -9.647 \times 10^3$ C.

$$F = \frac{1}{4\pi\epsilon_0} \frac{|q_1 q_2|}{r^2} = (8.988 \times 10^9 \text{ N} \cdot \text{m}^2/\text{C}^2) \frac{(9.647 \times 10^3 \text{ C})^2}{(0.0200 \text{ m})^2} = 2.09 \times 10^{21} \text{ N}$$

(b) $a = F/m$. Need the mass of 0.100 mol of Cl^- ions. For Cl, $M = 35.453 \times 10^{-3}$ kg/mol, so

$$m = (0.100 \text{ mol})(35.453 \times 10^{-3} \text{ kg/mol}) = 35.45 \times 10^{-4} \text{ kg. Then } a = \frac{F}{m} = \frac{2.09 \times 10^{21} \text{ N}}{35.45 \times 10^{-4} \text{ kg}} = 5.90 \times 10^{23} \text{ m/s}^2.$$

(c) EVALUATE: Is is not reasonable to have such a huge force. The net charges of objects are rarely larger than 1 μC; a charge of 10^4 C is immense. A small amount of material contains huge amounts of positive and negative charges.

21.79. **IDENTIFY:** Use Coulomb's law to calculate the forces between pairs of charges and sum these forces as vectors to find the net charge.
(a) SET UP: The forces are sketched in Figure 21.79a.

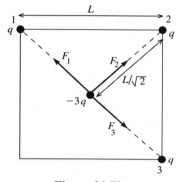

EXECUTE: $\vec{F}_1 + \vec{F}_3 = 0$, so the net force is $\vec{F} = \vec{F}_2$.

$$F = \frac{1}{4\pi\epsilon_0} \frac{q(3q)}{(L/\sqrt{2})^2} = \frac{6q^2}{4\pi\epsilon_0 L^2}, \text{ away from the vacant corner.}$$

Figure 21.79a

(b) SET UP: The forces are sketched in Figure 21.79b.

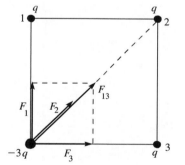

EXECUTE: $F_2 = \frac{1}{4\pi\epsilon_0} \frac{q(3q)}{(\sqrt{2}L)^2} = \frac{3q^2}{4\pi\epsilon_0 (2L^2)}$

$F_1 = F_3 = \frac{1}{4\pi\epsilon_0} \frac{q(3q)}{L^2} = \frac{3q^2}{4\pi\epsilon_0 L^2}$

The vector sum of F_1 and F_3 is $F_{13} = \sqrt{F_1^2 + F_3^2}$.

Figure 21.79b

$F_{13} = \sqrt{2}F_1 = \frac{3\sqrt{2}q^2}{4\pi\epsilon_0 L^2}$; $\vec{F}_{13}$ and $\vec{F}_2$ are in the same direction.

$F = F_{13} + F_2 = \frac{3q^2}{4\pi\epsilon_0 L^2}\left(\sqrt{2} + \frac{1}{2}\right)$, and is directed toward the center of the square.

EVALUATE: By symmetry the net force is along the diagonal of the square. The net force is only slightly larger when the $-3q$ charge is at the center. Here it is closer to the charge at point 2 but the other two forces cancel.

21.85. **IDENTIFY** and **SET UP:** Use the density of copper to calculate the number of moles and then the number of atoms. Calculate the net charge and then use Coulomb's law to calculate the force.

EXECUTE: **(a)** $m = \rho V = \rho\left(\frac{4}{3}\pi r^3\right) = (8.9 \times 10^3 \text{ kg/m}^3)\left(\frac{4}{3}\pi\right)(1.00 \times 10^{-3} \text{ m})^3 = 3.728 \times 10^{-5} \text{ kg}$

$n = m/M = (3.728 \times 10^{-5} \text{ kg})/(63.546 \times 10^{-3} \text{ kg/mol}) = 5.867 \times 10^{-4} \text{ mol}$

$N = nN_A = 3.5 \times 10^{20} \text{ atoms}$

(b) $N_e = (29)(3.5 \times 10^{20}) = 1.015 \times 10^{22}$ electrons and protons

$q_{net} = eN_e - (0.99900)eN_e = (0.100 \times 10^{-2})(1.602 \times 10^{-19} \text{ C})(1.015 \times 10^{22}) = 1.6 \text{ C}$

$F = k\dfrac{q^2}{r^2} = k\dfrac{(1.6 \text{ C})^2}{(1.00 \text{ m})^2} = 2.3 \times 10^{10} \text{ N}$

EVALUATE: The amount of positive and negative charge in even small objects is immense. If the charge of an electron and a proton weren't exactly equal, objects would have large net charges.

21.87. **IDENTIFY:** Eq. (21.3) gives the force exerted by the electric field. This force is constant since the electric field is uniform and gives the proton a constant acceleration. Apply the constant acceleration equations for the x- and y-components of the motion, just as for projectile motion.

(a) SET UP: The electric field is upward so the electric force on the positively charged proton is upward and has magnitude $F = eE$. Use coordinates where positive y is downward. Then applying $\sum \vec{F} = m\vec{a}$ to the proton gives that $a_x = 0$ and $a_y = -eE/m$. In these coordinates the initial velocity has components $v_x = +v_0 \cos\alpha$ and $v_y = +v_0 \sin\alpha$, as shown in Figure 21.87a.

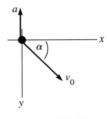

Figure 21.87a

EXECUTE: Finding h_{max}: At $y = h_{max}$ the y-component of the velocity is zero.

$v_y = 0,\ v_{0y} = v_0\sin\alpha,\ a_y = -eE/m,\ y - y_0 = h_{max} = ?$

$v_y^2 = v_{0y}^2 + 2a_y(y - y_0)$

$y - y_0 = \dfrac{v_y^2 - v_{0y}^2}{2a_y}$

$h_{max} = \dfrac{-v_0^2\sin^2\alpha}{2(-eE/m)} = \dfrac{mv_0^2\sin^2\alpha}{2eE}$

(b) Use the vertical motion to find the time t: $y - y_0 = 0,\ v_{0y} = v_0\sin\alpha,\ a_y = -eE/m,\ t = ?$

$y - y_0 = v_{0y}t + \frac{1}{2}a_y t^2$

With $y - y_0 = 0$ this gives $t = -\dfrac{2v_{0y}}{a_y} = -\dfrac{2(v_0\sin\alpha)}{-eE/m} = \dfrac{2mv_0\sin\alpha}{eE}$

Then use the x-component motion to find d: $a_x = 0,\ v_{0x} = v_0\cos\alpha,\ t = 2mv_0\sin\alpha/eE,\ x - x_0 = d = ?$

$x - x_0 = v_{0x}t + \frac{1}{2}a_x t^2$ gives $d = v_0\cos\alpha\left(\dfrac{2mv_0\sin\alpha}{eE}\right) = \dfrac{mv_0^2 2\sin\alpha\cos\alpha}{eE} = \dfrac{mv_0^2\sin 2\alpha}{eE}$

(c) The trajectory of the proton is sketched in Figure 21.87b.

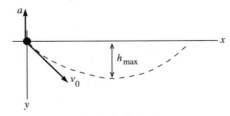

Figure 21.87b

(d) Use the expression in part (a): $h_{max} = \dfrac{\left[(4.00 \times 10^5 \text{ m/s})(\sin 30.0°)\right]^2 (1.673 \times 10^{-27} \text{ kg})}{2(1.602 \times 10^{-19} \text{ C})(500 \text{ N/C})} = 0.418 \text{ m}$

Use the expression in part (b): $d = \dfrac{\left(1.673\times10^{-27}\text{ kg}\right)\left(4.00\times10^{5}\text{ m/s}\right)^{2}\sin 60.0°}{\left(1.602\times10^{-19}\text{ C}\right)\left(500\text{ N/C}\right)} = 2.89\text{ m}$

EVALUATE: In part (a), $a_{y} = -eE/m = -4.8\times10^{10}\text{ m/s}^{2}$. This is much larger in magnitude than g, the acceleration due to gravity, so it is reasonable to ignore gravity. The motion is just like projectile motion, except that the acceleration is upward rather than downward and has a much different magnitude. $h_{\max}$ and d increase when α or v_{0} increase and decrease when E increases.

21.89. **IDENTIFY:** Divide the charge distribution into infinitesimal segments of length dx. Calculate E_{x} and E_{u} due to a segment and integrate to find the total field.

SET UP: The charge dQ of a segment of length dx is $dQ = (Q/a)dx$. The distance between a segment at x and the charge q is $a + r - x$. $(1-y)^{-1} \approx 1 + y$ when $|y| \ll 1$.

EXECUTE: (a) $dE_{x} = \dfrac{1}{4\pi\epsilon_{0}}\dfrac{dQ}{(a+r-x)^{2}}$ so $E_{x} = \dfrac{1}{4\pi\epsilon_{0}}\displaystyle\int_{0}^{a}\dfrac{Qdx}{a(a+r-x)^{2}} = \dfrac{1}{4\pi\epsilon_{0}}\dfrac{Q}{a}\left(\dfrac{1}{r} - \dfrac{1}{a+r}\right)$.

$a + r = x$, so $E_{x} = \dfrac{1}{4\pi\epsilon_{0}}\dfrac{Q}{a}\left(\dfrac{1}{x-a} - \dfrac{1}{x}\right)$. $E_{y} = 0$.

(b) $\vec{F} = q\vec{E} = \dfrac{1}{4\pi\epsilon_{0}}\dfrac{qQ}{a}\left(\dfrac{1}{x-a} - \dfrac{1}{x}\right)\hat{i}$.

EVALUATE: (c) For $x \gg a$, $F = \dfrac{kqQ}{ax}\left((1-a/x)^{-1} - 1\right) = \dfrac{kqQ}{ax}(1 + a/x + \cdots - 1) \approx \dfrac{kqQ}{x^{2}} \approx \dfrac{1}{4\pi\epsilon_{0}}\dfrac{qQ}{r^{2}}$. (Note that for

$x \gg a$, $r = x - a \approx x$.) The charge distribution looks like a point charge from far away, so the force takes the form of the force between a pair of point charges.

21.93. **IDENTIFY:** Apply Eq.(21.11).

SET UP: $\sigma = Q/A = Q/\pi R^{2}$. $(1+y^{2})^{-1/2} \approx 1 - y^{2}/2$, when $y^{2} \ll 1$.

EXECUTE: (a) $E = \dfrac{\sigma}{2\epsilon_{0}}\left[1 - \left(R^{2}/x^{2} + 1\right)^{-1/2}\right]$.

$E = \dfrac{4.00\text{ pC}/\pi(0.025\text{ m})^{2}}{2\epsilon_{0}}\left[1 - \left(\dfrac{(0.025\text{ m})^{2}}{(0.200\text{ m})^{2}} + 1\right)^{-1/2}\right] = 0.89\text{ N/C}$, in the $+x$ direction.

(b) For $x \gg R$ $E = \dfrac{\sigma}{2\epsilon_{0}}[1 - (1 - R^{2}/2x^{2} + \cdots)] \approx \dfrac{\sigma}{2\epsilon_{0}}\dfrac{R^{2}}{2x^{2}} = \dfrac{\sigma\pi R^{2}}{4\pi\epsilon_{0}x^{2}} = \dfrac{Q}{4\pi\epsilon_{0}x^{2}}$.

(c) The electric field of (a) is less than that of the point charge (0.90 N/C) since the first correction term to the point charge result is negative.

(d) For $x = 0.200\text{ m}$, the percent difference is $\dfrac{(0.90-0.89)}{0.89} = 0.01 = 1\%$. For $x = 0.100\text{ m}$,

$E_{\text{disk}} = 3.43\text{ N/C}$ and $E_{\text{point}} = 3.60\text{ N/C}$, so the percent difference is $\dfrac{(3.60-3.43)}{3.60} = 0.047 \approx 5\%$.

EVALUATE: The field of a disk becomes closer to the field of a point charge as the distance from the disk increases. At $x = 10.0\text{ cm}$, $R/x = 25\%$ and the percent difference between the field of the disk and the field of a point charge is 5%.

21.95. **IDENTIFY:** Find the resultant electric field due to the two point charges. Then use $\vec{F} = q\vec{E}$ to calculate the force on the point charge.

SET UP: Use the results of Problems 21.90 and 21.89.

EXECUTE: (a) The y-components of the electric field cancel, and the x-component from both charges, as given in

Problem 21.90, is $E_{x} = \dfrac{1}{4\pi\epsilon_{0}}\dfrac{-2Q}{a}\left(\dfrac{1}{y} - \dfrac{1}{(y^{2}+a^{2})^{1/2}}\right)$. Therefore, $\vec{F} = \dfrac{1}{4\pi\epsilon_{0}}\dfrac{-2Qq}{a}\left(\dfrac{1}{y} - \dfrac{1}{(y^{2}+a^{2})^{1/2}}\right)\hat{i}$. If $y \gg a$

$\vec{F} \approx \dfrac{1}{4\pi\epsilon_{0}}\dfrac{-2Qq}{ay}(1 - (1 - a^{2}/2y^{2} + \cdots))\hat{i} = -\dfrac{1}{4\pi\epsilon_{0}}\dfrac{Qqa}{y^{3}}\hat{i}$.

(b) If the point charge is now on the x-axis the two halves of the charge distribution provide different forces, though still along the x-axis, as given in Problem 21.89: $\vec{F}_+ = q\vec{E}_+ = \dfrac{1}{4\pi\epsilon_0}\dfrac{Qq}{a}\left(\dfrac{1}{x-a}-\dfrac{1}{x}\right)\hat{i}$

and $\vec{F}_- = q\,\vec{E}_- = -\dfrac{1}{4\pi\epsilon_0}\dfrac{Qq}{a}\left(\dfrac{1}{x}-\dfrac{1}{x+a}\right)\hat{i}$. Therefore, $\vec{F} = \vec{F}_+ + \vec{F}_- = \dfrac{1}{4\pi\epsilon_0}\dfrac{Qq}{a}\left(\dfrac{1}{x-a}-\dfrac{2}{x}+\dfrac{1}{x+a}\right)\hat{i}$. For $x \gg a$,

$\vec{F} \approx \dfrac{1}{4\pi\epsilon_0}\dfrac{Qq}{ax}\left(\left(1+\dfrac{a}{x}+\dfrac{a^2}{x^2}+\ldots\right)-2+\left(1-\dfrac{a}{x}+\dfrac{a^2}{x^2}-\ldots\right)\right)\hat{i} = \dfrac{1}{4\pi\epsilon_0}\dfrac{2Qqa}{x^3}\hat{i}$.

EVALUATE: If the charge distributed along the x-axis were all positive or all negative, the force would be proportional to $1/y^2$ in part (a) and to $1/x^2$ in part (b), when y or x is very large.

21.97. **IDENTIFY:** Divide the charge distribution into small segments, use the point charge formula for the electric field due to each small segment and integrate over the charge distribution to find the x and y components of the total field.

SET UP: Consider the small segment shown in Figure 21.97a.

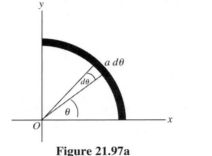

Figure 21.97a

EXECUTE: A small segment that subtends angle $d\theta$ has length $a\,d\theta$ and contains charge $dQ = \left(\dfrac{a\,d\theta}{\frac{1}{2}\pi a}\right)Q = \dfrac{2Q}{\pi}d\theta$.

($\frac{1}{2}\pi a$ is the total length of the charge distribution.)

The charge is negative, so the field at the origin is directed toward the small segment. The small segment is located at angle θ as shown in the sketch. The electric field due to dQ is shown in Figure 21.97b, along with its components.

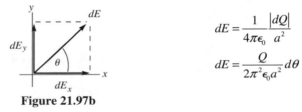

$dE = \dfrac{1}{4\pi\epsilon_0}\dfrac{|dQ|}{a^2}$

$dE = \dfrac{Q}{2\pi^2\epsilon_0 a^2}d\theta$

Figure 21.97b

$dE_x = dE\cos\theta = \left(Q/2\pi^2\epsilon_0 a^2\right)\cos\theta\,d\theta$

$E_x = \int dE_x = \dfrac{Q}{2\pi^2\epsilon_0 a^2}\int_0^{\pi/2}\cos\theta\,d\theta = \dfrac{Q}{2\pi^2\epsilon_0 a^2}\left(\sin\theta\big|_0^{\pi/2}\right) = \dfrac{Q}{2\pi^2\epsilon_0 a^2}$

$dE_y = dE\sin\theta = \left(Q/2\pi^2\epsilon_0 a^2\right)\sin\theta\,d\theta$

$E_y = \int dE_y = \dfrac{Q}{2\pi^2\epsilon_0 a^2}\int_0^{\pi/2}\sin\theta\,d\theta = \dfrac{Q}{2\pi^2\epsilon_0 a^2}\left(-\cos\theta\big|_0^{\pi/2}\right) = \dfrac{Q}{2\pi^2\epsilon_0 a^2}$

EVALUATE: Note that $E_x = E_y$, as expected from symmetry.

21.101. **IDENTIFY:** Each sheet produces an electric field that is independent of the distance from the sheet. The net field is the vector sum of the two fields.

SET UP: The formula for each field is $E = \sigma/2\epsilon_0$, and the net field is the vector sum of these,

$E_{net} = \dfrac{\sigma_B}{2\epsilon_0} \pm \dfrac{\sigma_A}{2\epsilon_0} = \dfrac{\sigma_B \pm \sigma_A}{2\epsilon_0}$, where we use the + or – sign depending on whether the fields are in the same or

opposite directions and σ_B and σ_A are the magnitudes of the surface charges.

EXECUTE: **(a)** The fields add and point to the left, giving $E_{net} = 1.19 \times 10^6$ N/C.

(b) The fields oppose and point to the left, so $E_{net} = 1.19 \times 10^5$ N/C.

(c) The fields oppose but now point to the right, giving $E_{net} = 1.19 \times 10^5$ N/C.

EVALUATE: We can simplify the calculations by sketching the fields and doing an algebraic solution first.

21.103. **IDENTIFY** and **SET UP:**   Example 21.12 gives the electric field due to one infinite sheet. Add the two fields as vectors.

EXECUTE: The electric field due to the first sheet, which is in the xy-plane, is $\vec{E}_1 = (\sigma/2\epsilon_0)\hat{k}$ for $z > 0$ and $\vec{E}_1 = -(\sigma/2\epsilon_0)\hat{k}$ for $z < 0$. We can write this as $\vec{E}_1 = (\sigma/2\epsilon_0)(z/|z|)\hat{k}$, since $z/|z| = +1$ for $z > 0$ and $z/|z| = -z/z = -1$ for $z < 0$. Similarly, we can write the electric field due to the second sheet as $\vec{E}_2 = -(\sigma/2\epsilon_0)(x/|x|)\hat{i}$, since its charge density is $-\sigma$. The net field is $\vec{E} = \vec{E}_1 + \vec{E}_2 = (\sigma/2\epsilon_0)\left(-(x/|x|)\hat{i} + (z/|z|)\hat{k}\right)$.

EVALUATE: The electric field is independent of the y-component of the field point since displacement in the $\pm y$- direction is parallel to both planes. The field depends on which side of each plane the field is located.

GAUSS'S LAW

22

22.1. **(a) IDENTIFY and SET UP:** $\Phi_E = \int E\cos\phi \, dA,$ where ϕ is the angle between the normal to the sheet $\hat{n}$ and the electric field $\vec{E}$.

EXECUTE: In this problem E and $\cos\phi$ are constant over the surface so

$$\Phi_E = E\cos\phi \int dA = E\cos\phi A = (14 \text{ N/C})(\cos 60°)(0.250 \text{ m}^2) = 1.8 \text{ N} \cdot \text{m}^2/\text{C}.$$

(b) EVALUATE: Φ_E is independent of the shape of the sheet as long as ϕ and E are constant at all points on the sheet.

(c) EXECUTE: (i) $\Phi_E = E\cos\phi A.$ Φ_E is largest for $\phi = 0°,$ so $\cos\phi = 1$ and $\Phi_E = EA.$

(ii) Φ_E is smallest for $\phi = 90°,$ so $\cos\phi = 0$ and $\Phi_E = 0.$

EVALUATE: Φ_E is 0 when the surface is parallel to the field so no electric field lines pass through the surface.

22.7. **(a) IDENTIFY:** Use Eq.(22.5) to calculate the flux through the surface of the cylinder.
SET UP: The line of charge and the cylinder are sketched in Figure 22.7.

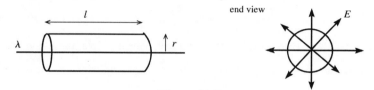

Figure 22.7

EXECUTE: The area of the curved part of the cylinder is $A = 2\pi rl.$

The electric field is parallel to the end caps of the cylinder, so $\vec{E} \cdot \vec{A} = 0$ for the ends and the flux through the cylinder end caps is zero.

The electric field is normal to the curved surface of the cylinder and has the same magnitude $E = \lambda/2\pi\epsilon_0 r$ at all points on this surface. Thus $\phi = 0°$ and

$$\Phi_E = EA\cos\phi = EA = (\lambda/2\pi\epsilon_0 r)(2\pi rl) = \frac{\lambda l}{\epsilon_0} = \frac{(6.00\times10^{-6} \text{ C/m})(0.400 \text{ m})}{8.854\times10^{-12} \text{ C}^2/\text{N}\cdot\text{m}^2} = 2.71\times10^5 \text{ N}\cdot\text{m}^2/\text{C}$$

(b) In the calculation in part (a) the radius r of the cylinder divided out, so the flux remains the same,
$\Phi_E = 2.71\times10^5 \text{ N}\cdot\text{m}^2/\text{C}.$

(c) $\Phi_E = \frac{\lambda l}{\epsilon_0} = \frac{(6.00\times10^{-6} \text{ C/m})(0.800 \text{ m})}{8.854\times10^{-12} \text{ C}^2/\text{N}\cdot\text{m}^2} = 5.42\times10^5 \text{ N}\cdot\text{m}^2/\text{C}$ (twice the flux calculated in parts (b) and (c)).

EVALUATE: The flux depends on the number of field lines that pass through the surface of the cylinder.

22.9. **IDENTIFY:** Apply the results in Example 21.10 for the field of a spherical shell of charge.

SET UP: Example 22.10 shows that $E = 0$ inside a uniform spherical shell and that $E = k\frac{|q|}{r^2}$ outside the shell.

EXECUTE: **(a)** $E = 0$

(b) $r = 0.060 \text{ m}$ and $E = (8.99\times10^9 \text{ N}\cdot\text{m}^2/\text{C}^2)\dfrac{15.0\times10^{-6} \text{ C}}{(0.060 \text{ m})^2} = 3.75\times10^7 \text{ N/C}$

(c) $r = 0.110 \text{ m}$ and $E = (8.99\times10^9 \text{ N}\cdot\text{m}^2/\text{C}^2)\dfrac{15.0\times10^{-6} \text{ C}}{(0.110 \text{ m})^2} = 1.11\times10^7 \text{ N/C}$

EVALUATE: Outside the shell the electric field is the same as if all the charge were concentrated at the center of the shell. But inside the shell the field is not the same as for a point charge at the center of the shell, inside the shell the electric field is zero.

22.13.   **(a) IDENTIFY** and **SET UP:**   It is rather difficult to calculate the flux directly from $\Phi = \oint \vec{E} \cdot d\vec{A}$ since the magnitude of $\vec{E}$ and its angle with $d\vec{A}$ varies over the surface of the cube. A much easier approach is to use Gauss's law to calculate the total flux through the cube. Let the cube be the Gaussian surface. The charge enclosed is the point charge.

EXECUTE: $\Phi_E = Q_{encl}/\epsilon_0 = \dfrac{9.60\times10^{-6}\ \text{C}}{8.854\times10^{-12}\ \text{C}^2/\text{N}\cdot\text{m}^2} = 1.084\times10^6\ \text{N}\cdot\text{m}^2/\text{C}.$

By symmetry the flux is the same through each of the six faces, so the flux through one face is

$\frac{1}{6}\left(1.084\times10^6\ \text{N}\cdot\text{m}^2/\text{C}\right) = 1.81\times10^5\ \text{N}\cdot\text{m}^2/\text{C}.$

(b) EVALUATE: In part (a) the size of the cube did not enter into the calculations. The flux through one face depends only on the amount of charge at the center of the cube. So the answer to (a) would not change if the size of the cube were changed.

22.15.   **IDENTIFY:**   The electric fields are produced by point charges.

SET UP: We use Coulomb's law, $E = \dfrac{1}{4\pi\epsilon_0}\dfrac{q}{r^2}$, to calculate the electric fields.

EXECUTE: **(a)** $E = \left(9.00\times10^9\ \text{N}\cdot\text{m}^2/\text{C}^2\right)\dfrac{5.00\times10^{-6}\ \text{C}}{(1.00\ \text{m})^2} = 4.50\times10^4\ \text{N/C}$

(b) $E = \left(9.00\times10^9\ \text{N}\cdot\text{m}^2/\text{C}^2\right)\dfrac{5.00\times10^{-6}\ \text{C}}{(7.00\ \text{m})^2} = 9.18\times10^2\ \text{N/C}$

(c) Every field line that enters the sphere on one side leaves it on the other side, so the net flux through the surface is zero.

EVALUATE: The flux would be zero no matter what shape the surface had, providing that no charge was inside the surface.

22.19.   **IDENTIFY** and **SET UP:**   Example 22.5 derived that the electric field just outside the surface of a spherical conductor that has net charge q is $E = \dfrac{1}{4\pi\epsilon_0}\dfrac{q}{R^2}$. Calculate q and from this the number of excess electrons.

EXECUTE: $q = \dfrac{R^2 E}{(1/4\pi\epsilon_0)} = \dfrac{(0.160\ \text{m})^2 (1150\ \text{N/C})}{8.988\times10^9\ \text{N}\cdot\text{m}^2/\text{C}^2} = 3.275\times10^{-9}\ \text{C}.$

Each electron has a charge of magnitude $e = 1.602\times10^{-19}\ \text{C}$, so the number of excess electrons needed is

$\dfrac{3.275\times10^{-9}\ \text{C}}{1.602\times10^{-19}\ \text{C}} = 2.04\times10^{10}.$

EVALUATE: The result we obtained for q is a typical value for the charge of an object. Such net charges correspond to a large number of excess electrons since the charge of each electron is very small.

22.21.   **IDENTIFY:**   Add the vector electric fields due to each line of charge. $E(r)$ for a line of charge is given by Example 22.6 and is directed toward a negative line of chage and away from a positive line.

SET UP: The two lines of charge are shown in Figure 22.21.

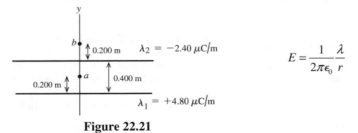

Figure 22.21

EXECUTE: **(a)** At point a, $\vec{E}_1$ and $\vec{E}_2$ are in the $+y$-direction (toward negative charge, away from positive charge).

$E_1 = (1/2\pi\epsilon_0)\left[\left(4.80\times10^{-6}\ \text{C/m}\right)/(0.200\ \text{m})\right] = 4.314\times10^5\ \text{N/C}$

$E_2 = (1/2\pi\epsilon_0)\left[\left(2.40\times10^{-6}\ \text{C/m}\right)/(0.200\ \text{m})\right] = 2.157\times10^5\ \text{N/C}$

$E = E_1 + E_2 = 6.47\times10^5\ \text{N/C}$, in the y-direction.

(b) At point b, $\vec{E}_1$ is in the $+y$-direction and $\vec{E}_2$ is in the $-y$-direction.

$$E_1 = (1/2\pi\epsilon_0)\left[(4.80\times10^{-6}\text{ C/m})/(0.600\text{ m})\right] = 1.438\times10^5\text{ N/C}$$

$$E_2 = (1/2\pi\epsilon_0)\left[(2.40\times10^{-6}\text{ C/m})/(0.200\text{ m})\right] = 2.157\times10^5\text{ N/C}$$

$$E = E_2 - E_1 = 7.2\times10^4\text{ N/C, in the }-y\text{-direction.}$$

EVALUATION: At point a the two fields are in the same direction and the magnitudes add. At point b the two fields are in opposite directions and the magnitudes subtract.

22.23. **IDENTIFY:** The electric field inside the conductor is zero, and all of its initial charge lies on its outer surface. The introduction of charge into the cavity induces charge onto the surface of the cavity, which induces an equal but opposite charge on the outer surface of the conductor. The net charge on the outer surface of the conductor is the sum of the positive charge initially there and the additional negative charge due to the introduction of the negative charge into the cavity.

(a) SET UP: First find the initial positive charge on the outer surface of the conductor using $q_i = \sigma A$, where A is the area of its outer surface. Then find the net charge on the surface after the negative charge has been introduced into the cavity. Finally use the definition of surface charge density.

EXECUTE: The original positive charge on the outer surface is

$$q_i = \sigma A = \sigma(4\pi r^2) = (6.37\times10^{-6}\text{ C/m}^2)4\pi(0.250\text{ m}^2) = 5.00\times10^{-6}\text{ C/m}^2$$

After the introduction of $-0.500\ \mu$C into the cavity, the outer charge is now

$$5.00\ \mu\text{C} - 0.500\ \mu\text{C} = 4.50\ \mu\text{C}$$

The surface charge density is now $\sigma = \dfrac{q}{A} = \dfrac{q}{4\pi r^2} = \dfrac{4.50\times10^{-6}\text{ C}}{4\pi(0.250\text{ m})^2} = 5.73\times10^{-6}\text{ C/m}^2$

(b) SET UP: Using Gauss's law, the electric field is $E = \dfrac{\Phi_E}{A} = \dfrac{q}{\epsilon_0 A} = \dfrac{q}{\epsilon_0 4\pi r^2}$

EXECUTE: Substituting numbers gives

$$E = \frac{4.50\times10^{-6}\text{ C}}{(8.85\times10^{-12}\text{ C}^2/\text{N}\cdot\text{m}^2)(4\pi)(0.250\text{ m})^2} = 6.47\times10^5\text{ N/C}.$$

(c) SET UP: We use Gauss's law again to find the flux. $\Phi_E = \dfrac{q}{\epsilon_0}$.

EXECUTE: Substituting numbers gives

$$\Phi_E = \frac{-0.500\times10^{-6}\text{ C}}{8.85\times10^{-12}\text{ C}^2/\text{N}\cdot\text{m}^2} = -5.65\times10^4\text{ N}\cdot\text{m}^2/\text{C}^2.$$

EVALUATE: The excess charge on the conductor is still $+5.00\ \mu$C, as it originally was. The introduction of the $-0.500\ \mu$C inside the cavity merely induced equal but opposite charges (for a net of zero) on the surfaces of the conductor.

22.29. **IDENTIFY:** Apply Gauss's law to a Gaussian surface and calculate E.

(a) SET UP: Consider the charge on a length l of the cylinder. This can be expressed as $q = \lambda l$. But since the surface area is $2\pi Rl$ it can also be expressed as $q = \sigma 2\pi Rl$. These two expressions must be equal, so $\lambda l = \sigma 2\pi Rl$ and $\lambda = 2\pi R\sigma$.

(b) Apply Gauss's law to a Gaussian surface that is a cylinder of length l, radius r, and whose axis coincides with the axis of the charge distribution, as shown in Figure 22.29.

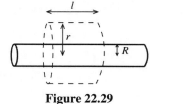

Figure 22.29

EXECUTE:
$$Q_{encl} = \sigma(2\pi Rl)$$
$$\Phi_E = 2\pi rlE$$

$\Phi_E = \dfrac{Q_{encl}}{\epsilon_0}$ gives $2\pi rlE = \dfrac{\sigma(2\pi Rl)}{\epsilon_0}$

$$E = \frac{\sigma R}{\epsilon_0 r}$$

(c) **EVALUATE:**   Example 22.6 shows that the electric field of an infinite line of charge is $E = \lambda / 2\pi\epsilon_0 r$. $\sigma = \dfrac{\lambda}{2\pi R}$,

so $E = \dfrac{\sigma R}{\epsilon_0 r} = \dfrac{R}{\epsilon_0 r}\left(\dfrac{\lambda}{2\pi R}\right) = \dfrac{\lambda}{2\pi\epsilon_0 r}$, the same as for an infinite line of charge that is along the axis of the cylinder.

22.31.   **IDENTIFY:**   Apply Gauss's law and conservation of charge.

SET UP: $E = 0$ in a conducting material.

EXECUTE: **(a)** Gauss's law says $+Q$ on inner surface, so $E = 0$ inside metal.

(b) The outside surface of the sphere is grounded, so no excess charge.

(c) Consider a Gaussian sphere with the $-Q$ charge at its center and radius less than the inner radius of the metal. This sphere encloses net charge $-Q$ so there is an electric field flux through it; there is electric field in the cavity.

(d) In an electrostatic situation $E = 0$ inside a conductor. A Gaussian sphere with the $-Q$ charge at its center and radius greater than the outer radius of the metal encloses zero net charge (the $-Q$ charge and the $+Q$ on the inner surface of the metal) so there is no flux through it and $E = 0$ outside the metal.

(e) No, $E = 0$ there. Yes, the charge has been shielded by the grounded conductor. There is nothing like positive and negative mass (the gravity force is always attractive), so this cannot be done for gravity.

EVALUATE: Field lines within the cavity terminate on the charges induced on the inner surface.

22.35.   **(a) IDENTIFY:**   Find the net flux through the parallelepiped surface and then use that in Gauss's law to find the net charge within. Flux out of the surface is positive and flux into the surface is negative.

SET UP: $\vec{E}_1$ gives flux out of the surface. See Figure 22.35a.

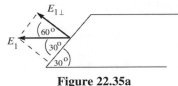

EXECUTE: $\Phi_1 = +E_{1\perp}A$

$A = (0.0600 \text{ m})(0.0500 \text{ m}) = 3.00\times10^{-3} \text{ m}^2$

$E_{1\perp} = E_1 \cos 60° = (2.50\times10^4 \text{ N/C})\cos 60°$

$E_{1\perp} = 1.25\times10^4 \text{ N/C}$

Figure 22.35a

$\Phi_{E_1} = +E_{1\perp}A = +(1.25\times10^4 \text{ N/C})(3.00\times10^{-3} \text{ m}^2) = 37.5 \text{ N}\cdot\text{m}^2/\text{C}$

SET UP: $\vec{E}_2$ gives flux into the surface. See Figure 22.35b.

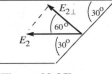

EXECUTE: $\Phi_2 = -E_{2\perp}A$

$A = (0.0600 \text{ m})(0.0500 \text{ m}) = 3.00\times10^{-3} \text{ m}^2$

$E_{2\perp} = E_2 \cos 60° = (7.00\times10^4 \text{ N/C})\cos 60°$

$E_{2\perp} = 3.50\times10^4 \text{ N/C}$

Figure 22.35b

$\Phi_{E_2} = -E_{2\perp}A = -(3.50\times10^4 \text{ N/C})(3.00\times10^{-3} \text{ m}^2) = -105.0 \text{ N}\cdot\text{m}^2/\text{C}$

The net flux is $\Phi_E = \Phi_{E_1} + \Phi_{E_2} = +37.5 \text{ N}\cdot\text{m}^2/\text{C} - 105.0 \text{ N}\cdot\text{m}^2/\text{C} = -67.5 \text{ N}\cdot\text{m}^2/\text{C}$.

The net flux is negative (inward), so the net charge enclosed is negative.

Apply Gauss's law: $\Phi_E = \dfrac{Q_{\text{encl}}}{\epsilon_0}$

$Q_{\text{encl}} = \Phi_E \epsilon_0 = (-67.5 \text{ N}\cdot\text{m}^2/\text{C})(8.854\times10^{-12} \text{ C}^2/\text{N}\cdot\text{m}^2) = -5.98\times10^{-10} \text{ C}$.

(b) EVALUATE: If there were no charge within the parallelpiped the net flux would be zero. This is not the case, so there is charge inside. The electric field lines that pass out through the surface of the parallelpiped must terminate on charges, so there also must be charges outside the parallelpiped.

22.37.   **(a) IDENTIFY:**   Apply Gauss's law to a Gaussian cylinder of length l and radius r, where $a < r < b$, and calculate E on the surface of the cylinder.

SET UP: The Gaussian surface is sketched in Figure 22.37a.

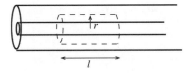

EXECUTE: $\Phi_E = E(2\pi rl)$

$Q_{\text{encl}} = \lambda l$ (the charge on the length l of the inner conductor that is inside the Gaussian surface).

Figure 22.37a

$\Phi_E = \dfrac{Q_{encl}}{\epsilon_0}$ gives $E(2\pi rl) = \dfrac{\lambda l}{\epsilon_0}$

$E = \dfrac{\lambda}{2\pi\epsilon_0 r}$. The enclosed charge is positive so the direction of $\vec{E}$ is radially outward.

(b) SET UP: Apply Gauss's law to a Gaussian cylinder of length l and radius r, where $r > c$, as shown in Figure 22.37b.

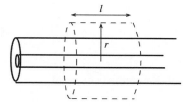

Figure 22.37b

EXECUTE: $\Phi_E = E(2\pi rl)$

$Q_{encl} = \lambda l$ (the charge on the length l of the inner conductor that is inside the Gaussian surface; the outer conductor carries no net charge).

$\Phi_E = \dfrac{Q_{encl}}{\epsilon_0}$ gives $E(2\pi rl) = \dfrac{\lambda l}{\epsilon_0}$

$E = \dfrac{\lambda}{2\pi\epsilon_0 r}$. The enclosed charge is positive so the direction of $\vec{E}$ is radially outward.

(c) $E = 0$ within a conductor. Thus $E = 0$ for $r < a$;

$E = \dfrac{\lambda}{2\pi\epsilon_0 r}$ for $a < r < b$; $E = 0$ for $b < r < c$;

$E = \dfrac{\lambda}{2\pi\epsilon_0 r}$ for $r > c$. The graph of E versus r is sketched in Figure 22.37c.

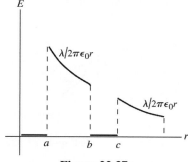

Figure 22.37c

EVALUATE: Inside either conductor $E = 0$. Between the conductors and outside both conductors the electric field is the same as for a line of charge with linear charge density λ lying along the axis of the inner conductor.

(d) IDENTIFY and SET UP: <u>inner surface:</u> Apply Gauss's law to a Gaussian cylinder with radius r, where $b < r < c$. We know E on this surface; calculate Q_{encl}.

EXECUTE: This surface lies within the conductor of the outer cylinder, where $E = 0$, so $\Phi_E = 0$. Thus by Gauss's law $Q_{encl} = 0$. The surface encloses charge λl on the inner conductor, so it must enclose charge $-\lambda l$ on the inner surface of the outer conductor. The charge per unit length on the inner surface of the outer cylinder is $-\lambda$.

<u>outer surface:</u> The outer cylinder carries no net charge. So if there is charge per unit length $-\lambda$ on its inner surface there must be charge per unit length $+\lambda$ on the outer surface.

EVALUATE: The electric field lines between the conductors originate on the surface charge on the outer surface of the inner conductor and terminate on the surface charges on the inner surface of the outer conductor. These surface charges are equal in magnitude (per unit length) and opposite in sign. The electric field lines outside the outer conductor originate from the surface charge on the outer surface of the outer conductor.

22.39.   **(a) IDENTIFY:**   Use Gauss's law to calculate $E(r)$.

(i) **SET UP:** $r < a$: Apply Gauss's law to a cylindrical Gaussian surface of length l and radius r, where $r < a$, as sketched in Figure 22.39a.

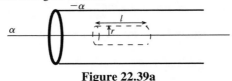

Figure 22.39a

EXECUTE: $\Phi_E = E(2\pi r l)$

$Q_{\text{encl}} = \alpha l$ (the charge on the length l of the line of charge)

$\Phi_E = \dfrac{Q_{\text{encl}}}{\epsilon_0}$ gives $E(2\pi r l) = \dfrac{\alpha l}{\epsilon_0}$

$E = \dfrac{\alpha}{2\pi\epsilon_0 r}$. The enclosed charge is positive so the direction of $\vec{E}$ is radially outward.

(ii) $a < r < b$: Points in this region are within the conducting tube, so $E = 0$.

(iii) **SET UP:** $r > b$: Apply Gauss's law to a cylindrical Gaussian surface of length l and radius r, where $r > b$, as sketched in Figure 22.39b.

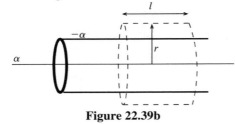

Figure 22.39b

EXECUTE: $\Phi_E = E(2\pi r l)$

$Q_{\text{encl}} = \alpha l$ (the charge on length l of the line of charge) $-\alpha l$ (the charge on length l of the tube) Thus $Q_{\text{encl}} = 0$.

$\Phi_E = \dfrac{Q_{\text{encl}}}{\epsilon_0}$ gives $E(2\pi r l) = 0$ and $E = 0$. The graph of E versus r is sketched in Figure 22.39c.

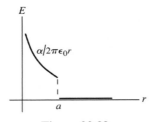

Figure 22.39c

(b) IDENTIFY: Apply Gauss's law to cylindrical surfaces that lie just outside the inner and outer surfaces of the tube. We know E so can calculate Q_{encl}.

(i) **SET UP:**   inner surface
Apply Gauss's law to a cylindrical Gaussian surface of length l and radius r, where $a < r < b$.

EXECUTE: This surface lies within the conductor of the tube, where $E = 0$, so $\Phi_E = 0$. Then by Gauss's law $Q_{\text{encl}} = 0$. The surface encloses charge αl on the line of charge so must enclose charge $-\alpha l$ on the inner surface of the tube. The charge per unit length on the inner surface of the tube is $-\alpha$.

(ii) outer surface
The net charge per unit length on the tube is $-\alpha$. We have shown in part (i) that this must all reside on the inner surface, so there is no net charge on the outer surface of the tube.

EVALUATE: For $r < a$ the electric field is due only to the line of charge. For $r > b$ the electric field of the tube is the same as for a line of charge along its axis. The fields of the line of charge and of the tube are equal in magnitude and opposite in direction and sum to zero. For $r < a$ the electric field lines originate on the line of charge and terminate on the surface charge on the inner surface of the tube. There is no electric field outside the tube and no surface charge on the outer surface of the tube.

22.41.   **IDENTIFY:**   First make a free-body diagram of the sphere. The electric force acts to the left on it since the electric field due to the sheet is horizontal. Since it hangs at rest, the sphere is in equilibrium so the forces on it add to zero, by Newton's first law. Balance horizontal and vertical force components separately.

SET UP: Call T the tension in the thread and E the electric field. Balancing horizontal forces gives $T \sin\theta = qE$. Balancing vertical forces we get $T \cos\theta = mg$. Combining these equations gives $\tan\theta = qE/mg$, which means that $\theta = \arctan(qE/mg)$. The electric field for a sheet of charge is $E = \sigma 2\epsilon_0$.

EXECUTE: Substituting the numbers gives us $E = \dfrac{\sigma}{2\epsilon_0} = \dfrac{2.50 \times 10^{-7} \text{ C/m}^2}{2\left(8.85 \times 10^{-12} \text{ C}^2/\text{N} \cdot \text{m}^2\right)} = 1.41 \times 10^4 \text{ N/C}$. Then

$$\theta = \arctan\left[\dfrac{\left(5.00 \times 10^{-8} \text{ C}\right)\left(1.41 \times 10^4 \text{ N/C}\right)}{\left(2.00 \times 10^{-2} \text{ kg}\right)\left(9.80 \text{ m/s}^2\right)}\right] = 19.8°$$

EVALUATE: Increasing the field, or decreasing the mass of the sphere, would cause the sphere to hang at a larger angle.

22.45.   **IDENTIFY:**   Apply Gauss's law to a spherical Gaussian surface with radius r. Calculate the electric field at the surface of the Gaussian sphere.

(a) SET UP: (i) $r < a$: The Gaussian surface is sketched in Figure 22.45a.

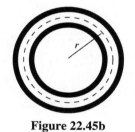

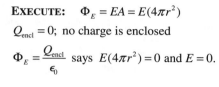

EXECUTE: $\Phi_E = EA = E(4\pi r^2)$

$Q_{\text{encl}} = 0$; no charge is enclosed

$\Phi_E = \dfrac{Q_{\text{encl}}}{\epsilon_0}$ says $E(4\pi r^2) = 0$ and $E = 0$.

Figure 22.45a

(ii) $a < r < b$: Points in this region are in the conductor of the small shell, so $E = 0$.
(iii) **SET UP:** $b < r < c$: The Gaussian surface is sketched in Figure 22.45b.
Apply Gauss's law to a spherical Gaussian surface with radius $b < r < c$.

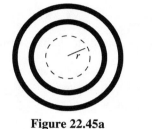

EXECUTE: $\Phi_E = EA = E(4\pi r^2)$
The Gaussian surface encloses all of the small shell and none of the large shell, so $Q_{\text{encl}} = +2q$.

Figure 22.45b

$\Phi_E = \dfrac{Q_{\text{encl}}}{\epsilon_0}$ gives $E(4\pi r^2) = \dfrac{2q}{\epsilon_0}$ so $E = \dfrac{2q}{4\pi\epsilon_0 r^2}$. Since the enclosed charge is positive the electric field is radially outward.

(iv) $c < r < d$: Points in this region are in the conductor of the large shell, so $E = 0$.
(v) **SET UP:** $r > d$: Apply Gauss's law to a spherical Gaussian surface with radius $r > d$, as shown in Figure 22.45c.

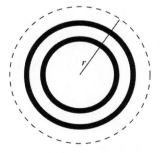

EXECUTE: $\Phi_E = EA = E\left(4\pi r^2\right)$
The Gaussian surface encloses all of the small shell and all of the large shell, so $Q_{\text{encl}} = +2q + 4q = 6q$.

Figure 22.45c

$\Phi_E = \dfrac{Q_{\text{encl}}}{\epsilon_0}$ gives $E\left(4\pi r^2\right) = \dfrac{6q}{\epsilon_0}$

$E = \dfrac{6q}{4\pi\epsilon_0 r^2}$. Since the enclosed charge is positive the electric field is radially outward.

The graph of E versus r is sketched in Figure 22.45d.

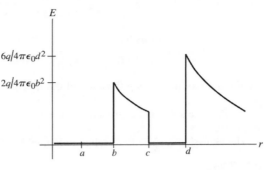

Figure 22.45d

(b) IDENTIFY and **SET UP:** Apply Gauss's law to a sphere that lies outside the surface of the shell for which we want to find the surface charge.

EXECUTE: (i) charge on inner surface of the small shell: Apply Gauss's law to a spherical Gaussian surface with radius $a < r < b$. This surface lies within the conductor of the small shell, where $E = 0$, so $\Phi_E = 0$. Thus by Gauss's law $Q_{encl} = 0$, so there is zero charge on the inner surface of the small shell.

(ii) charge on outer surface of the small shell: The total charge on the small shell is $+2q$. We found in part (i) that there is zero charge on the inner surface of the shell, so all $+2q$ must reside on the outer surface.

(iii) charge on inner surface of large shell: Apply Gauss's law to a spherical Gaussian surface with radius $c < r < d$. The surface lies within the conductor of the large shell, where $E = 0$, so $\Phi_E = 0$. Thus by Gauss's law $Q_{encl} = 0$. The surface encloses the $+2q$ on the small shell so there must be charge $-2q$ on the inner surface of the large shell to make the total enclosed charge zero.

(iv) charge on outer surface of large shell: The total charge on the large shell is $+4q$. We showed in part (iii) that the charge on the inner surface is $-2q$, so there must be $+6q$ on the outer surface.

EVALUATE: The electric field lines for $b < r < c$ originate from the surface charge on the outer surface of the inner shell and all terminate on the surface charge on the inner surface of the outer shell. These surface charges have equal magnitude and opposite sign. The electric field lines for $r > d$ originate from the surface charge on the outer surface of the outer sphere.

22.49. **IDENTIFY:** Use Gauss's law to find the electric field $\vec{E}$ produced by the shell for $r < R$ and $r > R$ and then use $\vec{F} = q\vec{E}$ to find the force the shell exerts on the point charge.

(a) SET UP: Apply Gauss's law to a spherical Gaussian surface that has radius $r > R$ and that is concentric with the shell, as sketched in Figure 22.49a.

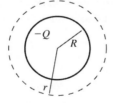

EXECUTE: $\Phi_E = E\left(4\pi r^2\right)$

$Q_{encl} = -Q$

Figure 22.49a

$\Phi_E = \dfrac{Q_{encl}}{\epsilon_0}$ gives $E\left(4\pi r^2\right) = \dfrac{-Q}{\epsilon_0}$

The magnitude of the field is $E = \dfrac{Q}{4\pi\epsilon_0 r^2}$ and it is directed toward the center of the shell. Then $F = qE = \dfrac{qQ}{4\pi\epsilon_0 r^2}$, directed toward the center of the shell. (Since q is positive, $\vec{E}$ and $\vec{F}$ are in the same direction.)

(b) SET UP: Apply Gauss's law to a spherical Gaussian surface that has radius $r < R$ and that is concentric with the shell, as sketched in Figure 22.49b.

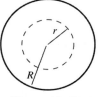

Figure 22.49b

EXECUTE: $\Phi_E = E\left(4\pi r^2\right)$

$Q_{encl} = 0$

$\Phi_E = \dfrac{Q_{encl}}{\epsilon_0}$ gives $E\left(4\pi r^2\right) = 0$

Then $E = 0$ so $F = 0$.

EVALUATE: Outside the shell the electric field and the force it exerts is the same as for a point charge $-Q$ located at the center of the shell. Inside the shell $E = 0$ and there is no force.

22.53.   **IDENTIFY:**   There is a force on each electron due to the other electron and a force due to the sphere of charge. Use Coulomb's law for the force between the electrons. Example 22.9 gives E inside a uniform sphere and Eq.(21.3) gives the force.

SET UP: The positions of the electrons are sketched in Figure 22.53a.

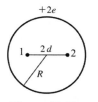

If the electrons are in equilibrium the net force on each one is zero.

Figure 22.53a

EXECUTE: Consider the forces on electron 2. There is a repulsive force F_1 due to the other electron, electron 1.

$$F_1 = \frac{1}{4\pi\epsilon_0}\frac{e^2}{\left(2d\right)^2}$$

The electric field inside the uniform distribution of positive charge is $E = \dfrac{Qr}{4\pi\epsilon_0 R^3}$ (Example 22.9), where $Q = +2e$.

At the position of electron 2, $r = d$. The force F_{cd} exerted by the positive charge distribution is $F_{cd} = eE = \dfrac{e\left(2e\right)d}{4\pi\epsilon_0 R^3}$ and is attractive.

The force diagram for electron 2 is given in Figure 22.53b.

$$\overset{F_{cd}}{\xleftarrow{\hspace{2cm}}}\bullet\overset{F_1}{\xrightarrow{\hspace{2cm}}}$$

Figure 22.53b

Net force equals zero implies $F_1 = F_{cd}$ and $\dfrac{1}{4\pi\epsilon_0}\dfrac{e^2}{4d^2} = \dfrac{2e^2 d}{4\pi\epsilon_0 R^3}$

Thus $\left(1/4d^2\right) = 2d/R^3$, so $d^3 = R^3/8$ and $d = R/2$.

EVALUATE: The electric field of the sphere is radially outward; it is zero at the center of the sphere and increases with distance from the center. The force this field exerts on one of the electrons is radially inward and increases as the electron is farther from the center. The force from the other electron is radially outward, is infinite when $d = 0$ and decreases as d increases. It is reasonable therefore for there to be a value of d for which these forces balance.

22.55.   **(a) IDENTIFY** and **SET UP:**   Consider the direction of the field for x slightly greater than and slightly less than zero. The slab is sketched in Figure 22.55a.

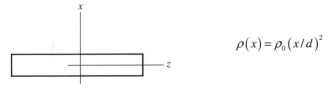

Figure 22.55a

EXECUTE: The charge distribution is symmetric about $x = 0$, so by symmetry $E(x) = E(-x)$. But for $x > 0$ the field is in the $+x$ direction and for $x < 0$ the field is in the $-x$ direction. At $x = 0$ the field can't be both in the $+x$ and $-x$ directions so must be zero. That is, $E_x(x) = -E_x(-x)$. At point $x = 0$ this gives $E_x(0) = -E_x(0)$ and this equation is satisfied only for $E_x(0) = 0$.

(b) IDENTIFY and **SET UP:** $|x| > d$ (outside the slab)

Apply Gauss's law to a cylindrical Gaussian surface whose axis is perpendicular to the slab and whose end caps have area A and are the same distance $|x| > d$ from $x = 0$, as shown in Figure 22.55b.

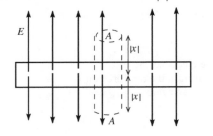

EXECUTE: $\Phi_E = 2EA$

Figure 22.55b

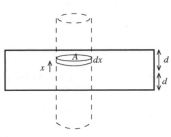

Figure 22.55c

To find Q_{encl} consider a thin disk at coordinate x and with thickness dx, as shown in Figure 22.55c. The charge within this disk is

$$dq = \rho\, dV = \rho A\, dx = \left(\rho_0 A / d^2\right) x^2\, dx.$$

The total charge enclosed by the Gaussian cylinder is

$$Q_{\text{encl}} = 2 \int_0^d dq = \left(2\rho_0 A / d^2\right) \int_0^d x^2\, dx = \left(2\rho_0 A / d^2\right)\left(d^3/3\right) = \tfrac{2}{3}\rho_0 A d.$$

Then $\Phi_E = \dfrac{Q_{\text{encl}}}{\epsilon_0}$ gives $2EA = 2\rho_0 A d / 3\epsilon_0$.

$$E = \rho_0 d / 3\epsilon_0$$

$\vec{E}$ is directed away from $x = 0$, so $\vec{E} = \left(\rho_0 d / 3\epsilon_0\right)\left(x/|x|\right)\hat{\imath}$.

IDENTIFY and **SET UP:** $|x| < d$ (inside the slab)

Apply Gauss's law to a cylindrical Gaussian surface whose axis is perpendicular to the slab and whose end caps have area A and are the same distance $|x| < d$ from $x = 0$, as shown in Figure 22.55d.

EXECUTE: $\Phi_E = 2EA$

Figure 22.55d

Q_{encl} is found as above, but now the integral on dx is only from 0 to x instead of 0 to d.

$$Q_{encl} = 2\int_0^x dq = \left(2\rho_0 A/d^2\right)\int_0^x x^2 dx = \left(2\rho_0 A/d^2\right)\left(x^3/3\right).$$

Then $\Phi_E = \dfrac{Q_{encl}}{\epsilon_0}$ gives $2EA = 2\rho_0 Ax^3/3\epsilon_0 d^2$.

$$E = \rho_0 x^3/3\epsilon_0 d^2$$

$\vec{E}$ is directed away from $x = 0$, so $\vec{E} = \left(\rho_0 x^3/3\epsilon_0 d^2\right)\hat{i}$.

EVALUATE: Note that $E = 0$ at $x = 0$ as stated in part (a). Note also that the expressions for $|x| > d$ and $|x| < d$ agree for $x = d$.

22.57. $\rho(r) = \rho_0(1 - r/R)$ for $r \le R$ where $\rho_0 = 3Q/\pi R^3$. $\rho(r) = 0$ for $r \ge R$

(a) IDENTIFY: The charge density varies with r inside the spherical volume. Divide the volume up into thin concentric shells, of radius r and thickness dr. Find the charge dq in each shell and integrate to find the total charge.
SET UP: The thin shell is sketched in Figure 22.57a.

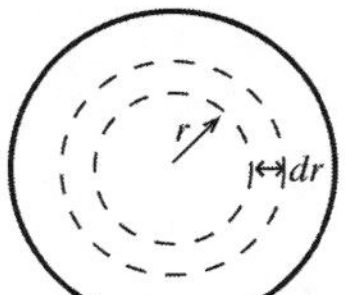

EXECUTE: The volume of such a shell is $dV = 4\pi r^2 dr$
The charge contained within the shell is
$dq = \rho(r)dV = 4\pi r^2 \rho_0(1 - r/R)dr$

Figure 22.57a

The total charge Q in the charge distribution is obtained by integrating dq over all such shells into which the sphere can be subdivided:

$$Q = \int dq = \int_0^R 4\pi r^2 \rho_0(1 - r/R)dr = 4\pi\rho_0 \int_0^R \left(r^2 - r^3/R\right)dr$$

$Q = 4\pi\rho_0\left[\dfrac{r^3}{3} - \dfrac{r^4}{4R}\right]_0^R = 4\pi\rho_0\left(\dfrac{R^3}{3} - \dfrac{R^4}{4R}\right) = 4\pi\rho_0\left(R^3/12\right) = 4\pi\left(3Q/\pi R^3\right)\left(R^3/12\right) = Q$, as was to be shown.

(b) IDENTIFY: Apply Gauss's law to a spherical surface of radius r, where $r > R$.
SET UP: The Gaussian surface is shown in Figure 22.57b.

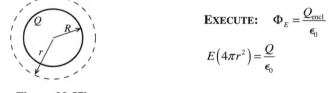

EXECUTE: $\Phi_E = \dfrac{Q_{encl}}{\epsilon_0}$

$$E\left(4\pi r^2\right) = \dfrac{Q}{\epsilon_0}$$

Figure 22.57b

$E = \dfrac{Q}{4\pi\epsilon_0 r^2}$; same as for point charge of charge Q.

(c) IDENTIFY: Apply Gauss's law to a spherical surface of radius r, where $r < R$:
SET UP: The Gaussian surface is shown in Figure 22.57c.

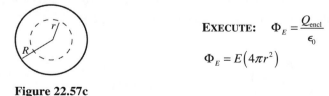

EXECUTE: $\Phi_E = \dfrac{Q_{encl}}{\epsilon_0}$

$$\Phi_E = E\left(4\pi r^2\right)$$

Figure 22.57c

The calculate the enclosed charge Q_{encl} use the same technique as in part (a), except integrate dq out to r rather than R. (We want the charge that is inside radius r.)

$$Q_{encl} = \int_0^r 4\pi r'^2 \rho_0 \left(1 - \frac{r'}{R}\right) dr' = 4\pi\rho_0 \int_0^r \left(r'^2 - \frac{r'^3}{R}\right) dr'$$

$$Q_{encl} = 4\pi\rho_0 \left[\frac{r'^3}{3} - \frac{r'^4}{4R}\right]_0^r = 4\pi\rho_0 \left(\frac{r^3}{3} - \frac{r^4}{4R}\right) = 4\pi\rho_0 r^3 \left(\frac{1}{3} - \frac{r}{4R}\right)$$

$$\rho_0 = \frac{3Q}{\pi R^3} \text{ so } Q_{encl} = 12Q \frac{r^3}{R^3}\left(\frac{1}{3} - \frac{r}{4R}\right) = Q\left(\frac{r^3}{R^3}\right)\left(4 - 3\frac{r}{R}\right).$$

Thus Gauss's law gives $E(4\pi r^2) = \dfrac{Q}{\epsilon_0}\left(\dfrac{r^3}{R^3}\right)\left(4 - 3\dfrac{r}{R}\right)$

$$E = \frac{Qr}{4\pi\epsilon_0 R^3}\left(4 - \frac{3r}{R}\right), \ r \le R$$

(d) The graph of E versus r is sketched in Figure 22.57d.

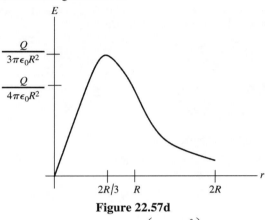

Figure 22.57d

(e) Where the electric field is a maximum, $\dfrac{dE}{dr} = 0$. Thus $\dfrac{d}{dr}\left(4r - \dfrac{3r^2}{R}\right) = 0$ so $4 - 6r/R = 0$ and $r = 2R/3$.

At this value of r, $E = \dfrac{Q}{4\pi\epsilon_0 R^3}\left(\dfrac{2R}{3}\right)\left(4 - \dfrac{3}{R}\dfrac{2R}{3}\right) = \dfrac{Q}{3\pi\epsilon_0 R^2}$

EVALUATE: Our expressions for $E(r)$ for $r < R$ and for $r > R$ agree at $r = R$. The results of part (e) for the value of r where $E(r)$ is a maximum agrees with the graph in part (d).

22.61.   **(a) IDENTIFY:** Use $\vec{E}(\vec{r})$ from Example (22.9) (inside the sphere) and relate the position vector of a point inside the sphere measured from the origin to that measured from the center of the sphere.

SET UP: For an insulating sphere of uniform charge density ρ and centered at the origin, the electric field inside the sphere is given by $E = Qr'/4\pi\epsilon_0 R^3$ (Example 22.9), where $\vec{r}'$ is the vector from the center of the sphere to the point where E is calculated.

But $\rho = 3Q/4\pi R^3$ so this may be written as $E = \rho r/3\epsilon_0$. And $\vec{E}$ is radially outward, in the direction of $\vec{r}'$, so $\vec{E} = \rho\vec{r}'/3\epsilon_0$.

For a sphere whose center is located by vector $\vec{b}$, a point inside the sphere and located by $\vec{r}$ is located by the vector $\vec{r}' = \vec{r} - \vec{b}$ relative to the center of the sphere, as shown in Figure 22.61.

EXECUTE: Thus $\vec{E} = \dfrac{\rho(\vec{r} - \vec{b})}{3\epsilon_0}$

Figure 22.61

EVALUATE: When $b = 0$ this reduces to the result of Example 22.9. When $\vec{r} = \vec{b}$, this gives $E = 0$, which is correct since we know that $E = 0$ at the center of the sphere.

(b) IDENTIFY: The charge distribution can be represented as a uniform sphere with charge density ρ and centered at the origin added to a uniform sphere with charge density $-\rho$ and centered at $\vec{r} = \vec{b}$.

SET UP: $\vec{E} = \vec{E}_{\text{uniform}} + \vec{E}_{\text{hole}}$, where $\vec{E}_{\text{uniform}}$ is the field of a uniformly charged sphere with charge density ρ and $\vec{E}_{\text{hole}}$ is the field of a sphere located at the hole and with charge density $-\rho$. (Within the spherical hole the net charge density is $+\rho - \rho = 0$.)

EXECUTE: $\vec{E}_{\text{uniform}} = \dfrac{\rho\vec{r}}{3\epsilon_0}$, where $\vec{r}$ is a vector from the center of the sphere.

$\vec{E}_{\text{hole}} = \dfrac{-\rho\left(\vec{r} - \vec{b}\right)}{3\epsilon_0}$, at points inside the hole.

Then $\vec{E} = \dfrac{\rho\vec{r}}{3\epsilon_0} + \left(\dfrac{-\rho\left(\vec{r} - \vec{b}\right)}{3\epsilon_0}\right) = \dfrac{\rho\vec{b}}{3\epsilon_0}$.

EVALUATE: $\vec{E}$ is independent of $\vec{r}$ so is uniform inside the hole. The direction of $\vec{E}$ inside the hole is in the direction of the vector $\vec{b}$, the direction from the center of the insulating sphere to the center of the hole.

22.63.   **IDENTIFY:**   The electric field at each point is the vector sum of the fields of the two charge distributions.

SET UP: Inside a sphere of uniform positive charge, $E = \dfrac{\rho r}{3\epsilon_0}$.

$\rho = \dfrac{Q}{\frac{4}{3}\pi R^3} = \dfrac{3Q}{4\pi R^3}$ so $E = \dfrac{Qr}{4\pi\epsilon_0 R^3}$, directed away from the center of the sphere. Outside a sphere of uniform positive charge, $E = \dfrac{Q}{4\pi\epsilon_0 r^2}$, directed away from the center of the sphere.

EXECUTE: **(a)** $x = 0$. This point is inside sphere 1 and outside sphere 2. The fields are shown in Figure 22.63a.

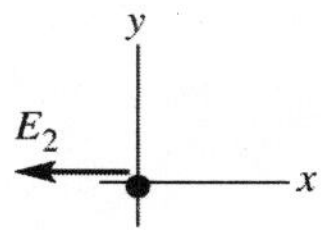

$E_1 = \dfrac{Qr}{4\pi\epsilon_0 R^3} = 0$, since $r = 0$.

Figure 22.63a

$E_2 = \dfrac{Q}{4\pi\epsilon_0 r^2}$ with $r = 2R$ so $E_2 = \dfrac{Q}{16\pi\epsilon_0 R^2}$, in the $-x$-direction.

Thus $\vec{E} = \vec{E}_1 + \vec{E}_2 = \dfrac{Q}{16\pi\epsilon_0 R^2}\hat{i}$.

(b) $x = R/2$. This point is inside sphere 1 and outside sphere 2. Each field is directed away from the center of the sphere that produces it. The fields are shown in Figure 22.63b.

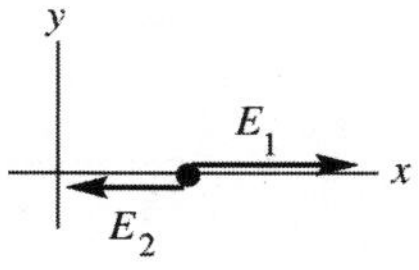

$E_1 = \dfrac{Qr}{4\pi\epsilon_0 R^3}$ with $r = R/2$ so

$E_1 = \dfrac{Q}{8\pi\epsilon_0 R^2}$

Figure 22.63b

$E_2 = \dfrac{Q}{4\pi\epsilon_0 r^2}$ with $r = 3R/2$ so $E_2 = \dfrac{Q}{9\pi\epsilon_0 R^2}$

$E = E_1 - E_2 = \dfrac{Q}{72\pi\epsilon_0 R^2}$, in the $+x$-direction and $\vec{E} = \dfrac{Q}{72\pi\epsilon_0 R^2}\hat{i}$

(c) $x = R$. This point is at the surface of each sphere. The fields have equal magnitudes and opposite directions, so $E = 0$.

(d) $x = 3R$. This point is outside both spheres. Each field is directed away from the center of the sphere that produces it. The fields are shown in Figure 22.63c.

$E_1 = \dfrac{Q}{4\pi\epsilon_0 r^2}$ with $r = 3R$ so

$E_1 = \dfrac{Q}{36\pi\epsilon_0 R^2}$

Figure 22.63c

$E_2 = \dfrac{Q}{4\pi\epsilon_0 r^2}$ with $r = R$ so $E_2 = \dfrac{Q}{4\pi\epsilon_0 R^2}$

$E = E_1 + E_2 = \dfrac{5Q}{18\pi\epsilon_0 R^2}$, in the $+x$-direction and $\vec{E} = \dfrac{5Q}{18\pi\epsilon_0 R^2}\hat{i}$

EVALUATE: The field of each sphere is radially outward from the center of the sphere. We must use the correct expression for $E(r)$ for each sphere, depending on whether the field point is inside or outside that sphere.

22.65. **IDENTIFY:** Let $-dQ$ be the electron charge contained within a spherical shell of radius r' and thickness dr'. Integrate r' from 0 to r to find the electron charge within a sphere of radius r. Apply Gauss's law to a sphere of radius r to find the electric field $E(r)$.

SET UP: The volume of the spherical shell is $dV = 4\pi r'^2\, dr'$.

EXECUTE: **(a)** $Q(r) = Q - \int \rho\, dV = Q - \dfrac{Q4\pi}{\pi a_0^3}\int e^{-2r/a_0} r^2\, dr = Q - \dfrac{4Q}{a_0^3}\int_0^r r'^2 e^{-2r'/a_0}\, dr'$.

$$Q(r) = Q - \dfrac{4Qe^{-\alpha r}}{a_0^3 \alpha^3}(2e^{\alpha r} - \alpha^2 r^2 - 2\alpha r - 2) = Qe^{-2r/a_0}[2(r/a_0)^2 + 2(r/a_0) + 1].$$

Note if $r \to \infty$, $Q(r) \to 0$; the total net charge of the atom is zero.

(b) The electric field is radially outward. Gauss's law gives $E(4\pi r^2) = \dfrac{Q(r)}{\epsilon_0}$ and

$E = \dfrac{kQe^{-2r/a_0}}{r^2}(2(r/a_0)^2 + 2(r/a_0) + 1)$.

(c) The graph of E versus r is sketched in Figure 22.65. What is plotted is the scaled E, equal to $E/E_{\text{pt charge}}$, versus scaled r, equal to r/a_0. $E_{\text{pt charge}} = \dfrac{kQ}{r^2}$ is the field of a point charge.

EVALUATE: As $r \to 0$, the field approaches that of the positive point charge (the proton). For increasing r the electric field falls to zero more rapidly than the $1/r^2$ dependence for a point charge.

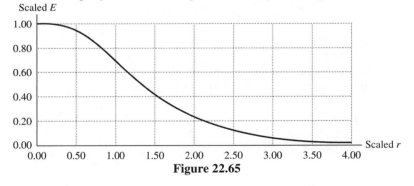

Figure 22.65

ELECTRIC POTENTIAL

23.1. **IDENTIFY:** Apply Eq.(23.2) to calculate the work. The electric potential energy of a pair of point charges is given by Eq.(23.9).

SET UP: Let the initial position of q_2 be point a and the final position be point b, as shown in Figure 23.1.

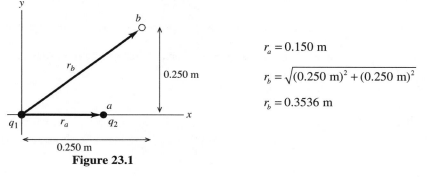

$r_a = 0.150 \text{ m}$

$r_b = \sqrt{(0.250 \text{ m})^2 + (0.250 \text{ m})^2}$

$r_b = 0.3536 \text{ m}$

Figure 23.1

EXECUTE: $W_{a \to b} = U_a - U_b$

$$U_a = \frac{1}{4\pi\epsilon_0} \frac{q_1 q_2}{r_a} = (8.988 \times 10^9 \text{ N} \cdot \text{m}^2 / \text{C}^2) \frac{(+2.40 \times 10^{-6} \text{ C})(-4.30 \times 10^{-6} \text{ C})}{0.150 \text{ m}}$$

$$U_a = -0.6184 \text{ J}$$

$$U_b = \frac{1}{4\pi\epsilon_0} \frac{q_1 q_2}{r_b} = (8.988 \times 10^9 \text{ N} \cdot \text{m}^2 / \text{C}^2) \frac{(+2.40 \times 10^{-6} \text{ C})(-4.30 \times 10^{-6} \text{ C})}{0.3536 \text{ m}}$$

$$U_b = -0.2623 \text{ J}$$

$$W_{a \to b} = U_a - U_b = -0.6184 \text{ J} - (-0.2623 \text{ J}) = -0.356 \text{ J}$$

EVALUATE: The attractive force on q_2 is toward the origin, so it does negative work on q_2 when q_2 moves to larger r.

23.5. **(a) IDENTIFY:** Use conservation of energy:

$$K_a + U_a + W_{\text{other}} = K_b + U_b$$

U for the pair of point charges is given by Eq.(23.9).
SET UP:

$v_a = 22.0 \text{ m/s}$

$v_b = ?$

a ◯ q_2 b ◯ q_2 ◯ q_1

$r_a = 0.800 \text{ m}$

$r_b = 0.400 \text{ m}$

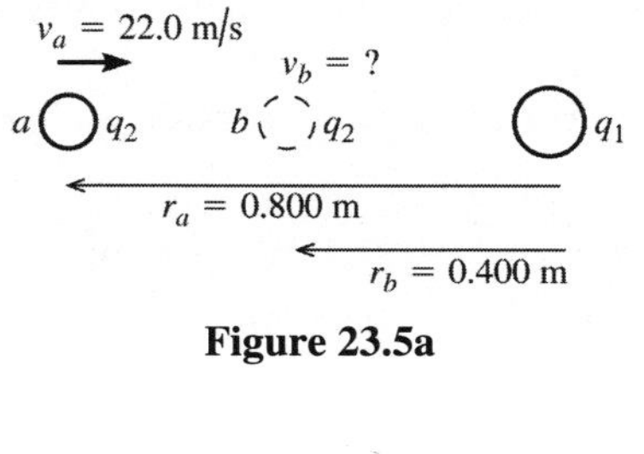

Let point a be where q_2 is 0.800 m from q_1 and point b be where q_2 is 0.400 m from q_1, as shown in Figure 23.5a.

Figure 23.5a

EXECUTE: Only the electric force does work, so $W_{other} = 0$ and $U = \dfrac{1}{4\pi\epsilon_0}\dfrac{q_1 q_2}{r}$.

$$K_a = \tfrac{1}{2}mv_a^2 = \tfrac{1}{2}(1.50\times10^{-3}\ \text{kg})(22.0\ \text{m/s})^2 = 0.3630\ \text{J}$$

$$U_a = \dfrac{1}{4\pi\epsilon_0}\dfrac{q_1 q_2}{r_a} = (8.988\times10^9\ \text{N}\cdot\text{m}^2/\text{C}^2)\dfrac{(-2.80\times10^{-6}\ \text{C})(-7.80\times10^{-6}\ \text{C})}{0.800\ \text{m}} = +0.2454\ \text{J}$$

$$K_b = \tfrac{1}{2}mv_b^2$$

$$U_b = \dfrac{1}{4\pi\epsilon_0}\dfrac{q_1 q_2}{r_b} = (8.988\times10^9\ \text{N}\cdot\text{m}^2/\text{C}^2)\dfrac{(-2.80\times10^{-6}\ \text{C})(-7.80\times10^{-6}\ \text{C})}{0.400\ \text{m}} = +0.4907\ \text{J}$$

The conservation of energy equation then gives $K_b = K_a + (U_a - U_b)$

$$\tfrac{1}{2}mv_b^2 = +0.3630\ \text{J} + (0.2454\ \text{J} - 0.4907\ \text{J}) = 0.1177\ \text{J}$$

$$v_b = \sqrt{\dfrac{2(0.1177\ \text{J})}{1.50\times10^{-3}\ \text{kg}}} = 12.5\ \text{m/s}$$

EVALUATE: The potential energy increases when the two positively charged spheres get closer together, so the kinetic energy and speed decrease.

(b) IDENTIFY: Let point c be where q_2 has its speed momentarily reduced to zero. Apply conservation of energy to points a and c: $K_a + U_a + W_{other} = K_c + U_c$.

SET UP: Points a and c are shown in Figure 23.5b.

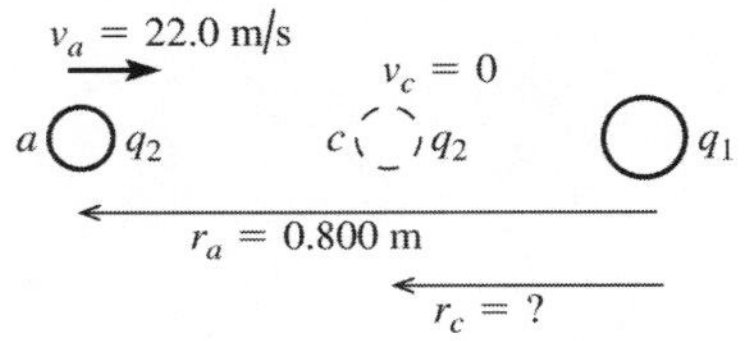

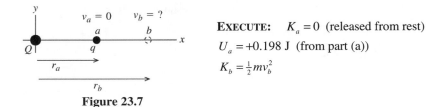

EXECUTE: $K_a = +0.3630\ \text{J}$ (from part (a))

$U_a = +0.2454\ \text{J}$ (from part (a))

Figure 23.5b

$K_c = 0$ (at distance of closest approach the speed is zero)

$$U_c = \dfrac{1}{4\pi\epsilon_0}\dfrac{q_1 q_2}{r_c}$$

Thus conservation of energy $K_a + U_a = U_c$ gives $\dfrac{1}{4\pi\epsilon_0}\dfrac{q_1 q_2}{r_c} = +0.3630\ \text{J} + 0.2454\ \text{J} = 0.6084\ \text{J}$

$$r_c = \dfrac{1}{4\pi\epsilon_0}\dfrac{q_1 q_2}{0.6084\ \text{J}} = (8.988\times10^9\ \text{N}\cdot\text{m}^2/\text{C}^2)\dfrac{(-2.80\times10^{-6}\ \text{C})(-7.80\times10^{-6}\ \text{C})}{+0.6084\ \text{J}} = 0.323\ \text{m}.$$

EVALUATE: $U \to \infty$ as $r \to 0$ so q_2 will stop no matter what its initial speed is.

23.7.   **(a) IDENTIFY and SET UP:** U is given by Eq.(23.9).

EXECUTE: $U = \dfrac{1}{4\varepsilon\pi_0}\dfrac{qq'}{r}$

$$U = (8.988\times10^9\ \text{N}\cdot\text{m}^2/\text{C}^2)\dfrac{(+4.60\times10^{-6}\ \text{C})(+1.20\times10^{-6}\ \text{C})}{0.250\ \text{m}} = +0.198\ \text{J}$$

EVALUATE: The two charges are both of the same sign so their electric potential energy is positive.

(b) IDENTIFY: Use conservation of energy: $K_a + U_a + W_{other} = K_b + U_b$

SET UP: Let point a be where q is released and point b be at its final position, as shown in Figure 23.7.

EXECUTE: $K_a = 0$ (released from rest)

$U_a = +0.198\ \text{J}$ (from part (a))

$K_b = \tfrac{1}{2}mv_b^2$

Figure 23.7

Only the electric force does work, so $W_{\text{other}} = 0$ and $U = \dfrac{1}{4\pi\epsilon_0} \dfrac{qQ}{r}$.

(i) $r_b = 0.500$ m

$$U_b = \frac{1}{4\pi\epsilon_0}\frac{qQ}{r} = (8.988\times10^9 \text{ N}\cdot\text{m}^2/\text{C}^2)\frac{(+4.60\times10^{-6}\text{ C})(+1.20\times10^{-6}\text{ C})}{0.500 \text{ m}} = +0.0992 \text{ J}$$

Then $K_a + U_a + W_{\text{other}} = K_b + U_b$ gives $K_b = U_a - U_b$ and $\frac{1}{2}mv_b^2 = U_a - U_b$ and

$$v_b = \sqrt{\frac{2(U_a - U_b)}{m}} = \sqrt{\frac{2(+0.198 \text{ J} - 0.0992 \text{ J})}{2.80\times10^{-4}\text{ kg}}} = 26.6 \text{ m/s.}$$

(ii) $r_b = 5.00$ m r_b is now ten times larger than in (i) so U_b is ten times smaller: $U_b = +0.0992 \text{ J}/10 = +0.00992$ J.

$$v_b = \sqrt{\frac{2(U_a - U_b)}{m}} = \sqrt{\frac{2(+0.198 \text{ J} - 0.00992 \text{ J})}{2.80\times10^{-4}\text{ kg}}} = 36.7 \text{ m/s.}$$

(iii) $r_b = 50.0$ m

r_b is now ten times larger than in (ii) so U_b is ten times smaller:

$$U_b = +0.00992 \text{ J}/10 = +0.000992 \text{ J.}$$

$$v_b = \sqrt{\frac{2(U_a - U_b)}{m}} = \sqrt{\frac{2(+0.198 \text{ J} - 0.000992 \text{ J})}{2.80\times10^{-4}\text{ kg}}} = 37.5 \text{ m/s.}$$

EVALUATE: The force between the two charges is repulsive and provides an acceleration to q. This causes the speed of q to increase as it moves away from Q.

23.11. **IDENTIFY:** Apply Eq.(23.2). The net work to bring the charges in from infinity is equal to the change in potential energy. The total potential energy is the sum of the potential energies of each pair of charges, calculated from Eq.(23.9).

SET UP: Let 1 be where all the charges are infinitely far apart. Let 2 be where the charges are at the corners of the triangle, as shown in Figure 23.11.

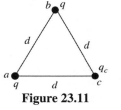

Let q_c be the third, unknown charge.

Figure 23.11

EXECUTE: $W = -\Delta U = -(U_2 - U_1)$

$$U_1 = 0$$

$$U_2 = U_{ab} + U_{ac} + U_{bc} = \frac{1}{4\pi\epsilon_0 d}(q^2 + 2qq_c)$$

Want $W = 0$, so $W = -(U_2 - U_1)$ gives $0 = -U_2$

$$0 = \frac{1}{4\pi\epsilon_0 d}(q^2 + 2qq_c)$$

$q^2 + 2qq_c = 0$ and $q_c = -q/2$.

EVALUATE: The potential energy for the two charges q is positive and for each q with q_c it is negative. There are two of the q, q_c terms so must have $q_c < q$.

23.13. **IDENTIFY:** $\vec{E}$ points from high potential to low potential. $\dfrac{W_{a\rightarrow b}}{q_0} = V_a - V_b$.

SET UP: The force on a positive test charge is in the direction of $\vec{E}$.

EXECUTE: V decreases in the eastward direction. A is east of B, so $V_B > V_A$. C is east of A, so $V_C < V_A$. The force on a positive test charge is east, so no work is done on it by the electric force when it moves due south (the force and displacement are perpendicular), and $V_D = V_A$.

EVALUATE: The electric potential is constant in a direction perpendicular to the electric field.

23.15.   **IDENTIFY** and **SET UP:**   Apply conservation of energy to points A and B.

EXECUTE: $K_A + U_A = K_B + U_B$

$U = qV$, so $K_A + qV_A = K_B + qV_B$

$K_B = K_A + q(V_A - V_B) = 0.00250 \text{ J} + (-5.00 \times 10^{-6} \text{ C})(200 \text{ V} - 800 \text{ V}) = 0.00550 \text{ J}$

$v_B = \sqrt{2K_B/m} = 7.42 \text{ m/s}$

EVALUATE: It is faster at B; a negative charge gains speed when it moves to higher potential.

23.17.   **IDENTIFY:**   Apply the equation that precedes Eq.(23.17): $W_{a \to b} = q' \int_a^b \vec{E} \cdot d\vec{l}$.

SET UP: Use coordinates where $+y$ is upward and $+x$ is to the right. Then $\vec{E} = E\hat{j}$ with $E = 4.00 \times 10^4$ N/C.

(a)   The path is sketched in Figure 23.17a.

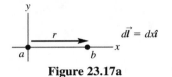

Figure 23.17a

EXECUTE: $\vec{E} \cdot d\vec{l} = (E\hat{j}) \cdot (dx\hat{i}) = 0$ so $W_{a \to b} = q' \int_a^b \vec{E} \cdot d\vec{l} = 0$.

EVALUATE: The electric force on the positive charge is upward (in the direction of the electric field) and does no work for a horizontal displacement of the charge.

(b) SET UP: The path is sketched in Figure 23.17b.

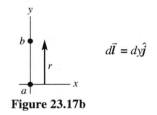

Figure 23.17b

EXECUTE: $\vec{E} \cdot d\vec{l} = (E\hat{j}) \cdot (dy\hat{j}) = E\,dy$

$$W_{a \to b} = q' \int_a^b \vec{E} \cdot d\vec{l} = q'E \int_a^b dy = q'E(y_b - y_a)$$

$y_b - y_a = +0.670$ m, positive since the displacement is upward and we have taken $+y$ to be upward.

$$W_{a \to b} = q'E(y_b - y_a) = (+28.0 \times 10^{-9} \text{ C})(4.00 \times 10^4 \text{ N/C})(+0.670 \text{ m}) = +7.50 \times 10^{-4} \text{ J}.$$

EVALUATE: The electric force on the positive charge is upward so it does positive work for an upward displacement of the charge.

(c) SET UP: The path is sketched in Figure 23.17c.

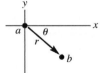

$y_a = 0$

$y_b = -r\sin\theta = -(2.60 \text{ m})\sin 45° = -1.838$ m

The vertical component of the 2.60 m displacement is 1.838 m downward.

Figure 23.17c

EXECUTE: $d\vec{l} = dx\hat{i} + dy\hat{j}$ (The displacement has both horizontal and vertical components.)

$\vec{E} \cdot d\vec{l} = (E\hat{j}) \cdot (dx\hat{i} + dy\hat{j}) = E\,dy$ (Only the vertical component of the displacement contributes to the work.)

$$W_{a \to b} = q' \int_a^b \vec{E} \cdot d\vec{l} = q'E \int_a^b dy = q'E(y_b - y_a)$$

$$W_{a \to b} = q'E(y_b - y_a) = (+28.0 \times 10^{-9} \text{ C})(4.00 \times 10^4 \text{ N/C})(-1.838 \text{ m}) = -2.06 \times 10^{-3} \text{ J}.$$

EVALUATE: The electric force on the positive charge is upward so it does negative work for a displacement of the charge that has a downward component.

23.21. **IDENTIFY:** $V = \dfrac{1}{4\pi\epsilon_0}\sum_i \dfrac{q_i}{r_i}$

SET UP: The locations of the changes and points A and B are sketched in Figure 23.21.

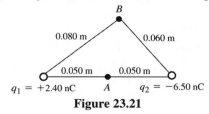

$q_1 = +2.40$ nC A $q_2 = -6.50$ nC

Figure 23.21

EXECUTE: **(a)** $V_A = \dfrac{1}{4\pi\epsilon_0}\left(\dfrac{q_1}{r_{A1}} + \dfrac{q_2}{r_{A2}}\right)$

$$V_A = (8.988\times10^9 \text{ N}\cdot\text{m}^2/\text{C}^2)\left(\dfrac{+2.40\times10^{-9} \text{ C}}{0.050 \text{ m}} + \dfrac{-6.50\times10^{-9} \text{ C}}{0.050 \text{ m}}\right) = -737 \text{ V}$$

(b) $V_B = \dfrac{1}{4\pi\epsilon_0}\left(\dfrac{q_1}{r_{B1}} + \dfrac{q_2}{r_{B2}}\right)$

$$V_B = (8.988\times10^9 \text{ N}\cdot\text{m}^2/\text{C}^2)\left(\dfrac{+2.40\times10^{-9} \text{ C}}{0.080 \text{ m}} + \dfrac{-6.50\times10^{-9} \text{ C}}{0.060 \text{ m}}\right) = -704 \text{ V}$$

(c) IDENTIFY and SET UP: Use Eq.(23.13) and the results of parts (a) and (b) to calculate W.

EXECUTE: $W_{B\to A} = q'(V_B - V_A) = (2.50\times10^{-9} \text{ C})(-704 \text{ V} - (-737 \text{ V})) = +8.2\times10^{-8} \text{ J}$

EVALUATE: The electric force does positive work on the positive charge when it moves from higher potential (point B) to lower potential (point A).

23.29. **(a) IDENTIFY and SET UP:** The direction of $\vec{E}$ is always from high potential to low potential so point b is at higher potential.

(b) Apply Eq.(23.17) to relate $V_b - V_a$ to E.

EXECUTE: $V_b - V_a = -\displaystyle\int_a^b \vec{E}\cdot d\vec{l} = \int_a^b E\,dx = E(x_b - x_a)$.

$$E = \dfrac{V_b - V_a}{x_b - x_a} = \dfrac{+240 \text{ V}}{0.90 \text{ m} - 0.60 \text{ m}} = 800 \text{ V/m}$$

(c) $W_{b\to a} = q(V_b - V_a) = (-0.200\times10^{-6} \text{ C})(+240 \text{ V}) = -4.80\times10^{-5} \text{ J}$.

EVALUATE: The electric force does negative work on a negative charge when the negative charge moves from high potential (point b) to low potential (point a).

23.31. **IDENTIFY and SET UP:** Apply conservation of energy, Eq.(23.3). Use Eq.(23.12) to express U in terms of V.

(a) EXECUTE: $K_1 + qV_1 = K_2 + qV_2$

$q(V_1 - V_2) = K_2 - K_1$; $q = -1.602\times10^{-19} \text{ C}$

$K_1 = \frac{1}{2}m_e v_1^2 = 4.099\times10^{-18} \text{ J}$; $K_2 = \frac{1}{2}m_e v_2^2 = 2.915\times10^{-17} \text{ J}$

$V_1 - V_2 = \dfrac{K_2 - K_1}{q} = -156 \text{ V}$

EVALUATE: The electron gains kinetic energy when it moves to higher potential.

(b) EXECUTE: Now $K_1 = 2.915\times10^{-17} \text{ J}$, $K_2 = 0$

$$V_1 - V_2 = \dfrac{K_2 - K_1}{q} = +182 \text{ V}$$

EVALUATE: The electron loses kinetic energy when it moves to lower potential.

23.33. **(a) IDENTIFY and SET UP:** The electric field on the ring's axis is calculated in Example 21.10. The force on the electron exerted by this field is given by Eq.(21.3).

EXECUTE: When the electron is on either side of the center of the ring, the ring exerts an attractive force directed toward the center of the ring. This restoring force produces oscillatory motion of the electron along the axis of the ring, with amplitude 30.0 cm. The force on the electron is *not* of the form $F = -kx$ so the oscillatory motion is not simple harmonic motion.

(b) IDENTIFY: Apply conservation of energy to the motion of the electron.

SET UP: $K_a + U_a = K_b + U_b$ with a at the initial position of the electron and b at the center of the ring. From

Example 23.11, $V = \dfrac{1}{4\pi\epsilon_0}\dfrac{Q}{\sqrt{x^2 + R^2}}$, where R is the radius of the ring.

EXECUTE: $x_a = 30.0$ cm, $x_b = 0$.

$K_a = 0$ (released from rest), $K_b = \frac{1}{2}mv^2$

Thus $\frac{1}{2}mv^2 = U_a - U_b$

And $U = qV = -eV$ so $v = \sqrt{\dfrac{2e(V_b - V_a)}{m}}$.

$$V_a = \frac{1}{4\pi\epsilon_0}\frac{Q}{\sqrt{x_a^2 + R^2}} = (8.988\times10^9 \text{ N}\cdot\text{m}^2/\text{C}^2)\frac{24.0\times10^{-9}\text{ C}}{\sqrt{(0.300\text{ m})^2 + (0.150\text{ m})^2}}$$

$$V_a = 643 \text{ V}$$

$$V_b = \frac{1}{4\pi\epsilon_0}\frac{Q}{\sqrt{x_b^2 + R^2}} = (8.988\times10^9 \text{ N}\cdot\text{m}^2/\text{C}^2)\frac{24.0\times10^{-9}\text{ C}}{0.150\text{ m}} = 1438 \text{ V}$$

$$v = \sqrt{\frac{2e(V_b - V_a)}{m}} = \sqrt{\frac{2(1.602\times10^{-19}\text{ C})(1438\text{ V} - 643\text{ V})}{9.109\times10^{-31}\text{ kg}}} = 1.67\times10^7 \text{ m/s}$$

EVALUATE: The positively charged ring attracts the negatively charged electron and accelerates it. The electron has its maximum speed at this point. When the electron moves past the center of the ring the force on it is opposite to its motion and it slows down.

23.35.   **IDENTIFY:**   The voltmeter measures the potential difference between the two points. We must relate this quantity to the linear charge density on the wire.

SET UP: For a very long (infinite) wire, the potential difference between two points is $\Delta V = \dfrac{\lambda}{2\pi\epsilon_0}\ln(r_b/r_a)$.

EXECUTE: **(a)** Solving for λ gives

$$\lambda = \frac{(\Delta V)2\pi\epsilon_0}{\ln(r_b/r_a)} = \frac{575 \text{ V}}{(18\times10^9 \text{ N}\cdot\text{m}^2/\text{C}^2)\ln\left(\dfrac{3.50\text{ cm}}{2.50\text{ cm}}\right)} = 9.49\times10^{-8} \text{ C/m}$$

(b) The meter will read less than 575 V because the electric field is weaker over this 1.00-cm distance than it was over the 1.00-cm distance in part (a).

(c) The potential difference is zero because both probes are at the same distance from the wire, and hence at the same potential.

EVALUATE: Since a voltmeter measures potential difference, we are actually given ΔV, even though that is not stated explicitly in the problem. We must also be careful when using the formula for the potential difference because each r is the distance from the *center* of the cylinder, not from the surface.

23.41.   **IDENTIFY** and **SET UP:**   Use the result of Example 23.9 to relate the electric field between the plates to the potential difference between them and their separation. The force this field exerts on the particle is given by Eq.(21.3). Use the equation that precedes Eq.(23.17) to calculate the work.

EXECUTE: **(a)** From Example 23.9, $E = \dfrac{V_{ab}}{d} = \dfrac{360\text{ V}}{0.0450\text{ m}} = 8000$ V/m

(b) $F = |q|E = (2.40\times10^{-9}\text{ C})(8000\text{ V/m}) = +1.92\times10^{-5}$ N

(c) The electric field between the plates is shown in Figure 23.41.

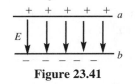

Figure 23.41

The plate with positive charge (plate a) is at higher potential. The electric field is directed from high potential toward low potential (or, $\vec{E}$ is from + charge toward − charge), so $\vec{E}$ points from a to b. Hence the force that $\vec{E}$ exerts on the positive charge is from a to b, so it does positive work.

$W = \displaystyle\int_a^b \vec{F}\cdot d\vec{l} = Fd$, where d is the separation between the plates.

$W = Fd = (1.92\times10^{-5}\text{ N})(0.0450\text{ m}) = +8.64\times10^{-7}$ J

(d) $V_a - V_b = +360$ V   (plate a is at higher potential)

$\Delta U = U_b - U_a = q(V_b - V_a) = (2.40 \times 10^{-9}$ C$)(-360$ V$) = -8.64 \times 10^{-7}$ J.

EVALUATE: We see that $W_{a \to b} = -(U_b - U_a) = U_a - U_b.$

23.43.   **IDENTIFY** and **SET UP:**   Consider the electric field outside and inside the shell and use that to deduce the potential.
EXECUTE: **(a)** The electric field outside the shell is the same as for a point charge at the center of the shell, so the potential outside the shell is the same as for a point charge:

$$V = \frac{q}{4\pi\epsilon_0 r} \text{ for } r > R.$$

The electric field is zero inside the shell, so no work is done on a test charge as it moves inside the shell and all points inside the shell are at the same potential as the surface of the shell: $V = \dfrac{q}{4\pi\epsilon_0 R}$ for $r \leq R.$

(b) $V = \dfrac{kq}{R}$ so $q = \dfrac{RV}{k} = \dfrac{(0.15 \text{ m})(-1200 \text{ V})}{k} = -20$ nC

(c) EVALUATE: No, the amount of charge on the sphere is very small. Since $U = qV$ the total amount of electric energy stored on the balloon is only $(20 \text{ nC})(1200 \text{ V}) = 2.4 \times 10^{-5}$ J.

23.47   **IDENTIFY** and **SET UP:**   Use Eq.(23.19) to calculate the components of $\vec{E}$.
EXECUTE: $V = Axy - Bx^2 + Cy$

(a) $E_x = -\dfrac{\partial V}{\partial x} = -Ay + 2Bx$

$E_y = -\dfrac{\partial V}{\partial y} = -Ax - C$

$E_z = -\dfrac{\partial V}{\partial z} = 0$

(b) $E = 0$ requires that $E_x = E_y = E_z = 0.$

$E_z = 0$ everywhere.

$E_y = 0$ at $x = -C/A.$

And E_x is also equal zero for this x, any value of z, and $y = 2Bx/A = (2B/A)(-C/A) = -2BC/A^2.$

EVALUATE: V doesn't depend on z so $E_z = 0$ everywhere.

23.49.   **IDENTIFY** and **SET UP:**   For a solid metal sphere or for a spherical shell, $V = \dfrac{kq}{r}$ outside the sphere and $V = \dfrac{kq}{R}$ at all points inside the sphere, where R is the radius of the sphere. When the electric field is radial, $E = -\dfrac{\partial V}{\partial r}.$

EXECUTE: **(a)** (i) $r < r_a$: This region is inside both spheres. $V = \dfrac{kq}{r_a} - \dfrac{kq}{r_b} = kq\left(\dfrac{1}{r_a} - \dfrac{1}{r_b}\right).$

(ii) $r_a < r < r_b$: This region is outside the inner shell and inside the outer shell. $V = \dfrac{kq}{r} - \dfrac{kq}{r_b} = kq\left(\dfrac{1}{r} - \dfrac{1}{r_b}\right).$

(iii) $r > r_b$: This region is outside both spheres and $V = 0$ since outside a sphere the potential is the same as for point charge. Therefore the potential is the same as for two oppositely charged point charges at the same location. These potentials cancel.

(b) $V_a = \dfrac{1}{4\pi\epsilon_0}\left(\dfrac{q}{r_a} - \dfrac{q}{r_b}\right)$ and $V_b = 0$, so $V_{ab} = \dfrac{1}{4\pi\epsilon_0}q\left(\dfrac{1}{r_a} - \dfrac{1}{r_b}\right).$

(c) Between the spheres $r_a < r < r_b$ and $V = kq\left(\dfrac{1}{r} - \dfrac{1}{r_b}\right)$. $E = -\dfrac{\partial V}{\partial r} = -\dfrac{q}{4\pi\epsilon_0}\dfrac{\partial}{\partial r}\left(\dfrac{1}{r} - \dfrac{1}{r_b}\right) = +\dfrac{1}{4\pi\epsilon_0}\dfrac{q}{r^2} = \dfrac{V_{ab}}{\left(\dfrac{1}{r_a} - \dfrac{1}{r_b}\right)}\dfrac{1}{r^2}.$

(d) From Equation (23.23): $E = 0$, since V is constant (zero) outside the spheres.

(e) If the outer charge is different, then outside the outer sphere the potential is no longer zero but is

$V = \dfrac{1}{4\pi\epsilon_0}\dfrac{q}{r} - \dfrac{1}{4\pi\epsilon_0}\dfrac{Q}{r} = \dfrac{1}{4\pi\epsilon_0}\dfrac{(q-Q)}{r}$. All potentials inside the outer shell are just shifted by an amount

$V = -\dfrac{1}{4\pi\epsilon_0}\dfrac{Q}{r_b}$. Therefore relative potentials within the shells are not affected. Thus (b) and (c) do not change.

However, now that the potential does vary outside the spheres, there is an electric field there:

$E = -\dfrac{\partial V}{\partial r} = -\dfrac{\partial}{\partial r}\left(\dfrac{kq}{r} + \dfrac{-kQ}{r}\right) = \dfrac{kq}{r^2}\left(1 - \dfrac{Q}{q}\right) = \dfrac{k}{r^2}(q - Q)$.

EVALUATE: In part (a) the potential is greater than zero for all $r < r_b$.

23.51.   IDENTIFY: Outside the cylinder it is equivalent to a line of charge at its center.

SET UP: The difference in potential between the surface of the cylinder (a distance R from the central axis) and a

general point a distance r from the central axis is given by $\Delta V = \dfrac{\lambda}{2\pi\epsilon_0}\ln(r/R)$.

EXECUTE: (a) The potential difference depends only on r, and not direction. Therefore all points at the same value of r will be at the same potential. Thus the equipotential surfaces are cylinders coaxial with the given cylinder.

(b) Solving $\Delta V = \dfrac{\lambda}{2\pi\epsilon_0}\ln(r/R)$ for r, gives $r = Re^{2\pi\epsilon_0\Delta V/\lambda}$.

For 10 V, the exponent is $(10\text{ V})/[(2 \times 9.00 \times 10^9\text{ N} \cdot \text{m}^2/\text{C}^2)(1.50 \times 10^{-9}\text{ C/m})] = 0.370$, which gives $r = (2.00\text{ cm})$
$e^{0.370} = 2.90\text{ cm}$. Likewise, the other radii are 4.20 cm (for 20 V) and 6.08 cm (for 30 V).

(c) $\Delta r_1 = 2.90\text{ cm} - 2.00\text{ cm} = 0.90\text{ cm}$; $\Delta r_2 = 4.20\text{ cm} - 2.90\text{ cm} = 1.30\text{ cm}$; $\Delta r_3 = 6.08\text{ cm} - 4.20\text{ cm} = 1.88\text{ cm}$

EVALUATE: As we can see, Δr increases, so the surfaces get farther apart. This is very different from a sheet of charge, where the surfaces are equally spaced planes.

23.53.   (a) IDENTIFY: Apply the work-energy theorem, Eq.(6.6).

SET UP: Points a and b are shown in Figure 23.53a.

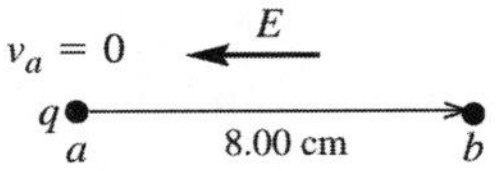

Figure 23.53a

EXECUTE: $W_{tot} = \Delta K = K_b - K_a = K_b = 4.35 \times 10^{-5}\text{ J}$

The electric force F_E and the additional force F both do work, so that $W_{tot} = W_{F_E} + W_F$.

$$W_{F_E} = W_{tot} - W_F = 4.35 \times 10^{-5}\text{ J} - 6.50 \times 10^{-5}\text{ J} = -2.15 \times 10^{-5}\text{ J}$$

EVALUATE: The forces on the charged particle are shown in Figure 23.53b.

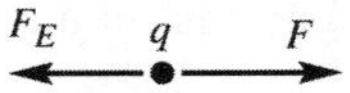

Figure 23.53b

The electric force is to the left (in the direction of the electric field since the particle has positive charge). The displacement is to the right, so the electric force does negative work. The additional force F is in the direction of the displacement, so it does positive work.

(b) IDENTIFY and SET UP: For the work done by the electric force, $W_{a\rightarrow b} = q(V_a - V_b)$

EXECUTE: $V_a - V_b = \dfrac{W_{a\rightarrow b}}{q} = \dfrac{-2.15 \times 10^{-5}\text{ J}}{7.60 \times 10^{-9}\text{ C}} = -2.83 \times 10^3\text{ V}$.

EVALUATE: The starting point (point a) is at 2.83×10^3 V lower potential than the ending point (point b). We know that $V_b > V_a$ because the electric field always points from high potential toward low potential.

(c) IDENTIFY: Calculate E from $V_a - V_b$ and the separation d between the two points.

SET UP: Since the electric field is uniform and directed opposite to the displacement $W_{a\rightarrow b} = -F_E d = -qEd$, where $d = 8.00$ cm is the displacement of the particle.

EXECUTE: $E = -\dfrac{W_{a\rightarrow b}}{qd} = -\dfrac{V_a - V_b}{d} = \dfrac{-2.83 \times 10^3\text{ V}}{0.0800\text{ m}} = 3.54 \times 10^4\text{ V/m}$.

EVALUATE: In part (a), W_{tot} is the total work done by both forces. In parts (b) and (c) $W_{a\rightarrow b}$ is the work done just by the electric force.

23.55.   **IDENTIFY** and **SET UP:**   Calculate the components of $\vec{E}$ from Eq.(23.19). Eq.(21.3) gives $\vec{F}$ from $\vec{E}$.
EXECUTE: **(a)** $V = Cx^{4/3}$

$$C = V/x^{4/3} = 240 \text{ V}/(13.0 \times 10^{-3} \text{ m})^{4/3} = 7.85 \times 10^4 \text{ V/m}^{4/3}$$

(b) $E_x = -\dfrac{\partial V}{\partial x} = -\dfrac{4}{3}Cx^{1/3} = -(1.05 \times 10^5 \text{ V/m}^{4/3})x^{1/3}$

The minus sign means that E_x is in the $-x$-direction, which says that $\vec{E}$ points from the positive anode toward the negative cathode.
(c) $\vec{F} = q\vec{E}$ so $F_x = -eE_x = \frac{4}{3}eCx^{1/3}$

Halfway between the electrodes means $x = 6.50 \times 10^{-3}$ m.

$$F_x = \tfrac{4}{3}(1.602 \times 10^{-19} \text{ C})(7.85 \times 10^4 \text{ V/m}^{4/3})(6.50 \times 10^{-3} \text{ m})^{1/3} = 3.13 \times 10^{-15} \text{ N}$$

F_x is positive, so the force is directed toward the positive anode.

EVALUATE: V depends only on x, so $E_y = E_z = 0$. $\vec{E}$ is directed from high potential (anode) to low potential (cathode). The electron has negative charge, so the force on it is directed opposite to the electric field.

23.59.   **(a) IDENTIFY:**   Use Eq.(23.10) for the electron and each proton.
SET UP: The positions of the particles are shown in Figure 23.59a.

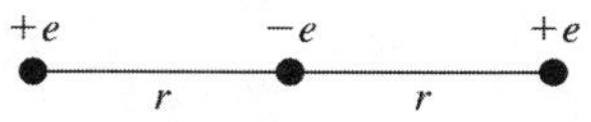

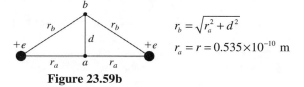

$r = (1.07 \times 10^{-10} \text{ m})/2 = 0.535 \times 10^{-10}$ m

Figure 23.59a

EXECUTE: The potential energy of interaction of the electron with each proton is
$U = \dfrac{1}{4\pi\epsilon_0}\dfrac{(-e^2)}{r}$,   so the total potential energy is

$$U = -\frac{2e^2}{4\pi\epsilon_0 r} = -\frac{2(8.988 \times 10^9 \text{ N} \cdot \text{m}^2/\text{C}^2)(1.60 \times 10^{-19} \text{ C})^2}{0.535 \times 10^{-10} \text{ m}} = -8.60 \times 10^{-18} \text{ J}$$

$$U = -8.60 \times 10^{-18} \text{ J}(1 \text{ eV}/1.602 \times 10^{-19} \text{ J}) = -53.7 \text{ eV}$$

EVALUATE: The electron and proton have charges of opposite signs, so the potential energy of the system is negative.
(b) IDENTIFY and **SET UP:**   The positions of the protons and points a and b are shown in Figure 23.59b.

b
r_b r_b
d
+e   +e
r_a a r_a

$r_b = \sqrt{r_a^2 + d^2}$

$r_a = r = 0.535 \times 10^{-10}$ m

Figure 23.59b

Apply $K_a + U_a + W_{\text{other}} = K_b + U_b$ with point a midway between the protons and point b where the electron instantaneously has $v = 0$ (at its maximum displacement d from point a).
EXECUTE: Only the Coulomb force does work, so $W_{\text{other}} = 0$.

$U_a = -8.60 \times 10^{-18}$ J (from part (a))

$K_a = \frac{1}{2}mv^2 = \frac{1}{2}(9.109 \times 10^{-31} \text{ kg})(1.50 \times 10^6 \text{ m/s})^2 = 1.025 \times 10^{-18}$ J

$K_b = 0$

$U_b = -2ke^2/r_b$

Then $U_b = K_a + U_a - K_b = 1.025 \times 10^{-18} \text{ J} - 8.60 \times 10^{-18} \text{ J} = -7.575 \times 10^{-18} \text{ J}$.

$$r_b = -\frac{2ke^2}{U_b} = -\frac{2(8.988 \times 10^9 \text{ N} \cdot \text{m}^2/\text{C}^2)(1.60 \times 10^{-19} \text{ C})^2}{-7.575 \times 10^{-18} \text{ J}} = 6.075 \times 10^{-11} \text{ m}$$

Then $d = \sqrt{r_b^2 - r_a^2} = \sqrt{(6.075 \times 10^{-11} \text{ m})^2 - (5.35 \times 10^{-11} \text{ m})^2} = 2.88 \times 10^{-11}$ m.

EVALUATE: The force on the electron pulls it back toward the midpoint. The transverse distance the electron moves is about 0.27 times the separation of the protons.

23.61.   **(a) IDENTIFY:**   The potential at any point is the sum of the potentials due to each of the two charged conductors.
SET UP: From Example 23.10, for a conducting cylinder with charge per unit length λ the potential outside the cylinder is given by $V = (\lambda/2\pi\epsilon_0)\ln(r_0/r)$ where r is the distance from the cylinder axis and r_0 is the distance from the axis for which we take $V = 0$. Inside the cylinder the potential has the same value as on the cylinder surface. The electric field is the same for a solid conducting cylinder or for a hollow conducting tube so this expression for V applies to both. This problem says to take $r_0 = b$.

EXECUTE: For the hollow tube of radius b and charge per unit length $-\lambda$: outside $V = -(\lambda/2\pi\epsilon_0)\ln(b/r)$; inside $V = 0$ since $V = 0$ at $r = b$.
For the metal cylinder of radius a and charge per unit length λ:
outside $V = (\lambda/2\pi\epsilon_0)\ln(b/r)$, inside $V = (\lambda/2\pi\epsilon_0)\ln(b/a)$, the value at $r = a$.
(i) $r < a$; inside both $V = (\lambda/2\pi\epsilon_0)\ln(b/a)$
(ii) $a < r < b$; outside cylinder, inside tube $V = (\lambda/2\pi\epsilon_0)\ln(b/r)$
(iii) $r > b$; outside both the potentials are equal in magnitude and opposite in sign so $V = 0$.
(b) For $r = a$, $V_a = (\lambda/2\pi\epsilon_0)\ln(b/a)$.
For $r = b$, $V_b = 0$.
Thus $V_{ab} = V_a - V_b = (\lambda/2\pi\epsilon_0)\ln(b/a)$.
(c) IDENTIFY and **SET UP:**   Use Eq.(23.23) to calculate E.

EXECUTE: $E = -\dfrac{\partial V}{\partial r} = -\dfrac{\lambda}{2\pi\epsilon_0}\dfrac{\partial}{\partial r}\ln\left(\dfrac{b}{r}\right) = -\dfrac{\lambda}{2\pi\epsilon_0}\left(\dfrac{r}{b}\right)\left(-\dfrac{b}{r^2}\right) = \dfrac{V_{ab}}{\ln(b/a)}\dfrac{1}{r}$.

(d) The electric field between the cylinders is due only to the inner cylinder, so V_{ab} is not changed,
$V_{ab} = (\lambda/2\pi\epsilon_0)\ln(b/a)$.
EVALUATE: The electric field is not uniform between the cylinders, so $V_{ab} \neq E(b - a)$.

23.63.   **IDENTIFY** and **SET UP:**   Use Eq.(21.3) to calculate $\vec{F}$ and then $\vec{F} = m\vec{a}$ gives $\vec{a}$.
EXECUTE: **(a)** $\vec{F}_E = q\vec{E}$. Since $q = -e$ is negative $\vec{F}_E$ and $\vec{E}$ are in opposite directions; $\vec{E}$ is upward so $\vec{F}_E$ is downward. The magnitude of F_E is
$$F_E = |q|E = eE = (1.602\times10^{-19}\ \text{C})(1.10\times10^3\ \text{N/C}) = 1.76\times10^{-16}\ \text{N}.$$
(b) Calculate the acceleration of the electron produced by the electric force:
$$a = \frac{F}{m} = \frac{1.76\times10^{-16}\ \text{N}}{9.109\times10^{-31}\ \text{kg}} = 1.93\times10^{14}\ \text{m/s}^2$$
EVALUATE: This is much larger than $g = 9.80\ \text{m/s}^2$, so the gravity force on the electron can be neglected. $\vec{F}_E$ is downward, so $\vec{a}$ is downward.
(c) IDENTIFY and **SET UP:**   The acceleration is constant and downward, so the motion is like that of a projectile. Use the horizontal motion to find the time and then use the time to find the vertical displacement.
EXECUTE: _x-component_
$v_{0x} = 6.50\times10^6\ \text{m/s};\ a_x = 0;\ x - x_0 = 0.060\ \text{m};\ t = ?$
$x - x_0 = v_{0x}t + \frac{1}{2}a_x t^2$ and the a_x term is zero, so
$t = \dfrac{x - x_0}{v_{0x}} = \dfrac{0.060\ \text{m}}{6.50\times10^6\ \text{m/s}} = 9.231\times10^{-9}\ \text{s}$
y-component
$v_{0y} = 0;\ a_y = 1.93\times10^{14}\ \text{m/s}^2;\ t = 9.231\times10^{-9}\ \text{m/s};\ y - y_0 = ?$
$y - y_0 = v_{0y}t + \frac{1}{2}a_y t^2$
$y - y_0 = \frac{1}{2}(1.93\times10^{14}\ \text{m/s}^2)(9.231\times10^{-9}\ \text{s})^2 = 0.00822\ \text{m} = 0.822\ \text{cm}$

(d) The velocity and its components as the electron leaves the plates are sketched in Figure 23.63.

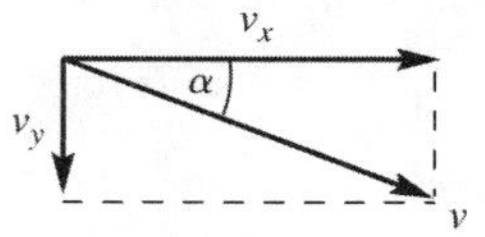

$v_x = v_{0x} = 6.50 \times 10^6$ m/s (since $a_x = 0$)

$v_y = v_{0y} + a_y t$

$v_y = 0 + (1.93 \times 10^{14} \text{ m/s}^2)(9.231 \times 10^{-9} \text{ s})$

$v_y = 1.782 \times 10^6$ m/s

Figure 23.63

$\tan \alpha = \dfrac{v_y}{v_x} = \dfrac{1.782 \times 10^6 \text{ m/s}}{6.50 \times 10^6 \text{ m/s}} = 0.2742$ so $\alpha = 15.3°$.

EVALUATE: The greater the electric field or the smaller the initial speed the greater the downward deflection.

(e) IDENTIFY and **SET UP:**   Consider the motion of the electron after it leaves the region between the plates. Outside the plates there is no electric field, so $a = 0$. (Gravity can still be neglected since the electron is traveling at such high speed and the times are small.) Use the horizontal motion to find the time it takes the electron to travel 0.120 m horizontally to the screen. From this time find the distance downward that the electron travels.

EXECUTE: x-component

$v_{0x} = 6.50 \times 10^6$ m/s; $a_x = 0$; $x - x_0 = 0.120$ m; $t = ?$

$x - x_0 = v_{0x}t + \frac{1}{2}a_x t^2$ and the a_x term is term is zero, so

$t = \dfrac{x - x_0}{v_{0x}} = \dfrac{0.120 \text{ m}}{6.50 \times 10^6 \text{ m/s}} = 1.846 \times 10^{-8}$ s

y-component

$v_{0y} = 1.782 \times 10^6$ m/s (from part (b)); $a_y = 0$; $t = 1.846 \times 10^{-8}$ m/s; $y - y_0 = ?$

$y - y_0 = v_{0y}t + \frac{1}{2}a_y t^2 = (1.782 \times 10^6 \text{ m/s})(1.846 \times 10^{-8} \text{ s}) = 0.0329$ m $= 3.29$ cm

EVALUATE: The electron travels downward a distance 0.822 cm while it is between the plates and a distance 3.29 cm while traveling from the edge of the plates to the screen. The total downward deflection is 0.822 cm $+$ 3.29 cm $=$ 4.11 cm.

The horizontal distance between the plates is half the horizontal distance the electron travels after it leaves the plates. And the vertical velocity of the electron increases as it travels between the plates, so it makes sense for it to have greater downward displacement during the motion after it leaves the plates.

about around 40 pC, the electric field is about 5000 V/m, and the electron speed would be about 3 million m/s.

23.65.   **(a) IDENTIFY** and **SET UP:**   Problem 23.61 derived that $E = \dfrac{V_{ab}}{\ln(b/a)}\dfrac{1}{r}$, where a is the radius of the inner cylinder (wire) and b is the radius of the outer hollow cylinder. The potential difference between the two cylinders is V_{ab}. Use this expression to calculate E at the specified r.

EXECUTE: Midway between the wire and the cylinder wall is at a radius of

$r = (a+b)/2 = (90.0 \times 10^{-6} \text{ m} + 0.140 \text{ m})/2 = 0.07004$ m.

$$E = \dfrac{V_{ab}}{\ln(b/a)}\dfrac{1}{r} = \dfrac{50.0 \times 10^3 \text{ V}}{\ln(0.140 \text{ m}/90.0 \times 10^{-6} \text{ m})(0.07004 \text{ m})} = 9.71 \times 10^4 \text{ V/m}$$

(b) IDENTIFY and **SET UP:**   The electric force is given by Eq.(21.3). Set this equal to ten times the weight of the particle and solve for $|q|$, the magnitude of the charge on the particle.

EXECUTE: $F_E = 10mg$

$$|q|E = 10mg \text{ and } |q| = \dfrac{10mg}{E} = \dfrac{10(30.0 \times 10^{-9} \text{ kg})(9.80 \text{ m/s}^2)}{9.71 \times 10^4 \text{ V/m}} = 3.03 \times 10^{-11} \text{ C}$$

EVALUATE: It requires only this modest net charge for the electric force to be much larger than the weight.

23.67.   **(a) IDENTIFY:**   Use $V_a - V_b = \displaystyle\int_a^b \vec{E} \cdot d\vec{l}.$

SET UP: From Problem 22.48, $E(r) = \dfrac{\lambda r}{2\pi\epsilon_0 R^2}$ for $r \leq R$ (inside the cylindrical charge distribution) and

$E(r) = \dfrac{\lambda r}{2\pi\epsilon_0 r}$ for $r \geq R$. Let $V = 0$ at $r = R$ (at the surface of the cylinder).

EXECUTE: $r > R$

Take point a to be at R and point b to be at r, where $r > R$. Let $d\vec{l} = d\vec{r}$. $\vec{E}$ and $d\vec{r}$ are both radially outward, so $\vec{E} \cdot d\vec{r} = E\,dr$. Thus $V_R - V_r = \int_R^r E\,dr$. Then $V_R = 0$ gives $V_r = -\int_R^r E\,dr$. In this interval $(r > R)$, $E(r) = \lambda/2\pi\epsilon_0 r$, so

$$V_r = -\int_R^r \frac{\lambda}{2\pi\epsilon_0 r}\,dr = -\frac{\lambda}{2\pi\epsilon_0}\int_R^r \frac{dr}{r} = -\frac{\lambda}{2\pi\epsilon_0}\ln\left(\frac{r}{R}\right).$$

EVALUATE: This expression gives $V_r = 0$ when $r = R$ and the potential decreases (becomes a negative number of larger magnitude) with increasing distance from the cylinder.

EXECUTE: $r < R$

Take point a at r, where $r < R$, and point b at R. $\vec{E} \cdot d\vec{r} = E\,dr$ as before. Thus $V_r - V_R = \int_r^R E\,dr$. Then $V_R = 0$ gives $V_r = \int_r^R E\,dr$. In this interval $(r < R)$, $E(r) = \lambda r/2\pi\epsilon_0 R^2$, so

$$V_r = \int_r^R \frac{\lambda}{2\pi\epsilon_0 R^2}\,dr = \frac{\lambda}{2\pi\epsilon_0 R^2}\int_r^R r\,dr = \frac{\lambda}{2\pi\epsilon_0 R^2}\left(\frac{R^2}{2} - \frac{r^2}{2}\right).$$

$$V_r = \frac{\lambda}{4\pi\epsilon_0}\left(1 - \left(\frac{r}{R}\right)^2\right).$$

EVALUATE: This expression also gives $V_r = 0$ when $r = R$. The potential is $\lambda/4\pi\epsilon_0$ at $r = 0$ and decreases with increasing r.

(b) EXECUTE: Graphs of V and E as functions of r are sketched in Figure 23.67.

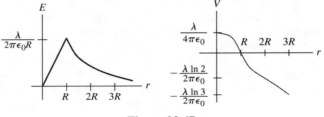

Figure 23.67

EVALUATE: E at any r is the negative of the slope of $V(r)$ at that r (Eq.23.23).

23.69.   **IDENTIFY:** $V_a - V_b = \int_a^b \vec{E} \cdot d\vec{l}$.

SET UP: From Example 21.10, we have: $E_x = \dfrac{1}{4\pi\epsilon_0}\dfrac{Qx}{(x^2 + a^2)^{3/2}}$. $\vec{E} \cdot d\vec{l} = E_x\,dx$. Let $a = \infty$ so $V_a = 0$.

EXECUTE: $V = -\dfrac{Q}{4\pi\epsilon_0}\int_\infty^x \dfrac{x'}{(x'^2 + a^2)^{3/2}}\,dx' = \dfrac{Q}{4\pi\epsilon_0}u^{-1/2}\Big|_{u=\infty}^{u=x^2+a^2} = \dfrac{1}{4\pi\epsilon_0}\dfrac{Q}{\sqrt{x^2 + a^2}}$.

EVALUATE: Our result agrees with Eq.(23.16) in Example 23.11.

23.71.   **IDENTIFY:**   We must integrate to find the total energy because the energy to bring in more charge depends on the charge already present.

SET UP: If ρ is the uniform volume charge density, the charge of a spherical shell or radius r and thickness dr is $dq = \rho\,4\pi r^2\,dr$, and $\rho = Q/(4/3\,\pi R^3)$. The charge already present in a sphere of radius r is $q = \rho(4/3\,\pi r^3)$. The energy to bring the charge dq to the surface of the charge q is $V\,dq$, where V is the potential due to q, which is $q/4\pi\epsilon_0 r$.

EXECUTE: The total energy to assemble the entire sphere of radius R and charge Q is sum (integral) of the tiny increments of energy.

$$U = \int V\,dq = \int \frac{q}{4\pi\epsilon_0 r}\,dq = \int_0^R \frac{\rho\frac{4}{3}\pi r^3}{4\pi\epsilon_0 r}\left(\rho\,4\pi r^2\,dr\right) = \frac{3}{5}\left(\frac{1}{4\pi\epsilon_0}\frac{Q^2}{R}\right)$$

where we have substituted $\rho = Q/(4/3\,\pi R^3)$ and simplified the result.

EVALUATE: For a point-charge, $R \to 0$ so $U \to \infty$, which means that a point-charge should have infinite self-energy. This suggests that either point-charges are impossible, or that our present treatment of physics is not adequate at the extremely small scale, or both.

23.73.   **IDENTIFY:**   Problem 23.70 shows that $V_r = \dfrac{Q}{8\pi\epsilon_0 R}(3 - r^2/R^2)$ for $r \le R$ and $V_r = \dfrac{Q}{4\pi\epsilon_0 r}$ for $r \ge R$.

SET UP: $V_0 = \dfrac{3Q}{8\pi\epsilon_0 R}, V_R = \dfrac{Q}{4\pi\epsilon_0 R}$

EXECUTE: (a) $V_0 - V_R = \dfrac{Q}{8\pi\epsilon_0 R}$

(b) If $Q > 0$, V is higher at the center. If $Q < 0$, V is higher at the surface.

EVALUATE: For $Q > 0$ the electric field is radially outward, $\vec{E}$ is directed toward lower potential, so V is higher at the center. If $Q < 0$, the electric field is directed radially inward and V is higher at the surface.

23.77.   **IDENTIFY:**   Use Eq.(23.17) to calculate V_{ab}.

SET UP: From Problem 22.43, for $R \le r \le 2R$ (between the sphere and the shell) $E = Q/4\pi\epsilon_0 r^2$
Take a at R and b at $2R$.

EXECUTE: $V_{ab} = V_a - V_b = \displaystyle\int_R^{2R} E\,dr = \dfrac{Q}{4\pi\epsilon_0}\int_R^{2R}\dfrac{dr}{r^2} = \dfrac{Q}{4\pi\epsilon_0}\left[-\dfrac{1}{r}\right]_R^{2R} = \dfrac{Q}{4\pi\epsilon_0}\left(\dfrac{1}{R} - \dfrac{1}{2R}\right)$

$$V_{ab} = \dfrac{Q}{8\pi\epsilon_0 R}$$

EVALUATE: The electric field is radially outward and points in the direction of decreasing potential, so the sphere is at higher potential than the shell.

23.79.   **IDENTIFY:**   Slice the rod into thin slices and use Eq.(23.14) to calculate the potential due to each slice. Integrate over the length of the rod to find the total potential at each point.

(a) **SET UP:**   An infinitesimal slice of the rod and its distance from point P are shown in Figure 23.79a.

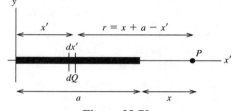

Figure 23.79a

Use coordinates with the origin at the left-hand end of the rod and one axis along the rod. Call the axes x' and y' so as not to confuse them with the distance x given in the problem.

EXECUTE: Slice the charged rod up into thin slices of width dx'. Each slice has charge $dQ = Q(dx'/a)$ and a distance $r = x + a - x'$ from point P. The potential at P due to the small slice dQ is

$$dV = \dfrac{1}{4\pi\epsilon_0}\left(\dfrac{dQ}{r}\right) = \dfrac{1}{4\pi\epsilon_0}\dfrac{Q}{a}\left(\dfrac{dx'}{x+a-x'}\right).$$

Compute the total V at P due to the entire rod by integrating dV over the length of the rod ($x' = 0$ to $x' = a$):

$$V = \int dV = \dfrac{Q}{4\pi\epsilon_0 a}\int_0^a \dfrac{dx'}{(x+a-x')} = \dfrac{Q}{4\pi\epsilon_0 a}[-\ln(x+a-x')]_0^a = \dfrac{Q}{4\pi\epsilon_0 a}\ln\left(\dfrac{x+a}{x}\right).$$

EVALUATE: As $x \to \infty$, $V \to \dfrac{Q}{4\pi\epsilon_0 a}\ln\left(\dfrac{x}{x}\right) = 0$.

(b) **SET UP:**   An infinitesimal slice of the rod and its distance from point R are shown in Figure 23.79b.

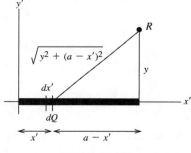

Figure 23.79b

$dQ = (Q/a)dx'$ as in part (a)

Each slice dQ is a distance $r = \sqrt{y^2 + (a - x')^2}$ from point R.

EXECUTE: The potential dV at R due to the small slice dQ is

$$dV = \frac{1}{4\pi\epsilon_0}\left(\frac{dQ}{r}\right) = \frac{1}{4\pi\epsilon_0}\frac{Q}{a}\frac{dx'}{\sqrt{y^2 + (a - x')^2}}.$$

$$V = \int dV = \frac{Q}{4\pi\epsilon_0 a}\int_0^a \frac{dx'}{\sqrt{y^2 + (a - x')^2}}.$$

In the integral make the change of variable $u = a - x'$; $du = -dx'$

$$V = -\frac{Q}{4\pi\epsilon_0 a}\int_a^0 \frac{du}{\sqrt{y^2 + u^2}} = -\frac{Q}{4\pi\epsilon_0 a}\left[\ln(u + \sqrt{y^2 + u^2})\right]_a^0$$

$$V = -\frac{Q}{4\pi\epsilon_0 a}[\ln y - \ln(a + \sqrt{y^2 + a^2})] = \frac{Q}{4\pi\epsilon_0 a}\left[\ln\left(\frac{a + \sqrt{a^2 + y^2}}{y}\right)\right].$$

(The expression for the integral was found in appendix B.)

EVALUATE: As $y \to \infty$, $V \to \dfrac{Q}{4\pi\epsilon_0 a}\ln\left(\dfrac{y}{y}\right) = 0.$

(c) SET UP: *part (a):* $V = \dfrac{Q}{4\pi\epsilon_0 a}\ln\left(\dfrac{x + a}{x}\right) = \dfrac{Q}{4\pi\epsilon_0 a}\ln\left(1 + \dfrac{a}{x}\right).$

From Appendix B, $\ln(1 + u) = u - u^2/2\ldots$, so $\ln(1 + a/x) = a/x - a^2/2x^2$ and this becomes a/x when x is large.

EXECUTE: Thus $V \to \dfrac{Q}{4\pi\epsilon_0 a}\left(\dfrac{a}{x}\right) = \dfrac{Q}{4\pi\epsilon_0 a}.$ For large x, V becomes the potential of a point charge.

part (b): $V = \dfrac{Q}{4\pi\epsilon_0 a}\left[\ln\left(\dfrac{a + \sqrt{a^2 + y^2}}{y}\right)\right] = \dfrac{Q}{4\pi\epsilon_0 a}\ln\left(\dfrac{a}{y} + \sqrt{1 + \dfrac{a^2}{y^2}}\right).$

From Appendix B, $\sqrt{1 + a^2/y^2} = (1 + a^2/y^2)^{1/2} = 1 + a^2/2y^2 + \ldots$

Thus $a/y + \sqrt{1 + a^2/y^2} \to 1 + a/y + a^2/2y^2 + \ldots \to 1 + a/y.$ And then using $\ln(1 + u) \approx u$ gives

$$V \to \frac{Q}{4\pi\epsilon_0 a}\ln(1 + a/y) \to \frac{Q}{4\pi\epsilon_0 a}\left(\frac{a}{y}\right) = \frac{Q}{4\pi\epsilon_0 y}.$$

EVALUATE: For large y, V becomes the potential of a point charge.

23.81.   **(a) IDENTIFY and SET UP:**   The potential at the surface of a charged conducting sphere is given by Example 23.8:

$V = \dfrac{1}{4\pi\epsilon_0}\dfrac{q}{R}.$ For spheres A and B this gives

$$V_A = \frac{Q_A}{4\pi\epsilon_0 R_A} \text{ and } V_B = \frac{Q_B}{4\pi\epsilon_0 R_B}.$$

EXECUTE: $V_A = V_B$ gives $Q_A/4\pi\epsilon_0 R_A = Q_B/4\pi\epsilon_0 R_B$ and $Q_B/Q_A = R_B/R_A$. And then $R_A = 3R_B$ implies $Q_B/Q_A = 1/3.$

(b) IDENTIFY and SET UP: The electric field at the surface of a charged conducting sphere is given in Example 22.5:

$$E = \frac{1}{4\pi\epsilon_0}\frac{|q|}{R^2}.$$

EXECUTE: For spheres A and B this gives

$$E_A = \frac{|Q_A|}{4\pi\epsilon_0 R_A^2} \text{ and } E_B = \frac{|Q_B|}{4\pi\epsilon_0 R_B^2}.$$

$$\frac{E_B}{E_A} = \left(\frac{|Q_B|}{4\pi\epsilon_0 R^2{}_B}\right)\left(\frac{4\pi\epsilon_0 R_A^2}{|Q_A|}\right) = |Q_B/Q_A|(R_A/R_B)^2 = (1/3)(3)^2 = 3.$$

EVALUATE: The sphere with the larger radius needs more net charge to produce the same potential. We can write $E = V/R$ for a sphere, so with equal potentials the sphere with the smaller R has the larger V.

23.83.   **IDENTIFY** and **SET UP:**   The potential at the surface is given by Example 23.8 and the electric field at the surface is given by Example 22.5. The charge initially on sphere 1 spreads between the two spheres such as to bring them to the same potential.

EXECUTE: **(a)** $E_1 = \dfrac{1}{4\pi\epsilon_0}\dfrac{Q_1}{R_1^2}$, $V_1 = \dfrac{1}{4\pi\epsilon_0}\dfrac{Q_1}{R_1} = R_1 E_1$

(b) Two conditions must be met:

1) Let q_1 and q_2 be the final potentials of each sphere. Then $q_1 + q_2 = Q_1$ (charge conservation)

2) Let V_1 and V_2 be the final potentials of each sphere. All points of a conductor are at the same potential, so $V_1 = V_2$.

$V_1 = V_2$ requires that $\dfrac{1}{4\pi\epsilon_0}\dfrac{q_1}{R_1} = \dfrac{1}{4\pi\epsilon_0}\dfrac{q_2}{R_2}$ and then $q_1/R_1 = q_2/R_2$

$$q_1 R_2 = q_2 R_1 = (Q_1 - q_1)R_1$$

This gives $q_1 = (R_1/[R_1 + R_2])Q_1$ and $q_2 = Q_1 - q_1 = Q_1(1 - R_1/[R_1 + R_2]) = Q_1(R_2/[R_1 + R_2])$

(c) $V_1 = \dfrac{1}{4\pi\epsilon_0}\dfrac{q_1}{R_1} = \dfrac{Q_1}{4\pi\epsilon_0(R_1 + R_2)}$ and $V_2 = \dfrac{1}{4\pi\epsilon_0}\dfrac{q_2}{R_2} = \dfrac{Q_1}{4\pi\epsilon_0(R_1 + R_2)}$, which equals V_1 as it should.

(d) $E_1 = \dfrac{V_1}{R_1} = \dfrac{Q_1}{4\pi\epsilon_0 R_1(R_1 + R_2)}$. $E_2 = \dfrac{V_2}{R_2} = \dfrac{Q_1}{4\pi\epsilon_0 R_2(R_1 + R_2)}$.

EVALUATE: Part (a) says $q_2 = q_1(R_2/R_1)$. The sphere with the larger radius needs more charge to produce the same potential at its surface. When $R_1 = R_2$, $q_1 = q_2 = Q_1/2$. The sphere with the larger radius has the smaller electric field at its surface.

CAPACITANCE AND DIELECTRICS

24.3. **IDENTIFY** and **SET UP:** It is a parallel-plate air capacitor, so we can apply the equations of Sections 24.1.

EXECUTE: **(a)** $C = \dfrac{Q}{V_{ab}}$ so $V_{ab} = \dfrac{Q}{C} = \dfrac{0.148 \times 10^{-6} \text{ C}}{245 \times 10^{-12} \text{ F}} = 604$ V

(b) $C = \dfrac{\epsilon_0 A}{d}$ so $A = \dfrac{Cd}{\epsilon_0} = \dfrac{\left(245 \times 10^{-12} \text{ F}\right)\left(0.328 \times 10^{-3} \text{ m}\right)}{8.854 \times 10^{-12} \text{ C}^2/\text{N} \cdot \text{m}^2} = 9.08 \times 10^{-3} \text{ m}^2 = 90.8 \text{ cm}^2$

(c) $V_{ab} = Ed$ so $E = \dfrac{V_{ab}}{d} = \dfrac{604 \text{ V}}{0.328 \times 10^{-3} \text{ m}} = 1.84 \times 10^6$ V/m

(d) $E = \dfrac{\sigma}{\epsilon_0}$ so $\sigma = E\epsilon_0 = \left(1.84 \times 10^6 \text{ V/m}\right)\left(8.854 \times 10^{-12} \text{ C}^2/\text{N} \cdot \text{m}^2\right) = 1.63 \times 10^{-5}$ C/m^2

EVALUATE: We could also calculate σ directly as Q/A. $\sigma = \dfrac{Q}{A} = \dfrac{0.148 \times 10^{-6} \text{ C}}{9.08 \times 10^{-3} \text{ m}^2} = 1.63 \times 10^{-5}$ C/m^2, which checks.

24.5. **IDENTIFY:** $C = \dfrac{Q}{V_{ab}}$. $C = \dfrac{\epsilon_0 A}{d}$.

SET UP: When the capacitor is connected to the battery, $V_{ab} = 12.0$ V .

EXECUTE: **(a)** $Q = CV_{ab} = (10.0 \times 10^{-6} \text{ F})(12.0 \text{ V}) = 1.20 \times 10^{-4} \text{ C} = 120 \ \mu\text{C}$

(b) When d is doubled C is halved, so Q is halved. $Q = 60 \ \mu\text{C}$.

(c) If r is doubled, A increases by a factor of 4. C increases by a factor of 4 and Q increases by a factor of 4. $Q = 480 \ \mu\text{C}$.

EVALUATE: When the plates are moved apart, less charge on the plates is required to produce the same potential difference. With the separation of the plates constant, the electric field must remain constant to produce the same potential difference. The electric field depends on the surface charge density, σ. To produce the same σ, more charge is required when the area increases.

24.11. **IDENTIFY** and **SET UP:** Use the expression for C/L derived in Example 24.4. Then use Eq.(24.1) to calculate Q.

EXECUTE: **(a)** From Example 24.4, $\dfrac{C}{L} = \dfrac{2\pi\epsilon_0}{\ln\left(r_b / r_a\right)}$

$\dfrac{C}{L} = \dfrac{2\pi\left(8.854 \times 10^{-12} \text{ C}^2/\text{N} \cdot \text{m}^2\right)}{\ln\left(3.5 \text{ mm}/1.5 \text{ mm}\right)} = 6.57 \times 10^{-11} \text{ F/m} = 66 \text{ pF/m}$

(b) $C = \left(6.57 \times 10^{-11} \text{ F/m}\right)\left(2.8 \text{ m}\right) = 1.84 \times 10^{-10} \text{ F}.$

$Q = CV = \left(1.84 \times 10^{-10} \text{ F}\right)\left(350 \times 10^{-3} \text{ V}\right) = 6.4 \times 10^{-11} \text{ C} = 64 \text{ pC}$

The conductor at higher potential has the positive charge, so there is +64 pC on the inner conductor and −64 pC on the outer conductor.

EVALUATE: C depends only on the dimensions of the capacitor. Q and V are proportional.

24.13. **IDENTIFY:** We can use the definition of capacitance to find the capacitance of the capacitor, and then relate the capacitance to geometry to find the inner radius.

(a) SET UP: By the definition of capacitance, $C = Q/V$.

EXECUTE: $C = \dfrac{Q}{V} = \dfrac{3.30 \times 10^{-9} \text{ C}}{2.20 \times 10^2 \text{ V}} = 1.50 \times 10^{-11} \text{ F} = 15.0 \text{ pF}$

(b) SET UP: The capacitance of a spherical capacitor is $C = 4\pi\epsilon_0 \dfrac{r_a r_b}{r_b - r_a}$.

EXECUTE: Solve for r_a and evaluate using $C = 15.0$ pF and $r_b = 4.00$ cm, giving $r_a = 3.09$ cm.

(c) SET UP: We can treat the inner sphere as a point-charge located at its center and use Coulomb's law,

$$E = \frac{1}{4\pi\epsilon_0} \frac{q}{r^2}.$$

EXECUTE: $E = \dfrac{\left(9.00\times10^9 \text{ N}\cdot\text{m}^2/\text{C}^2\right)\left(3.30\times10^{-9} \text{ C}\right)}{\left(0.0309 \text{ m}\right)^2} = 3.12\times10^4 \text{ N/C}$

EVALUATE: Outside the capacitor, the electric field is zero because the charges on the spheres are equal in magnitude but opposite in sign.

24.15.   **IDENTIFY:**   Replace series and parallel combinations of capacitors by their equivalents. In each equivalent network apply the rules for Q and V for capacitors in series and parallel; start with the simplest network and work back to the original circuit.

SET UP: Do parts **(a)** and **(b)** together. The capacitor network is drawn in Figure 24.15a.

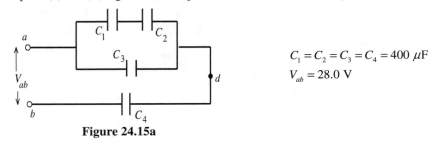

$C_1 = C_2 = C_3 = C_4 = 400 \ \mu\text{F}$

$V_{ab} = 28.0$ V

Figure 24.15a

EXECUTE: Simplify the circuit by replacing the capacitor combinations by their equivalents: C_1 and C_2 are in series and are equivalent to C_{12} (Figure 24.15b).

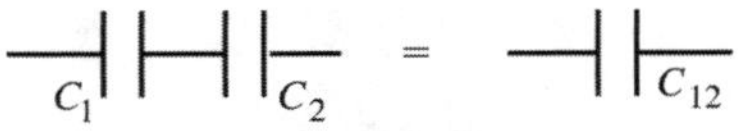

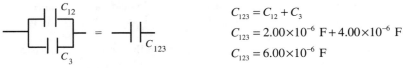

$\dfrac{1}{C_{12}} = \dfrac{1}{C_1} + \dfrac{1}{C_2}$

Figure 24.15b

$C_{12} = \dfrac{C_1 C_2}{C_1 + C_2} = \dfrac{\left(4.00\times10^{-6} \text{ F}\right)\left(4.00\times10^{-6} \text{ F}\right)}{4.00\times10^{-6} \text{ F} + 4.00\times10^{-6} \text{ F}} = 2.00\times10^{-6} \text{ F}$

C_{12} and C_3 are in parallel and are equivalent to C_{123} (Figure 24.15c).

$C_{123} = C_{12} + C_3$

$C_{123} = 2.00\times10^{-6} \text{ F} + 4.00\times10^{-6} \text{ F}$

$C_{123} = 6.00\times10^{-6} \text{ F}$

Figure 24.15c

C_{123} and C_4 are in series and are equivalent to C_{1234} (Figure 24.15d).

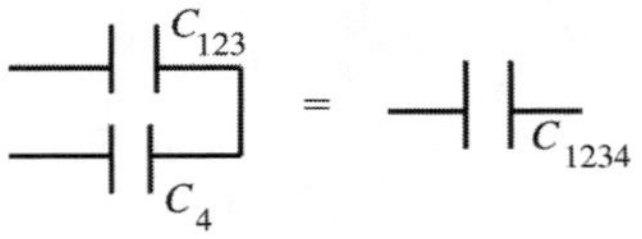

$\dfrac{1}{C_{1234}} = \dfrac{1}{C_{123}} + \dfrac{1}{C_4}$

Figure 24.15d

$C_{1234} = \dfrac{C_{123} C_4}{C_{123} + C_4} = \dfrac{\left(6.00\times10^{-6} \text{ F}\right)\left(4.00\times10^{-6} \text{ F}\right)}{6.00\times10^{-6} \text{ F} + 4.00\times10^{-6} \text{ F}} = 2.40\times10^{-6} \text{ F}$

The circuit is equivalent to the circuit shown in Figure 24.15e.

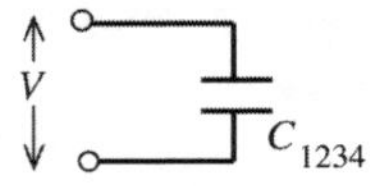

$V_{1234} = V = 28.0$ V

$Q_{1234} = C_{1234} V = \left(2.40\times10^{-6} \text{ F}\right)\left(28.0 \text{ V}\right) = 67.2 \ \mu\text{C}$

Figure 24.15e

Now build back up the original circuit, step by step. C_{1234} represents C_{123} and C_4 in series (Figure 24.15f).

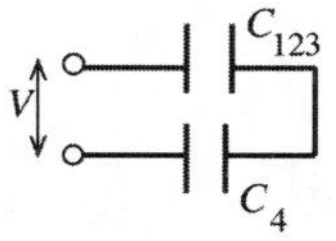

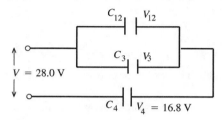

$Q_{123} = Q_4 - Q_{1234} = 67.2\ \mu C$
(charge same for capacitors in series)

Figure 24.15f

Then $V_{123} = \dfrac{Q_{123}}{C_{123}} = \dfrac{67.2\ \mu C}{6.00\ \mu F} = 11.2\ V$

$V_4 = \dfrac{Q_4}{C_4} = \dfrac{67.2\ \mu C}{4.00\ \mu F} = 16.8\ V$

Note that $V_4 + V_{123} = 16.8\ V + 11.2\ V = 28.0\ V$, as it should.

Next consider the circuit as written in Figure 24.15g.

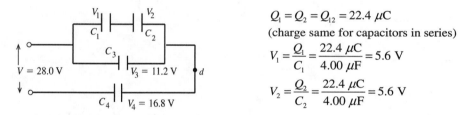

$V_3 = V_{12} = 28.0\ V - V_4$

$V_3 = 11.2\ V$

$Q_3 = C_3 V_3 = (4.00\ \mu F)(11.2\ V)$

$Q_3 = 44.8\ \mu C$

$Q_{12} - C_{12} V_{12} = (2.00\ \mu F)(11.2\ V)$

$Q_{12} = 22.4\ \mu C$

Figure 24.15g

Finally, consider the original circuit, as shown in Figure 24.15h.

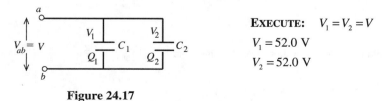

$Q_1 = Q_2 = Q_{12} = 22.4\ \mu C$
(charge same for capacitors in series)

$V_1 = \dfrac{Q_1}{C_1} = \dfrac{22.4\ \mu C}{4.00\ \mu F} = 5.6\ V$

$V_2 = \dfrac{Q_2}{C_2} = \dfrac{22.4\ \mu C}{4.00\ \mu F} = 5.6\ V$

Figure 24.15h

Note that $V_1 + V_2 = 11.2\ V$, which equals V_3 as it should.

Summary: $Q_1 = 22.4\ \mu C$, $V_1 = 5.6\ V$

$Q_2 = 22.4\ \mu C$, $V_2 = 5.6\ V$

$Q_3 = 44.8\ \mu C$, $V_3 = 11.2\ V$

$Q_4 = 67.2\ \mu C$, $V_4 = 16.8\ V$

(c) $V_{ad} = V_3 = 11.2\ V$

EVALUATE: $V_1 + V_2 + V_4 = V$, or $V_3 + V_4 = V$. $Q_1 = Q_2$, $Q_1 + Q_3 = Q_4$ and $Q_4 = Q_{1234}$.

24.17.   **IDENTIFY:**   The two capacitors are in parallel so the voltage is the same on each, and equal to the applied voltage V_{ab}.

SET UP: Do parts **(a)** and **(b)** together. The network is sketched in Figure 24.17.

EXECUTE: $V_1 = V_2 = V$

$V_1 = 52.0\ V$

$V_2 = 52.0\ V$

Figure 24.17

$C = Q/V$ so $Q = CV$

$Q_1 = C_1 V_1 = (3.00\ \mu F)(52.0\ V) = 156\ \mu C$. $Q_2 = C_2 V_2 = (5.00\ \mu F)(52.0\ V) = 260\ \mu C$.

EVALUATE: To produce the same potential difference, the capacitor with the larger C has the larger Q.

24.19.   **IDENTIFY** and **SET UP:**   Use the rules for V for capacitors in series and parallel: for capacitors in parallel the voltages are the same and for capacitors in series the voltages add.

EXECUTE: $V_1 = Q_1 / C_1 = (150 \ \mu\text{C})/(3.00 \ \mu\text{F}) = 50 \text{ V}$

C_1 and C_2 are in parallel, so $V_2 = 50 \text{ V}$

$V_3 = 120 \text{ V} - V_1 = 70 \text{ V}$

EVALUATE: Now that we know the voltages, we could also calculate Q for the other two capacitors.

24.21.   **IDENTIFY** and **SET UP:** $C = \dfrac{\epsilon_0 A}{d}$. For two capacitors in parallel, $C_{eq} = C_1 + C_2$.

EXECUTE: $C_{eq} = C_1 + C_2 = \dfrac{\epsilon_0 A_1}{d} + \dfrac{\epsilon_0 A_2}{d} = \dfrac{\epsilon_0 (A_1 + A_2)}{d}$. So the combined capacitance for two capacitors in parallel is

that of a single capacitor of their combined area $(A_1 + A_2)$ and common plate separation d.

EVALUATE: C_{eq} is larger than either C_1 or C_2.

24.25.   **IDENTIFY** and **SET UP:**   The energy density is given by Eq.(24.11): $u = \frac{1}{2}\epsilon_0 E^2$. Use $V = Ed$ to solve for E.

EXECUTE: Calculate E: $E = \dfrac{V}{d} = \dfrac{400 \text{ V}}{5.00 \times 10^{-3} \text{ m}} = 8.00 \times 10^4 \text{ V/m}$.

Then $u = \frac{1}{2}\epsilon_0 E^2 = \frac{1}{2}\left(8.854 \times 10^{-12} \text{ C}^2/\text{N} \cdot \text{m}^2\right)\left(8.00 \times 10^4 \text{ V/m}\right)^2 = 0.0283 \text{ J/m}^3$

EVALUATE: E is smaller than the value in Example 24.8 by about a factor of 6 so u is smaller by about a factor of $6^2 = 36$.

24.29.   **IDENTIFY** and **SET UP:**   Combine Eqs. (24.9) and (24.2) to write the stored energy in terms of the separation between the plates.

EXECUTE: **(a)** $U = \dfrac{Q^2}{2C}$; $C = \dfrac{\epsilon_0 A}{x}$ so $U = \dfrac{x Q^2}{2\epsilon_0 A}$

(b) $x \rightarrow x + dx$ gives $U = \dfrac{(x + dx)Q^2}{2\epsilon_0 A}$

$dU = \dfrac{(x + dx)Q^2}{2\epsilon_0 A} - \dfrac{x Q^2}{2\epsilon_0 A} = \left(\dfrac{Q^2}{2\epsilon_0 A}\right) dx$

(c) $dW = F \, dx = dU$, so $F = \dfrac{Q^2}{2\epsilon_0 A}$

(d) EVALUATE: The capacitor plates and the field between the plates are shown in Figure 24.29a.

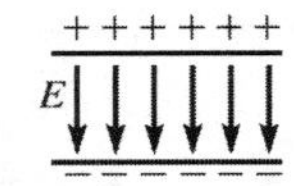

$E = \dfrac{\sigma}{\epsilon_0} = \dfrac{Q}{\epsilon_0 A}$

$F = \frac{1}{2} QE$, not QE

Figure 24.29a

The reason for the difference is that E is the field due to <u>both</u> plates. If we consider the positive plate only and calculate its electric field using Gauss's law (Figure 24.29b):

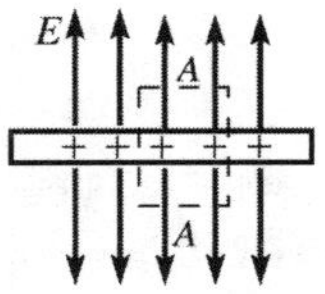

$\oint \vec{E} \cdot d\vec{A} = \dfrac{Q_{encl}}{\epsilon_0}$

$2EA = \dfrac{\sigma A}{\epsilon_0}$

$E = \dfrac{\sigma}{2\epsilon_0} = \dfrac{Q}{2\epsilon_0 A}$

Figure 24.29b

The force this field exerts on the other plate, that has charge $-Q$, is $F = \dfrac{Q^2}{2\epsilon_0 A}$.

24.33.   **IDENTIFY:**   The two capacitors are in parallel. $C_{eq} = C_1 + C_2$. $C = \dfrac{Q}{V}$. $U = \frac{1}{2}CV^2$.

SET UP: For capacitors in parallel, the voltages are the same and the charges add.

EXECUTE: **(a)** $C_{eq} = C_1 + C_2 = 35 \text{ nF} + 75 \text{ nF} = 110 \text{ nF}$. $Q_{tot} = C_{eq}V = (110 \times 10^{-9} \text{ F})(220 \text{ V}) = 24.2 \ \mu\text{C}$

(b) $V = 220 \text{ V}$ for each capacitor.

35 nF: $Q_{35} = C_{35}V = (35 \times 10^{-9} \text{ F})(220 \text{ V}) = 7.7 \ \mu\text{C}$; 75 nF: $Q_{75} = C_{75}V = (75 \times 10^{-9} \text{ F})(220 \text{ V}) = 16.5 \ \mu\text{C}$. Note that $Q_{35} + Q_{75} = Q_{tot}$.

(c) $U_{tot} = \frac{1}{2}C_{eq}V^2 = \frac{1}{2}(110 \times 10^{-9} \text{ F})(220 \text{ V})^2 = 2.66 \text{ mJ}$

(d) 35 nF: $U_{35} = \frac{1}{2}C_{35}V^2 = \frac{1}{2}(35 \times 10^{-9} \text{ F})(220 \text{ V})^2 = 0.85 \text{ mJ}$;

75 nF: $U_{75} = \frac{1}{2}C_{75}V^2 = \frac{1}{2}(75 \times 10^{-9} \text{ F})(220 \text{ V})^2 = 1.81 \text{ mJ}$. Since V is the same the capacitor with larger C stores more energy.

(e) 220 V for each capacitor.

EVALUATE: The capacitor with the larger C has the larger Q.

24.37. **IDENTIFY:** Use the rules for series and for parallel capacitors to express the voltage for each capacitor in terms of the applied voltage. Express U, Q, and E in terms of the capacitor voltage.

SET UP: Le the applied voltage be V. Let each capacitor have capacitance C. $U = \frac{1}{2}CV^2$ for a single capacitor with voltage V.

EXECUTE: (a) series

Voltage across each capacitor is $V/2$. The total energy stored is $U_s = 2\left(\frac{1}{2}C[V/2]^2\right) = \frac{1}{4}CV^2$

parallel

Voltage across each capacitor is V. The total energy stored is $U_p = 2\left(\frac{1}{2}CV^2\right) = CV^2$

$U_p = 4U_s$

(b) $Q = CV$ for a single capacitor with voltage V. $Q_s = 2\left(C[V/2]\right) = CV$; $Q_p = 2(CV) = 2CV$; $Q_p = 2Q_s$

(c) $E = V/d$ for a capacitor with voltage V. $E_s = V/2d$; $E_p = V/d$; $E_p = 2E_s$

EVALUATE: The parallel combination stores more energy and more charge since the voltage for each capacitor is larger for parallel. More energy stored and larger voltage for parallel means larger electric field in the parallel case.

24.39. **IDENTIFY and SET UP:** Q is constant so we can apply Eq.(24.14). The charge density on each surface of the dielectric is given by Eq.(24.16).

EXECUTE: $E = \frac{E_0}{K}$ so $K = \frac{E_0}{E} = \frac{3.20 \times 10^5 \text{ V/m}}{2.50 \times 10^5 \text{ V/m}} = 1.28$

(a) $\sigma_i = \sigma(1 - 1/K)$

$\sigma = \epsilon_0 E_0 = (8.854 \times 10^{-12} \text{ C}^2/\text{N} \cdot \text{m}^2)(3.20 \times 10^5 \text{ N/C}) = 2.833 \times 10^{-6} \text{ C/m}^2$

$\sigma_i = (2.833 \times 10^{-6} \text{ C/m}^2)(1 - 1/1.28) = 6.20 \times 10^{-7} \text{ C/m}^2$

(b) As calculated above, $K = 1.28$.

EVALUATE: The surface charges on the dielectric produce an electric field that partially cancels the electric field produced by the charges on the capacitor plates.

24.41. **IDENTIFY and SET UP:** For a parallel-plate capacitor with a dielectric we can use the equation $C = K\epsilon_0 A/d$.

Minimum A means smallest possible d. d is limited by the requirement that E be less than 1.60×10^7 V/m when V is as large as 5500 V.

EXECUTE: $V = Ed$ so $d = \frac{V}{E} = \frac{5500 \text{ V}}{1.60 \times 10^7 \text{ V/m}} = 3.44 \times 10^{-4} \text{ m}$

Then $A = \frac{Cd}{K\epsilon_0} = \frac{(1.25 \times 10^{-9} \text{ F})(3.44 \times 10^{-4} \text{ m})}{(3.60)(8.854 \times 10^{-12} \text{ C}^2/\text{N} \cdot \text{m}^2)} = 0.0135 \text{ m}^2$.

EVALUATE: The relation $V = Ed$ applies with or without a dielectric present. A would have to be larger if there were no dielectric.

24.45. **(a) IDENTIFY and SET UP:** Since the capacitor remains connected to the power supply the potential difference doesn't change when the dielectric is inserted. Use Eq.(24.9) to calculate V and combine it with Eq.(24.12) to obtain a relation between the stored energies and the dielectric constant and use this to calculate K.

EXECUTE: Before the dielectric is inserted $U_0 = \frac{1}{2}C_0V^2$ so $V = \sqrt{\frac{2U_0}{C_0}} = \sqrt{\frac{2(1.85 \times 10^{-5} \text{ J})}{360 \times 10^{-9} \text{ F}}} = 10.1 \text{ V}$

(b) $K = C/C_0$

$U_0 = \frac{1}{2}C_0V^2$, $U = \frac{1}{2}CV^2$ so $C/C_0 = U/U_0$

$K = \frac{U}{U_0} = \frac{1.85 \times 10^{-5} \text{ J} + 2.32 \times 10^{-5} \text{ J}}{1.85 \times 10^{-5} \text{ J}} = 2.25$

EVALUATE: K increases the capacitance and then from $U = \frac{1}{2}CV^2$, with V constant an increase in C gives an increase in U.

24.49.   **IDENTIFY:**   Apply Eq.(24.23) to calculate E. $V = Ed$ and $C = Q/V$ apply whether there is a dielectric between the plates or not.
(a) **SET UP:**   Apply Eq.(24.23) to the dashed surface in Figure 24.49:

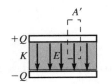

EXECUTE: $\oint K\vec{E} \cdot d\vec{A} = \dfrac{Q_{\text{encl-free}}}{\epsilon_0}$

$\oint K\vec{E} \cdot d\vec{A} = KEA'$

since $E = 0$ outside the plates
$Q_{\text{encl-free}} = \sigma A' = (Q/A)A'$

Figure 24.49

Thus $KEA' = \dfrac{(Q/A)A'}{\epsilon_0}$ and $E = \dfrac{Q}{\epsilon_0 AK}$

(b) $V = Ed = \dfrac{Qd}{\epsilon_0 AK}$

(c) $C = \dfrac{Q}{V} = \dfrac{Q}{(Qd/\epsilon_0 AK)} = K\dfrac{\epsilon_0 A}{d} = KC_0$.

EVALUATE: Our result shows that $K = C/C_0$, which is Eq.(24.12).

24.51.   **IDENTIFY** and **SET UP:**   If the capacitor remains connected to the battery, the battery keeps the potential difference between the plates constant by changing the charge on the plates.

EXECUTE: (a) $C = \dfrac{\epsilon_0 A}{d}$

$C = \dfrac{(8.854 \times 10^{-12}\ \text{C}^2/\text{N} \cdot \text{m}^2)(0.16\ \text{m})^2}{9.4 \times 10^{-3}\ \text{m}} = 2.4 \times 10^{-11}\ \text{F} = 24\ \text{pF}$

(b) Remains connected to the battery says that V stays 12 V. $Q = CV = (2.4 \times 10^{-11}\ \text{F})(12\ \text{V}) = 2.9 \times 10^{-10}\ \text{C}$

(c) $E = \dfrac{V}{d} = \dfrac{12\ \text{V}}{9.4 \times 10^{-3}\ \text{m}} = 1.3 \times 10^3\ \text{V/m}$

(d) $U = \tfrac{1}{2}QV = \tfrac{1}{2}(2.9 \times 10^{-10}\ \text{C})(12.0\ \text{V}) = 1.7 \times 10^{-9}\ \text{J}$

EVALUATE: Increasing the separation decreases C. With V constant, this means that Q decreases and U decreases. Q decreases and $E = Q/\epsilon_0 A$ so E decreases. We come to the same conclusion from $E = V/d$.

24.55.   **IDENTIFY:**   Example 24.4 shows that $C = \dfrac{2\pi\epsilon_0 L}{\ln(r_b/r_a)}$ for a cylindrical capacitor.

SET UP: $\ln(1 + x) \approx x$ when x is small. The area of each conductor is approximately $A = 2\pi r_a L$.

EXECUTE: (a) $d \ll r_a$: $C = \dfrac{2\pi\epsilon_0 L}{\ln(r_b/r_a)} = \dfrac{2\pi\epsilon_0 L}{\ln((d + r_a)/r_a)} = \dfrac{2\pi\epsilon_0 L}{\ln(1 + d/r_a)} \approx \dfrac{2\pi r_a L \epsilon_0}{d} = \dfrac{\epsilon_0 A}{d}$

EVALUATE: (b) At the scale of part (a) the cylinders appear to be flat, and so the capacitance should appear like that of flat plates.

24.57.   **IDENTIFY:**   Simplify the network by replacing series and parallel combinations by their equivalent. The stored energy in a capacitor is $U = \tfrac{1}{2}CV^2$.

SET UP: For capacitors in series the voltages add and the charges are the same; $\dfrac{1}{C_{\text{eq}}} = \dfrac{1}{C_1} + \dfrac{1}{C_2} + \cdots$. For capacitors in parallel the voltages are the same and the charges add; $C_{\text{eq}} = C_1 + C_2 + \cdots$ $C = \dfrac{Q}{V}$. $U = \tfrac{1}{2}CV^2$.

EXECUTE: (a) Find C_{eq} for the network by replacing each series or parallel combination by its equivalent. The successive simplified circuits are shown in Figure 24.57a–c.
$U_{\text{tot}} = \tfrac{1}{2}C_{\text{eq}}V^2 = \tfrac{1}{2}(2.19 \times 10^{-6}\ \text{F})(12.0\ \text{V})^2 = 1.58 \times 10^{-4}\ \text{J} = 158\ \mu\text{J}$

(b) From Figure 24.57c, $Q_{\text{tot}} = C_{\text{eq}}V = (2.19 \times 10^{-6}\ \text{F})(12.0\ \text{V}) = 2.63 \times 10^{-5}\ \text{C}$. From Figure 24.57b, $Q_{\text{tot}} = 2.63 \times 10^{-5}\ \text{C}$.

$V_{4.8} = \dfrac{Q_{4.8}}{C_{4.8}} = \dfrac{2.63 \times 10^{-5}\ \text{C}}{4.80 \times 10^{-6}\ \text{F}} = 5.48\ \text{V}$. $U_{4.8} = \tfrac{1}{2}CV^2 = \tfrac{1}{2}(4.80 \times 10^{-6}\ \text{F})(5.48\ \text{V})^2 = 7.21 \times 10^{-5}\ \text{J} = 72.1\ \mu\text{J}$

This one capacitor stores nearly half the total stored energy.

EVALUATE: $U = \dfrac{Q^2}{2C}$. For capacitors in series the capacitor with the smallest C stores the greatest amount of energy.

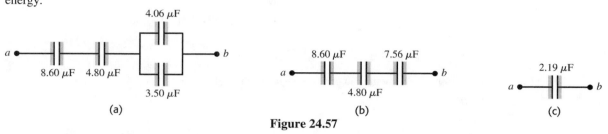

Figure 24.57

24.59. **(a) IDENTIFY:**   Replace series and parallel combinations of capacitors by their equivalents.
SET UP: The network is sketched in Figure 24.59a.

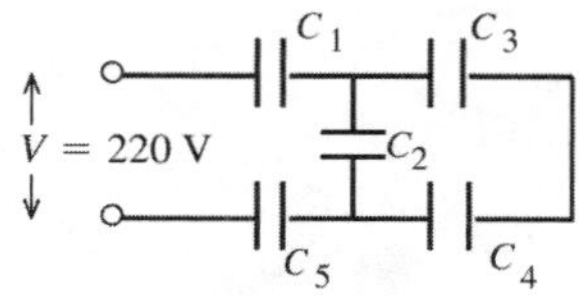

$C_1 = C_5 = 8.4 \ \mu F$

$C_2 = C_3 = C_4 = 4.2 \ \mu F$

Figure 24.59a

EXECUTE: Simplify the circuit by replacing the capacitor combinations by their equivalents: C_3 and C_4 are in series and can be replaced by C_{34} (Figure 24.59b):

$$\frac{1}{C_{34}} = \frac{1}{C_3} + \frac{1}{C_4}$$

$$\frac{1}{C_{34}} = \frac{C_3 + C_4}{C_3 C_4}$$

Figure 24.59b

$$C_{34} = \frac{C_3 C_4}{C_3 + C_4} = \frac{(4.2 \ \mu F)(4.2 \ \mu F)}{4.2 \ \mu F + 4.2 \ \mu F} = 2.1 \ \mu F$$

C_2 and C_{34} are in parallel and can be replaced by their equivalent (Figure 24.59c):

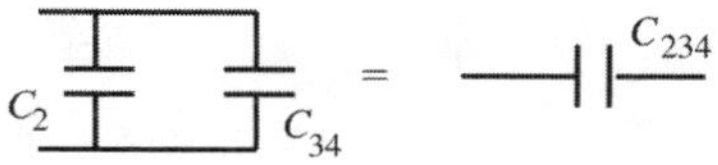

$$C_{234} = C_2 + C_{34}$$

$$C_{234} = 4.2 \ \mu F + 2.1 \ \mu F$$

$$C_{234} = 6.3 \ \mu F$$

Figure 24.59c

C_1, C_5 and C_{234} are in series and can be replaced by C_{eq} (Figure 24.59d):

$$\frac{1}{C_{eq}} = \frac{1}{C_1} + \frac{1}{C_5} + \frac{1}{C_{234}}$$

$$\frac{1}{C_{eq}} = \frac{2}{8.4 \ \mu F} + \frac{1}{6.3 \ \mu F}$$

$$C_{eq} = 2.5 \ \mu F$$

Figure 24.59d

EVALUATE: For capacitors in series the equivalent capacitor is smaller than any of those in series. For capacitors in parallel the equivalent capacitance is larger than any of those in parallel.
(b) IDENTIFY and SET UP: In each equivalent network apply the rules for Q and V for capacitors in series and parallel; start with the simplest network and work back to the original circuit.

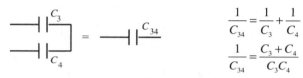
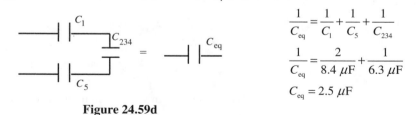

EXECUTE: The equivalent circuit is drawn in Figure 24.59e.

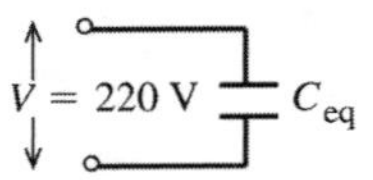

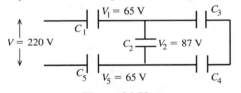

$$Q_{eq} = C_{eq}V$$
$$Q_{eq} = (2.5\ \mu F)(220\ V) = 550\ \mu C$$

Figure 24.59e

$Q_1 = Q_5 = Q_{234} = 550\ \mu C$ (capacitors in series have same charge)

$$V_1 = \frac{Q_1}{C_1} = \frac{550\ \mu C}{8.4\ \mu F} = 65\ V$$

$$V_5 = \frac{Q_5}{C_5} = \frac{550\ \mu C}{8.4\ \mu F} = 65\ V$$

$$V_{234} = \frac{Q_{234}}{C_{234}} = \frac{550\ \mu C}{6.3\ \mu F} = 87\ V$$

Now draw the network as in Figure 24.59f.

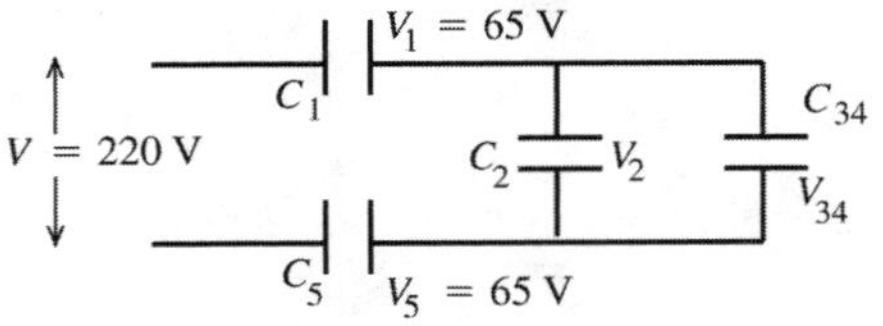

$V_2 = V_{34} = V_{234} = 87\ V$
capacitors in parallel have the same potential

Figure 24.59f

$$Q_2 = C_2 V_2 = (4.2\ \mu F)(87\ V) = 370\ \mu C$$
$$Q_{34} = C_{34}V_{34} = (2.1\ \mu F)(87\ V) = 180\ \mu C$$

Finally, consider the original circuit (Figure 24.59g).

$Q_3 = Q_4 = Q_{34} = 180\ \mu C$
capacitors in series have the same charge

Figure 24.59g

$$V_3 = \frac{Q_3}{C_3} = \frac{180\ \mu C}{4.2\ \mu F} = 43\ V$$

$$V_4 = \frac{Q_4}{C_4} = \frac{180\ \mu C}{4.2\ \mu F} = 43\ V$$

Summary: $Q_1 = 550\ \mu C$, $V_1 = 65\ V$

$Q_2 = 370\ \mu C$, $V_2 = 87\ V$

$Q_3 = 180\ \mu C$, $V_3 = 43\ V$

$Q_4 = 180\ \mu C$, $V_4 = 43\ V$

$Q_5 = 550\ \mu C$, $V_5 = 65\ V$

EVALUATE: $V_3 + V_4 = V_2$ and $V_1 + V_2 + V_5 = 220\ V$ (apart from some small rounding error)

$Q_1 = Q_2 + Q_3$ and $Q_5 = Q_2 + Q_4$

24.61.   **(a) IDENTIFY:**   Replace the three capacitors in series by their equivalent. The charge on the equivalent capacitor equals the charge on each of the original capacitors.
SET UP: The three capacitors can be replaced by their equivalent as shown in Figure 24.61a.

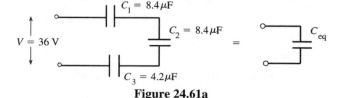

Figure 24.61a

EXECUTE: $C_3 = C_1/2$ so $\dfrac{1}{C_{eq}} = \dfrac{1}{C_1} + \dfrac{1}{C_2} + \dfrac{1}{C_3} = \dfrac{4}{8.4\ \mu F}$ and $C_{eq} = 8.4\ \mu F/4 = 2.1\ \mu F$

$Q = C_{eq}V = (2.1\ \mu F)(36\ V) = 76\ \mu C$

The three capacitors are in series so they each have the same charge: $Q_1 = Q_2 = Q_3 = 76\ \mu C$

EVALUATE: The equivalent capacitance for capacitors in series is smaller than each of the original capacitors.
(b) IDENTIFY and SET UP: Use $U = \frac{1}{2}QV$. We know each Q and we know that $V_1 + V_2 + V_3 = 36$ V.
EXECUTE: $U = \frac{1}{2}Q_1V_1 + \frac{1}{2}Q_2V_2 + \frac{1}{2}Q_3V_3$

But $Q_1 = Q_2 = Q_3 = Q$ so $U = \frac{1}{2}Q(V_1 + V_2 + V_3)$

But also $V_1 + V_2 + V_3 = V = 36$ V, so $U = \frac{1}{2}QV = \frac{1}{2}(76\ \mu C)(36\ V) = 1.4\times10^{-3}$ J.

EVALUATE: We could also use $U = Q^2/2C$ and calculate U for each capacitor.

(c) IDENTIFY: The charges on the plates redistribute to make the potentials across each capacitor the same.
SET UP: The capacitors before and after they are connected are sketched in Figure 24.61b.

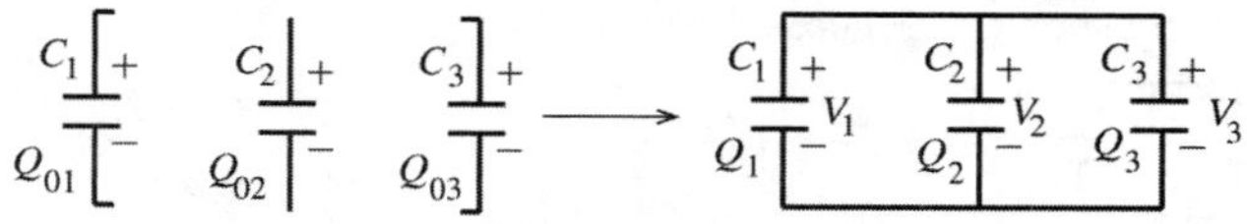

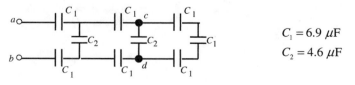

Figure 24.61b

EXECUTE: The total positive charge that is available to be distributed on the upper plates of the three capacitors is
$Q_0 = Q_{01} + Q_{02} + Q_{03} = 3(76\ \mu C) = 228\ \mu C$. Thus $Q_1 + Q_2 + Q_3 = 228\ \mu C$. After the circuit is completed the charge distributes to make $V_1 = V_2 = V_3$. $V = Q/C$ and $V_1 = V_2$ so $Q_1/C_1 = Q_2/C_2$ and then $C_1 = C_2$ says $Q_1 = Q_2$. $V_1 = V_3$ says $Q_1/C_1 = Q_3/C_3$ and $Q_1 = Q_3(C_1/C_3) = Q_3(8.4\ \mu F/4.2\ \mu F) = 2Q_3$

Using $Q_2 = Q_1$ and $Q_1 = 2Q_3$ in the above equation gives $2Q_3 + 2Q_3 + Q_3 = 228\ \mu C$.

$5Q_3 = 228\ \mu C$ and $Q_3 = 45.6\ \mu C$, $Q_1 = Q_2 = 91.2\ \mu C$

Then $V_1 = \dfrac{Q_1}{C_1} = \dfrac{91.2\ \mu C}{8.4\ \mu F} = 11$ V, $V_2 = \dfrac{Q_2}{C_2} = \dfrac{91.2\ \mu C}{8.4\ \mu F} = 11$ V, and $V_3 = \dfrac{Q_3}{C_3} = \dfrac{45.6\ \mu C}{4.2\ \mu F} = 11$ V.

The voltage across each capacitor in the parallel combination is 11 V.
(d) $U = \frac{1}{2}Q_1V_1 + \frac{1}{2}Q_2V_2 + \frac{1}{2}Q_3V_3$.

But $V_1 = V_2 = V_3$ so $U = \frac{1}{2}V_1(Q_1 + Q_2 + Q_3) = \frac{1}{2}(11\ V)(228\ \mu C) = 1.3\times10^{-3}$ J.

EVALUATE: This is less than the original energy of 1.4×10^{-3} J. The stored energy has decreased, as in Example 24.7.

24.63.   **IDENTIFY:**   Replace series and parallel combinations of capacitors by their equivalents. In each equivalent network apply the rules for Q and V for capacitors in series and parallel; start with the simplest network and work back to the original circuit.
(a) SET UP: The network is sketched in Figure 24.63a.

$C_1 = 6.9\ \mu F$

$C_2 = 4.6\ \mu F$

Figure 24.63a

EXECUTE: Simplify the network by replacing the capacitor combinations by their equivalents. Make the replacement shown in Figure 24.63b.

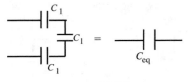

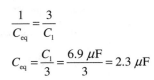

$$\frac{1}{C_{eq}} = \frac{3}{C_1}$$

$$C_{eq} = \frac{C_1}{3} = \frac{6.9 \ \mu\text{F}}{3} = 2.3 \ \mu\text{F}$$

Figure 24.63b

Next make the replacement shown in Figure 24.63c.

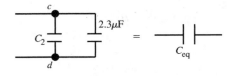

$$C_{eq} = 2.3 \ \mu\text{F} + C_2$$

$$C_{eq} = 2.3 \ \mu\text{F} + 4.6 \ \mu\text{F} = 6.9 \ \mu\text{F}$$

Figure 24.63c

Make the replacement shown in Figure 24.63d.

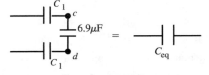

$$\frac{1}{C_{eq}} = \frac{2}{C_1} + \frac{1}{6.9 \ \mu\text{F}} = \frac{3}{6.9 \ \mu\text{F}}$$

$$C_{eq} = 2.3 \ \mu\text{F}$$

Figure 24.63d

Make the replacement shown in Figure 24.63e.

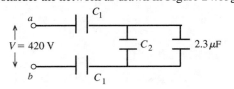

$$C_{eq} = C_2 + 2.3 \ \mu\text{F} = 4.6 \ \mu\text{F} + 2.3 \ \mu\text{F}$$

$$C_{eq} = 6.9 \ \mu\text{F}$$

Figure 24.63e

Make the replacement shown in Figure 24.63f.

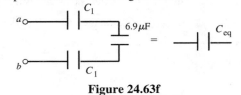

$$\frac{1}{C_{eq}} = \frac{2}{C_1} + \frac{1}{6.9 \ \mu\text{F}} = \frac{3}{6.9 \ \mu\text{F}}$$

$$C_{eq} = 2.3 \ \mu\text{F}$$

Figure 24.63f

(b) Consider the network as drawn in Figure 24.63g.

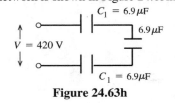

From part (a) 2.3 μF is the equivalent capacitance of the rest of the network.

Figure 24.63g

The equivalent network is shown in Figure 24.63h.

The capacitors are in series, so all three capacitors have the same Q.

Figure 24.63h

But here all three have the same C, so by $V = Q/C$ all three must have the same V. The three voltages must add to 420 V, so each capacitor has $V = 140$ V. The 6.9 μF to the right is the equivalent of C_2 and the 2.3 μF capacitor in parallel, so $V_2 = 140$ V. (Capacitors in parallel have the same potential difference.) Hence

$Q_1 = C_1V_1 = (6.9\ \mu\text{F})(140\ \text{V}) = 9.7 \times 10^{-4}$ C and $Q_2 = C_2V_2 = (4.6\ \mu\text{F})(140\ \text{V}) = 6.4 \times 10^{-4}$ C.

(c) From the potentials deduced in part (b) we have the situation shown in Figure 24.63i.

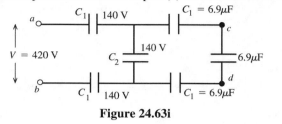

Figure 24.63i

The three right-most capacitors are in series and therefore have the same charge. But their capacitances are also equal, so by $V = Q/C$ they each have the same potential difference. Their potentials must sum to 140 V, so the potential across each is 47 V and $V_{cd} = 47$ V.

EVALUATE: In each capacitor network the rules for combining V for capacitors in series and parallel are obeyed. Note that $V_{cd} < V$, in fact $V - 2(140\ \text{V}) - 2(47\ \text{V}) = V_{cd}$.

24.65. **(a) IDENTIFY** and **SET UP:** Q is constant. $C = KC_0$; use Eq.(24.1) to relate the dielectric constant K to the ratio of the voltages without and with the dielectric.
EXECUTE: With the dielectric: $V = Q/C = Q/(KC_0)$
without the dielectric: $V_0 = Q/C_0$
$V_0/V = K$, so $K = (45.0\ \text{V})/(11.5\ \text{V}) = 3.91$

EVALUATE: Our analysis agrees with Eq.(24.13).
(b) IDENTIFY: The capacitor can be treated as equivalent to two capacitors C_1 and C_2 in parallel, one with area $2A/3$ and air between the plates and one with area $A/3$ and dielectric between the plates.
SET UP: The equivalent network is shown in Figure 24.65.

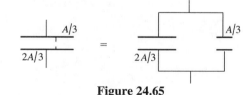

Figure 24.65

EXECUTE: Let $C_0 = \epsilon_0 A/d$ be the capacitance with only air between the plates. $C_1 = KC_0/3$, $C_2 = 2C_0/3$; $C_{eq} = C_1 + C_2 = (C_0/3)(K+2)$

$$V = \frac{Q}{C_{eq}} = \frac{Q}{C_0}\left(\frac{3}{K+2}\right) = V_0\left(\frac{3}{K+2}\right) = (45.0\ \text{V})\left(\frac{3}{5.91}\right) = 22.8\ \text{V}$$

EVALUATE: The voltage is reduced by the dielectric. The voltage reduction is less when the dielectric doesn't completely fill the volume between the plates.

24.67. **(a) IDENTIFY:** The conductor can be at some potential V, where $V = 0$ far from the conductor. This potential depends on the charge Q on the conductor so we can define $C = Q/V$ where C will not depend on V or Q.
(b) SET UP: Use the expression for the potential at the surface of the sphere in the analysis in part (a).
EXECUTE: For any point on a solid conducting sphere $V = Q/4\pi\epsilon_0 R$ if $V = 0$ at $r \to \infty$.

$C = \dfrac{Q}{V} = Q\left(\dfrac{4\pi\epsilon_0 R}{Q}\right) = 4\pi\epsilon_0 R$

(c) $C = 4\pi\epsilon_0 R = 4\pi(8.854 \times 10^{-12}\ \text{F/m})(6.38 \times 10^6\ \text{m}) = 7.10 \times 10^{-4}$ F $= 710\ \mu$F.

EVALUATE: The capacitance of the earth is about seven times larger than the largest capacitances in this range. The capacitance of the earth is quite small, in view of its large size.

24.69. **IDENTIFY:** We model the earth as a spherical capacitor.

SET UP: The capacitance of the earth is $C = 4\pi\epsilon_0 \dfrac{r_a r_b}{r_b - r_a}$ and, the charge on it is $Q = CV$, and its stored energy is

$U = \frac{1}{2}CV^2$.

EXECUTE: **(a)** $C = \dfrac{1}{9.00 \times 10^9 \text{ N} \cdot \text{m}^2/\text{C}^2} \dfrac{(6.38 \times 10^6 \text{ m})(6.45 \times 10^6 \text{ m})}{6.45 \times 10^6 \text{ m} - 6.38 \times 10^6 \text{ m}} = 6.5 \times 10^{-2} \text{ F}$

(b) $Q = CV = (6.54 \times 10^{-2} \text{ F})(350,000 \text{ V}) = 2.3 \times 10^4 \text{ C}$

(c) $U = \frac{1}{2}CV^2 = \frac{1}{2}(6.54 \times 10^{-2} \text{ F})(350,000 \text{ V})^2 = 4.0 \times 10^9 \text{ J}$

EVALUATE: While the capacitance of the earth is larger than ordinary laboratory capacitors, capacitors much larger than this, such as 1 F, are readily available.

24.71. **IDENTIFY:** $C = Q/V$, so we need to calculate the effect of the dielectrics on the potential difference between the plates.

SET UP: Let the potential of the positive plate be V_a, the potential of the negative plate be V_c, and the potential midway between the plates where the dielectrics meet be V_b, as shown in Figure 24.71.

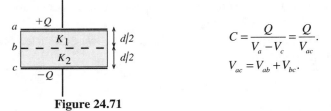

$C = \dfrac{Q}{V_a - V_c} = \dfrac{Q}{V_{ac}}.$

$V_{ac} = V_{ab} + V_{bc}.$

Figure 24.71

EXECUTE: The electric field in the absence of any dielectric is $E_0 = \dfrac{Q}{\epsilon_0 A}$. In the first dielectric the electric field is

reduced to $E_1 = \dfrac{E_0}{K_1} = \dfrac{Q}{K_1 \epsilon_0 A}$ and $V_{ab} = E_1\left(\dfrac{d}{2}\right) = \dfrac{Qd}{K_1 2\epsilon_0 A}$. In the second dielectric the electric field is reduced to

$E_2 = \dfrac{E_0}{K_2} = \dfrac{Q}{K_2 \epsilon_0 A}$ and $V_{bc} = E_2\left(\dfrac{d}{2}\right) = \dfrac{Qd}{K_2 2\epsilon_0 A}$. Thus $V_{ac} = V_{ab} + V_{bc} = \dfrac{Qd}{K_1 2\epsilon_0 A} + \dfrac{Qd}{K_2 2\epsilon_0 A} = \dfrac{Qd}{2\epsilon_0 A}\left(\dfrac{1}{K_1} + \dfrac{1}{K_2}\right).$

$V_{ac} = \dfrac{Qd}{2\epsilon_0 A}\left(\dfrac{K_1 + K_2}{K_1 K_2}\right)$. This gives $C = \dfrac{Q}{V_{ac}} = Q\left(\dfrac{2\epsilon_0 A}{Qd}\right)\left(\dfrac{K_1 K_2}{K_1 + K_2}\right) = \dfrac{2\epsilon_0 A}{d}\left(\dfrac{K_1 K_2}{K_1 + K_2}\right).$

EVALUATE: An equivalent way to calculate C is to consider the capacitor to be two in series, one with dielectric constant K_1 and the other with dielectric constant K_2 and both with plate separation $d/2$. (Can imagine inserting a thin conducting plate between the dielectric slabs.)

$C_1 = K_1 \dfrac{\epsilon_0 A}{d/2} = 2K_1 \dfrac{\epsilon_0 A}{d}$

$C_2 = K_2 \dfrac{\epsilon_0 A}{d/2} = 2K_2 \dfrac{\epsilon_0 A}{d}$

Since they are in series the total capacitance C is given by $\dfrac{1}{C} = \dfrac{1}{C_1} + \dfrac{1}{C_2}$ so $C = \dfrac{C_1 C_2}{C_1 + C_2} = \dfrac{2\epsilon_0 A}{d}\left(\dfrac{K_1 K_2}{K_1 + K_2}\right)$

CURRENT, RESISTANCE, AND ELECTROMOTIVE FORCE

25.5. **IDENTIFY** and **SET UP:** Use Eq. (25.3) to calculate the drift speed and then use that to find the time to travel the length of the wire.

EXECUTE: **(a)** Calculate the drift speed v_d:

$$J = \frac{I}{A} = \frac{I}{\pi r^2} = \frac{4.85\ \text{A}}{\pi\left(1.025\times10^{-3}\ \text{m}\right)^2} = 1.469\times10^6\ \text{A/m}^2$$

$$v_d = \frac{J}{n|q|} = \frac{1.469\times10^6\ \text{A/m}^2}{\left(8.5\times10^{28}/\text{m}^3\right)\left(1.602\times10^{-19}\ \text{C}\right)} = 1.079\times10^{-4}\ \text{m/s}$$

$$t = \frac{L}{v_d} = \frac{0.710\ \text{m}}{1.079\times10^{-4}\ \text{m/s}} = 6.58\times10^3\ \text{s} = 110\ \text{min.}$$

(b) $v_d = \dfrac{I}{\pi r^2 n|q|}$

$$t = \frac{L}{v_d} = \frac{\pi r^2 n|q|L}{I}$$

t is proportional to r^2 and hence to d^2 where $d = 2r$ is the wire diameter.

$$t = \left(6.58\times10^3\ \text{s}\right)\left(\frac{4.12\ \text{mm}}{2.05\ \text{mm}}\right)^2 = 2.66\times10^4\ \text{s} = 440\ \text{min.}$$

(c) EVALUATE: The drift speed is proportional to the current density and therefore it is inversely proportional to the square of the diameter of the wire. Increasing the diameter by some factor decreases the drift speed by the square of that factor.

25.7. **IDENTIFY** and **SET UP:** Apply Eq. (25.1) to find the charge dQ in time dt. Integrate to find the total charge in the whole time interval.

EXECUTE: **(a)** $dQ = I\ dt$

$$Q = \int_0^{8.0\ \text{s}}\left(55\ \text{A} - \left(0.65\ \text{A/s}^2\right)t^2\right)dt = \left[\left(55\ \text{A}\right)t - \left(0.217\ \text{A/s}^2\right)t^3\right]_0^{8.0\ \text{s}}$$

$$Q = \left(55\ \text{A}\right)\left(8.0\ \text{s}\right) - \left(0.217\ \text{A/s}^2\right)\left(8.0\ \text{s}\right)^3 = 330\ \text{C}$$

(b) $I = \dfrac{Q}{t} = \dfrac{330\ \text{C}}{8.0\ \text{s}} = 41\ \text{A}$

EVALUATE: The current decreases from 55 A to 13.4 A during the interval. The decrease is not linear and the average current is not equal to (55A + 13.4 A) / 2.

25.9. **IDENTIFY:** The number of moles of silver atoms is the mass of 1.00 m^3 divided by the atomic mass of silver. There are $N_A = 6.023\times10^{23}$ atoms per mole.

SET UP: For silver, density $= 10.5\times10^3\ \text{kg/m}^3$ and the atomic mass is $M = 107.868\times10^{-3}\ \text{kg/mol}$.

EXECUTE: Consider 1 m^3 of silver. $m = (\text{density})V = 10.5\times10^3\ \text{kg}$. $n = m/M = 9.734\times10^4\ \text{mol}$ and the number of atoms is $N = nN_A = 5.86\times10^{28}$ atoms. If there is one free electron per atom, there are 5.86×10^{28} free electrons$/\text{m}^3$. This agrees with the value given in Exercise 25.2.

EVALUATE: Our result verifies that for silver there is approximately one free electron per atom. Exercise 25.6 showed that for copper there is also one free electron per atom.

25.11.   **IDENTIFY:**   First use Ohm's law to find the resistance at 20.0°C; then calculate the resistivity from the resistance. Finally use the dependence of resistance on temperature to calculate the temperature coefficient of resistance.

   SET UP: Ohm's law is $R = V/I$, $R = \rho L/A$, $R = R_0[1 + \alpha(T - T_0)]$, and the radius is one-half the diameter.

   EXECUTE: **(a)** At 20.0°C, $R = V/I = (15.0 \text{ V})/(18.5 \text{ A}) = 0.811 \ \Omega$. Using $R = \rho L/A$ and solving for ρ gives $\rho = RA/L = R\pi(D/2)^2/L = (0.811 \ \Omega)\pi[(0.00500 \text{ m})/2]^2/(1.50 \text{ m}) = 1.06 \times 10^{-6} \ \Omega \cdot \text{m}$.

   (b) At 92.0°C, $R = V/I = (15.0 \text{ V})/(17.2 \text{ A}) = 0.872 \ \Omega$. Using $R = R_0[1 + \alpha(T - T_0)]$ with T_0 taken as 20.0°C, we have $0.872 \ \Omega = (0.811 \ \Omega)[1 + \alpha(92.0°C - 20.0°C)]$. This gives $\alpha = 0.00105 \ (\text{C}°)^{-1}$

   EVALUATE: The results are typical of ordinary metals.

25.15.   **(a) IDENTIFY:**   Start with the definition of resistivity and use its dependence on temperature to find the electric field.

   SET UP: $E = \rho J = \rho_{20}[1 + \alpha(T - T_0)]\dfrac{I}{\pi r^2}$

   EXECUTE: $E = (5.25 \times 10^{-8} \ \Omega \cdot \text{m})[1 + (0.0045/\text{C}°)(120°C - 20°C)](12.5 \text{ A})/[\pi(0.000500 \text{ m})^2] = 1.21 \text{ V/m}$.

   (Note that the resistivity at 120°C turns out to be $7.61 \times 10^{-8} \ \Omega \cdot \text{m}$.)

   EVALUATE: This result is fairly large because tungsten has a larger resisitivity than copper.

   (b) IDENTIFY: Relate resistance and resistivity.

   SET UP: $R = \rho L/A = \rho L/\pi r^2$

   EXECUTE: $R = (7.61 \times 10^{-8} \ \Omega \cdot \text{m})(0.150 \text{ m})/[\pi(0.000500 \text{ m})^2] = 0.0145 \ \Omega$

   EVALUATE: Most metals have very low resistance.

   (c) IDENTIFY: The potential difference is proportional to the length of wire.

   SET UP: $V = EL$

   EXECUTE: $V = (1.21 \text{ V/m})(0.150 \text{ m}) = 0.182 \text{ V}$

   EVALUATE: We could also calculate $V = IR = (12.5 \text{ A})(0.0145 \ \Omega) = 0.181 \text{ V}$, in agreement with part (c).

25.19.   **IDENTIFY** and **SET UP:**   Use Eq. (25.10) to calculate A. Find the volume of the wire and use the density to calculate the mass.

   EXECUTE: Find the volume of one of the wires:

   $R = \dfrac{\rho L}{A}$ so $A = \dfrac{\rho L}{R}$ and

   $\text{volume} = AL = \dfrac{\rho L^2}{R} = \dfrac{(1.72 \times 10^{-8} \Omega \cdot \text{m})(3.50 \text{ m})^2}{0.125 \ \Omega} = 1.686 \times 10^{-6} \text{ m}^3$

   $m = (\text{density})V = (8.9 \times 10^3 \text{ kg/m}^3)(1.686 \times 10^{-6} \text{ m}^3) = 15 \text{ g}$

   EVALUATE: The mass we calculated is reasonable for a wire.

25.21.   **IDENTIFY:** $R = \dfrac{\rho L}{A}$.

   SET UP: $L = 1.80 \text{ m}$, the length of one side of the cube. $A = L^2$.

   EXECUTE: $R = \dfrac{\rho L}{A} = \dfrac{\rho L}{L^2} = \dfrac{\rho}{L} = \dfrac{2.75 \times 10^{-8} \ \Omega \cdot \text{m}}{1.80 \text{ m}} = 1.53 \times 10^{-8} \ \Omega$

   EVALUATE: The resistance is very small because A is very much larger than the typical value for a wire.

25.25.   **IDENTIFY** and **SET UP:**   Eq. (25.5) relates the electric field that is given to the current density. $V = EL$ gives the potential difference across a length L of wire and Eq. (25.11) allows us to calculate R.

   EXECUTE: **(a)** Eq. (25.5): $\rho = E/J$ so $J = E/\rho$

   From Table 25.1 the resistivity for gold is $2.44 \times 10^{-8} \ \Omega \cdot \text{m}$.

   $J = \dfrac{E}{\rho} = \dfrac{0.49 \text{ V/m}}{2.44 \times 10^{-8} \ \Omega \cdot \text{m}} = 2.008 \times 10^7 \text{ A/m}^2$

   $I = JA = J\pi r^2 = (2.008 \times 10^7 \text{ A/m}^2)\pi(0.41 \times 10^{-3} \text{ m})^2 = 11 \text{ A}$

   (b) $V = EL = (0.49 \text{ V/m})(6.4 \text{ m}) = 3.1 \text{ V}$

   (c) We can use Ohm's law (Eq. (25.11)): $V = IR$.

   $R = \dfrac{V}{I} = \dfrac{3.1 \text{ V}}{11 \text{ A}} = 0.28 \ \Omega$

   EVALUATE: We can also calculate R from the resistivity and the dimensions of the wire (Eq. 25.10):

   $R = \dfrac{\rho L}{A} = \dfrac{\rho L}{\pi r^2} = \dfrac{(2.44 \times 10^{-8} \ \Omega \cdot \text{m})(6.4 \text{ m})}{\pi(0.42 \times 10^{-3} \text{ m})^2} = 0.28 \ \Omega$, which checks.

25.27.   **IDENTIFY:**   Apply $R = R_0\left[1 + \alpha\left(T - T_0\right)\right]$ to calculate the resistance at the second temperature.

(a) **SET UP:** $\alpha = 0.0004\ \left(\text{C}°\right)^{-1}$ (Table 25.1). Let T_0 be 0.0°C and T be 11.5°C.

EXECUTE: $R_0 = \dfrac{R}{1 + \alpha\left(T - T_0\right)} = \dfrac{100.0\ \Omega}{1 + \left(0.0004\ \left(\text{C}°\right)^{-1}\left(11.5\ \text{C}°\right)\right)} = 99.54\ \Omega$

(b) **SET UP:** $\alpha = -0.0005\ \left(\text{C}°\right)^{-1}$ (Table 25.2). Let $T_0 = 0.0°$C and $T = 25.8°$C.

EXECUTE: $R = R_0\left[1 + \alpha\left(T - T_0\right)\right] = 0.0160\ \Omega\left[1 + \left(-0.0005\ \left(\text{C}°\right)^{-1}\right)\left(25.8\ \text{C}°\right)\right] = 0.0158\ \Omega$

EVALUATE: Nichrome, like most metallic conductors, has a positive α and its resistance increases with temperature. For carbon, α is negative and its resistance decreases as T increases.

25.29.   **IDENTIFY** and **SET UP:**   Apply $R = \dfrac{\rho L}{A}$ to determine the effect of increasing A and L.

EXECUTE: (a) If 120 strands of wire are placed side by side, we are effectively increasing the area of the current carrier by 120. So the resistance is smaller by that factor: $R = (5.60\times10^{-6}\ \Omega)/120 = 4.67\times10^{-8}\ \Omega$.

(b) If 120 strands of wire are placed end to end, we are effectively increasing the length of the wire by 120, and so $R = (5.60\times10^{-6}\ \Omega)120 = 6.72\times10^{-4}\ \Omega$.

EVALUATE: Placing the strands side by side decreases the resistance and placing them end to end increases the resistance.

25.31.   **IDENTIFY:**   Use $R = \dfrac{\rho L}{A}$ to calculate R and then apply $V = IR$. $P = VI$ and energy $= Pt$

SET UP: For copper, $\rho = 1.72\times10^{-8}\ \Omega\cdot\text{m}$. $A = \pi r^2$, where $r = 0.050$ m.

EXECUTE: (a) $R = \dfrac{\rho L}{A} = \dfrac{(1.72\times10^{-8}\ \Omega\cdot\text{m})(100\times10^{3}\text{m})}{\pi(0.050\ \text{m})^2} = 0.219\ \Omega$. $V = IR = (125\ \text{A})(0.219\ \Omega) = 27.4\ \text{V}$.

(b) $P = VI = (27.4\ \text{V})(125\ \text{A}) = 3422\ \text{W} = 3422\ \text{J/s}$ and energy $= Pt = (3422\ \text{J/s})(3600\ \text{s}) = 1.23\times10^{7}\ \text{J}$.

EVALUATE: The rate of electrical energy loss in the cable is large, over 3 kW.

25.33.   **IDENTIFY:** $V = \mathcal{E} - Ir$.

SET UP: The graph gives $V = 9.0$ V when $I = 0$ and $I = 2.0$ A when $V = 0$.

EXECUTE: (a) $\mathcal{E}$ is equal to the terminal voltage when the current is zero. From the graph, this is 9.0 V.

(b) When the terminal voltage is zero, the potential drop across the internal resistance is just equal in magnitude to the internal emf, so $rI = \mathcal{E}$, which gives $r = \mathcal{E}/I = (9.0\ \text{V})/(2.0\ \text{A}) = 4.5\ \Omega$.

EVALUATE: The terminal voltage decreases as the current through the battery increases.

25.37.   **IDENTIFY:**   The voltmeter reads the potential difference V_{ab} between the terminals of the battery.

SET UP: open circuit $I = 0$. The circuit is sketched in Figure 25.37a.

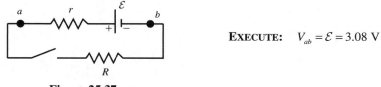

EXECUTE: $V_{ab} = \mathcal{E} = 3.08$ V

Figure 25.37a

SET UP: switch closed The circuit is sketched in Figure 35.37b.

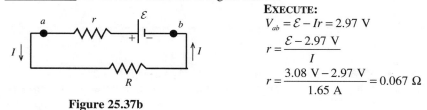

EXECUTE:
$V_{ab} = \mathcal{E} - Ir = 2.97$ V
$r = \dfrac{\mathcal{E} - 2.97\ \text{V}}{I}$
$r = \dfrac{3.08\ \text{V} - 2.97\ \text{V}}{1.65\ \text{A}} = 0.067\ \Omega$

Figure 25.37b

And $V_{ab} = IR$ so $R = \dfrac{V_{ab}}{I} = \dfrac{2.97\ \text{V}}{1.65\ \text{A}} = 1.80\ \Omega$.

EVALUATE: When current flows through the battery there is a voltage drop across its internal resistance and its terminal voltage V is less than its emf.

25.39.   **(a) IDENTIFY** and **SET UP:**   Assume that the current is clockwise. The circuit is sketched in Figure 25.39a.

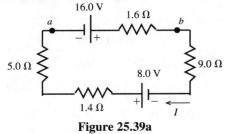

Figure 25.39a

Add up the potential rises and drops as travel clockwise around the circuit.

EXECUTE: $16.0 \text{ V} - I(1.6 \text{ }\Omega) - I(9.0 \text{ }\Omega) + 8.0 \text{ V} - I(1.4 \text{ }\Omega) - I(5.0 \text{ }\Omega) = 0$

$$I = \frac{16.0 \text{ V} + 8.0 \text{ V}}{9.0 \text{ }\Omega + 1.4 \text{ }\Omega + 5.0 \text{ }\Omega + 1.6 \text{ }\Omega} = \frac{24.0 \text{ V}}{17.0 \text{ }\Omega} = 1.41 \text{ A, clockwise}$$

EVALUATE: The 16.0 V battery drives the current clockwise more strongly than the 8.0 V battery does in the opposite direction.

(b) IDENTIFY and **SET UP:**   Start at point a and travel through the battery to point b, keeping track of the potential changes. At point b the potential is V_b.

EXECUTE: $V_a + 16.0 \text{ V} - I(1.6 \text{ }\Omega) = V_b$

$V_a - V_b = -16.0 \text{ V} + (1.41 \text{ A})(1.6 \text{ }\Omega)$

$V_{ab} = -16.0 \text{ V} + 2.3 \text{ V} = -13.7 \text{ V}$ (point a is at lower potential; it is the negative terminal)

EVALUATE: Could also go counterclockwise from a to b:

$V_a + (1.41 \text{ A})(5.0 \text{ }\Omega) + (1.41 \text{ A})(1.4 \text{ }\Omega) - 8.0 \text{ V} + (1.41 \text{ A})(9.0 \text{ }\Omega) = V_b$

$V_{ab} = -13.7 \text{ V}$, which checks.

(c) IDENTIFY and **SET UP:**   State at point a and travel through the battery to point c, keeping track of the potential changes.

EXECUTE: $V_a + 16.0 \text{ V} - I(1.6 \text{ }\Omega) - I(9.0 \text{ }\Omega) = V_c$

$V_a - V_c = -16.0 \text{ V} + (1.41 \text{ A})(1.6 \text{ }\Omega + 9.0 \text{ }\Omega)$

$V_{ac} = -16.0 \text{ V} + 15.0 \text{ V} = -1.0 \text{ V}$ (point a is at lower potential than point c)

EVALUATE: Could also go counterclockwise from a to c:

$V_a + (1.41 \text{ A})(5.0 \text{ }\Omega) + (1.41 \text{ A})(1.4 \text{ }\Omega) - 8.0 \text{ V} = V_c$

$V_{ac} = -1.0 \text{ V}$, which checks.

(d) Call the potential zero at point a. Travel clockwise around the circuit. The graph is sketched in Figure 25.39b.

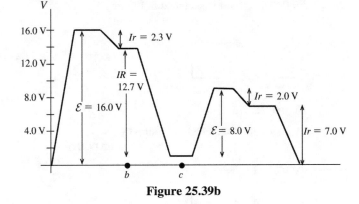

Figure 25.39b

25.45.   **IDENTIFY:**   A "100-W" European bulb dissipates 100 W when used across 220 V.

(a) SET UP: Take the ratio of the power in the US to the power in Europe, as in the alternative method for problem 25.44, using $P = V^2/R$.

EXECUTE: $\dfrac{P_{\text{US}}}{P_{\text{E}}} = \dfrac{V_{\text{US}}^2/R}{V_{\text{E}}^2/R} = \left(\dfrac{V_{\text{US}}}{V_{\text{E}}}\right)^2 = \left(\dfrac{120 \text{ V}}{220 \text{ V}}\right)^2$. This gives $P_{\text{US}} = (100 \text{ W})\left(\dfrac{120 \text{ V}}{220 \text{ V}}\right)^2 = 29.8 \text{ W}$.

(b) SET UP: Use $P = IV$ to find the current.

EXECUTE: $I = P/V = (29.8 \text{ W})/(120 \text{ V}) = 0.248 \text{ A}$

EVALUATE: The bulb draws considerably less power in the U.S., so it would be much dimmer than in Europe.

25.47.   **IDENTIFY** and **SET UP:**   By definition $p = \dfrac{P}{LA}$. Use $P = VI$, $E = VL$ and $I = JA$ to rewrite this expression in terms of the specified variables.

EXECUTE: **(a)** E is related to V and J is related to I, so use $P = VI$. This gives $p = \dfrac{VI}{LA}$

$\dfrac{V}{L} = E$ and $\dfrac{I}{A} = J$ so $p = EJ$

(b) J is related to I and ρ is related to R, so use $P = IR^2$. This gives $p = \dfrac{I^2 R}{LA}$.

$I = JA$ and $R = \dfrac{\rho L}{A}$ so $p = \dfrac{J^2 A^2 \rho L}{LA^2} \rho J^2$

(c) E is related to V and ρ is related to R, so use $P = V^2 / R$. This gives $p = \dfrac{V^2}{RLA}$.

$V = EL$ and $R = \dfrac{\rho L}{A}$ so $p = \dfrac{E^2 L^2}{LA}\left(\dfrac{A}{\rho L}\right) = \dfrac{E^2}{\rho}$.

EVALUATE: For a given material (ρ constant), p is proportional to J^2 or to E^2.

25.49.   **(a) IDENTIFY** and **SET UP:** $P = VI$ and energy = (power) × (time).
EXECUTE: $P = VI = (12\ \text{V})(60\ \text{A}) = 720\ \text{W}$

The battery can provide this for 1.0 h, so the energy the battery has stored is
$U = Pt = (720\ \text{W})(3600\ \text{s}) = 2.6 \times 10^6\ \text{J}$

(b) IDENTIFY and **SET UP:**   For gasoline the heat of combustion is $L_c = 46 \times 10^6\ \text{J/kg}$. Solve for the mass m required to supply the energy calculated in part (a) and use density $\rho = m/V$ to calculate V.

EXECUTE: The mass of gasoline that supplies $2.6 \times 10^6\ \text{J}$ is $m = \dfrac{2.6 \times 10^6\ \text{J}}{46 \times 10^6\ \text{J/kg}} = 0.0565\ \text{kg}$.

The volume of this mass of gasoline is
$V = \dfrac{m}{\rho} = \dfrac{0.0565\ \text{kg}}{900\ \text{kg/m}^3} = 6.3 \times 10^{-5}\ \text{m}^3 \left(\dfrac{1000\ \text{L}}{1\ \text{m}^3}\right) = 0.063\ \text{L}$

(c) IDENTIFY and **SET UP:**   Energy = (power) × (time); the energy is that calculated in part (a).
EXECUTE: $U = Pt$, $t = \dfrac{U}{P} = \dfrac{2.6 \times 10^6\ \text{J}}{450\ \text{W}} = 5800\ \text{s} = 97\ \text{min} = 1.6\ \text{h}$.

EVALUATE: The battery discharges at a rate of 720 W (for 60 A) and is charged at a rate of 450 W, so it takes longer to charge than to discharge.

25.51.   **IDENTIFY:**   Some of the power generated by the internal emf of the battery is dissipated across the battery's internal resistance, so it is not available to the bulb.
SET UP: Use $P = I^2 R$ and take the ratio of the power dissipated in the internal resistance r to the total power.

EXECUTE: $\dfrac{P_r}{P_{\text{Total}}} = \dfrac{I^2 r}{I^2 (r + R)} = \dfrac{r}{r + R} = \dfrac{3.5\ \Omega}{28.5\ \Omega} = 0.123 = 12.3\%$

EVALUATE: About 88% of the power of the battery goes to the bulb. The rest appears as heat in the internal resistance.

25.53.   **IDENTIFY:**   Solve for the current I in the circuit. Apply Eq. (25.17) to the specified circuit elements to find the rates of energy conversion.
SET UP: The circuit is sketched in Figure 25.53.

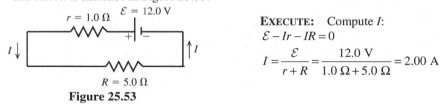

EXECUTE: Compute I:
$\mathcal{E} - Ir - IR = 0$

$I = \dfrac{\mathcal{E}}{r + R} = \dfrac{12.0\ \text{V}}{1.0\ \Omega + 5.0\ \Omega} = 2.00\ \text{A}$

Figure 25.53

(a) The rate of conversion of chemical energy to electrical energy in the emf of the battery is
$P = \mathcal{E}I = (12.0\ \text{V})(2.00\ \text{A}) = 24.0\ \text{W}$.

(b) The rate of dissipation of electrical energy in the internal resistance of the battery is
$P = I^2 r = (2.00\ \text{A})^2 (1.0\ \Omega) = 4.0\ \text{W}$.

(c) The rate of dissipation of electrical energy in the external resistor R is $P = I^2 R = (2.00 \text{ A})^2 (5.0 \ \Omega) = 20.0 \text{ W}$.

EVALUATE: The rate of production of electrical energy in the circuit is 24.0 W. The total rate of consumption of electrical energy in the circuit is 4.00 W + 20.0 W = 24.0 W. Equal rate of production and consumption of electrical energy are required by energy conservation.

25.57.   (a) **IDENTIFY** and **SET UP:**   Use $R = \dfrac{\rho L}{A}$.

EXECUTE: $\rho = \dfrac{RA}{L} = \dfrac{(0.104 \ \Omega)\pi(1.25 \times 10^{-3} \text{ m})^2}{14.0 \text{ m}} = 3.65 \times 10^{-8} \Omega \cdot \text{m}$

EVALUATE: This value is similar to that for good metallic conductors in Table 25.1.
(b) **IDENTIFY** and **SET UP:**   Use $V = EL$ to calculate E and then Ohm's law gives I.
EXECUTE: $V = EL = (1.28 \text{ V/m})(14.0 \text{ m}) = 17.9 \text{ V}$

$I = \dfrac{V}{R} = \dfrac{17.9 \text{ V}}{0.104 \ \Omega} = 172 \text{ A}$

EVALUATE: We could do the calculation another way:

$E = \rho J$ so $J = \dfrac{E}{\rho} = \dfrac{1.28 \text{ V/m}}{3.65 \times 10^{-8} \ \Omega \cdot \text{m}} = 3.51 \times 10^7 \text{ A/m}^2$

$I = JA = (3.51 \times 10^7 \text{ A/m}^2)\pi(1.25 \times 10^{-3} \text{ m})^2 = 172 \text{ A}$, which checks

(c) **IDENTIFY** and **SET UP:**   Calculate $J = I/A$ or $J = E/\rho$ and then use Eq. (25.3) for the target variable v_d.
EXECUTE: $J = n|q|v_d = nev_d$

$v_d = \dfrac{J}{ne} = \dfrac{3.51 \times 10^7 \text{ A/m}^2}{(8.5 \times 10^{28} \text{ m}^{-3})(1.602 \times 10^{-19} \text{ C})} = 2.58 \times 10^{-3} \text{ m/s} = 2.58 \text{ mm/s}$

EVALUATE: Even for this very large current the drift speed is small.

25.59.   **IDENTIFY** and **SET UP:**   With the voltmeter connected across the terminals of the battery there is no current through the battery and the voltmeter reading is the battery emf; $\mathcal{E} = 12.6 \text{ V}$.
With a wire of resistance R connected to the battery current I flows and $\mathcal{E} - Ir - IR = 0$, where r is the internal resistance of the battery. Apply this equation to each piece of wire to get two equations in the two unknowns.
EXECUTE: Call the resistance of the 20.0-m piece R_1; then the resistance of the 40.0-m piece is $R_2 = 2R_1$.
$\mathcal{E} - I_1 r - I_1 R_1 = 0$; $12.6 \text{ V} - (7.00 \text{ A})r - (7.00 \text{ A})R_1 = 0$
$\mathcal{E} - I_2 r - I_2(2R_1) = 0$; $12.6 \text{ V} - (4.20 \text{ A})r - (4.20 \text{ A})(2R_1) = 0$

Solving these two equations in two unknowns gives $R_1 = 1.20 \ \Omega$. This is the resistance of 20.0 m, so the resistance of one meter is $\left[1.20 \ \Omega/(20.0 \text{ m}) \right](1.00 \text{ m}) = 0.060 \ \Omega$

EVALUATE: We can also solve for r and we get $r = 0.600 \ \Omega$. When measuring small resistances, the internal resistance of the battery has a large effect.

25.61.   **IDENTIFY:**   Conservation of charge requires that the current be the same in both sections of the wire.

$E = \rho J = \dfrac{\rho I}{A}$. For each section, $V = IR = JAR = \left(\dfrac{EA}{\rho} \right)\left(\dfrac{\rho L}{A} \right) = EL$. The voltages across each section add.

SET UP: $A = (\pi/4)D^2$, where D is the diameter.

EXECUTE: (a) The current must be the same in both sections of the wire, so the current in the thin end is 2.5 mA.

(b) $E_{1.6\text{mm}} = \rho J = \dfrac{\rho I}{A} = \dfrac{(1.72 \times 10^{-8} \ \Omega \cdot \text{m})(2.5 \times 10^{-3} \text{ A})}{(\pi/4)(1.6 \times 10^{-3} \text{ m})^2} = 2.14 \times 10^{-5} \text{ V/m}$.

(c) $E_{0.8\text{mm}} = \rho J = \dfrac{\rho I}{A} = \dfrac{(1.72 \times 10^{-8} \ \Omega \cdot \text{m})(2.5 \times 10^{-3} \text{ A})}{(\pi/4)(0.80 \times 10^{-3} \text{ m})^2} = 8.55 \times 10^{-5} \text{ V/m}$. This is $4E_{1.6\text{mm}}$.

(d) $V = E_{1.6\text{mm}}L_{1.6 \text{ mm}} + E_{0.8 \text{ mm}}L_{0.8 \text{ mm}}$. $V = (2.14 \times 10^{-5} \text{ V/m})(1.20 \text{ m}) + (8.55 \times 10^{-5} \text{ V/m})(1.80 \text{ m}) = 1.80 \times 10^{-4} \text{ V}$.

EVALUATE: The currents are the same but the current density is larger in the thinner section and the electric field is larger there.

25.63. (a) **IDENTIFY:** Apply Eq. (25.10) to calculate the resistance of each thin disk and then integrate over the truncated cone to find the total resistance.
SET UP:

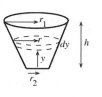

Figure 25.63

EXECUTE: The radius of a truncated cone a distance y above the bottom is given by

$$r = r_2 + (y/h)(r_1 - r_2) = r_2 + y\beta$$

with $\beta = (r_1 - r_2)/h$

Consider a thin slice a distance y above the bottom. The slice has thickness dy and radius r. The resistance of the slice is

$$dR = \frac{\rho \, dy}{A} = \frac{\rho \, dy}{\pi r^2} = \frac{\rho \, dy}{\pi (r_2 + \beta y)^2}$$

The total resistance of the cone if obtained by integrating over these thin slices:

$$R = \int dR = \frac{\rho}{\pi} \int_0^h \frac{dy}{(r_2 + \beta y)^2} = \frac{\rho}{\pi}\left[-\frac{1}{\beta}(r_2 + y\beta)^{-1} \right]_0^h = -\frac{\rho}{\pi\beta}\left[\frac{1}{r_2 + h\beta} - \frac{1}{r_2} \right]$$

But $r_2 + h\beta = r_1$

$$R = \frac{\rho}{\pi\beta}\left[\frac{1}{r_2} - \frac{1}{r_1} \right] = \frac{\rho}{\pi}\left(\frac{h}{r_1 - r_2} \right)\left(\frac{r_1 - r_2}{r_1 r_2} \right) = \frac{\rho h}{\pi r_1 r_2}$$

(b) **EVALUATE:** Let $r_1 = r_2 = r$. Then $R = \rho h/\pi r^2 = \rho L/A$ where $A = \pi r^2$ and $L = h$. This agrees with Eq. (25.10).

25.65. **IDENTIFY** and **SET UP:** Use $E = \rho J$ to calculate the current density between the plates. Let A be the area of each plate; then $I = JA$.

EXECUTE: $J = \dfrac{E}{\rho}$ and $E = \dfrac{\sigma}{K\epsilon_0} = \dfrac{Q}{KA\epsilon_0}$

Thus $J = \dfrac{Q}{KA\epsilon_0 \rho}$ and $I = JA = \dfrac{Q}{K\epsilon_0 \rho}$, as was to be shown.

EVALUATE: $C = K\epsilon_0 A/d$ and $V = Q/C = Qd/K\epsilon_0 A$ so the result can also be written as $I = VA/d\rho$. The resistance of the dielectric is $R = V/I = d\rho/A$, which agrees with Eq. (25.10).

25.69. **IDENTIFY:** In each case write the terminal voltage in terms of $\mathcal{E}$, I, and r. Since I is known, this gives two equations in the two unknowns $\mathcal{E}$ and r.
SET UP: The battery with the 1.50 A current is sketched in Figure 25.69a.

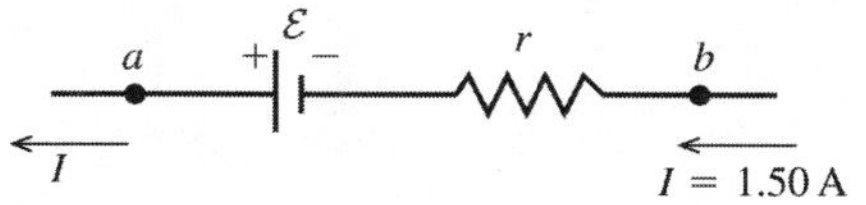

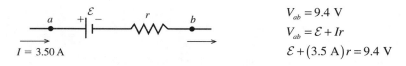

$V_{ab} = 8.4$ V

$V_{ab} = \mathcal{E} - Ir$

$\mathcal{E} - (1.50 \text{ A})r = 8.4$ V

Figure 25.69a

The battery with the 3.50 A current is sketched in Figure 25.69b.

$V_{ab} = 9.4$ V

$V_{ab} = \mathcal{E} + Ir$

$\mathcal{E} + (3.5 \text{ A})r = 9.4$ V

Figure 25.69b

EXECUTE: (a) Solve the first equation for $\mathcal{E}$ and use that result in the second equation:

$\mathcal{E} = 8.4$ V $+ (1.50 \text{ A})r$

8.4 V $+ (1.50 \text{ A})r + (3.50 \text{ A})r = 9.4$ V

$(5.00 \text{ A})r = 1.0$ V so $r = \dfrac{1.0 \text{ V}}{5.00 \text{ A}} = 0.20\ \Omega$

(b) Then $\mathcal{E} = 8.4 \text{ V} + (1.50 \text{ A})r = 8.4 \text{ V} + (1.50 \text{ A})(0.20 \ \Omega) = 8.7 \text{ V}$

EVALUATE: When the current passes through the emf in the direction from $-$ to $+$, the terminal voltage is less than the emf and when it passes through from $+$ to $-$, the terminal voltage is greater than the emf.

25.73.   **IDENTIFY:**   Set the sum of the potential rises and drops around the circuit equal to zero and solve for I.
   SET UP: The circuit is sketched in Figure 25.73.

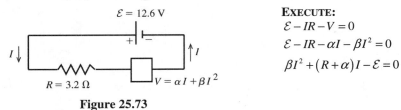

$$\mathcal{E} = 12.6 \text{ V}$$

EXECUTE:
$\mathcal{E} - IR - V = 0$
$\mathcal{E} - IR - \alpha I - \beta I^2 = 0$
$\beta I^2 + (R + \alpha)I - \mathcal{E} = 0$

$R = 3.2 \ \Omega$ $V = \alpha I + \beta I^2$

Figure 25.73

The quadratic formula gives $I = (1/2\beta)\left[-(R+\alpha) \pm \sqrt{(R+\alpha)^2 + 4\beta\mathcal{E}}\right]$

I must be positive, so take the $+$ sign

$$I = (1/2\beta)\left[-(R+\alpha) + \sqrt{(R+\alpha)^2 + 4\beta\mathcal{E}}\right]$$

$I = -2.692 \text{ A} + 4.116 \text{ A} = 1.42 \text{ A}$

EVALUATE: For this I the voltage across the thermistor is 8.0 V. The voltage across the resistor must then be $12.6 \text{ V} - 8.0 \text{ V} = 4.6 \text{ V}$, and this agrees with Ohm's law for the resistor.

25.75.   **IDENTIFY:**   The ammeter acts as a resistance in the circuit loop. Set the sum of the potential rises and drops around the circuit equal to zero.
   (a) SET UP: The circuit with the ammeter is sketched in Figure 25.75a.

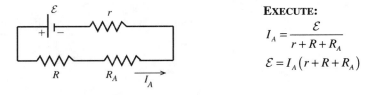

EXECUTE:
$$I_A = \frac{\mathcal{E}}{r + R + R_A}$$
$$\mathcal{E} = I_A(r + R + R_A)$$

Figure 25.75a

SET UP: The circuit with the ammeter removed is sketched in Figure 25.75b.

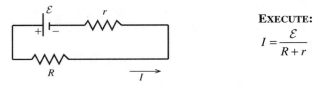

EXECUTE:
$$I = \frac{\mathcal{E}}{R + r}$$

Figure 25.75b

Combining the two equations gives
$$I = \left(\frac{1}{R+r}\right)I_A(r + R + R_A) = I_A\left(1 + \frac{R_A}{r+R}\right)$$

(b) Want $I_A = 0.990I$. Use this in the result for part (a).

$$I = 0.990I\left(1 + \frac{R_A}{r+R}\right)$$

$$0.010 = 0.990\left(\frac{R_A}{r+R}\right)$$

$R_A = (r + R)(0.010/0.990) = (0.45 \ \Omega + 3.80 \ \Omega)(0.010/0.990) = 0.0429 \ \Omega$

(c) $I - I_A = \dfrac{\mathcal{E}}{r+R} - \dfrac{\mathcal{E}}{r+R+R_A}$

$$I - I_A = \mathcal{E}\left(\frac{r + R + R_A - r - R}{(r+R)(r+R+R_A)}\right) = \frac{\mathcal{E}R_A}{(r+R)(r+R+R_A)}.$$

EVALUATE: The difference between I and I_A increases as R_A increases. If R_A is larger than the value calculated in part (b) then I_A differs from I by more than 1.0%.

25.79.   **(a) IDENTIFY:**   Set the sum of the potential rises and drops around the circuit equal to zero and solve for the resulting equation for the current I. Apply Eq. (25.17) to each circuit element to find the power associated with it.

SET UP: The circuit is sketched in Figure 25.79.

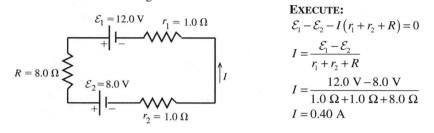

EXECUTE:

$$\mathcal{E}_1 - \mathcal{E}_2 - I(r_1 + r_2 + R) = 0$$

$$I = \frac{\mathcal{E}_1 - \mathcal{E}_2}{r_1 + r_2 + R}$$

$$I = \frac{12.0 \text{ V} - 8.0 \text{ V}}{1.0 \text{ }\Omega + 1.0 \text{ }\Omega + 8.0 \text{ }\Omega}$$

$$I = 0.40 \text{ A}$$

Figure 25.79

(b) $P = I^2 R + I^2 r_1 + I^2 r_2 = I^2 (R + r_1 + r_2) = (0.40 \text{ A})^2 (8.0 \text{ }\Omega + 1.0 \text{ }\Omega + 1.0 \text{ }\Omega)$

$P = 1.6 \text{ W}$

(c) Chemical energy is converted to electrical energy in a battery when the current goes through the battery from the negative to the positive terminal, so the electrical energy of the charges increases as the current passes through. This happens in the 12.0 V battery, and the rate of production of electrical energy is

$P = \mathcal{E}_1 I = (12.0 \text{ V})(0.40 \text{ A}) = 4.8 \text{ W}.$

(d) Electrical energy is converted to chemical energy in a battery when the current goes through the battery from the positive to the negative terminal, so the electrical energy of the charges decreases as the current passes through. This happens in the 8.0 V battery, and the rate of consumption of electrical energy is

$P = \mathcal{E}_2 I = (8.0 \text{ V})(0.40 \text{ V}) = 3.2 \text{ W}.$

(e) EVALUATE: Total rate of production of electrical energy = 4.8 W. Total rate of consumption of electrical energy = 1.6 W + 3.2 W = 4.8 W, which equals the rate of production, as it must.

25.81.   **IDENTIFY** and **SET UP:**   The terminal voltage is $V_{ab} = \mathcal{E} - Ir = IR$, where R is the resistance connected to the battery. During the charging the terminal voltage is $V_{ab} = \mathcal{E} + Ir$. $P = VI$ and energy is $E = Pt$. $I^2 r$ is the rate at which energy is dissipated in the internal resistance of the battery.

EXECUTE: **(a)** $V_{ab} = \mathcal{E} + Ir = 12.0 \text{ V} + (10.0 \text{ A})(0.24 \text{ }\Omega) = 14.4 \text{ V}.$

(b) $E = Pt = IVt = (10 \text{ A})(14.4 \text{ V})(5)(3600 \text{ s}) = 2.59 \times 10^6 \text{ J}.$

(c) $E_{\text{diss}} = P_{\text{diss}} t = I^2 rt = (10 \text{ A})^2 (0.24 \text{ }\Omega)(5)(3600 \text{ s}) = 4.32 \times 10^5 \text{ J}.$

(d) Discharged at 10 A: $I = \dfrac{\mathcal{E}}{r + R} \Rightarrow R = \dfrac{\mathcal{E} - Ir}{I} = \dfrac{12.0 \text{ V} - (10 \text{ A})(0.24 \text{ }\Omega)}{10 \text{ A}} = 0.96 \text{ }\Omega.$

(e) $E = Pt = IVt = (10 \text{ A})(9.6 \text{ V})(5)(3600 \text{ s}) = 1.73 \times 10^6 \text{ J}.$

(f) Since the current through the internal resistance is the same as before, there is the same energy dissipated as in (c): $E_{\text{diss}} = 4.32 \times 10^5 \text{ J}.$

(g) Part of the energy originally supplied was stored in the battery and part was lost in the internal resistance. So the stored energy was less than what was supplied during charging. Then when discharging, even more energy is lost in the internal resistance, and only what is left is dissipated by the external resistor.

DIRECT-CURRENT CIRCUITS

26.5. **IDENTIFY:** The equivalent resistance will vary for the different connections because the series-parallel combinations vary, and hence the current will vary.
SET UP: First calculate the equivalent resistance using the series-parallel formulas, then use Ohm's law ($V = RI$) to find the current.
EXECUTE: **(a)** $1/R = 1/(15.0 \ \Omega) + 1/(30.0 \ \Omega)$ gives $R = 10.0 \ \Omega$. $I = V/R = (35.0 \text{ V})/(10.0 \ \Omega) = 3.50$ A.
(b) $1/R = 1/(10.0 \ \Omega) + 1/(35.0 \ \Omega)$ gives $R = 7.78 \ \Omega$. $I = (35.0 \text{ V})/(7.78 \ \Omega) = 4.50$ A
(c) $1/R = 1/(20.0 \ \Omega) + 1/(25.0 \ \Omega)$ gives $R = 11.11 \ \Omega$, so $I = (35.0 \text{ V})/(11.11 \ \Omega) = 3.15$ A.
(d) From part (b), the resistance of the triangle alone is $7.78 \ \Omega$. Adding the 3.00-Ω internal resistance of the battery gives an equivalent resistance for the circuit of $10.78 \ \Omega$. Therefore the current is $I = (35.0 \text{ V})/(10.78 \ \Omega) = 3.25$ A
EVALUATE: It makes a big difference how the triangle is connected to the battery.

26.7. **IDENTIFY:** First do as much series-parallel reduction as possible.
SET UP: The 45.0-Ω and 15.0-Ω resistors are in parallel, so first reduce them to a single equivalent resistance. Then find the equivalent series resistance of the circuit.
EXECUTE: $1/R_p = 1/(45.0 \ \Omega) + 1/(15.0 \ \Omega)$ and $R_p = 11.25 \ \Omega$. The total equivalent resistance is $18.0 \ \Omega + 11.25 \ \Omega + 3.26 \ \Omega = 32.5 \ \Omega$. Ohm's law gives $I = (25.0 \text{ V})/(32.5 \ \Omega) = 0.769$ A.
EVALUATE: The circuit appears complicated until we realize that the 45.0-Ω and 15.0-Ω resistors are in parallel.

26.9. **IDENTIFY:** For a series network, the current is the same in each resistor and the sum of voltages for each resistor equals the battery voltage. The equivalent resistance is $R_{eq} = R_1 + R_2 + R_3$. $P = I^2 R$.
SET UP: Let $R_1 = 1.60 \ \Omega$, $R_2 = 2.40 \ \Omega$, $R_3 = 4.80 \ \Omega$.
EXECUTE: **(a)** $R_{eq} = 1.60 \ \Omega + 2.40 \ \Omega + 4.80 \ \Omega = 8.80 \ \Omega$

(b) $I = \dfrac{V}{R_{eq}} = \dfrac{28.0 \text{ V}}{8.80 \ \Omega} = 3.18$ A

(c) $I = 3.18$ A , the same as for each resistor.
(d) $V_1 = IR_1 = (3.18 \text{ A})(1.60 \ \Omega) = 5.09$ V . $V_2 = IR_2 = (3.18 \text{ A})(2.40 \ \Omega) = 7.63$ V .

$V_3 = IR_3 = (3.18 \text{ A})(4.80 \ \Omega) = 15.3$ V . Note that $V_1 + V_2 + V_3 = 28.0$ V .
(e) $P_1 = I^2 R_1 = (3.18 \text{ A})^2 (1.60 \ \Omega) = 16.2$ W . $P_2 = I^2 R_2 = (3.18 \text{ A})^2 (2.40 \ \Omega) = 24.3$ W .

$P_3 = I^2 R_3 = (3.18 \text{ A})^2 (4.80 \ \Omega) = 48.5$ W .

(f) Since $P = I^2 R$ and the current is the same for each resistor, the resistor with the greatest R dissipates the greatest power.
EVALUATE: When resistors are connected in parallel, the resistor with the smallest R dissipates the greatest power.

26.13. **IDENTIFY:** In both circuits, with and without R_4, replace series and parallel combinations of resistors by their equivalents. Calculate the currents and voltages in the equivalent circuit and infer from this the currents and voltages in the original circuit. Use $P = I^2 R$ to calculate the power dissipated in each bulb.
(a) SET UP: The circuit is sketched in Figure 26.13a.

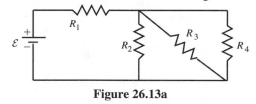

Figure 26.13a

EXECUTE: R_2, R_3, and R_4 are in parallel, so their equivalent resistance R_{eq} is given by $\dfrac{1}{R_{eq}} = \dfrac{1}{R_2} + \dfrac{1}{R_3} + \dfrac{1}{R_4}$

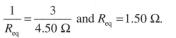
$\dfrac{1}{R_{eq}} = \dfrac{3}{4.50 \ \Omega}$ and $R_{eq} = 1.50 \ \Omega$.

The equivalent circuit is drawn in Figure 26.13b.

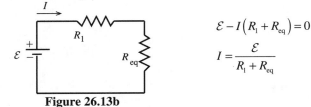

$$\mathcal{E} - I\left(R_1 + R_{eq}\right) = 0$$

$$I = \frac{\mathcal{E}}{R_1 + R_{eq}}$$

Figure 26.13b

$$I = \frac{9.00 \text{ V}}{4.50 \ \Omega + 1.50 \ \Omega} = 1.50 \text{ A and } I_1 = 1.50 \text{ A}$$

Then $V_1 = I_1 R_1 = (1.50 \text{ A})(4.50 \ \Omega) = 6.75 \text{ V}$

$I_{eq} = 1.50$ A, $V_{eq} = I_{eq}R_{eq} = (1.50 \text{ A})(1.50 \ \Omega) = 2.25$ V

For resistors in parallel the voltages are equal and are the same as the voltage across the equivalent resistor, so
$V_2 = V_3 = V_4 = 2.25$ V.

$$I_2 = \frac{V_2}{R_2} = \frac{2.25 \text{ V}}{4.50 \ \Omega} = 0.500 \text{ A}, I_3 = \frac{V_3}{R_3} = 0.500 \text{ A}, I_4 = \frac{V_4}{R_4} = 0.500 \text{ A}$$

EVALUATE: Note that $I_2 + I_3 + I_4 = 1.50$ A, which is I_{eq}. For resistors in parallel the currents add and their sum
is the current through the equivalent resistor.

(b) SET UP: $P = I^2 R$

EXECUTE: $P_1 = (1.50 \text{ A})^2 (4.50 \ \Omega) = 10.1$ W

$P_2 = P_3 = P_4 = (0.500 \text{ A})^2 (4.50 \ \Omega) = 1.125$ W, which rounds to 1.12 W. R_1 glows brightest.

EVALUATE: Note that $P_2 + P_3 + P_4 = 3.37$ W. This equals $P_{eq} = I_{eq}^2 R_{eq} = (1.50 \text{ A})^2 (1.50 \ \Omega) = 3.37$ W, the power
dissipated in the equivalent resistor.

(c) SET UP: With R_4 removed the circuit becomes the circuit in Figure 26.13c.

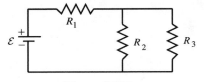

EXECUTE: R_2 and R_3 are in parallel and
their equivalent resistance R_{eq} is given by

$$\frac{1}{R_{eq}} = \frac{1}{R_2} + \frac{1}{R_3} = \frac{2}{4.50 \ \Omega} \text{ and } R_{eq} = 2.25 \ \Omega$$

Figure 26.13c

The equivalent circuit is shown in Figure 26.13d.

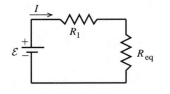

$$\mathcal{E} - I\left(R_1 + R_{eq}\right) = 0$$

$$I = \frac{\mathcal{E}}{R_1 + R_{eq}}$$

$$I = \frac{9.00 \text{ V}}{4.50 \ \Omega + 2.25 \ \Omega} = 1.333 \text{ A}$$

Figure 26.13d

$I_1 = 1.33$ A, $V_1 = I_1 R_1 = (1.333 \text{ A})(4.50 \ \Omega) = 6.00$ V

$I_{eq} = 1.33$ A, $V_{eq} = I_{eq}R_{eq} = (1.333 \text{ A})(2.25 \ \Omega) = 3.00$ V and $V_2 = V_3 = 3.00$ V.

$$I_2 = \frac{V_2}{R_2} = \frac{3.00 \text{ V}}{4.50 \ \Omega} = 0.667 \text{ A}, I_3 = \frac{V_3}{R_3} = 0.667 \text{ A}$$

(d) SET UP: $P = I^2 R$

EXECUTE: $P_1 = (1.333 \text{ A})^2 (4.50 \ \Omega) = 8.00$ W

$P_2 = P_3 = (0.667 \text{ A})^2 (4.50 \ \Omega) = 2.00$ W.

(e) EVALUATE: When R_4 is removed, P_1 decreases and P_2 and P_3 increase. Bulb R_1 glows less brightly and bulbs
R_2 and R_3 glow more brightly. When R_4 is removed the equivalent resistance of the circuit increases and the current
through R_1 decreases. But in the parallel combination this current divides into two equal currents rather than three,
so the currents through R_2 and R_3 increase. Can also see this by noting that with R_4 removed and less current
through R_1 the voltage drop across R_1 is less so the voltage drop across R_2 and across R_3 must become larger.

26.15.   **IDENTIFY:**   Apply Ohm's law to each resistor.
SET UP: For resistors in parallel the voltages are the same and the currents add. For resistors in series the currents are the same and the voltages add.
EXECUTE: The current through 2.00-Ω resistor is 6.00 A. Current through 1.00-Ω resistor also is 6.00 A and the voltage is 6.00 V. Voltage across the 6.00-Ω resistor is 12.0 V + 6.0 V = 18.0 V. Current through the 6.00-Ω resistor is (18.0 V)/(6.00 Ω) = 3.00 A. The battery emf is 18.0 V.
EVALUATE: The current through the battery is 6.00 A + 3.00 A = 9.00 A. The equivalent resistor of the resistor network is 2.00 Ω, and this equals (18.0 V)/(9.00 A).

26.17.   **IDENTIFY:**   For resistors in series, the voltages add and the current is the same. For resistors in parallel, the voltages are the same and the currents add. $P = I^2R$.
(a) SET UP: The circuit is sketched in Figure 26.17a.

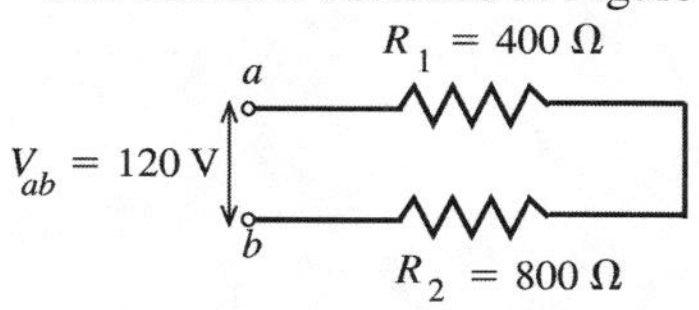

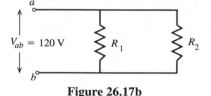

Figure 26.17a

For resistors in series the current is the same through each.

EXECUTE: $R_{eq} = R_1 + R_2 = 1200 \ \Omega$. $I = \dfrac{V}{R_{eq}} = \dfrac{120 \text{ V}}{1200 \ \Omega} = 0.100$ A. This is the current drawn from the line.

(b) $P_1 = I_1^2 R_1 = (0.100 \text{ A})^2 (400 \ \Omega) = 4.0$ W

$P_2 = I_2^2 R_2 = (0.100 \text{ A})^2 (800 \ \Omega) = 8.0$ W

(c) $P_{out} = P_1 + P_2 = 12.0$ W, the total power dissipated in both bulbs. Note that

$P_{in} = V_{ab}I = (120 \text{ V})(0.100 \text{ A}) = 12.0$ W, the power delivered by the potential source, equals P_{out}.

(d) SET UP: The circuit is sketched in Figure 26.17b.

For resistors in parallel the voltage across each resistor is the same.

Figure 26.17b

EXECUTE: $I_1 = \dfrac{V_1}{R_1} = \dfrac{120 \text{ V}}{400 \ \Omega} = 0.300$ A, $I_2 = \dfrac{V_2}{R_2} = \dfrac{120 \text{ V}}{800 \ \Omega} = 0.150$ A

EVALUATE: Note that each current is larger than the current when the resistors are connected in series.
(e) EXECUTE: $P_1 = I_1^2 R_1 = (0.300 \text{ A})^2 (400 \ \Omega) = 36.0$ W

$P_2 = I_2^2 R_2 = (0.150 \text{ A})^2 (800 \ \Omega) = 18.0$ W

(f) $P_{out} = P_1 + P_2 = 54.0$ W

EVALUATE: Note that the total current drawn from the line is $I = I_1 + I_2 = 0.450$ A. The power input from the line is $P_{in} = V_{ab}I = (120 \text{ V})(0.450 \text{ A}) = 54.0$ W, which equals the total power dissipated by the bulbs.
(g) The bulb that is dissipating the most power glows most brightly. For the series connection the currents are the same and by $P = I^2R$ the bulb with the larger R has the larger P; the 800 Ω bulb glows more brightly. For the parallel combination the voltages are the same and by $P = V^2/R$ the bulb with the smaller R has the larger P; the 400 Ω bulb glows more brightly.
(h) The total power output P_{out} equals $P_{in} = V_{ab}I$, so P_{out} is larger for the parallel connection where the current drawn from the line is larger (because the equivalent resistance is smaller.)

26.19.   **IDENTIFY and SET UP:**   Replace series and parallel combinations of resistors by their equivalents until the circuit is reduced to a single loop. Use the loop equation to find the current through the 20.0 Ω resistor. Set $P = I^2R$ for the 20.0 Ω resistor equal to the rate Q/t at which heat goes into the water and set $Q = mc\Delta T$.

EXECUTE: Replace the network by the equivalent resistor, as shown in Figure 26.19.

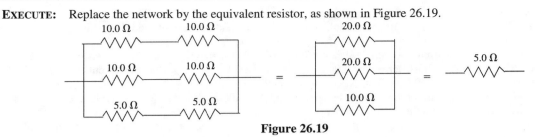

Figure 26.19

$30.0 \text{ V} - I(20.0 \text{ }\Omega + 5.0 \text{ }\Omega + 5.0 \text{ }\Omega) = 0; \ I = 1.00 \text{ A}$

For the 20.0-Ω resistor thermal energy is generated at the rate $P = I^2 R = 20.0$ W. $Q = Pt$ and $Q = mc\Delta T$ gives

$$t = \frac{mc\Delta T}{P} = \frac{(0.100 \text{ kg})(4190 \text{ J/kg} \cdot \text{ K})(48.0 \text{ C}^\circ)}{20.0 \text{ W}} = 1.01 \times 10^3 \text{ s}$$

EVALUATE: The battery is supplying heat at the rate $P = \mathcal{E}I = 30.0$ W. In the series circuit, more energy is dissipated in the larger resistor (20.0 Ω) than in the smaller ones (5.00 Ω).

26.21.   **IDENTIFY:**   Apply Kirchhoff's point rule at point a to find the current through R. Apply Kirchhoff's loop rule to loops (1) and (2) shown in Figure 26.21a to calculate R and $\mathcal{E}$. Travel around each loop in the direction shown.

(a) SET UP:

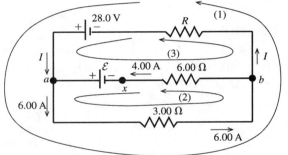

Figure 26.21a

EXECUTE: Apply Kirchhoff's point rule to point a: $\sum I = 0$ so $I + 4.00 \text{ A} - 6.00 \text{ A} = 0$

$I = 2.00$ A (in the direction shown in the diagram).

(b) Apply Kirchhoff's loop rule to loop (1): $-(6.00 \text{ A})(3.00 \text{ }\Omega) - (2.00 \text{ A})R + 28.0 \text{ V} = 0$

$-18.0 \text{ V} - (2.00 \text{ }\Omega)R + 28.0 \text{ V} = 0$

$$R = \frac{28.0 \text{ V} - 18.0 \text{ V}}{2.00 \text{ A}} = 5.00 \text{ }\Omega$$

(c) Apply Kirchhoff's loop rule to loop (2): $-(6.00 \text{ A})(3.00 \text{ }\Omega) - (4.00 \text{ A})(6.00 \text{ }\Omega) + \mathcal{E} = 0$

$\mathcal{E} = 18.0 \text{ V} + 24.0 \text{ V} = 42.0 \text{ V}$

EVALUATE: Can check that the loop rule is satisfied for loop (3), as a check of our work:

$28.0 \text{ V} - \mathcal{E} + (4.00 \text{ A})(6.00 \text{ }\Omega) - (2.00 \text{ A})R = 0$

$28.0 \text{ V} - 42.0 \text{ V} + 24.0 \text{ V} - (2.00 \text{ A})(5.00 \text{ }\Omega) = 0$

$52.0 \text{ V} = 42.0 \text{ V} + 10.0 \text{ V}$

$52.0 \text{ V} = 52.0 \text{ V}$, so the loop rule is satisfied for this loop.

(d) IDENTIFY: If the circuit is broken at point x there can be no current in the 6.00 Ω resistor. There is now only a single current path and we can apply the loop rule to this path.

SET UP: The circuit is sketched in Figure 26.21b.

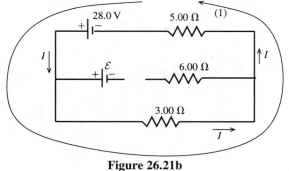

Figure 26.21b

EXECUTE: $+28.0 \text{ V} - (3.00 \ \Omega)I - (5.00 \ \Omega)I = 0$

$I = \dfrac{28.0 \text{ V}}{8.00 \ \Omega} = 3.50 \text{ A}$

EVALUATE: Breaking the circuit at x removes the 42.0 V emf from the circuit and the current through the 3.00 Ω resistor is reduced.

26.23.   **IDENTIFY:**   Apply the junction rule at points a, b, c and d to calculate the unknown currents. Then apply the loop rule to three loops to calculate $\mathcal{E}_1$, $\mathcal{E}_2$ and R.

(a) SET UP: The circuit is sketched in Figure 26.23.

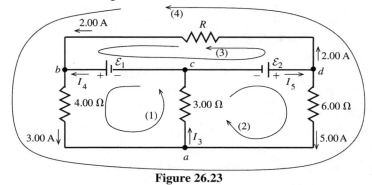

Figure 26.23

EXECUTE: Apply the junction rule to point a: $3.00 \text{ A} + 5.00 \text{ A} - I_3 = 0$

$I_3 = 8.00 \text{ A}$

Apply the junction rule to point b: $2.00 \text{ A} + I_4 - 3.00 \text{ A} = 0$

$I_4 = 1.00 \text{ A}$

Apply the junction rule to point c: $I_3 - I_4 - I_5 = 0$

$I_5 = I_3 - I_4 = 8.00 \text{ A} - 1.00 \text{ A} = 7.00 \text{ A}$

EVALUATE: As a check, apply the junction rule to point d: $I_5 - 2.00 \text{ A} - 5.00 \text{ A} = 0$

$I_5 = 7.00 \text{ A}$

(b) EXECUTE: Apply the loop rule to loop (1): $\mathcal{E}_1 - (3.00 \text{ A})(4.00 \ \Omega) - I_3 (3.00 \ \Omega) = 0$

$\mathcal{E}_1 = 12.0 \text{ V} + (8.00 \text{ A})(3.00 \ \Omega) = 36.0 \text{ V}$

Apply the loop rule to loop (2): $\mathcal{E}_2 - (5.00 \text{ A})(6.00 \ \Omega) - I_3 (3.00 \ \Omega) = 0$

$\mathcal{E}_2 = 30.0 \text{ V} + (8.00 \text{ A})(3.00 \ \Omega) = 54.0 \text{ V}$

(c) Apply the loop rule to loop (3): $-(2.00 \text{ A})R - \mathcal{E}_1 + \mathcal{E}_2 = 0$

$R = \dfrac{\mathcal{E}_2 - \mathcal{E}_1}{2.00 \text{ A}} = \dfrac{54.0 \text{ V} - 36.0 \text{ V}}{2.00 \text{ A}} = 9.00 \ \Omega$

EVALUATE: Apply the loop rule to loop (4) as a check of our calculations:

$-(2.00 \text{ A})R - (3.00 \text{ A})(4.00 \ \Omega) + (5.00 \text{ A})(6.00 \ \Omega) = 0$

$-(2.00 \text{ A})(9.00 \ \Omega) - 12.0 \text{ V} + 30.0 \text{ V} = 0$

$-18.0 \text{ V} + 18.0 \text{ V} = 0$

26.25. **IDENTIFY:** Apply the junction rule to reduce the number of unknown currents. Apply the loop rule to two loops to obtain two equations for the unknown currents I_1 and I_2

(a) **SET UP:** The circuit is sketched in Figure 26.25.

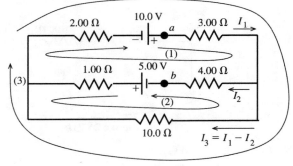

Figure 26.25

Let I_1 be the current in the $3.00\ \Omega$ resistor and I_2 be the current in the $4.00\ \Omega$ resistor and assume that these currents are in the directions shown. Then the current in the $10.0\ \Omega$ resistor is $I_3 = I_1 - I_2$, in the direction shown, where we have used Kirchhoff's point rule to relate I_3 to I_1 and I_2. If we get a negative answer for any of these currents we know the current is actually in the opposite direction to what we have assumed. Three loops and directions to travel around the loops are shown in the circiut diagram. Apply Kirchhoff's loop rule to each loop.

EXECUTE: <u>loop (1)</u>
$+10.0\ \text{V} - I_1(3.00\ \Omega) - I_2(4.00\ \Omega) + 5.00\ \text{V} - I_2(1.00\ \Omega) - I_1(2.00\ \Omega) = 0$

$15.00\ \text{V} - (5.00\ \Omega)I_1 - (5.00\ \Omega)I_2 = 0$

$3.00\ \text{A} - I_1 - I_2 = 0$

<u>loop (2)</u>
$+5.00\ \text{V} - I_2(1.00\ \Omega) + (I_1 - I_2)10.0\ \Omega - I_2(4.00\ \Omega) = 0$

$5.00\ \text{V} + (10.0\ \Omega)I_1 - (15.0\ \Omega)I_2 = 0$

$1.00\ \text{A} + 2.00I_1 - 3.00I_2 = 0$

The first equation says $I_2 = 3.00\ \text{A} - I_1$.

Use this in the second equation: $1.00\ \text{A} + 2.00I_1 - 9.00\ \text{A} + 3.00I_1 = 0$

$5.00I_1 = 8.00\ \text{A},\ I_1 = 1.60\ \text{A}$

Then $I_2 = 3.00\ \text{A} - I_1 = 3.00\ \text{A} - 1.60\ \text{A} = 1.40\ \text{A}$.

$I_3 = I_1 - I_2 = 1.60\ \text{A} - 1.40\ \text{A} = 0.20\ \text{A}$

EVALUATE: Loop (3) can be used as a check.
$+10.0\ \text{V} - (1.60\ \text{A})(3.00\ \Omega) - (0.20\ \text{A})(10.00\ \Omega) - (1.60\ \text{A})(2.00\ \Omega) = 0$

$10.0\ \text{V} = 4.8\ \text{V} + 2.0\ \text{V} + 3.2\ \text{V}$

$10.0\ \text{V} = 10.0\ \text{V}$

We find that with our calculated currents the loop rule is satisfied for loop (3). Also, all the currents came out to be positive, so the current directions in the circuit diagram are correct.

(b) **IDENTIFY and SET UP:** To find $V_{ab} = V_a - V_b$ start at point b and travel to point a. Many different routes can be taken from b to a and all must yield the same result for V_{ab}.

EXECUTE: Travel through the $4.00\ \Omega$ resistor and then through the $3.00\ \Omega$ resistor:
$V_b + I_2(4.00\ \Omega) + I_1(3.00\ \Omega) = V_a$

$V_a - V_b = (1.40\ \text{A})(4.00\ \Omega) + (1.60\ \text{A})(3.00\ \Omega) = 5.60\ \text{V} + 4.8\ \text{V} = 10.4\ \text{V}$ (point a is at higher potential than point b)

EVALUATE: Alternatively, travel through the $5.00\ \text{V}$ emf, the $1.00\ \Omega$ resistor, the $2.00\ \Omega$ resistor, and the $10.0\ \text{V}$ emf.
$V_b + 5.00\ \text{V} - I_2(1.00\ \Omega) - I_1(2.00\ \Omega) + 10.0\ \text{V} = V_a$

$V_a - V_b = 15.0\ \text{V} - (1.40\ \text{A})(1.00\ \Omega) - (1.60\ \text{A})(2.00\ \Omega) = 15.0\ \text{V} - 1.40\ \text{V} - 3.20\ \text{V} = 10.4\ \text{V}$, the same as before.

26.29. (a) **IDENTIFY:** With the switch open, we have a series circuit with two batteries.
SET UP: Take a loop to find the current, then use Ohm's law to find the potential difference between a and b.
EXECUTE: Taking the loop: $I = (40.0\ \text{V})/(175\ \Omega) = 0.229\ \text{A}$. The potential difference between a and b is
$V_b - V_a = +15.0\ \text{V} - (75.0\ \Omega)(0.229\ \text{A}) = -2.14\ \text{V}$.
EVALUATE: The minus sign means that a is at a higher potential than b.

(b) IDENTIFY: With the switch closed, the ammeter part of the circuit divides the original circuit into two circuits. We can apply Kirchhoff's rules to both parts.

SET UP: Take loops around the left and right parts of the circuit, and then look at the current at the junction.

EXECUTE: The left-hand loop gives $I_{100} = (25.0 \text{ V})/(100.0 \text{ }\Omega) = 0.250$ A. The right-hand loop gives $I_{75} = (15.0 \text{ V})/(75.0 \text{ }\Omega) = 0.200$ A. At the junction just above the switch we have $I_{100} = 0.250$ A (in) and $I_{75} = 0.200$ A (out) , so $I_A = 0.250$ A $- 0.200$ A $= 0.050$ A, downward. The voltmeter reads zero because the potential difference across it is zero with the switch closed.

EVALUATE: The ideal ammeter acts like a short circuit, making a and b at the same potential. Hence the voltmeter reads zero.

26.31.   **IDENTIFY:**   To construct an ammeter, add a shunt resistor in parallel with the galvanometer coil. To construct a voltmeter, add a resistor in series with the galvanometer coil.

SET UP: The full-scale deflection current is 500 μA and the coil resistance is 25.0 Ω .

EXECUTE: **(a)** For a 20-mA ammeter, the two resistances are in parallel and the voltages across each are the same. $V_c = V_s$ gives $I_c R_c = I_s R_s$. $\left(500 \times 10^{-6} \text{ A}\right)\left(25.0 \text{ }\Omega\right) = \left(20 \times 10^{-3} \text{ A} - 500 \times 10^{-6} \text{ A}\right) R_s$ and $R_s = 0.641 \text{ }\Omega$.

(b) For a 500-mV voltmeter, the resistances are in series and the current is the same through each: $V_{ab} = I\left(R_c + R_s\right)$

and $R_s = \dfrac{V_{ab}}{I} - R_c = \dfrac{500 \times 10^{-3} \text{ V}}{500 \times 10^{-6} \text{ A}} - 25.0 \text{ }\Omega = 975 \text{ }\Omega$.

EVALUATE: The equivalent resistance of the voltmeter is $R_{eq} = R_s + R_c = 1000 \text{ }\Omega$. The equivalent resistance of the ammeter is given by $\dfrac{1}{R_{eq}} = \dfrac{1}{R_{sh}} + \dfrac{1}{R_c}$ and $R_{eq} = 0.625 \text{ }\Omega$. The voltmeter is a high-resistance device and the ammeter is a low-resistance device.

26.33.   **IDENTIFY:**   The meter introduces resistance into the circuit, which affects the current through the 5.00-kΩ resistor and hence the potential drop across it.

SET UP: Use Ohm's law to find the current through the 5.00-kΩ resistor and then the potential drop across it.

EXECUTE: **(a)** The parallel resistance with the voltmeter is 3.33 kΩ, so the total equivalent resistance across the battery is 9.33 kΩ, giving $I = (50.0 \text{ V})/(9.33 \text{ k}\Omega) = 5.36$ mA. Ohm's law gives the potential drop across the 5.00-kΩ resistor: $V_{5 \text{ k}\Omega} = (3.33 \text{ k}\Omega)(5.36 \text{ mA}) = 17.9$ V

(b) The current in the circuit is now $I = (50.0 \text{ V})/(11.0 \text{ k}\Omega) = 4.55$ mA. $V_{5 \text{ k}\Omega} = (5.00 \text{ k}\Omega)(4.55 \text{ mA}) = 22.7$ V.

(c) % error = (22.7 V $-$ 17.9 V)/(22.7 V) = 0.214 = 21.4%. (We carried extra decimal places for accuracy since we had to subtract our answers.)

EVALUATE: The presence of the meter made a very large percent error in the reading of the "true" potential across the resistor.

26.41.   **IDENTIFY:**   The capacitors, which are in parallel, will discharge exponentially through the resistors.

SET UP: Since V is proportional to Q, V must obey the same exponential equation as Q, $V = V_0 e^{-t/RC}$. The current is $I = (V_0/R) e^{-t/RC}$.

EXECUTE: **(a)** Solve for time when the potential across each capacitor is 10.0 V:

$$t = - RC \ln(V/V_0) = -(80.0 \text{ }\Omega)(35.0 \text{ }\mu\text{F}) \ln(10/45) = 4210 \text{ }\mu\text{s} = 4.21 \text{ ms}$$

(b) $I = (V_0/R) e^{-t/RC}$. Using the above values, with $V_0 = 45.0$ V, gives $I = 0.125$ A.

EVALUATE: Since the current and the potential both obey the same exponential equation, they are both reduced by the same factor (0.222) in 4.21 ms.

26.43.   **IDENTIFY and SET UP:**   Apply the loop rule. The voltage across the resistor depends on the current through it and the voltage across the capacitor depends on the charge on its plates.

EXECUTE: $\mathcal{E} - V_R - V_C = 0$

$\mathcal{E} = 120$ V, $V_R = IR = \left(0.900 \text{ A}\right)\left(80.0 \text{ }\Omega\right) = 72$ V, so $V_C = 48$ V

$Q = CV = \left(4.00 \times 10^{-6} \text{ F}\right)\left(48 \text{ V}\right) = 192 \text{ }\mu\text{C}$

EVALUATE: The initial charge is zero and the final charge is $C\mathcal{E} = 480 \text{ }\mu\text{C}$. Since current is flowing at the instant considered in the problem the capacitor is still being charged and its charge has not reached its final value.

26.47.   **IDENTIFY:**   In both cases, simplify the complicated circuit by eliminating the appropriate circuit elements. The potential across an uncharged capacitor is initially zero, so it behaves like a short circuit. A fully charged capacitor allows no current to flow through it.

(a) SET UP: Just after closing the switch, the uncharged capacitors all behave like short circuits, so any resistors in parallel with them are eliminated from the circuit.

EXECUTE: The equivalent circuit consists of 50 Ω and 25 Ω in parallel, with this combination in series with 75 Ω, 15 Ω, and the 100-V battery. The equivalent resistance is 90 Ω + 16.7 Ω = 106.7 Ω, which gives $I = (100 \text{ V})/(106.7 \text{ }\Omega) = 0.937$ A.

(b) SET UP: Long after closing the switch, the capacitors are essentially charged up and behave like open circuits since no charge can flow through them. They effectively eliminate any resistors in series with them since no current can flow through these resistors.

EXECUTE: The equivalent circuit consists of resistances of 75 Ω, 15 Ω, and three 25-Ω resistors, all in series with the 100-V battery, for a total resistance of 165 Ω. Therefore $I = (100 \text{ V})/(165 \text{ Ω}) = 0.606$ A

EVALUATE: The initial and final behavior of the circuit can be calculated quite easily using simple series-parallel circuit analysis. Intermediate times would require much more difficult calculations!

26.49. **IDENTIFY:** For each circuit apply the loop rule to relate the voltages across the circuit elements.

(a) SET UP: With the switch in position 2 the circuit is the charging circuit shown in Figure 26.49a.

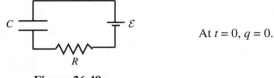

At $t = 0$, $q = 0$.

Figure 26.49a

EXECUTE: The charge q on the capacitor is given as a function of time by Eq.(26.12):

$q = C\mathcal{E}\left(1 - e^{-t/RC}\right)$

$Q_f = C\mathcal{E} = \left(1.50 \times 10^{-5} \text{ F}\right)\left(18.0 \text{ V}\right) = 2.70 \times 10^{-4}$ C.

$RC = \left(980 \text{ Ω}\right)\left(1.50 \times 10^{-5} \text{ F}\right) = 0.0147$ s

Thus, at $t = 0.0100$ s, $q = \left(2.70 \times 10^{-4} \text{ C}\right)\left(1 - e^{-(0.0100 \text{ s})/(0.0147 \text{ s})}\right) = 133$ μC.

(b) $v_C = \dfrac{q}{C} = \dfrac{133 \ \mu\text{C}}{1.50 \times 10^{-5} \text{ F}} = 8.87$ V

The loop rule says $\mathcal{E} - v_C - v_R = 0$

$v_R = \mathcal{E} - v_C = 18.0 \text{ V} - 8.87 \text{ V} = 9.13$ V

(c) SET UP: Throwing the switch back to position 1 produces the discharging circuit shown in Figure 26.49b.

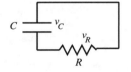

The initial charge Q_0 is the charge calculated in part (b), $Q_0 = 133$ μC.

Figure 26.49b

EXECUTE: $v_C = \dfrac{q}{C} = \dfrac{133 \ \mu\text{C}}{1.50 \times 10^{-5} \text{ F}} = 8.87$ V, the same as just before the switch is thrown. But now

$v_C - v_R = 0$, so $v_R = v_C = 8.87$ V.

(d) SET UP: In the discharging circuit the charge on the capacitor as a function of time is given by Eq.(26.16):

$q = Q_0 e^{-t/RC}$.

EXECUTE: $RC = 0.0147$ s, the same as in part (a). Thus at $t = 0.0100$ s, $q = (133 \ \mu\text{C})e^{-((0.0100 \text{ s})/(0.0147 \text{ s}))} = 67.4$ μC.

EVALUATE: $t = 10.0$ ms is less than one time constant, so at the instant described in part (a) the capacitor is not fully charged; its voltage (8.87 V) is less than the emf. There is a charging current and a voltage drop across the resistor. In the discharging circuit the voltage across the capacitor starts at 8.87 V and decreases. After $t = 10.0$ ms it has decreased to $v_C = q/C = 4.49$ V.

26.51. **IDENTIFY and SET UP:** The heater and hair dryer are in parallel so the voltage across each is 120 V and the current through the fuse is the sum of the currents through each appliance. As the power consumed by the dryer increases the current through it increases. The maximum power setting is the highest one for which the current through the fuse is less than 20 A.

EXECUTE: Find the current through the heater. $P = VI$ so $I = P/V = 1500 \text{ W}/120 \text{ V} = 12.5$ A. The maximum total current allowed is 20 A, so the current through the dryer must be less than $20 \text{ A} - 12.5 \text{ A} = 7.5$ A. The power dissipated by the dryer if the current has this value is $P = VI = (120 \text{ V})(7.5 \text{ A}) = 900$ W. For P at this value or larger the circuit breaker trips.

EVALUATE: $P = V^2/R$ and for the dryer V is a constant 120 V. The higher power settings correspond to a smaller resistance R and larger current through the device.

26.53.   **IDENTIFY** and **SET UP:**   Ohm's law and Eq.(25.18) can be used to calculate I and P given V and R. Use Eq.(25.12) to calculate the resistance at the higher temperature.

(a) EXECUTE: When the heater element is first turned on it is at room temperature and has resistance $R = 20\ \Omega$.

$$I = \frac{V}{R} = \frac{120\ \text{V}}{20\ \Omega} = 6.0\ \text{A}$$

$$P = \frac{V^2}{R} = \frac{(120\ \text{V})^2}{20\ \Omega} = 720\ \text{W}$$

(b) Find the resistance $R(T)$ of the element at the operating temperature of $280°\text{C}$.

Take $T_0 = 23.0°\text{C}$ and $R_0 = 20\ \Omega$. Eq.(25.12) gives

$$R(T) = R_0\left(1 + \alpha(T - T_0)\right) = 20\ \Omega\left(1 + \left(2.8 \times 10^{-3}\ (\text{C}°)^{-1}\right)(280°\text{C} - 23.0°\text{C})\right) = 34.4\ \Omega.$$

$$I = \frac{V}{R} = \frac{120\ \text{V}}{34.4\ \Omega} = 3.5\ \text{A}$$

$$P = \frac{V^2}{R} = \frac{(120\ \text{V})^2}{34.4\ \Omega} = 420\ \text{W}$$

EVALUATE: When the temperature increases, R increases and I and P decrease. The changes are substantial.

26.57.   **IDENTIFY:**   The terminal voltage of the battery depends on the current through it and therefore on the equivalent resistance connected to it. The power delivered to each bulb is $P = I^2R$, where I is the current through it.

SET UP: The terminal voltage of the source is $\mathcal{E} - Ir$.

EXECUTE: **(a)** The equivalent resistance of the two bulbs is $1.0\ \Omega$. This equivalent resistance is in series with the internal resistance of the source, so the current through the battery is $I = \dfrac{V}{R_{\text{total}}} = \dfrac{8.0\ \text{V}}{1.0\ \Omega + 0.80\ \Omega} = 4.4\ \text{A}$ and the current through each bulb is 2.2 A. The voltage applied to each bulb is $\mathcal{E} - Ir = 8.0\ \text{V} - (4.4\ \text{A})(0.80\ \Omega) = 4.4\ \text{V}$. Therefore, $P_{\text{bulb}} = I^2R = (2.2\ \text{A})^2(2.0\ \Omega) = 9.7\ \text{W}$.

(b) If one bulb burns out, then $I = \dfrac{V}{R_{\text{total}}} = \dfrac{8.0\ \text{V}}{2.0\ \Omega + 0.80\ \Omega} = 2.9\ \text{A}$. . The current through the remaining bulb is 2.9 A, and $P = I^2R = (2.9\ \text{A})^2\ (2.0\ \Omega) = 16.3\ \text{W}$. The remaining bulb is brighter than before, because it is consuming more power.

EVALUATE: In Example 26.2 the internal resistance of the source is negligible and the brightness of the remaining bulb doesn't change when one burns out.

26.59.   **IDENTIFY:**   The ohmmeter reads the equivalent resistance between points a and b. Replace series and parallel combinations by their equivalent.

SET UP: For resistors in parallel, $\dfrac{1}{R_{\text{eq}}} = \dfrac{1}{R_1} + \dfrac{1}{R_2}$. For resistors in series, $R_{\text{eq}} = R_1 + R_2$

EXECUTE: Circuit (a): The $75.0\ \Omega$ and $40.0\ \Omega$ resistors are in parallel and have equivalent resistance $26.09\ \Omega$. The $25.0\ \Omega$ and $50.0\ \Omega$ resistors are in parallel and have an equivalent resistance of $16.67\ \Omega$. The equivalent network is given in Figure 26.59a. $\dfrac{1}{R_{\text{eq}}} = \dfrac{1}{100.0\ \Omega} + \dfrac{1}{23.05\ \Omega}$, so $R_{\text{eq}} = 18.7\ \Omega$.

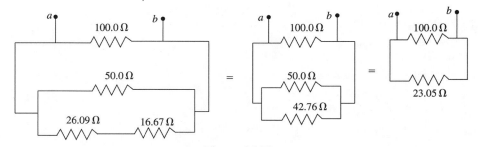

Figure 26.59a

Circuit (b): The $30.0\ \Omega$ and $45.0\ \Omega$ resistors are in parallel and have equivalent resistance $18.0\ \Omega$. The equivalent network is given in Figure 26.59b. $\dfrac{1}{R_{eq}} = \dfrac{1}{10.0\ \Omega} + \dfrac{1}{30.3\ \Omega}$, so $R_{eq} = 7.5\ \Omega$.

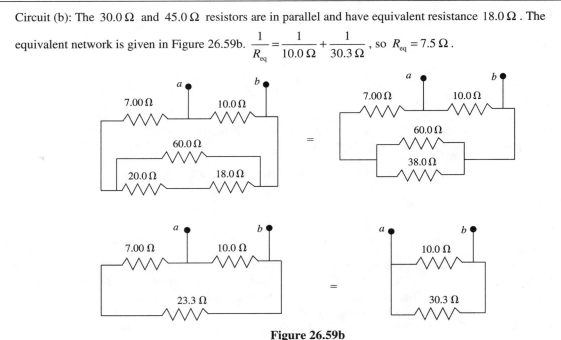

Figure 26.59b

EVALUATE: In circuit (a) the resistance along one path between a and b is $100.0\ \Omega$, but that is not the equivalent resistance between these points. A similar comment can be made about circuit (b).

26.61. IDENTIFY: Apply the junction rule to express the currents through the $5.00\ \Omega$ and $8.00\ \Omega$ resistors in terms of I_1, I_2 and I_3. Apply the loop rule to three loops to get three equations in the three unknown currents.

SET UP: The circuit is sketched in Figure 26.61.

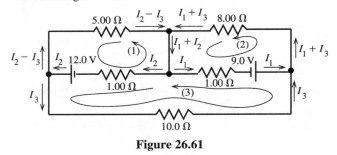

Figure 26.61

The current in each branch has been written in terms of I_1, I_2 and I_3 such that the junction rule is satisfied at each junction point.

EXECUTE: Apply the loop rule to loop (1).

$-12.0\ \text{V} + I_2(1.00\ \Omega) + (I_2 - I_3)(5.00\ \Omega) = 0$

$I_2(6.00\ \Omega) - I_3(5.00\ \Omega) = 12.0\ \text{V}$ eq.(1)

Apply the loop rule to loop (2).

$-I_1(1.00\ \Omega) + 9.00\ \text{V} - (I_1 + I_3)(8.00\ \Omega) = 0$

$I_1(9.00\ \Omega) + I_3(8.00\ \Omega) = 9.00\ \text{V}$ eq.(2)

Apply the loop rule to loop (3).

$-I_3(10.0\ \Omega) - 9.00\ \text{V} + I_1(1.00\ \Omega) - I_2(1.00\ \Omega) + 12.0\ \text{V} = 0$

$-I_1(1.00\ \Omega) + I_2(1.00\ \Omega) + I_3(10.0\ \Omega) = 3.00\ \text{V}$ eq.(3)

Eq.(1) gives $I_2 = 2.00\ \text{A} + \frac{5}{6}I_3$; eq.(2) gives $I_1 = 1.00\ \text{A} - \frac{8}{9}I_3$

Using these results in eq.(3) gives $-(1.00\ \text{A} - \frac{8}{9}I_3)(1.00\ \Omega) + (2.00\ \text{A} + \frac{5}{6}I_3)(1.00\ \Omega) + I_3(10.0\ \Omega) = 3.00\ \text{V}$

$\left(\frac{16+15+180}{18}\right)I_3 = 2.00\ \text{A}$; $I_3 = \frac{18}{211}(2.00\ \text{A}) = 0.171\ \text{A}$

Then $I_2 = 2.00\ \text{A} + \frac{5}{6}I_3 = 2.00\ \text{A} + \frac{5}{6}(0.171\ \text{A}) = 2.14\ \text{A}$ and $I_1 = 1.00\ \text{A} - \frac{8}{9}I_3 = 1.00\ \text{A} - \frac{8}{9}(0.171\ \text{A}) = 0.848\ \text{A}$.

EVALUATE: We could check that the loop rule is satisfied for a loop that goes through the 5.00 Ω, 8.00 Ω and 10.0 Ω resistors. Going around the loop clockwise: $-(I_2-I_3)(5.00\ \Omega)+(I_1+I_3)(8.00\ \Omega)+I_3(10.0\ \Omega)=$ $-9.85\ \text{V}+8.15\ \text{V}+1.71\ \text{V}$, which does equal zero, apart from rounding.

26.63.   **IDENTIFY** and **SET UP:**   The circuit is sketched in Figure 26.63.

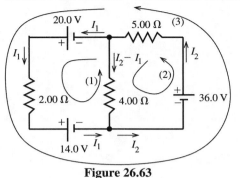

Two unknown currents I_1 (through the 2.00 Ω resistor) and I_2 (through the 5.00 Ω resistor) are labeled on the circuit diagram. The current through the 4.00 Ω resistor has been written as I_2-I_1 using the junction rule.

Figure 26.63

Apply the loop rule to loops (1) and (2) to get two equations for the unknown currents, I_1 and I_2. Loop (3) can then be used to check the results.

EXECUTE: loop (1): $+20.0\ \text{V}-I_1(2.00\ \Omega)-14.0\ \text{V}+(I_2-I_1)(4.00\ \Omega)=0$

$6.00I_1-4.00I_2=6.00\ \text{A}$

$3.00I_1-2.00I_2=3.00\ \text{A}$ eq.(1)

loop (2): $+36.0\ \text{V}-I_2(5.00\ \Omega)-(I_2-I_1)(4.00\ \Omega)=0$

$-4.00I_1+9.00I_2=36.0\ \text{A}$ eq.(2)

Solving eq. (1) for I_1 gives $I_1=1.00\ \text{A}+\frac{2}{3}I_2$

Using this in eq.(2) gives $-4.00\left(1.00\ \text{A}+\frac{2}{3}I_2\right)+9.00I_2=36.0\ \text{A}$

$\left(-\frac{8}{3}+9.00\right)I_2=40.0\ \text{A and }I_2=6.32\ \text{A.}$

Then $I_1=1.00\ \text{A}+\frac{2}{3}I_2=1.00\ \text{A}+\frac{2}{3}(6.32\ \text{A})=5.21\ \text{A.}$

In summary then

Current through the 2.00 Ω resistor: $I_1=5.21\ \text{A.}$

Current through the 5.00 Ω resistor: $I_2=6.32\ \text{A.}$

Current through the 4.00 Ω resistor: $I_2-I_1=6.32\ \text{A}-5.21\ \text{A}=1.11\ \text{A.}$

EVALUATE: Use loop (3) to check. $+20.0\ \text{V}-I_1(2.00\ \Omega)-14.0\ \text{V}+36.0\ \text{V}-I_2(5.00\ \Omega)=0$

$(5.21\ \text{A})(2.00\ \Omega)+(6.32\ \text{A})(5.00\ \Omega)=42.0\ \text{V}$

$10.4\ \text{V}+31.6\ \text{V}=42.0\ \text{V}$, so the loop rule is satisfied for this loop.

26.65.   **(a) IDENTIFY:**   Break the circuit between points a and b means no current in the middle branch that contains the 3.00 Ω resistor and the 10.0 V battery. The circuit therefore has a single current path. Find the current, so that potential drops across the resistors can be calculated. Calculate V_{ab} by traveling from a to b, keeping track of the potential changes along the path taken.

SET UP: The circuit is sketched in Figure 26.65a.

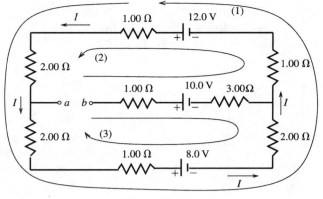

Figure 26.65a

EXECUTE: Apply the loop rule to loop (1).

$+12.0 \text{ V} - I(1.00 \text{ }\Omega + 2.00 \text{ }\Omega + 2.00 \text{ }\Omega + 1.00 \text{ }\Omega) - 8.0 \text{ V} - I(2.00 \text{ }\Omega + 1.00 \text{ }\Omega) = 0$

$$I = \frac{12.0 \text{ V} - 8.0 \text{ V}}{9.00 \text{ }\Omega} = 0.4444 \text{ A}.$$

To find V_{ab} start at point b and travel to a, adding up the potential rises and drops. Travel on path (2) shown on the diagram. The $1.00 \text{ }\Omega$ and $3.00 \text{ }\Omega$ resistors in the middle branch have no current through them and hence no voltage across them. Therefore, $V_b - 10.0 \text{ V} + 12.0 \text{ V} - I(1.00 \text{ }\Omega + 1.00 \text{ }\Omega + 2.00 \text{ }\Omega) = V_a$; thus

$V_a - V_b = 2.0 \text{ V} - (0.4444 \text{ A})(4.00 \text{ }\Omega) = +0.22 \text{ V}$ (point a is at higher potential)

EVALUATE: As a check on this calculation we also compute V_{ab} by traveling from b to a on path (3).

$V_b - 10.0 \text{ V} + 8.0 \text{ V} + I(2.00 \text{ }\Omega + 1.00 \text{ }\Omega + 2.00 \text{ }\Omega) = V_a$

$V_{ab} = -2.00 \text{ V} + (0.4444 \text{ A})(5.00 \text{ }\Omega) = +0.22 \text{ V}$, which checks.

(b) IDENTIFY and SET UP: With points a and b connected by a wire there are three current branches, as shown in Figure 26.65b.

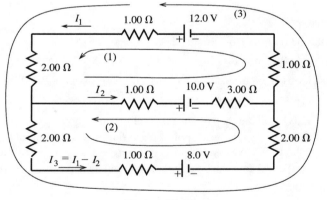

Figure 26.65b

The junction rule has been used to write the third current (in the 8.0 V battery) in terms of the other currents. Apply the loop rule to loops (1) and (2) to obtain two equations for the two unknowns I_1 and I_2.

EXECUTE: Apply the loop rule to loop (1).

$12.0 \text{ V} - I_1(1.00 \text{ }\Omega) - I_1(2.00 \text{ }\Omega) - I_2(1.00 \text{ }\Omega) - 10.0 \text{ V} - I_2(3.00 \text{ }\Omega) - I_1(1.00 \text{ }\Omega) = 0$

$2.0 \text{ V} - I_1(4.00 \text{ }\Omega) - I_2(4.00 \text{ }\Omega) = 0$

$(2.00 \text{ }\Omega)I_1 + (2.00 \text{ }\Omega)I_2 = 1.0 \text{ V}$ eq.(1)

Apply the loop rule to loop (2).

$-(I_1 - I_2)(2.00 \text{ }\Omega) - (I_1 - I_2)(1.00 \text{ }\Omega) - 8.0 \text{ V} - (I_1 - I_2)(2.00 \text{ }\Omega) + I_2(3.00 \text{ }\Omega) + 10.0 \text{ V} + I_2(1.00 \text{ }\Omega) = 0$

$2.0 \text{ V} - (5.00 \text{ }\Omega)I_1 + (9.00 \text{ }\Omega)I_2 = 0$ eq.(2)

Solve eq.(1) for I_2 and use this to replace I_2 in eq.(2).

$I_2 = 0.50 \text{ A} - I_1$

$2.0 \text{ V} - (5.00 \ \Omega)I_1 + (9.00 \ \Omega)(0.50 \text{ A} - I_1) = 0$

$(14.0 \ \Omega)I_1 = 6.50 \text{ V}$ so $I_1 = (6.50 \text{ V})/(14.0 \ \Omega) = 0.464 \text{ A}$

$I_2 = 0.500 \text{ A} - 0.464 \text{ A} = 0.036 \text{ A}.$

The current in the 12.0 V battery is $I_1 = 0.464 \text{ A}.$

EVALUATE: We can apply the loop rule to loop (3) as a check.

$+12.0 \text{ V} - I_1(1.00 \ \Omega + 2.00 \ \Omega + 1.00 \ \Omega) - (I_1 - I_2)(2.00 \ \Omega + 1.00 \ \Omega + 2.00 \ \Omega) - 8.0 \text{ V} = 4.0 \text{ V} - 1.86 \text{ V} - 2.14 \text{ V} = 0,$

as it should.

26.67.   **IDENTIFY:**   In Figure 26.67, points a and c are at the same potential and points d and b are at the same potential, so we can calculate V_{ab} by calculating V_{cd}. We know the current through the resistor that is between points c and d. We thus can calculate the terminal voltage of the 24.0 V battery without calculating the current through it.

SET UP:

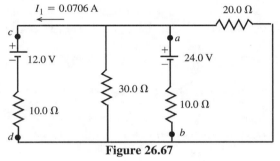

Figure 26.67

EXECUTE: $V_d + I_1(10.0 \ \Omega) + 12.0 \text{ V} = V_c$

$V_c - V_d = 12.7 \text{ V}; \ V_a - V_b = V_c - V_d = 12.7 \text{ V}$

EVALUATE: The voltage across each parallel branch must be the same. The current through the 24.0 V battery must be $(24.0 \text{ V} - 12.7 \text{ V})/(10.0 \ \Omega) = 1.13 \text{ A}$ in the direction b to a.

26.69.   **IDENTIFY** and **SET UP:**   Simplify the circuit by replacing the parallel networks of resistors by their equivalents. In this simplified circuit apply the loop and junction rules to find the current in each branch.

EXECUTE: The 20.0-Ω and 30.0-Ω resistors are in parallel and have equivalent resistance 12.0 Ω. The two resistors R are in parallel and have equivalent resistance $R/2$. The circuit is equivalent to the circuit sketched in Figure 26.69.

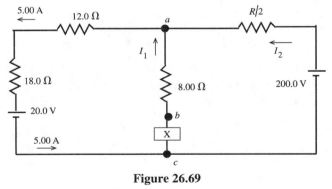

Figure 26.69

(a) Calculate V_{ca} by traveling along the branch that contains the 20.0 V battery, since we know the current in that branch.

$V_a - (5.00 \text{ A})(12.0 \ \Omega) - (5.00 \text{ A})(18.0 \ \Omega) - 20.0 \text{ V} = V_c$

$V_a - V_c = 20.0 \text{ V} + 90.0 \text{ V} + 60.0 \text{ V} = 170.0 \text{ V}$

$V_b - V_a = V_{ab} = 16.0 \text{ V}$

$X - V_{ba} = 170.0 \text{ V}$ so $X = 186.0 \text{ V}$, with the upper terminal +

(b) $I_1 = (16.0\ \text{V})/(8.0\ \Omega) = 2.00\ \text{A}$

The junction rule applied to point a gives $I_2 + I_1 = 5.00\ \text{A}$, so $I_2 = 3.00\ \text{A}$. The current through the 200.0 V battery is in the direction from the $-$ to the $+$ terminal, as shown in the diagram.

(c) $200.0\ \text{V} - I_2(R/2) = 170.0\ \text{V}$

$(3.00\ \text{A})(R/2) = 30.0\ \text{V}$ so $R = 20.0\ \Omega$

EVALUATE: We can check the loop rule by going clockwise around the outer circuit loop. This gives $+20.0\ \text{V} + (5.00\ \text{A})(18.0\ \Omega + 12.0\ \Omega) + (3.00\ \text{A})(10.0\ \Omega) - 200.0\ \text{V} = 20.0\ \text{V} + 150.0\ \text{V} + 30.0\ \text{V} - 200.0\ \text{V}$, which does equal zero.

26.71. **IDENTIFY and SET UP:**   For part (a) use that the full emf is across each resistor. In part (b), calculate the power dissipated by the equivalent resistance, and in this expression express R_1 and R_2 in terms of P_1, P_2 and $\mathcal{E}$.

EXECUTE: $P_1 = \mathcal{E}^2 / R_1$ so $R_1 = \mathcal{E}^2 / P_1$

$P_2 = \mathcal{E}^2 / R_2$ so $R_2 = \mathcal{E}^2 / P_2$

(a) When the resistors are connected in parallel to the emf, the voltage across each resistor is $\mathcal{E}$ and the power dissipated by each resistor is the same as if only the one resistor were connected. $P_{\text{tot}} = P_1 + P_2$

(b) When the resistors are connected in series the equivalent resistance is $R_{\text{eq}} = R_1 + R_2$

$$P_{\text{tot}} = \frac{\mathcal{E}^2}{R_1 + R_2} = \frac{\mathcal{E}^2}{\mathcal{E}^2 / P_1 + \mathcal{E}^2 / P_2} = \frac{P_1 P_2}{P_1 + P_2}$$

EVALUATE: The result in part (b) can be written as $\dfrac{1}{P_{\text{tot}}} = \dfrac{1}{P_1} + \dfrac{1}{P_2}$. Our results are that for parallel the powers add and that for series the reciprocals of the power add. This is opposite the result for combining resistance. Since $P = \mathcal{E}^2 / R$ tells us that P is proportional to $1/R$, this makes sense.

26.73. **(a) IDENTIFY and SET UP:**   The circuit is sketched in Figure 26.73a.

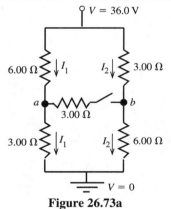

With the switch open there is no current through it and there are only the two currents I_1 and I_2 indicated in the sketch.

Figure 26.73a

The potential drop across each parallel branch is 36.0 V. Use this fact to calculate I_1 and I_2. Then travel from point a to point b and keep track of the potential rises and drops in order to calculate V_{ab}.

EXECUTE: $-I_1(6.00\ \Omega + 3.00\ \Omega) + 36.0\ \text{V} = 0$

$I_1 = \dfrac{36.0\ \text{V}}{6.00\ \Omega + 3.00\ \Omega} = 4.00\ \text{A}$

$-I_2(3.00\ \Omega + 6.00\ \Omega) + 36.0\ \text{V} = 0$

$I_2 = \dfrac{36.0\ \text{V}}{3.00\ \Omega + 6.00\ \Omega} = 4.00\ \text{A}$

To calculate $V_{ab} = V_a - V_b$ start at point b and travel to point a, adding up all the potential rises and drops along the way. We can do this by going from b up through the 3.00 Ω resistor:

$V_b + I_2(3.00\ \Omega) - I_1(6.00\ \Omega) = V_a$

$V_a - V_b = (4.00\ \text{A})(3.00\ \Omega) - (4.00\ \text{A})(6.00\ \Omega) = 12.0\ \text{V} - 24.0\ \text{V} = -12.0\ \text{V}$

$V_{ab} = -12.0\ \text{V}$ (point a is 12.0 V lower in potential than point b)

EVALUATE: Alternatively, we can go from point b down through the 6.00 Ω resistor.

$V_b - I_2(6.00\ \Omega) + I_1(3.00\ \Omega) = V_a$

$V_a - V_b = -(4.00\ \text{A})(6.00\ \Omega) + (4.00\ \text{A})(3.00\ \Omega) = -24.0\ \text{V} + 12.0\ \text{V} = -12.0\ \text{V}$, which checks.

(b) IDENTIFY: Now there are multiple current paths, as shown in Figure 26.73b. Use junction rule to write the current in each branch in terms of three unknown currents I_1, I_2, and I_3. Apply the loop rule to three loops to get three equations for the three unknowns. The target variable is I_3, the current through the switch. R_{eq} is calculated from $V = IR_{eq}$, where I is the total current that passes through the network.

SET UP:

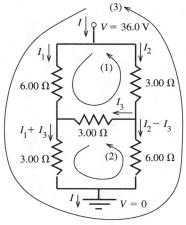

Figure 26.73b

The three unknown currents I_1, I_2, and I_3 are labeled on Figure 26.73b.

EXECUTE: Apply the loop rule to loops (1), (2), and (3).

loop (1): $-I_1(6.00 \ \Omega) + I_3(3.00 \ \Omega) + I_2(3.00 \ \Omega) = 0$

$I_2 = 2I_1 - I_3$ eq.(1)

loop (2): $-(I_1 + I_3)(3.00 \ \Omega) + (I_2 - I_3)(6.00 \ \Omega) - I_3(3.00 \ \Omega) = 0$

$6I_2 - 12I_3 - 3I_1 = 0$ so $2I_2 - 4I_3 - I_1 = 0$

Use eq(1) to replace I_2:

$4I_1 - 2I_3 - 4I_3 - I_1 = 0$

$3I_1 = 6I_3$ and $I_1 = 2I_3$ eq.(2)

loop (3) (This loop is completed through the battery [not shown], in the direction from the $-$ to the $+$ terminal.):

$-I_1(6.00 \ \Omega) - (I_1 + I_3)(3.00 \ \Omega) + 36.0 \ \text{V} = 0$

$9I_1 + 3I_3 = 36.0 \ \text{A}$ and $3I_1 + I_3 = 12.0 \ \text{A}$ eq.(3)

Use eq.(2) in eq.(3) to replace I_1:

$3(2I_3) + I_3 = 12.0 \ \text{A}$

$I_3 = 12.0 \ \text{A}/7 = 1.71 \ \text{A}$

$I_1 = 2I_3 = 3.42 \ \text{A}$

$I_2 = 2I_1 - I_3 = 2(3.42 \ \text{A}) - 1.71 \ \text{A} = 5.13 \ \text{A}$

The current through the switch is $I_3 = 1.71 \ \text{A}$.

(c) From the results in part (a) the current through the battery is $I = I_1 + I_2 = 3.42 \ \text{A} + 5.13 \ \text{A} = 8.55 \ \text{A}$. The equivalent circuit is a single resistor that produces the same current through the 36.0 V battery, as shown in Figure 26.73c.

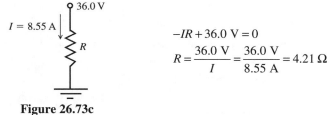

$-IR + 36.0 \ \text{V} = 0$

$R = \dfrac{36.0 \ \text{V}}{I} = \dfrac{36.0 \ \text{V}}{8.55 \ \text{A}} = 4.21 \ \Omega$

Figure 26.73c

EVALUATE: With the switch open (part a), point b is at higher potential than point a, so when the switch is closed the current flows in the direction from b to a. With the switch closed the circuit cannot be simplified using series and parallel combinations but there is still an equivalent resistance that represents the network.

26.77.   **IDENTIFY:**   Connecting the voltmeter between point *b* and ground gives a resistor network and we can solve for the current through each resistor. The voltmeter reading equals the potential drop across the 200 kΩ resistor.

SET UP: For resistors in parallel, $\dfrac{1}{R_{eq}} = \dfrac{1}{R_1} + \dfrac{1}{R_2}$. For resistors in series, $R_{eq} = R_1 + R_2$.

EXECUTE: **(a)** $R_{eq} = 100 \text{ k}\Omega + \left(\dfrac{1}{200 \text{ k}\Omega} + \dfrac{1}{50 \text{ k}\Omega} \right)^{-1} = 140 \text{ k}\Omega$. The total current is $I = \dfrac{0.400 \text{ kV}}{140 \text{ k}\Omega} = 2.86 \times 10^{-3}$ A.

The voltage across the 200 kΩ resistor is $V_{200k\Omega} = IR = (2.86 \times 10^{-3} \text{ A}) \left(\dfrac{1}{200 \text{ k}\Omega} + \dfrac{1}{50 \text{ k}\Omega} \right)^{-1} = 114.4$ V.

(b) If $V_R = 5.00 \times 10^6 \ \Omega$, then we carry out the same calculations as above to find $R_{eq} = 292 \text{ k}\Omega$, $I = 1.37 \times 10^{-3}$ A and $V_{200k\Omega} = 263$ V.

(c) If $V_R = \infty$, then we find $R_{eq} = 300 \text{ k}\Omega$, $I = 1.33 \times 10^{-3}$ A and $V_{200k\Omega} = 266$ V.

EVALUATE: When a voltmeter of finite resistance is connected to a circuit, current flows through the voltmeter and the presence of the voltmeter alters the currents and voltages in the original circuit. The effect of the voltmeter on the circuit decreases as the resistance of the voltmeter increases.

26.81.   **IDENTIFY** and **SET UP:**   Without the meter, the circuit consists of the two resistors in series. When the meter is connected, its resistance is added to the circuit in parallel with the resistor it is connected across.

(a) EXECUTE: $I = I_1 = I_2$

$I = \dfrac{90.0 \text{ V}}{R_1 + R_2} = \dfrac{90.0 \text{ V}}{224 \ \Omega + 589 \ \Omega} = 0.1107$ A

$V_1 = I_1 R_1 = (0.1107 \text{ A})(224 \ \Omega) = 24.8$ V; $V_2 = I_2 R_2 = (0.1107 \text{ A})(589 \ \Omega) = 65.2$ V

(b) SET UP: The resistor network is sketched in Figure 26.81a.

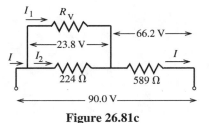

The voltmeter reads the potential difference across its terminals, which is 23.8 V. If we can find the current I_1 through the voltmeter then we can use Ohm's law to find its resistance.

Figure 26.81c

EXECUTE: The voltage drop across the 589 Ω resistor is $90.0 \text{ V} - 23.8 \text{ V} = 66.2$ V, so

$I = \dfrac{V}{R} = \dfrac{66.2 \text{ V}}{589 \ \Omega} = 0.1124$ A. The voltage drop across the 224 Ω resistor is 23.8 V, so $I_2 = \dfrac{V}{R} = \dfrac{23.8 \text{ V}}{224 \ \Omega} = 0.1062$ A.

Then $I = I_1 + I_2$ gives $I_1 = I - I_2 = 0.1124 \text{ A} - 0.1062 \text{ A} = 0.0062$ A. $R_V = \dfrac{V}{I_1} = \dfrac{23.8 \text{ V}}{0.0062 \text{ A}} = 3840 \ \Omega$

(c) SET UP: The circuit with the voltmeter connected is sketched in Figure 26.81b.

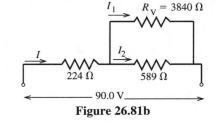

Figure 26.81b

EXECUTE: Replace the two resistors in parallel by their equivalent, as shown in Figure 26.81c.

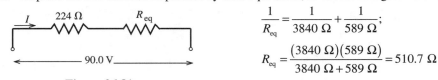

$\dfrac{1}{R_{eq}} = \dfrac{1}{3840 \ \Omega} + \dfrac{1}{589 \ \Omega}$;

$R_{eq} = \dfrac{(3840 \ \Omega)(589 \ \Omega)}{3840 \ \Omega + 589 \ \Omega} = 510.7 \ \Omega$

Figure 26.81c

$I = \dfrac{90.0 \text{ V}}{224 \ \Omega + 510.7 \ \Omega} = 0.1225$ A

The potential drop across the 224 Ω resistor then is $IR = (0.1225 \text{ A})(224 \text{ }\Omega) = 27.4 \text{ V},$ so the potential drop across the 589 Ω resistor and across the voltmeter (what the voltmeter reads) is $90.0 \text{ V} - 27.4 \text{ V} = 62.6 \text{ V}.$

(d) EVALUATE: No, any real voltmeter will draw some current and thereby reduce the current through the resistance whose voltage is being measured. Thus the presence of the voltmeter connected in parallel with the resistance lowers the voltage drop across that resistance. The resistance of the voltmeter is only about a factor of ten larger than the resistances in the circuit, so the voltmeter has a noticeable effect on the circuit.

26.83. **IDENTIFY:** Apply the loop rule to the circuit. The initial current determines R. We can then use the time constant to calculate C.

SET UP: The circuit is sketched in Figure 26.83.

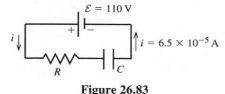

Initially, the charge of the capacitor is zero, so by $v = q/C$ the voltage across the capacitor is zero.

Figure 26.83

EXECUTE: The loop rule therefore gives $\mathcal{E} - iR = 0$ and $R = \dfrac{\mathcal{E}}{i} = \dfrac{110 \text{ V}}{6.5 \times 10^{-5} \text{ A}} = 1.7 \times 10^{6} \text{ }\Omega$

The time constant is given by $\tau = RC$ (Eq.26.14), so $C = \dfrac{\tau}{R} = \dfrac{6.2 \text{ s}}{1.7 \times 10^{6} \text{ }\Omega} = 3.6 \text{ }\mu\text{F}.$

EVALUATE: The resistance is large so the initial current is small and the time constant is large.

26.87. **IDENTIFY and SET UP:** For parts (a) and (b) evaluate the integrals as specified in the problem. The current as a function of time is given by Eq.(26.13) $i = \dfrac{\mathcal{E}}{R} e^{-t/RC}.$ The energy stored in the capacitor is given by $Q^2/2C.$

EXECUTE: **(a)** $P = \mathcal{E}i$

The total energy supplied by the battery is $\int_0^\infty P\,dt = \int_0^\infty \mathcal{E}i\,dt = (\mathcal{E}^2/R)\int_0^\infty e^{-t/RC}\,dt = (\mathcal{E}^2/R)\left[-RCe^{-t/RC}\right]_0^\infty = C\mathcal{E}^2.$

(b) $P = i^2 R$

The total energy dissipated in the resistor is

$\int_0^\infty P\,dt = \int_0^\infty i^2 R\,dt = (\mathcal{E}^2/R)\int_0^\infty e^{-2t/RC}\,dt = (\mathcal{E}^2/R)\left[-(RC/2)e^{-2t/RC}\right]_0^\infty = \frac{1}{2}C\mathcal{E}^2.$

(c) The final charge on the capacitor is $Q = C\mathcal{E}.$ The energy stored is $U = Q^2/(2C) = \frac{1}{2}C\mathcal{E}^2.$ The final energy stored in the capacitor $\left(\frac{1}{2}C\mathcal{E}^2\right) =$ total energy supplied by the battery $\left(C\mathcal{E}^2\right) -$ energy dissipated in the resistor $\left(\frac{1}{2}C\mathcal{E}^2\right)$

(d) EVALUATE: $\frac{1}{2}$ of the energy supplied by the battery is stored in the capacitor. This fraction is independent of R. The other $\frac{1}{2}$ of the energy supplied by the battery is dissipated in the resistor. When R is small the current initially is large but dies away quickly. When R is large the current initially is small but lasts longer.

27

MAGNETIC FIELD AND MAGNETIC FORCES

27.1. **IDENTIFY** and **SET UP:** Apply Eq.(27.2) to calculate $\vec{F}$. Use the cross products of unit vectors from Section 1.10.

EXECUTE: $\vec{v} = \left(+4.19 \times 10^4 \text{ m/s}\right)\hat{i} + \left(-3.85 \times 10^4 \text{ m/s}\right)\hat{j}$

(a) $\vec{B} = \left(1.40 \text{ T}\right)\hat{i}$

$\vec{F} = q\vec{v} \times \vec{B} = \left(-1.24 \times 10^{-8} \text{ C}\right)\left(1.40 \text{ T}\right)\left[\left(4.19 \times 10^4 \text{ m/s}\right)\hat{i} \times \hat{i} - \left(3.85 \times 10^4 \text{ m/s}\right)\hat{j} \times \hat{i}\right]$

$\hat{i} \times \hat{i} = 0, \hat{j} \times \hat{i} = -\hat{k}$

$\vec{F} = \left(-1.24 \times 10^{-8} \text{ C}\right)\left(1.40 \text{ T}\right)\left(-3.85 \times 10^4 \text{ m/s}\right)\left(-\hat{k}\right) = \left(-6.68 \times 10^{-4} \text{ N}\right)\hat{k}$

EVALUATE: The directions of $\vec{v}$ and $\vec{B}$ are shown in Figure 27.1a.

 The right-hand rule gives that $\vec{v} \times \vec{B}$ is directed out of the paper (+z-direction). The charge is negative so $\vec{F}$ is opposite to $\vec{v} \times \vec{B}$;

Figure 27.1a

$\vec{F}$ is in the $-z$- direction. This agrees with the direction calculated with unit vectors.

(b) EXECUTE: $\vec{B} = \left(1.40 \text{ T}\right)\hat{k}$

$\vec{F} = q\vec{v} \times \vec{B} = \left(-1.24 \times 10^{-8} \text{ C}\right)\left(1.40 \text{ T}\right)\left[\left(+4.19 \times 10^4 \text{ m/s}\right)\hat{i} \times \hat{k} - \left(3.85 \times 10^4 \text{ m/s}\right)\hat{j} \times \hat{k}\right]$

$\hat{i} \times \hat{k} = -\hat{j}, \hat{j} \times \hat{k} = \hat{i}$

$\vec{F} = \left(-7.27 \times 10^{-4} \text{ N}\right)\left(-\hat{j}\right) + \left(6.68 \times 10^{-4} \text{ N}\right)\hat{i} = \left[\left(6.68 \times 10^{-4} \text{ N}\right)\hat{i} + \left(7.27 \times 10^{-4} \text{ N}\right)\hat{j}\right]$

EVALUATE: The directions of $\vec{v}$ and $\vec{B}$ are shown in Figure 27.1b.

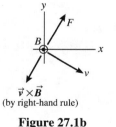 The direction of $\vec{F}$ is opposite to $\vec{v} \times \vec{B}$ since q is negative. The direction of $\vec{F}$ computed from the right-hand rule agrees qualitatively with the direction calculated with unit vectors.

Figure 27.1b

27.7. **IDENTIFY:** Apply $\vec{F} = q\vec{v} \times \vec{B}$.

SET UP: $\vec{v} = v_y \hat{j}$, with $v_y = -3.80 \times 10^3 \text{ m/s}$. $F_x = +7.60 \times 10^{-3} \text{ N}$, $F_y = 0$, and $F_z = -5.20 \times 10^{-3} \text{ N}$.

EXECUTE: **(a)** $F_x = q(v_y B_z - v_z B_y) = qv_y B_z$.

$B_z = F_x / qv_y = (7.60 \times 10^{-3} \text{ N}) / ([7.80 \times 10^{-6} \text{ C}](-3.80 \times 10^3 \text{ m/s})] = -0.256 \text{ T}$

$F_y = q(v_z B_x - v_x B_z) = 0$, which is consistent with $\vec{F}$ as given in the problem. There is no force component along the direction of the velocity.

$F_z = q(v_x B_y - v_y B_x) = -qv_y B_x$. $B_x = -F_z / qv_y = -0.175 \text{ T}$.

(b) B_y is not determined. No force due to this component of $\vec{B}$ along $\vec{v}$; measurement of the force tells us nothing about B_y.

(c) $\vec{B} \cdot \vec{F} = B_x F_x + B_y F_y + B_z F_z = (-0.175 \text{ T})(+7.60 \times 10^{-3} \text{ N}) + (-0.256 \text{ T})(-5.20 \times 10^{-3} \text{ N})$

$\vec{B} \cdot \vec{F} = 0$. $\vec{B}$ and $\vec{F}$ are perpendicular (angle is $90°$).

EVALUATE: The force is perpendicular to both $\vec{v}$ and $\vec{B}$, so $\vec{v} \cdot \vec{F}$ is also zero.

27.9.   **IDENTIFY:**   Apply $\vec{F} = q\vec{v} \times \vec{B}$ to the force on the proton and to the force on the electron. Solve for the components of $\vec{B}$.

SET UP: $\vec{F}$ is perpendicular to both $\vec{v}$ and $\vec{B}$. Since the force on the proton is in the $+y$-direction, $B_y = 0$ and $\vec{B} = B_x \hat{i} + B_z \hat{k}$. For the proton, $\vec{v} = (1.50 \text{ km/s})\hat{i}$.

EXECUTE: **(a)** For the proton, $\vec{F} = q(1.50 \times 10^3 \text{ m/s})\hat{i} \times (B_x \hat{i} + B_z \hat{k}) = q(1.50 \times 10^3 \text{ m/s})B_z(-\hat{j})$. $\vec{F} = (2.25 \times 10^{-16} \text{ N})\hat{j}$,

so $B_z = -\dfrac{2.25 \times 10^{-16} \text{ N}}{(1.60 \times 10^{-19} \text{ C})(1.50 \times 10^3 \text{ m/s})} = -0.938 \text{ T}$. The force on the proton is independent of B_x. For the

electron, $\vec{v} = (4.75 \text{ km/s})(-\hat{k})$. $\vec{F} = q\vec{v} \times \vec{B} = (-e)(4.75 \times 10^3 \text{ m/s})(-\hat{k}) \times (B_x \hat{i} + B_z \hat{k}) = +e(4.75 \times 10^3 \text{ m/s})B_x \hat{j}$.

The magnitude of the force is $F = e(4.75 \times 10^3 \text{ m/s})|B_x|$. Since $F = 8.50 \times 10^{-16} \text{ N}$,

$|B_x| = \dfrac{8.50 \times 10^{-16} \text{ N}}{(1.60 \times 10^{-19} \text{ C})(4.75 \times 10^3 \text{ m/s})} = 1.12 \text{ T}$. $B_x = \pm 1.12 \text{ T}$. The sign of B_x is not determined by measuring

the magnitude of the force on the electron. $B = \sqrt{B_x^2 + B_z^2} = \sqrt{(\pm 1.12 \text{ T}) + (-0.938 \text{ T})^2} = 1.46 \text{ T}$.

$\tan\theta = \dfrac{B_z}{B_x} = \dfrac{-0.938 \text{ T}}{\pm 1.12 \text{ T}}$. $\theta = \pm 40°$. $\vec{B}$ is in the xz-plane and is either at $40°$ from the $+x$-direction toward the

$-z$-direction or $40°$ from the $-x$-direction toward the $-z$-direction.

(b) $\vec{B} = B_x \hat{i} + B_z \hat{k}$. $\vec{v} = (3.2 \text{ km/s})(-\hat{j})$.

$\vec{F} = q\vec{v} \times \vec{B} = (-e)(3.2 \text{ km/s})(-\hat{j}) \times (B_x \hat{i} + B_z \hat{k}) = e(3.2 \times 10^3 \text{ m/s})(B_x(-\hat{k}) + B_z \hat{i})$.

$\vec{F} = e(3.2 \times 10^3 \text{ m/s})(-[\pm 1.12 \text{ T}]\hat{k} - [0.938 \text{ T}]\hat{i}) = -(4.80 \times 10^{-16} \text{ N})\hat{i} \pm (5.73 \times 10^{-16} \text{ N})\hat{k}$

$F = \sqrt{F_x^2 + F_z^2} = 7.47 \times 10^{-16} \text{ N}$. $\tan\theta = \dfrac{F_z}{F_x} = \dfrac{\pm 5.73 \times 10^{-16} \text{ N}}{-4.80 \times 10^{-16} \text{ N}}$. $\theta = \pm 50.0°$. The force is in the xz-plane and is

directed at $50.0°$ from the $-x$-axis toward either the $+z$ or $-z$ axis, depending on the sign of B_x.

EVALUATE: If the direction of the force on the first electron were measured, then the sign of B_x would be determined.

27.11.   **IDENTIFY and SET UP:** $\Phi_B = \int \vec{B} \cdot d\vec{A}$

Circular area in the xy-plane, so $A = \pi r^2 = \pi(0.0650 \text{ m})^2 = 0.01327 \text{ m}^2$ and $d\vec{A}$ is in the z-direction. Use Eq.(1.18) to calculate the scalar product.

EXECUTE: **(a)** $\vec{B} = (0.230 \text{ T})\hat{k}$; $\vec{B}$ and $d\vec{A}$ are parallel $(\phi = 0°)$ so $\vec{B} \cdot d\vec{A} = B \, dA$.

B is constant over the circular area so $\Phi_B = \int \vec{B} \cdot d\vec{A} = \int B \, dA = B \int dA = BA = (0.230 \text{ T})(0.01327 \text{ m}^2) = 3.05 \times 10^{-3} \text{ Wb}$

(b) The directions of $\vec{B}$ and $d\vec{A}$ are shown in Figure 27.11a.

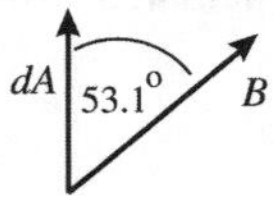

$\vec{B} \cdot d\vec{A} = B\cos\phi \, dA$
with $\phi = 53.1°$

Figure 27.11a

B and ϕ are constant over the circular area so $\Phi_B = \int \vec{B} \cdot d\vec{A} = \int B\cos\phi \, dA = B\cos\phi \int dA = B\cos\phi A$

$\Phi_B = (0.230 \text{ T})\cos 53.1°(0.01327 \text{ m}^2) = 1.83 \times 10^{-3} \text{ Wb}$

(c) The directions of $\vec{B}$ and $d\vec{A}$ are shown in Figure 27.11b.

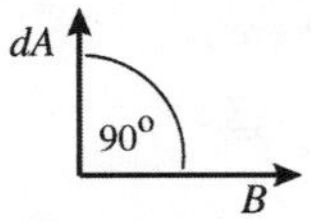

$\vec{B} \cdot d\vec{A} = 0$ since $d\vec{A}$ and $\vec{B}$ are perpendicular $(\phi = 90°)$

$\Phi_B = \int \vec{B} \cdot d\vec{A} = 0$.

Figure 27.11b

EVALUATE: Magnetic flux is a measure of how many magnetic field lines pass through the surface. It is maximum when $\vec{B}$ is perpendicular to the plane of the loop (part a) and is zero when $\vec{B}$ is parallel to the plane of the loop (part c).

27.15.   **(a) IDENTIFY:**   Apply Eq.(27.2) to relate the magnetic force $\vec{F}$ to the directions of $\vec{v}$ and $\vec{B}$. The electron has negative charge so $\vec{F}$ is opposite to the direction of $\vec{v} \times \vec{B}$. For motion in an arc of a circle the acceleration is toward the center of the arc so $\vec{F}$ must be in this direction. $a = v^2/R$.
SET UP:

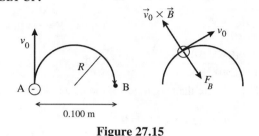

As the electron moves in the semicircle, its velocity is tangent to the circular path. The direction of $\vec{v}_0 \times \vec{B}$ at a point along the path is shown in Figure 27.15.

Figure 27.15

EXECUTE: For circular motion the acceleration of the electron $\vec{a}_{rad}$ is directed in toward the center of the circle. Thus the force $\vec{F}_B$ exerted by the magnetic field, since it is the only force on the electron, must be radially inward. Since q is negative, $\vec{F}_B$ is opposite to the direction given by the right-hand rule for $\vec{v}_0 \times \vec{B}$. Thus $\vec{B}$ is directed into the page. Apply Newton's 2nd law to calculate the magnitude of $\vec{B}$: $\sum \vec{F} = m\vec{a}$ gives $\sum F_{rad} = ma$

$$F_B = m(v^2/R)$$
$$F_B = |q|vB\sin\phi = |q|vB, \text{ so } |q|vB = m(v^2/R)$$
$$B = \frac{mv}{|q|R} = \frac{(9.109 \times 10^{-31} \text{ kg})(1.41 \times 10^6 \text{ m/s})}{(1.602 \times 10^{-19} \text{ C})(0.050 \text{ m})} = 1.60 \times 10^{-4} \text{ T}$$

(b) IDENTIFY and SET UP: The speed of the electron as it moves along the path is constant. ($\vec{F}_B$ changes the direction of $\vec{v}$ but not its magnitude.) The time is given by the distance divided by v_0.

EXECUTE: The distance along the semicircular path is πR, so $t = \dfrac{\pi R}{v_0} = \dfrac{\pi(0.050 \text{ m})}{1.41 \times 10^6 \text{ m/s}} = 1.11 \times 10^{-7} \text{ s}$

EVALUATE: The magnetic field required increases when v increases or R decreases and also depends on the mass to charge ratio of the particle.

27.17.   **IDENTIFY and SET UP:**   Use conservation of energy to find the speed of the ball when it reaches the bottom of the shaft. The right-hand rule gives the direction of $\vec{F}$ and Eq.(27.1) gives its magnitude. The number of excess electrons determines the charge of the ball.
EXECUTE: $q = (4.00 \times 10^8)(-1.602 \times 10^{-19} \text{ C}) = -6.408 \times 10^{-11} \text{ C}$

speed at bottom of shaft: $\frac{1}{2}mv^2 = mgy$; $v = \sqrt{2gy} = 49.5$ m/s

$\vec{v}$ is downward and $\vec{B}$ is west, so $\vec{v} \times \vec{B}$ is north. Since $q < 0$, $\vec{F}$ is south.

$$F = |q|vB\sin\theta = (6.408 \times 10^{-11} \text{ C})(49.5 \text{ m/s})(0.250 \text{ T})\sin 90° = 7.93 \times 10^{-10} \text{ N}$$

EVALUATE: Both the charge and speed of the ball are relatively small so the magnetic force is small, much less than the gravity force of 1.5 N.

27.21.   **(a) IDENTIFY and SET UP:**   Apply Newton's 2nd law, with $a = v^2/R$ since the path of the particle is circular.
EXECUTE: $\sum \vec{F} = m\vec{a}$ says $|q|vB = m(v^2/R)$

$$v = \frac{|q|BR}{m} = \frac{(1.602 \times 10^{-19} \text{ C})(2.50 \text{ T})(6.96 \times 10^{-3} \text{ m})}{3.34 \times 10^{-27} \text{ kg}} = 8.35 \times 10^5 \text{ m/s}$$

(b) IDENTIFY and SET UP: The speed is constant so $t = $ distance/v.

EXECUTE: $t = \dfrac{\pi R}{v} = \dfrac{\pi(6.96 \times 10^{-3} \text{ m})}{8.35 \times 10^5 \text{ m/s}} = 2.62 \times 10^{-8} \text{ s}$

(c) IDENTIFY and SET UP: kinetic energy gained = electric potential energy lost
EXECUTE: $\frac{1}{2}mv^2 = |q|V$

$$V = \frac{mv^2}{2|q|} = \frac{(3.34 \times 10^{-27} \text{ kg})(8.35 \times 10^5 \text{ m/s})^2}{2(1.602 \times 10^{-19} \text{ C})} = 7.27 \times 10^3 \text{ V} = 7.27 \text{ kV}$$

EVALUATE: The deutron has a much larger mass to charge ratio than an electron so a much larger B is required for the same v and R. The deutron has positive charge so gains kinetic energy when it goes from high potential to low potential.

27.25.   **IDENTIFY:**   When a particle of charge $-e$ is accelerated through a potential difference of magnitude V, it gains kinetic energy eV. When it moves in a circular path of radius R, its acceleration is $\dfrac{v^2}{R}$.

SET UP: An electron has charge $q = -e = -1.60\times10^{-19}$ C and mass 9.11×10^{-31} kg.

EXECUTE: $\tfrac{1}{2}mv^2 = eV$ and $v = \sqrt{\dfrac{2eV}{m}} = \sqrt{\dfrac{2(1.60\times10^{-19}\ \text{C})(2.00\times10^3\ \text{V})}{9.11\times10^{-31}\ \text{kg}}} = 2.65\times10^7$ m/s. $\vec{F} = m\vec{a}$ gives

$|q|vB\sin\phi = m\dfrac{v^2}{R}$. $\phi = 90°$ and $B = \dfrac{mv}{|q|R} = \dfrac{(9.11\times10^{-31}\ \text{kg})(2.65\times10^7\ \text{m/s})}{(1.60\times10^{-19}\ \text{C})(0.180\ \text{m})} = 8.38\times10^{-4}$ T.

EVALUATE: The smaller the radius of the circular path, the larger the magnitude of the magnetic field that is required.

27.27.   **(a) IDENTIFY** and **SET UP:**   Eq.(27.4) gives the total force on the proton. At $t = 0$,

$\vec{F} = q\vec{v}\times\vec{B} = q\left(v_x\hat{i} + v_z\hat{k}\right)\times B_x\hat{i} = qv_zB_x\hat{j}$. $\vec{F} = \left(1.60\times10^{-19}\ \text{C}\right)\left(2.00\times10^5\ \text{m/s}\right)\left(0.500\ \text{T}\right)\hat{j} = \left(1.60\times10^{-14}\ \text{N}\right)\hat{j}$.

(b) Yes. The electric field exerts a force in the direction of the electric field, since the charge of the proton is positive and there is a component of acceleration in this direction.

(c) EXECUTE: In the plane perpendicular to $\vec{B}$ (the yz-plane) the motion is circular. But there is a velocity component in the direction of $\vec{B}$, so the motion is a helix. The electric field in the $+\hat{i}$ direction exerts a force in the $+\hat{i}$ direction. This force produces an acceleration in the $+\hat{i}$ direction and this causes the pitch of the helix to vary. The force does not affect the circular motion in the yz-plane, so the electric field does not affect the radius of the helix.

(d) IDENTIFY and **SET UP:**   Eq.(27.12) and $T = 2\pi/\omega$ to calculate the period of the motion. Calculate a_x produced by the electric force and use a constant acceleration equation to calculate the displacement in the x-direction in time $T/2$.

EXECUTE: Calculate the period T: $\omega = |q|B/m$

$T = \dfrac{2\pi}{\omega} = \dfrac{2\pi m}{|q|B} = \dfrac{2\pi\left(1.67\times10^{-27}\ \text{kg}\right)}{\left(1.60\times10^{-19}\ \text{C}\right)\left(0.500\ \text{T}\right)} = 1.312\times10^{-7}$ s. Then $t = T/2 = 6.56\times10^{-8}$ s. $v_{0x} = 1.50\times10^5$ m/s

$a_x = \dfrac{F_x}{m} = \dfrac{\left(1.60\times10^{-19}\ \text{C}\right)\left(2.00\times10^4\ \text{V/m}\right)}{1.67\times10^{-27}\ \text{kg}} = +1.916\times10^{12}$ m/s^2

$x - x_0 = v_{0x}t + \tfrac{1}{2}a_x t^2$

$x - x_0 = \left(1.50\times10^5\ \text{m/s}\right)\left(6.56\times10^{-8}\ \text{s}\right) + \tfrac{1}{2}\left(1.916\times10^{12}\ \text{m/s}^2\right)\left(6.56\times10^{-8}\ \text{s}\right)^2 = 1.40$ cm

EVALUATE: The electric and magnetic fields are in the same direction but produce forces that are in perpendicular directions to each other.

27.29.   **IDENTIFY:**   For the alpha particles to emerge from the plates undeflected, the magnetic force on them must exactly cancel the electric force. The battery produces an electric field between the plates, which acts on the alpha particles.

SET UP: First use energy conservation to find the speed of the alpha particles as they enter the plates: $qV = 1/2\ mv^2$. The electric field between the plates due to the battery is $E = V_b/d$. For the alpha particles not to be deflected, the magnetic force must cancel the electric force, so $qvB = qE$, giving $B = E/v$.

EXECUTE: Solve for the speed of the alpha particles just as they enter the region between the plates. Their charge is $2e$.

$$v_\alpha = \sqrt{\dfrac{2(2e)V}{m}} = \sqrt{\dfrac{4\left(1.60\times10^{-19}\ \text{C}\right)(1750\,\text{V})}{6.64\times10^{-27}\ \text{kg}}} = 4.11\times10^5\ \text{m/s}$$

The electric field between the plates, produced by the battery, is

$$E = V_b/d = (150\ \text{V})/(0.00820\ \text{m}) = 18{,}300\ \text{V}$$

The magnetic force must cancel the electric force:

$$B = E/v_\alpha = (18{,}300\ \text{V})/(4.11\times10^5\ \text{m/s}) = 0.0445\ \text{T}$$

The magnetic field is perpendicular to the electric field. If the charges are moving to the right and the electric field points upward, the magnetic field is out of the page.

EVALUATE: The sign of the charge of the alpha particle does not enter the problem, so negative charges of the same magnitude would also not be deflected.

27.35.   **IDENTIFY:**   Apply $F = IlB\sin\phi$.

SET UP: Label the three segments in the field as a, b, and c. Let x be the length of segment a. Segment b has length 0.300 m and segment c has length $0.600 \text{ cm} - x$. Figure 27.35a shows the direction of the force on each segment. For each segment, $\phi = 90°$. The total force on the wire is the vector sum of the forces on each segment.

EXECUTE: $F_a = IlB = (4.50 \text{ A})x(0.240 \text{ T})$. $F_c = (4.50 \text{ A})(0.600 \text{ m} - x)(0.240 \text{ T})$. Since $\vec{F}_a$ and $\vec{F}_c$ are in the same direction their vector sum has magnitude $F_{ac} = F_a + F_c = (4.50 \text{ A})(0.600 \text{ m})(0.240 \text{ T}) = 0.648 \text{ N}$ and is directed toward the bottom of the page in Figure 27.35a. $F_b = (4.50 \text{ A})(0.300 \text{ m})(0.240 \text{ T}) = 0.324 \text{ N}$ and is directed to the right. The vector addition diagram for $\vec{F}_{ac}$ and $\vec{F}_b$ is given in Figure 27.35b.

$F = \sqrt{F_{ac}^2 + F_b^2} = \sqrt{(0.648 \text{ N})^2 + (0.324 \text{ N})^2} = 0.724 \text{ N}$. $\tan\theta = \dfrac{F_{ac}}{F_b} = \dfrac{0.648 \text{ N}}{0.324 \text{ N}}$ and $\theta = 63.4°$. The net force has magnitude 0.724 N and its direction is specified by $\theta = 63.4°$ in Figure 27.35b.

EVALUATE: All three current segments are perpendicular to the magnetic field, so $\phi = 90°$ for each in the force equation. The direction of the force on a segment depends on the direction of the current for that segment.

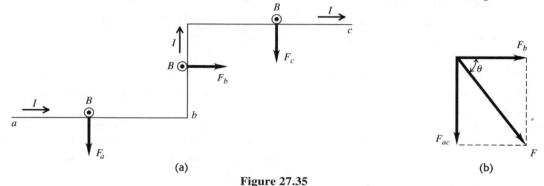

Figure 27.35

27.39.   **IDENTIFY** and **SET UP:**   The magnetic force is given by Eq.(27.19). $F_I = mg$ when the bar is just ready to levitate. When I becomes larger, $F_I > mg$ and $F_I - mg$ is the net force that accelerates the bar upward. Use Newton's 2nd law to find the acceleration.

(a) EXECUTE: $IlB = mg$, $I = \dfrac{mg}{lB} = \dfrac{(0.750 \text{ kg})(9.80 \text{ m/s}^2)}{(0.500 \text{ m})(0.450 \text{ T})} = 32.67 \text{ A}$

$\mathcal{E} = IR = (32.67 \text{ A})(25.0 \ \Omega) = 817 \text{ V}$

(b) $R = 2.0 \ \Omega$, $I = \mathcal{E}/R = (816.7 \text{ V})/(2.0 \ \Omega) = 408 \text{ A}$

$F_I = IlB = 92 \text{ N}$

$a = (F_I - mg)/m = 113 \text{ m/s}^2$

EVALUATE: I increases by over an order of magnitude when R changes to $F_I >> mg$ and a is an order of magnitude larger than g.

27.41.   **IDENTIFY:**   Apply $\vec{F} = I\vec{l} \times \vec{B}$ to each segment of the conductor: the straight section parallel to the x axis, the semicircular section and the straight section that is perpendicular to the plane of the figure in Example 27.8.

SET UP: $\vec{B} = B_x\hat{i}$. The force is zero when the current is along the direction of $\vec{B}$.

EXECUTE: **(a)** The force on the straight section along the $-x$-axis is zero. For the half of the semicircle at negative x the force is out of the page. For the half of the semicircle at positive x the force is into the page. The net force on the semicircular section is zero. The force on the straight section that is perpendicular to the plane of the figure is in the $-y$-direction and has magnitude $F = ILB$. The total magnetic force on the conductor is ILB, in the $-y$-direction.

EVALUATE: **(b)** If the semicircular section is replaced by a straight section along the x-axis, then the magnetic force on that straight section would be zero, the same as it is for the semicircle.

27.45.   **IDENTIFY:**   The magnetic field exerts a torque on the current-carrying coil, which causes it to turn. We can use the rotational form of Newton's second law to find the angular acceleration of the coil.

SET UP: The magnetic torque is given by $\vec{\tau} = \vec{\mu} \times \vec{B}$, and the rotational form of Newton's second law is $\sum\tau = I\alpha$. The magnetic field is parallel to the plane of the loop.

EXECUTE: **(a)** The coil rotates about axis A_2 because the only torque is along top and bottom sides of the coil.

(b) To find the moment of inertia of the coil, treat the two 1.00-m segments as point-masses (since all the points in them are 0.250 m from the rotation axis) and the two 0.500-m segments as thin uniform bars rotated about their centers. Since the coil is uniform, the mass of each segment is proportional to its fraction of the total perimeter of the coil. Each 1.00-m segment is 1/3 of the total perimeter, so its mass is $(1/3)(210 \text{ g}) = 70 \text{ g} = 0.070 \text{ kg}$. The mass of each 0.500-m segment is half this amount, or 0.035 kg. The result is

$$I = 2(0.070 \text{ kg})(0.250 \text{ m})^2 + 2\tfrac{1}{12}(0.035 \text{ kg})(0.500 \text{ m})^2 = 0.0102 \text{ kg} \cdot \text{m}^2$$

The torque is

$$|\vec{\tau}| = |\vec{\mu} \times \vec{B}| = IAB \sin 90° = (2.00 \text{ A})(0.500 \text{ m})(1.00 \text{ m})(3.00 \text{ T}) = 3.00 \text{ N} \cdot \text{m}$$

Using the above values, the rotational form of Newton's second law gives

$$\alpha = \frac{\tau}{I} = 290 \text{ rad/s}^2$$

EVALUATE: This angular acceleration will not continue because the torque changes as the coil turns.

27.47.   **IDENTIFY and SET UP:**   The potential energy is given by Eq.(27.27): $U = \vec{\mu} \cdot \vec{B}$. The scalar product depends on the angle between $\vec{\mu}$ and $\vec{B}$.

EXECUTE: For $\vec{\mu}$ and $\vec{B}$ parallel, $\phi = 0°$ and $\vec{\mu} \cdot \vec{B} = \mu B \cos\phi = \mu B$. For $\vec{\mu}$ and $\vec{B}$ antiparallel, $\phi = 180°$ and $\vec{\mu} \cdot \vec{B} = \mu B \cos\phi = -\mu B$.

$U_1 = +\mu B,\ U_2 = -\mu B$

$\Delta U = U_2 - U_1 = -2\mu B = -2(1.45 \text{ A} \cdot \text{m}^2)(0.835 \text{ T}) = -2.42 \text{ J}$

EVALUATE: U is maximum when $\vec{\mu}$ and $\vec{B}$ are antiparallel and minimum when they are parallel. When the coil is rotated as specified its magnetic potential energy decreases.

27.49.   **IDENTIFY:**   The circuit consists of two parallel branches with the potential difference of 120 V applied across each. One branch is the rotor, represented by a resistance R_r and an induced emf that opposes the applied potential. Apply the loop rule to each parallel branch and use the junction rule to relate the currents through the field coil and through the rotor to the 4.82 A supplied to the motor.

SET UP: The circuit is sketched in Figure 27.49.

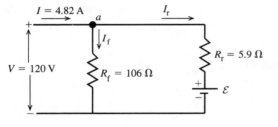

$\mathcal{E}$ is the induced emf developed by the motor. It is directed so as to oppose the current through the rotor.

Figure 27.49

EXECUTE: **(a)** The field coils and the rotor are in parallel with the applied potential difference V, so $V = I_f R_f$.

$I_f = \dfrac{V}{R_f} = \dfrac{120 \text{ V}}{106 \text{ }\Omega} = 1.13 \text{ A}.$

(b) Applying the junction rule to point a in the circuit diagram gives $I - I_f - I_r = 0$.

$I_r = I - I_f = 4.82 \text{ A} - 1.13 \text{ A} = 3.69 \text{ A}.$

(c) The potential drop across the rotor, $I_r R_r + \mathcal{E}$, must equal the applied potential difference $V: V = I_r R_r + \mathcal{E}$

$\mathcal{E} = V - I_r R_r = 120 \text{ V} - (3.69 \text{ A})(5.9 \text{ }\Omega) = 98.2 \text{ V}$

(d) The mechanical power output is the electrical power input minus the rate of dissipation of electrical energy in the resistance of the motor:

electrical power input to the motor

$P_{in} = IV = (4.82 \text{ A})(120 \text{ V}) = 578 \text{ W}$

electrical power loss in the two resistances

$P_{loss} = I_f^2 R_f + I_f^2 R = (1.13 \text{ A})^2 (106 \text{ }\Omega) + (3.69 \text{ A})^2 (5.9 \text{ }\Omega) = 216 \text{ W}$

mechanical power output

$P_{out} = P_{in} - P_{loss} = 578 \text{ W} - 216 \text{ W} = 362 \text{ W}$

The mechanical power output is the power associated with the induced emf $\mathcal{E}$

$P_{out} = P_{\mathcal{E}} = \mathcal{E} I_r = (98.2 \text{ V})(3.69 \text{ A}) = 362 \text{ W}$, which agrees with the above calculation.

EVALUATE: The induced emf reduces the amount of current that flows through the rotor. This motor differs from the one described in Example 27.12. In that example the rotor and field coils are connected in series and in this problem they are in parallel.

27.51.   **IDENTIFY:**   The drift velocity is related to the current density by Eq.(25.4). The electric field is determined by the requirement that the electric and magnetic forces on the current-carrying charges are equal in magnitude and opposite in direction.

(a) SET UP: The section of the silver ribbon is sketched in Figure 27.51a.

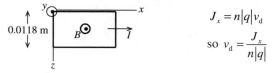

$$J_x = n|q|v_d$$

$$\text{so } v_d = \frac{J_x}{n|q|}$$

Figure 27.51a

EXECUTE: $J_x = \dfrac{I}{A} = \dfrac{I}{y_1 z_1} = \dfrac{120 \text{ A}}{(0.23\times 10^{-3} \text{ m})(0.0118 \text{ m})} = 4.42\times 10^7 \text{ A/m}^2$

$v_d = \dfrac{J_x}{n|q|} = \dfrac{4.42\times 10^7 \text{ A/m}^2}{\left(5.85\times 10^{28}/\text{m}^3\right)\left(1.602\times 10^{-19} \text{ C}\right)} = 4.7\times 10^{-3} \text{ m/s} = 4.7 \text{ mm/s}$

(b) magnitude of $\vec{E}$

$|q|E_z = |q|v_d B_y$

$E_z = v_d B_y = (4.7\times 10^{-3} \text{ m/s})(0.95 \text{ T}) = 4.5\times 10^{-3} \text{ V/m}$

direction of $\vec{E}$

The drift velocity of the electrons is in the opposite direction to the current, as shown in Figure 27.51b.

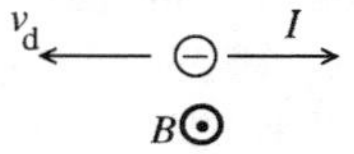

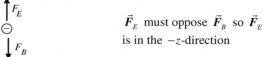

$\vec{v}\times\vec{B}\uparrow$

$\vec{F}_B = q\vec{v}\times\vec{B} = -e\vec{v}\times\vec{B}\downarrow$

Figure 27.51b

The directions of the electric and magnetic forces on an electron in the ribbon are shown in Figure 27.51c.

$\vec{F}_E$ must oppose $\vec{F}_B$ so $\vec{F}_E$
is in the $-z$-direction

Figure 27.51c

$\vec{F}_E = q\vec{E} = -e\vec{E}$ so $\vec{E}$ is opposite to the direction of $\vec{F}_E$ and thus $\vec{E}$ is in the $+z$-direction.

(c) The Hall emf is the potential difference between the two edges of the strip (at $z = 0$ and $z = z_1$) that results from the electric field calculated in part (b). $\mathcal{E}_{\text{Hall}} = Ez_1 = (4.5\times 10^{-3} \text{ V/m})(0.0118 \text{ m}) = 53 \ \mu\text{V}$

EVALUATE: Even though the current is quite large the Hall emf is very small. Our calculated Hall emf is more than an order of magnitude larger than in Example 27.13. In this problem the magnetic field and current density are larger than in the example, and this leads to a larger Hall emf.

27.53.   **(a) IDENTIFY:**   Use Eq.(27.2) to relate $\vec{v}$, $\vec{B}$, and $\vec{F}$.

SET UP: The directions of $\vec{v}_1$ and $\vec{F}_1$ are shown in Figure 27.53a.

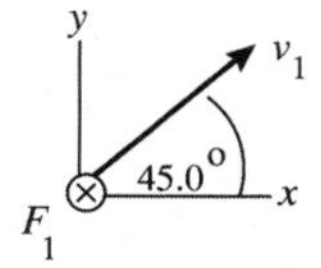

$\vec{F} = q\vec{v}\times\vec{B}$ says that $\vec{F}$ is perpendicular to $\vec{v}$ and $\vec{B}$. The information given here means that $\vec{B}$ can have no z-component.

Figure 27.53a

The directions of $\vec{v}_2$ and $\vec{F}_2$ are shown in Figure 27.53b.

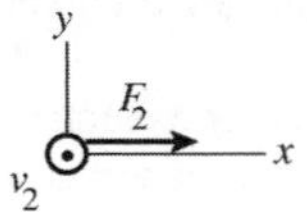

$\vec{F}$ is perpendicular to $\vec{v}$ and $\vec{B}$, so $\vec{B}$ can have no x-component.

Figure 27.53b

Both pieces of information taken together say that $\vec{B}$ is in the y-direction; $\vec{B} = B_y\hat{\boldsymbol{j}}$.

EXECUTE: Use the information given about $\vec{F}_2$ to calculate F_y: $\vec{F}_2 = F_2\hat{\boldsymbol{i}}$, $\vec{v}_2 = v_2\hat{\boldsymbol{k}}$, $\vec{B} = B_y\hat{\boldsymbol{j}}$.

$\vec{F}_2 = q\vec{v}_2 \times \vec{B}$ says $F_2\hat{\boldsymbol{i}} = qv_2B_y\hat{\boldsymbol{k}} \times \hat{\boldsymbol{j}} = qv_2B_y(-\hat{\boldsymbol{i}})$ and $F_2 = -qv_2B_y$

$B_y = -F_2/(qv_2) = -F_2/(qv_1)$. $\vec{B}$ has the magnitude $F_2/(qv_1)$ and is in the $-y$-direction.

(b) $F_1 = qvB\sin\phi = qv_1|B_y|/\sqrt{2} = F_2/\sqrt{2}$

EVALUATE: $v_1 = v_2$. $\vec{v}_2$ is perpendicular to $\vec{B}$ whereas only the component of $\vec{v}_1$ perpendicular to $\vec{B}$ contributes to the force, so it is expected that $F_2 > F_1$, as we found.

27.55. **IDENTIFY:**   The sum of the magnetic, electrical, and gravitational forces must be zero to aim at and hit the target.
SET UP: The magnetic field must point to the left when viewed in the direction of the target for no net force. The net force is zero, so $\sum F = F_B - F_E - mg = 0$ and $qvB - qE - mg = 0$.
EXECUTE: Solving for B gives

$$B = \frac{qE + mg}{qv} = \frac{(2500\times10^{-6}\,\text{C})(27.5\,\text{N/C}) + (0.0050\,\text{kg})(9.80\,\text{m/s}^2)}{(2500\times10^{-6}\,\text{C})(12.8\,\text{m/s})} = 3.7\,\text{T}$$

The direction should be perpendicular to the initial velocity of the coin.
EVALUATE: This is a very strong magnetic field, but achievable in some labs.

27.57. **(a) IDENTIFY** and **SET UP:**   The maximum radius of the orbit determines the maximum speed v of the protons. Use Newton's 2nd law and $a_c = v^2/R$ for circular motion to relate the variables. The energy of the particle is the kinetic energy $K = \frac{1}{2}mv^2$.

EXECUTE: $\sum \vec{F} = m\vec{a}$ gives $|q|vB = m(v^2/R)$

$v = \dfrac{|q|BR}{m} = \dfrac{(1.60\times10^{-19}\,\text{C})(0.85\,\text{T})(0.40\,\text{m})}{1.67\times10^{-27}\,\text{kg}} = 3.257\times10^7\,\text{m/s}$. The kinetic energy of a proton moving with this

speed is $K = \frac{1}{2}mv^2 = \frac{1}{2}(1.67\times10^{-27}\,\text{kg})(3.257\times10^7\,\text{m/s})^2 = 8.9\times10^{-13}\,\text{J} = 5.6\,\text{MeV}$

(b) The time for one revolution is the period $T = \dfrac{2\pi R}{v} = \dfrac{2\pi(0.40\,\text{m})}{3.257\times10^7\,\text{m/s}} = 7.7\times10^{-8}\,\text{s}$

(c) $K = \frac{1}{2}mv^2 = \frac{1}{2}m\left(\dfrac{|q|BR}{m}\right)^2 = \frac{1}{2}\dfrac{|q|^2 B^2 R^2}{m}$. Or, $B = \dfrac{\sqrt{2Km}}{|q|R}$. B is proportional to $\sqrt{K}$, so if K is increased by a

factor of 2 then B must be increased by a factor of $\sqrt{2}$. $B = \sqrt{2}(0.85\,\text{T}) = 1.2\,\text{T}$.

(d) $v = \dfrac{|q|BR}{m} = \dfrac{(3.20\times10^{-19}\,\text{C})(0.85\,\text{T})(0.40\,\text{m})}{6.65\times10^{-27}\,\text{kg}} = 1.636\times10^7\,\text{m/s}$

$K = \frac{1}{2}mv^2 = \frac{1}{2}(6.65\times10^{-27}\,\text{kg})(1.636\times10^7\,\text{m/s})^2 = 8.9\times10^{-13}\,\text{J} = 5.5\,\text{MeV}$, the same as the maximum energy for protons.

EVALUATE: We can see that the maximum energy must be approximately the same as follows: From part (c),

$K = \frac{1}{2}m\left(\dfrac{|q|BR}{m}\right)^2$. For alpha particles $|q|$ is larger by a factor of 2 and m is larger by a factor of 4 (approximately).

Thus $|q|^2/m$ is unchanged and K is the same.

27.61. **IDENTIFY** and **SET UP:**   Use Eq.(27.2) to relate $q, \vec{v}, \vec{B}$ and $\vec{F}$. The force $\vec{F}$ and $\vec{a}$ are related by Newton's 2nd law.
$\vec{B} = -(0.120\,\text{T})\hat{\boldsymbol{k}}, \vec{v} = (1.05\times10^6\,\text{m/s})(-3\hat{\boldsymbol{i}} + 4\hat{\boldsymbol{j}} + 12\hat{\boldsymbol{k}})$, $F = 1.25\,\text{N}$
(a) EXECUTE: $\vec{F} = q\vec{v} \times \vec{B}$
$\vec{F} = q(-0.120\,\text{T})(1.05\times10^6\,\text{m/s})(-3\hat{\boldsymbol{i}} \times \hat{\boldsymbol{k}} + 4\hat{\boldsymbol{j}} \times \hat{\boldsymbol{k}} + 12\hat{\boldsymbol{k}} \times \hat{\boldsymbol{k}})$
$\hat{\boldsymbol{i}} \times \hat{\boldsymbol{k}} = -\hat{\boldsymbol{j}}, \hat{\boldsymbol{j}} \times \hat{\boldsymbol{k}} = \hat{\boldsymbol{i}}, \hat{\boldsymbol{k}} \times \hat{\boldsymbol{k}} = 0$
$\vec{F} = -q(1.26\times10^5\,\text{N/C})(+3\hat{\boldsymbol{j}} + 4\hat{\boldsymbol{i}}) = -q(1.26\times10^5\,\text{N/C})(+4\hat{\boldsymbol{i}} + 3\hat{\boldsymbol{j}})$

The magnitude of the vector $+4\hat{i}+3\hat{j}$ is $\sqrt{3^2+4^2}=5$. Thus $F=-q(1.26\times10^5 \text{ N/C})(5)$.

$$q=-\frac{F}{5(1.26\times10^5 \text{ N/C})}=-\frac{1.25 \text{ N}}{5(1.26\times10^5 \text{ N/C})}=-1.98\times10^{-6} \text{ C}$$

(b) $\sum\vec{F}=m\vec{a}$ so $\vec{a}=\vec{F}/m$

$$\vec{F}=-q(1.26\times10^5 \text{ N/C})(+4\hat{i}+3\hat{j})=-(-1.98\times10^{-6} \text{ C})(1.26\times10^5 \text{ N/C})(+4\hat{i}+3\hat{j})=+0.250 \text{ N}(+4\hat{i}+3\hat{j})$$

Then $\vec{a}=\vec{F}/m=\left(\dfrac{0.250 \text{ N}}{2.58\times10^{-15} \text{ kg}}\right)(+4\hat{i}+3\hat{j})=(9.69\times10^{13} \text{ m/s}^2)(+4\hat{i}+3\hat{j})$

(c) IDENTIFY and **SET UP:** $\vec{F}$ is in the xy-plane, so in the z-direction the particle moves with constant speed 12.6×10^6 m/s. In the xy-plane the force $\vec{F}$ causes the particle to move in a circle, with $\vec{F}$ directed in towards the center of the circle.

EXECUTE: $\sum\vec{F}=m\vec{a}$ gives $F=m(v^2/R)$ and $R=mv^2/F$

$$v^2=v_x^2+v_y^2=(-3.15\times10^6 \text{ m/s})^2+(+4.20\times10^6 \text{ m/s})^2=2.756\times10^{13} \text{ m}^2/\text{s}^2$$

$$F=\sqrt{F_x^2+F_y^2}=(0.250 \text{ N})\sqrt{4^2+3^2}=1.25 \text{ N}$$

$$R=\frac{mv^2}{F}=\frac{(2.58\times10^{-15} \text{ kg})(2.756\times10^{13} \text{ m}^2/\text{s}^2)}{1.25 \text{ N}}=0.0569 \text{ m}=5.69 \text{ cm}$$

(d) IDENTIFY and **SET UP:** By Eq.(27.12) the cyclotron frequency is $f=\omega/2\pi=v/2\pi R$.

EXECUTE: The circular motion is in the xy-plane, so $v=\sqrt{v_x^2+v_y^2}=5.25\times10^6$ m/s.

$$f=\frac{v}{2\pi R}=\frac{5.25\times10^6 \text{ m/s}}{2\pi(0.0569 \text{ m})}=1.47\times10^7 \text{ Hz, and } \omega=2\pi f=9.23\times10^7 \text{ rad/s}$$

(e) IDENTIFY and **SET UP** Compare t to the period T of the circular motion in the xy-plane to find the x and y coordinates at this t. In the z-direction the particle moves with constant speed, so $z=z_0+v_z t$.

EXECUTE: The period of the motion in the xy-plane is given by $T=\dfrac{1}{f}=\dfrac{1}{1.47\times10^7 \text{ Hz}}=6.80\times10^{-8}$ s

In $t=2T$ the particle has returned to the same x and y coordinates. The z-component of the motion is motion with a constant velocity of $v_z=+12.6\times10^6$ m/s. Thus $z=z_0+v_z t=0+(12.6\times10^6 \text{ m/s})(2)(6.80\times10^{-8} \text{ s})=+1.71$ m. The coordinates at $t=2T$ are $x=R, y=0, z=+1.71$ m.

EVALUATE: The circular motion is in the plane perpendicular to $\vec{B}$. The radius of this motion gets smaller when B increases and it gets larger when v increases. There is no magnetic force in the direction of $\vec{B}$ so the particle moves with constant velocity in that direction. The superposition of circular motion in the xy-plane and constant speed motion in the z-direction is a helical path.

27.65. **IDENTIFY:** For the velocity selector, $E=vB$. For the circular motion in the field B', $R=\dfrac{mv}{|q|B'}$.

SET UP: $B=B'=0.701$ T.

EXECUTE: $v=\dfrac{E}{B}=\dfrac{1.88\times10^4 \text{ N/C}}{0.701 \text{ T}}=2.68\times10^4$ m/s. $R=\dfrac{mv}{qB'}$, so

$$R_{82}=\frac{82(1.66\times10^{-27} \text{ kg})(2.68\times10^4 \text{ m/s})}{(1.60\times10^{-19} \text{ C})(0.701 \text{ T})}=0.0325 \text{ m.}$$

$$R_{84}=\frac{84(1.66\times10^{-27} \text{ kg})(2.68\times10^4 \text{ m/s})}{(1.60\times10^{-19} \text{ C})(0.701 \text{ T})}=0.0333 \text{ m.}$$

$$R_{86}=\frac{86(1.66\times10^{-27} \text{ kg})(2.68\times10^4 \text{ m/s})}{(1.60\times10^{-19} \text{ C})(0.701 \text{ T})}=0.0341 \text{ m.}$$

The distance between two adjacent lines is $\Delta R=1.6$ mm.

EVALUATE: The distance between the ^{82}Kr line and the ^{84}Kr line is 1.6 mm and the distance between the ^{84}Kr line and the ^{86}Kr line is 1.6 mm. Adjacent lines are equally spaced since the ^{82}Kr versus ^{84}Kr and ^{84}Kr versus ^{86}Kr mass differences are the same.

27.67.   IDENTIFY: The force exerted by the magnetic field is given by Eq.(27.19). The net force on the wire must be zero.
SET UP: For the wire to remain at rest the force exerted on it by the magnetic field must have a component directed up the incline. To produce a force in this direction, the current in the wire must be directed from right to left in Figure 27.61 in the textbook. Or, viewing the wire from its left-hand end the directions are shown in Figure 27.67a.

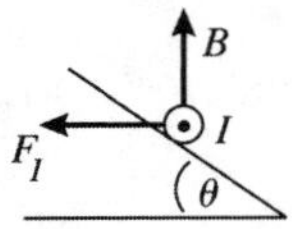

Figure 27.67a

The free-body diagram for the wire is given in Figure 27.67b.

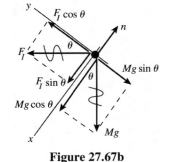

Figure 27.67b

EXECUTE: $\sum F_y = 0$

$F_I \cos\theta - Mg\sin\theta = 0$

$F_I = ILB\sin\phi$

$\phi = 90°$ since $\vec{B}$ is perpendicular to the current direction.

Thus $(ILB)\cos\theta - Mg\sin\theta = 0$ and $I = \dfrac{Mg\tan\theta}{LB}$

EVALUATE: The magnetic and gravitational forces are in perpendicular directions so their components parallel to the incline involve different trig functions. As the tilt angle θ increases there is a larger component of Mg down the incline and the component of F_I up the incline is smaller; I must increase with θ to compensate. As $\theta \to 0$, $I \to 0$ and as $\theta \to 90°, I \to \infty$.

27.73.   IDENTIFY: Apply $F = IlB\sin\phi$ to calculate the force on each segment of the wire that is in the magnetic field. The net force is the vector sum of the forces on each segment.
SET UP: The direction of the magnetic force on each current segment in the field is shown in Figure 27.73. By symmetry, $F_a = F_b$. $\vec{F}_a$ and $\vec{F}_b$ are in opposite directions so their vector sum is zero. The net force equals F_c. For F_c, $\phi = 90°$ and $l = 0.450$ m.
EXECUTE: $F_c = IlB = (6.00 \text{ A})(0.450 \text{ m})(0.666 \text{ T}) = 1.80$ N . The net force is 1.80 N, directed to the left.
EVALUATE: The shape of the region of uniform field doesn't matter, as long as all of segment c is in the field and as long as the lengths of the portions of segments a and b that are in the field are the same.

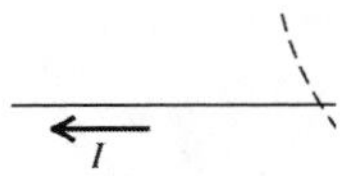

Figure 27.73

27.75.   IDENTIFY: For the loop to be in equilibrium the net torque on it must be zero. Use Eq.(27.26) to calculate the torque due to the magnetic field and use Eq.(10.3) for the torque due to the gravity force.
SET UP: See Figure 27.75a.

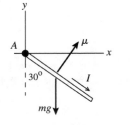

Use $\sum\tau_A = 0$, where point A is at the origin.

Figure 27.75a

EXECUTE: See Figure 27.75b.

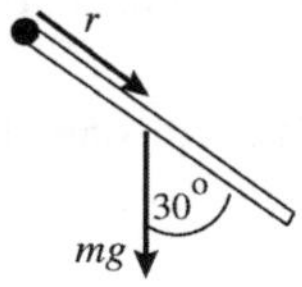

Figure 27.75b

$\tau_{mg} = mgr \sin\phi = mg(0.400 \text{ m})\sin 30.0°$

The torque is clockwise; $\vec{\tau}_{mg}$ is directed into the paper.

For the loop to be in equilibrium the torque due to $\vec{B}$ must be counterclockwise (opposite to $\vec{\tau}_{mg}$) and it must be that $\tau_B = \tau_{mg}$. See Figure 27.75c.

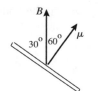

Figure 27.75c

$\vec{\tau}_B = \vec{\mu}\times\vec{B}$. For this torque to be counterclockwise ($\vec{\tau}_B$ directed out of the paper), $\vec{B}$ must be in the $+y$-direction.

$\tau_B = \mu B\sin\phi = IAB\sin 60.0°$

$\tau_B = \tau_{mg}$ gives $IAB\sin 60.0° = mg(0.0400 \text{ m})\sin 30.0°$

$m = (0.15 \text{ g/cm})2(8.00 \text{ cm} + 6.00 \text{ cm}) = 4.2 \text{ g} = 4.2\times 10^{-3} \text{ kg}$

$A = (0.800 \text{ m})(0.0600 \text{ m}) = 4.80\times 10^{-3} \text{ m}^2$

$B = \dfrac{mg(0.0400 \text{ m})(\sin 30.0°)}{IA\sin 60.0°}$

$B = \dfrac{(4.2\times 10^{-3} \text{ kg})(9.80 \text{ m/s}^2)(0.0400 \text{ m})\sin 30.0°}{(8.2 \text{ A})(4.80\times 10^{-3} \text{ m}^2)\sin 60.0°} = 0.024 \text{ T}$

EVALUATE: As the loop swings up the torque due to $\vec{B}$ decreases to zero and the torque due to mg increases from zero, so there must be an orientation of the loop where the net torque is zero.

27.77.   **IDENTIFY:**   Use Eq.(27.20) to calculate the force and then the torque on each small section of the rod and integrate to find the total magnetic torque. At equilibrium the torques from the spring force and from the magnetic force cancel. The spring force depends on the amount x the spring is stretched and then $U = \frac{1}{2}kx^2$ gives the energy stored in the spring.

(a) SET UP:

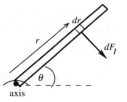

Divide the rod into infinitesimal sections of length dr, as shown in Figure 27.77.

Figure 27.77

EXECUTE: The magnetic force on this section is $dF_I = IBdr$ and is perpendicular to the rod. The torque $d\tau$ due to the force on this section is $d\tau = rdF_I = IBr\,dr$. The total torque is $\int d\tau = IB\int_0^l r\,dr = \frac{1}{2}Il^2B = 0.0442 \text{ N}\cdot\text{m}$, clockwise.

(b) SET UP: F_I produces a clockwise torque so the spring force must produce a counterclockwise torque. The spring force must be to the left; the spring is stretched.

EXECUTE: Find x, the amount the spring is stretched:

$\sum\tau = 0$, axis at hinge, counterclockwise torques positive

$(kx)l\sin 53° - \frac{1}{2}Il^2B = 0$

$x = \dfrac{IlB}{2k\sin 53.0°} = \dfrac{(6.50 \text{ A})(0.200 \text{ m})(0.340 \text{ T})}{2(4.80 \text{ N/m})\sin 53.0°} = 0.05765 \text{ m}$

$U = \frac{1}{2}kx^2 = 7.98\times 10^{-3} \text{ J}$

EVALUATE: The magnetic torque calculated in part (a) is the same torque calculated from a force diagram in which the total magnetic force $F_I = IlB$ acts at the center of the rod. We didn't include a gravity torque since the problem said the rod had negligible mass.

27.79.   **IDENTIFY:**   Use Eq.(27.20) to calculate the force on a short segment of the coil and integrate over the entire coil to find the total force.

SET UP: See Figure 27.79a.

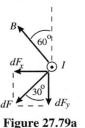

Figure 27.79a

Consider the force $d\vec{F}$ on a short segment dl at the left-hand side of the coil, as viewed in Figure 27.69 in the textbook. The current at this point is directed out of the page. $d\vec{F}$ is perpendicular both to $\vec{B}$ and to the direction of I.

See Figure 27.79b.

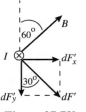

Figure 27.79b

Consider also the force $d\vec{F}'$ on a short segment on the opposite side of the coil, at the right-hand side of the coil in Figure 27.69 in the textbook. The current at this point is directed into the page.

The two sketches show that the x-components cancel and that the y-components add. This is true for all pairs of short segments on opposite sides of the coil. The net magnetic force on the coil is in the y-direction and its magnitude is given by $F = \int dF_y$.

EXECUTE: $dF = Idl\,B\sin\phi$. But $\vec{B}$ is perpendicular to the current direction so $\phi = 90°$.

$dF_y = dF\cos 30.0 = IB\cos 30.0°\,dl$

$F = \int dF_y = IB\cos 30.0° \int dl$

But $\int dl = N(2\pi r)$, the total length of wire in the coil.

$F = IB\cos 30.0° N(2\pi r) = (0.950\text{ A})(0.200\text{ T})(\cos 30.0°)(50)2\pi(0.0078\text{ m}) = 0.444\text{ N}$ and $\vec{F} = -(0.444\text{ N})\hat{j}$

EVALUATE: The magnetic field makes a constant angle with the plane of the coil but has a different direction at different points around the circumference of the coil so is not uniform. The net force is proportional to the magnitude of the current and reverses direction when the current reverses direction.

27.81.   **IDENTIFY:**   Apply $d\vec{F} = Id\vec{l} \times \vec{B}$ to each side of the loop.

SET UP: For each side of the loop, $d\vec{l}$ is parallel to that side of the loop and is in the direction of I. Since the loop is in the xy-plane, $z = 0$ at the loop and $B_y = 0$ at the loop.

EXECUTE: **(a)** The magnetic field lines in the yz-plane are sketched in Figure 27.81.

(b) Side 1, that runs from (0,0) to (0,L): $\vec{F} = \int_0^L Id\vec{l} \times \vec{B} = I\int_0^L \frac{B_0 y\,dy}{L}\hat{i} = \frac{1}{2}B_0L\hat{i}$.

Side 2, that runs from (0,L) to (L,L): $\vec{F} = \int_{0,y=L}^L Id\vec{l} \times \vec{B} = I\int_{0,y=L}^L \frac{B_0 y\,dx}{L}\hat{j} = -IB_0L\hat{j}$.

Side 3, that runs from (L,L) to (L,0): $\vec{F} = \int_{L,x=L}^0 Id\vec{l} \times \vec{B} = I\int_{L,x=L}^0 \frac{B_0 y\,dy}{L}(-\hat{i}) = -\frac{1}{2}IB_0L\hat{i}$.

Side 4, that runs from (L,0) to (0,0): $\vec{F} = \int_{L,y=0}^0 Id\vec{l} \times \vec{B} = I\int_{L,y=0}^0 \frac{B_0 y\,dx}{L}\hat{j} = 0$.

(c) The sum of all forces is $\vec{F}_{\text{total}} = -IB_0L\hat{j}$.

EVALUATE: The net force on sides 1 and 3 is zero. The force on side 4 is zero, since $y = 0$ and $z = 0$ at that side and therefore $B = 0$ there. The net force on the loop equals the force on side 2.

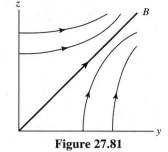

Figure 27.81

27.83.   **IDENTIFY:**   While the ends of the wire are in contact with the mercury and current flows in the wire, the magnetic field exerts an upward force and the wire has an upward acceleration. After the ends leave the mercury the electrical connection is broken and the wire is in free-fall.

(a) SET UP: After the wire leaves the mercury its acceleration is g, downward. The wire travels upward a total distance of 0.350 m from its initial position. Its ends lose contact with the mercury after the wire has traveled 0.025 m, so the wire travels upward 0.325 m after it leaves the mercury. Consider the motion of the wire after it leaves the mercury. Take $+y$ to be upward and take the origin at the position of the wire as it leaves the mercury.

$a_y = -9.80$ m/s^2, $y - y_0 = +0.325$ m, $v_y = 0$ (at maximum height), $v_{0y} = ?$

$$v_y^2 = v_{0y}^2 + 2a_y\left(y - y_0\right)$$

EXECUTE: $v_{0y} = \sqrt{-2a_y(y - y_0)} = \sqrt{-2(-9.80 \text{ m/s}^2)(0.325 \text{ m})} = 2.52$ m/s

(b) SET UP: Now consider the motion of the wire while it is in contact with the mercury. Take $+y$ to be upward and the origin at the initial position of the wire. Calculate the acceleration: $y - y_0 = +0.025$ m, $v_{0y} = 0$ (starts from rest), $v_y = +2.52$ m/s (from part (a)), $a_y = ?$

$$v_y^2 = v_{0y}^2 + 2a_y(y - y_0)$$

EXECUTE: $a_y = \dfrac{v_y^2}{2(y - y_0)} = \dfrac{(2.52 \text{ m/s})^2}{2(0.025 \text{ m})} = 127$ m/s^2

SET UP: The free-body diagram for the wire is given in Figure 27.83.

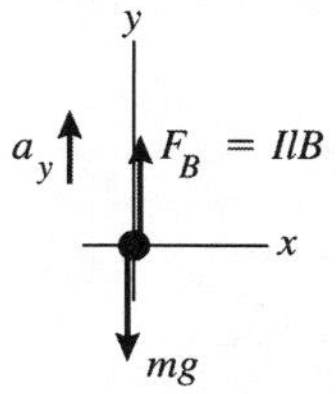

$a_y \uparrow$ $F_B = IlB$ — x

$\downarrow mg$

Figure 27.83

EXECUTE: $\sum F_y = ma_y$

$F_B - mg = ma_y$

$IlB = m\left(g + a_y\right)$

$I = \dfrac{m\left(g + a_y\right)}{lB}$

l is the length of the horizontal section of the wire; $l = 0.150$ m

$$I = \frac{(5.40 \times 10^{-5} \text{ kg})(9.80 \text{ m/s}^2 + 127 \text{ m/s}^2)}{(0.150 \text{ m})(0.00650 \text{ T})} = 7.58 \text{ A}$$

(c) IDENTIFY and SET UP: Use Ohm's law.

EXECUTE: $V = IR$ so $R = \dfrac{V}{I} = \dfrac{1.50 \text{ V}}{7.58 \text{ A}} = 0.198 \ \Omega$

EVALUATE: The current is large and the magnetic force provides a large upward acceleration. During this upward acceleration the wire moves a much shorter distance as it gains speed than the distance is moves while in free-fall with a much smaller acceleration, as it loses the speed it gained. The large current means the resistance of the wire must be small.

27.85.   **(a) IDENTIFY:**   Use Eq.(27.27) to relate U, μ and $\vec{B}$ and use Eq.(27.26) to relate $\vec{\tau}$, $\vec{\mu}$ and $\vec{B}$. We also know that $B_0^2 = B_x^2 + B_y^2 + B_z^2$. This gives three equations for the three components of $\vec{B}$.

SET UP: The loop and current are shown in Figure 27.85.

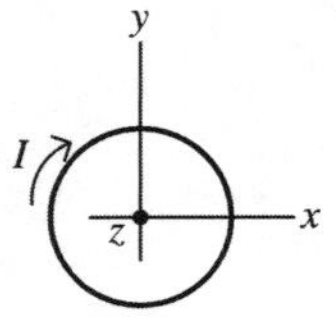

$\vec{\mu}$ is into the plane of the paper, in the $-z$-direction

Figure 27.85

$\vec{\mu} = -\mu\hat{k} = -IA\hat{k}$

(b) EXECUTE: $\vec{\tau} = D(+4\hat{i} - 3\hat{j})$, where $D > 0$.

$\vec{\mu} = -IA\hat{k}$, $\vec{B} = B_x\hat{i} + B_y\hat{j} + B_y\hat{k}$

$\vec{\tau} = \vec{\mu} \times \vec{B} = (-IA)(B_x\hat{k} \times \hat{i} + B_y\hat{k} \times \hat{j} + B_z\hat{k} \times \hat{k}) = IAB_y\hat{i} - IAB_x\hat{j}$

Compare this to the expression given for $\vec{\tau}$: $IAB_y = 4D$ so $B_y = 4D/IA$ and $-IAB_x = -3D$ so $B_x = 3D/IA$

B_z doesn't contribute to the torque since $\vec{\mu}$ is along the z-direction. But $B = B_0$ and $B_x^2 + B_y^2 + B_z^2 = B_0^2$; with $B_0 = 13D/IA$. Thus $B_z = \pm\sqrt{B_0^2 - B_x^2 - B_y^2} = \pm(D/IA)\sqrt{169 - 9 - 16} = \pm12(D/IA)$

That $U = -\vec{\mu} \cdot \vec{B}$ is negative determines the sign of B_z: $U = -\vec{\mu} \cdot \vec{B} = -(-IA\hat{k}) \cdot (B_x\hat{i} + B_y\hat{j} + B_z\hat{k}) = +IAB_z$

So U negative says that B_z is negative, and thus $B_z = -12D/IA$.

EVALUATE: $\vec{\mu}$ is along the z-axis so only B_x and B_y contribute to the torque. B_x produces a y-component of $\vec{\tau}$ and B_y produces an x-component of $\vec{\tau}$. Only B_z affects U, and U is negative when $\vec{\mu}$ and $\vec{B}_z$ are parallel.

SOURCES OF MAGNETIC FIELD

28.1. **IDENTIFY** and **SET UP:** Use Eq.(28.2) to calculate $\vec{B}$ at each point.

$$\vec{B} = \frac{\mu_0}{4\pi} \frac{q\vec{v} \times \hat{r}}{r^2} = \frac{\mu_0}{4\pi} \frac{q\vec{v} \times \vec{r}}{r^3}, \text{ since } \hat{r} = \frac{\vec{r}}{r}.$$

$\vec{v} = (8.00 \times 10^6 \text{ m/s})\hat{j}$ and $\vec{r}$ is the vector from the charge to the point where the field is calculated.

EXECUTE: **(a)** $\vec{r} = (0.500 \text{ m})\hat{i}$, $r = 0.500$ m

$\vec{v} \times \vec{r} = vr\hat{j} \times \hat{i} = -vr\hat{k}$

$$\vec{B} = -\frac{\mu_0}{4\pi} \frac{qv}{r^2}\hat{k} = -(1 \times 10^{-7} \text{ T} \cdot \text{m/A})\frac{(6.00 \times 10^{-6} \text{ C})(8.00 \times 10^6 \text{ m/s})}{(0.500 \text{ m})^2}\hat{k}$$

$\vec{B} = -(1.92 \times 10^{-5} \text{ T})\hat{k}$

(b) $\vec{r} = -(0.500 \text{ m})\hat{j}$, $r = 0.500$ m

$\vec{v} \times \vec{r} = -vr\hat{j} \times \hat{j} = 0$ and $\vec{B} = 0$.

(c) $\vec{r} = (0.500 \text{ m})\hat{k}$, $r = 0.500$ m

$\vec{v} \times \vec{r} = vr\hat{j} \times \hat{k} = vr\hat{i}$

$$\vec{B} = (1 \times 10^{-7} \text{ T} \cdot \text{m/A})\frac{(6.00 \times 10^{-6} \text{ C})(8.00 \times 10^6 \text{ m/s})}{(0.500 \text{ m})^2}\hat{i} = +(1.92 \times 10^{-5} \text{ T})\hat{i}$$

(d) $\vec{r} = -(0.500 \text{ m})\hat{j} + (0.500 \text{ m})\hat{k}$, $r = \sqrt{(0.500 \text{ m})^2 + (0.500 \text{ m})^2} = 0.7071$ m

$\vec{v} \times \vec{r} = v(0.500 \text{ m})(-\hat{j} \times \hat{j} + \hat{j} \times \hat{k}) = (4.00 \times 10^6 \text{ m}^2/\text{s})\hat{i}$

$$\vec{B} = (1 \times 10^{-7} \text{ T} \cdot \text{m/A})\frac{(6.00 \times 10^{-6} \text{ C})(4.00 \times 10^6 \text{ m/s})}{(0.7071 \text{ m})^3}\hat{i} = +(6.79 \times 10^{-6} \text{ T})\hat{i}$$

EVALUATE: At each point $\vec{B}$ is perpendicular to both $\vec{v}$ and $\vec{r}$. $B = 0$ along the direction of $\vec{v}$.

28.5. **IDENTIFY:** Apply $\vec{B} = \frac{\mu_0}{4\pi} \frac{q\vec{v} \times \vec{r}}{r^3}$.

SET UP: Since the charge is at the origin, $\vec{r} = x\hat{i} + y\hat{j} + z\hat{k}$.

EXECUTE: **(a)** $\vec{v} = v\hat{i}, \vec{r} = r\hat{i}$; $\vec{v} \times \vec{r} = 0, B = 0$.

(b) $\vec{v} = v\hat{i}, \vec{r} = r\hat{j}$; $\vec{v} \times \vec{r} = vr\hat{k}$, $r = 0.500$ m.

$$B = \left(\frac{\mu_0}{4\pi}\right)\frac{|q|v}{r^2} = \frac{(1.0 \times 10^{-7} \text{ N} \cdot \text{s}^2/\text{C}^2)(4.80 \times 10^{-6} \text{ C})(6.80 \times 10^5 \text{ m/s})}{(0.500 \text{ m})^2} = 1.31 \times 10^{-6} \text{ T}.$$

q is negative, so $\vec{B} = -(1.31 \times 10^{-6} \text{ T})\hat{k}$.

(c) $\vec{v} = v\hat{i}, \vec{r} = (0.500 \text{ m})(\hat{i} + \hat{j})$; $\vec{v} \times \vec{r} = (0.500 \text{ m})v\hat{k}$, $r = 0.7071$ m.

$$B = \left(\frac{\mu_0}{4\pi}\right)(|q||\vec{v} \times \vec{r}|/r^3) = \frac{(1.0 \times 10^{-7} \text{ N} \cdot \text{s}^2/\text{C}^2)(4.80 \times 10^{-6} \text{ C})(0.500 \text{ m})(6.80 \times 10^5 \text{ m/s})}{(0.7071 \text{ m})^3}.$$

$B = 4.62 \times 10^{-7}$ T. $\vec{B} = -(4.62 \times 10^{-7} \text{ T})\hat{k}.$

(d) $\vec{v} = v\hat{i}, \vec{r} = r\hat{k}; \;\; \vec{v} \times \vec{r} = -vr\hat{j}, r = 0.500$ m

$$B = \left(\frac{\mu_0}{4\pi}\right)\frac{|q|v}{r^2} = \frac{(1.0\times10^{-7}\ \text{N}\cdot\text{s}^2/\text{C}^2)(4.80\times10^{-6}\ \text{C})(6.80\times10^5\ \text{m/s})}{(0.500\ \text{m})^2} = 1.31\times10^{-6}\ \text{T}.$$

$$\vec{B} = \left(1.31\times10^{-6}\ \text{T}\right)\hat{j}.$$

EVALUATE: In each case, $\vec{B}$ is perpendicular to both $\vec{r}$ and $\vec{v}$.

28.7. **IDENTIFY:** Apply $\vec{B} = \dfrac{\mu_0}{4\pi}\dfrac{q\vec{v}\times\vec{r}}{r^3}$. For the magnetic force on q', use $\vec{F}_B = q'\vec{v}\times\vec{B}_q$ and for the magnetic force on

q use $\vec{F}_B = q\vec{v}\times\vec{B}_{q'}$.

SET UP: In part (a), $r = d$ and $\dfrac{|\vec{v}\times\vec{r}|}{r^3} = \dfrac{v}{d^2}$.

EXECUTE: **(a)** $q' = -q$; $B_q = \dfrac{\mu_0 qv}{4\pi d^2}$, into the page; $B_{q'} = \dfrac{\mu_0 qv'}{4\pi d^2}$, out of the page.

(i) $v' = \dfrac{v}{2}$ gives $B = \dfrac{\mu_0 qv}{4\pi d^2}\left(1-\tfrac{1}{2}\right) = \dfrac{\mu_0 qv}{4\pi(2d^2)}$, into the page. (ii) $v' = v$ gives $B = 0$.

(iii) $v' = 2v$ gives $B = \dfrac{\mu_0 qv}{4\pi d^2}$, out of the page.

(b) The force that q exerts on q' is given by $\vec{F} = q'\vec{v}\times\vec{B}_q$, so $F = \dfrac{\mu_0 q^2 v'v}{4\pi(2d)^2}$. $\vec{B}_q$ is into the page, so the force on

q' is toward q. The force that q' exerts on q is toward q'. The force between the two charges is attractive.

(c) $F_B = \dfrac{\mu_0 q^2 vv'}{4\pi(2d)^2}$, $F_C = \dfrac{q^2}{4\pi\epsilon_0(2d)^2}$ so $\dfrac{F_B}{F_C} = \mu_0\epsilon_0 vv' = \mu_0\epsilon_0(3.00\times10^5\ \text{m/s})^2 = 1.00\times10^{-6}$.

EVALUATE: When charges of opposite sign move in opposite directions, the force between them is attractive. For the values specified in part (c), the magnetic force between the two charges is much smaller in magnitude than the Coulomb force between them.

28.11. **IDENTIFY** and **SET UP:** The magnetic field produced by an infinitesimal current element is given by Eq.(28.6).

$d\vec{B} = \dfrac{\mu_0}{4\pi}\dfrac{I\vec{l}\times\hat{r}}{r^2}$ As in Example 28.2 use this equation for the finite 0.500-mm segment of wire since the

$\Delta l = 0.500$ mm length is much smaller than the distances to the field points.

$$\vec{B} = \frac{\mu_0}{4\pi}\frac{I\Delta\vec{l}\times\hat{r}}{r^2} = \frac{\mu_0}{4\pi}\frac{I\Delta\vec{l}\times\vec{r}}{r^3}$$

I is in the $+z$-direction, so $\Delta\vec{l} = \left(0.500\times10^{-3}\ \text{m}\right)\hat{k}$

EXECUTE: **(a)** Field point is at $x = 2.00$ m, $y = 0$, $z = 0$ so the vector $\vec{r}$ from the source point (at the origin) to the field point is $\vec{r} = (2.00\ \text{m})\hat{i}$.

$$\Delta\vec{l}\times\vec{r} = \left(0.500\times10^{-3}\ \text{m}\right)(2.00\ \text{m})\hat{k}\times\hat{i} = +\left(1.00\times10^{-3}\ \text{m}^2\right)\hat{j}$$

$$\vec{B} = \frac{\left(1\times10^{-7}\ \text{T}\cdot\text{m/A}\right)(4.00\ \text{A})\left(1.00\times10^{-3}\ \text{m}^2\right)}{(2.00\ \text{m})^3}\hat{j} = \left(5.00\times10^{-11}\ \text{T}\right)\hat{j}$$

(b) $\vec{r} = (2.00\ \text{m})\hat{j}$, $r = 2.00$ m.

$$\Delta\vec{l}\times\vec{r} = \left(0.500\times10^{-3}\ \text{m}\right)(2.00\ \text{m})\hat{k}\times\hat{j} = -\left(1.00\times10^{-3}\ \text{m}^2\right)\hat{i}$$

$$\vec{B} = -\frac{\left(1\times10^{-7}\ \text{T}\cdot\text{m/A}\right)(4.00\ \text{A})\left(1.00\times10^{-3}\ \text{m}^2\right)}{(2.00\ \text{m})^3}\hat{i} = -\left(5.00\times10^{-11}\ \text{T}\right)\hat{i}$$

(c) $\vec{r} = (2.00\ \text{m})\left(\hat{i}+\hat{j}\right)$, $r = \sqrt{2}(2.00\ \text{m})$.

$$\Delta\vec{l}\times\vec{r} = \left(0.500\times10^{-3}\ \text{m}\right)(2.00\ \text{m})\hat{k}\times\left(\hat{i}+\hat{j}\right) = \left(1.00\times10^{-3}\ \text{m}^2\right)\left(\hat{j}-\hat{i}\right)$$

$$\vec{B} = \frac{\left(1\times10^{-7}\ \text{T}\cdot\text{m/A}\right)(4.00\ \text{A})\left(1.00\times10^{-3}\ \text{m}^2\right)}{\left[\sqrt{2}(2.00\ \text{m})\right]^3}\left(\hat{j}-\hat{i}\right) = \left(-1.77\times10^{-11}\ \text{T}\right)\left(\hat{i}-\hat{j}\right)$$

(d) $\vec{r} = (2.00 \text{ m})\hat{k}$, $r = 2.00$ m.

$\Delta\vec{l} \times \vec{r} = (0.500 \times 10^{-3} \text{ m})(2.00 \text{ m})\hat{k} \times \hat{k} = 0$; $\vec{B} = 0$.

EVALUATE: At each point $\vec{B}$ is perpendicular to both $\vec{r}$ and $\Delta\vec{l}$. $B = 0$ along the length of the wire.

28.13. **IDENTIFY:** A current segment creates a magnetic field.

SET UP: The law of Biot and Savart gives $dB = \dfrac{\mu_0}{4\pi}\dfrac{Idl\sin\phi}{r^2}$. Both fields are into the page, so their magnitudes add.

EXECUTE: Applying the Biot and Savart law, where $r = \tfrac{1}{2}\sqrt{(3.00 \text{ cm})^2 + (3.00 \text{ cm})^2} = 2.121$ cm, we have

$$dB = 2\frac{4\pi \times 10^{-7} \text{ T} \cdot \text{m/A}}{4\pi}\frac{(28.0 \text{ A})(0.00200 \text{ m})\sin 45.0°}{(0.02121 \text{ m})^2} = 1.76 \times 10^{-5} \text{ T, into the paper.}$$

EVALUATE: Even though the two wire segments are at right angles, the magnetic fields they create are in the same direction.

28.17. **IDENTIFY:** The long current-carrying wire produces a magnetic field.

SET UP: The magnetic field due to a long wire is $B = \dfrac{\mu_0 I}{2\pi r}$.

EXECUTE: First solve for the current, then substitute the numbers using the above equation.
(a) Solving for the current gives

$$I = 2\pi r B/\mu_0 = 2\pi(0.0200 \text{ m})(1.00 \times 10^{-4} \text{ T})/(4\pi \times 10^{-7} \text{ T} \cdot \text{m/A}) = 10.0 \text{ A}$$

(b) The earth's horizontal field points northward, so at all points directly above the wire the field of the wire would point northward.
(c) At all points directly east of the wire, its field would point northward.

EVALUATE: Even though the Earth's magnetic field is rather weak, it requires a fairly large current to cancel this field.

28.19. **IDENTIFY:** The total magnetic field is the vector sum of the constant magnetic field and the wire's magnetic field.

SET UP: For the wire, $B_{\text{wire}} = \dfrac{\mu_0 I}{2\pi r}$ and the direction of B_{wire} is given by the right-hand rule that is illustrated in Figure 28.6 in the textbook. $\vec{B}_0 = (1.50 \times 10^{-6} \text{ T})\hat{i}$.

EXECUTE: **(a)** At $(0, 0, 1 \text{ m})$, $\vec{B} = \vec{B}_0 - \dfrac{\mu_0 I}{2\pi r}\hat{i} = (1.50 \times 10^{-6} \text{ T})\hat{i} - \dfrac{\mu_0(8.00 \text{ A})}{2\pi(1.00 \text{ m})}\hat{i} = -(1.0 \times 10^{-7} \text{ T})\hat{i}$.

(b) At $(1 \text{ m}, 0, 0)$, $\vec{B} = \vec{B}_0 + \dfrac{\mu_0 I}{2\pi r}\hat{k} = (1.50 \times 10^{-6} \text{ T})\hat{i} + \dfrac{\mu_0(8.00 \text{ A})}{2\pi(1.00 \text{ m})}\hat{k}$.

$\vec{B} = (1.50 \times 10^{-6} \text{ T})\hat{i} + (1.6 \times 10^{-6} \text{ T})\hat{k} = 2.19 \times 10^{-6}$ T, at $\theta = 46.8°$ from x to z.

(c) At $(0, 0, -0.25 \text{ m})$, $\vec{B} = \vec{B}_0 + \dfrac{\mu_0 I}{2\pi r}\hat{i} = (1.50 \times 10^{-6} \text{ T})\hat{i} + \dfrac{\mu_0(8.00 \text{ A})}{2\pi(0.25 \text{ m})}\hat{i} = (7.9 \times 10^{-6} \text{ T})\hat{i}$.

EVALUATE: At point c the two fields are in the same direction and their magnitudes add. At point a they are in opposite directions and their magnitudes subtract. At point b the two fields are perpendicular.

28.21. **IDENTIFY:** $B = \dfrac{\mu_0 I}{2\pi r}$. The direction of $\vec{B}$ is given by the right-hand rule in Section 20.7.

SET UP: Call the wires a and b, as indicated in Figure 28.21. The magnetic fields of each wire at points P_1 and P_2 are shown in Figure 28.21a. The fields at point 3 are shown in Figure 28.21b.
EXECUTE: **(a)** At P_1, $B_a = B_b$ and the two fields are in opposite directions, so the net field is zero.

(b) $B_a = \dfrac{\mu_0 I}{2\pi r_a}$. $B_b = \dfrac{\mu_0 I}{2\pi r_b}$. $\vec{B}_a$ and $\vec{B}_b$ are in the same direction so

$$B = B_a + B_b = \frac{\mu_0 I}{2\pi}\left(\frac{1}{r_a} + \frac{1}{r_b}\right) = \frac{(4\pi \times 10^{-7} \text{ T} \cdot \text{m/A})(4.00 \text{ A})}{2\pi}\left[\frac{1}{0.300 \text{ m}} + \frac{1}{0.200 \text{ m}}\right] = 6.67 \times 10^{-6} \text{ T}$$

$\vec{B}$ has magnitude 6.67 μT and is directed toward the top of the page.

(c) In Figure 28.21b, $\vec{B}_a$ is perpendicular to $\vec{r}_a$ and $\vec{B}_b$ is perpendicular to $\vec{r}_b$. $\tan\theta = \dfrac{5 \text{ cm}}{20 \text{ cm}}$ and $\theta = 14.04°$.

$r_a = r_b = \sqrt{(0.200 \text{ m})^2 + (0.050 \text{ m})^2} = 0.206$ m and $B_a = B_b$.

$$B = B_a\cos\theta + B_b\cos\theta = 2B_a\cos\theta = 2\left(\frac{\mu_0 I}{2\pi r_a}\right)\cos\theta = \frac{2(4\pi \times 10^{-7} \text{ T} \cdot \text{m/A})(4.0 \text{ A})\cos 14.04°}{2\pi(0.206 \text{ m})} = 7.54 \ \mu\text{T}$$

B has magnitude 7.53 μT and is directed to the left.

EVALUATE: At points directly to the left of both wires the net field is directed toward the bottom of the page.

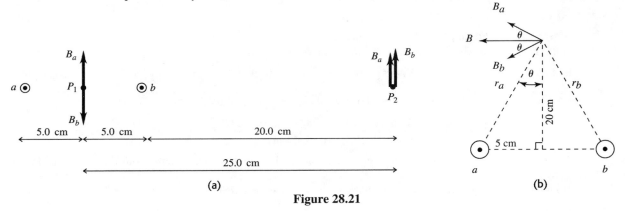

(a)

Figure 28.21

(b)

28.25.   **IDENTIFY:**   Apply Eq.(28.11).
SET UP: Two parallel conductors carrying current in the same direction attract each other. Parallel conductors carrying currents in opposite directions repel each other.

EXECUTE: (a) $F = \dfrac{\mu_0 I_1 I_2 L}{2\pi r} = \dfrac{\mu_0 (5.00 \text{ A})(2.00 \text{ A})(1.20 \text{ m})}{2\pi (0.400 \text{ m})} = 6.00 \times 10^{-6}$ N, and the force is repulsive since the

currents are in opposite directions.

(b) Doubling the currents makes the force increase by a factor of four to $F = 2.40 \times 10^{-5}$ N.

EVALUATE: Doubling the current in a wire doubles the magnetic field of that wire. For fixed magnetic field, doubling the current in a wire doubles the force that the magnetic field exerts on the wire.

28.29.   **IDENTIFY:**   The wire *CD* rises until the upward force F_I due to the currents balances the downward force of gravity.
SET UP: The forces on wire *CD* are shown in Figure 28.29.

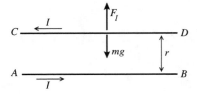

Currents in opposite directions so the force is repulsive and F_I is upward, as shown.

Figure 28.29

Eq.(28.11) says $F_I = \dfrac{\mu_0 I^2 L}{2\pi h}$ where *L* is the length of wire *CD* and *h* is the distance between the wires.

EXECUTE: $mg = \lambda L g$

Thus $F_I - mg = 0$ says $\dfrac{\mu_0 I^2 L}{2\pi h} = \lambda L g$ and $h = \dfrac{\mu_0 I^2}{2\pi g \lambda}$.

EVALUATE: The larger *I* is or the smaller λ is, the larger *h* will be.

28.31.   **IDENTIFY:**   Calculate the magnetic field vector produced by each wire and add these fields to get the total field.
SET UP: First consider the field at *P* produced by the current I_1 in the upper semicircle of wire. See Figure 28.31a.

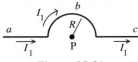

Consider the three parts of this wire
a: long straight section,
b: semicircle
c: long, straight section

Figure 28.31a

Apply the Biot-Savart law $d\vec{B} = \dfrac{\mu_0}{4\pi} \dfrac{I d\vec{l} \times \hat{r}}{r^2} = \dfrac{\mu_0}{4\pi} \dfrac{I d\vec{l} \times \vec{r}}{r^3}$ to each piece.

EXECUTE: part *a* See Figure 28.31b.

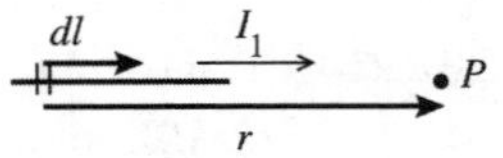

$d\vec{l} \times \vec{r} = 0$,
so $dB = 0$

Figure 28.31b

The same is true for all the infinitesimal segments that make up this piece of the wire, so $B = 0$ for this piece.

part c See Figure 28.31c.

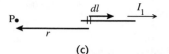

$d\vec{l} \times \vec{r} = 0$,

so $dB = 0$ and $B = 0$ for this piece.

(c)

Figure 28.31c

part b See Figure 28.31d.

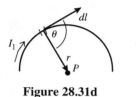

$d\vec{l} \times \vec{r}$ is directed into the paper for all infinitesimal segments that make up this semicircular piece, so $\vec{B}$ is directed into the paper and $B = \int dB$ (the vector sum of the $d\vec{B}$ is obtained by adding their magnitudes since they are in the same direction).

Figure 28.31d

$\left| d\vec{l} \times \vec{r} \right| = r\, dl \sin\theta$. The angle θ between $d\vec{l}$ and $\vec{r}$ is $90°$ and $r = R$, the radius of the semicircle. Thus $\left| d\vec{l} \times \vec{r} \right| = R\, dl$

$$dB = \frac{\mu_0}{4\pi} \frac{I \left| d\vec{l} \times \vec{r} \right|}{r^3} = \frac{\mu_0 I_1}{4\pi} \frac{R}{R^3} dl = \left(\frac{\mu_0 I_1}{4\pi R^2} \right) dl$$

$$B = \int dB = \left(\frac{\mu_0 I_1}{4\pi R^2} \right) \int dl = \left(\frac{\mu_0 I_1}{4\pi R^2} \right) (\pi R) = \frac{\mu_0 I_1}{4R}$$

(We used that $\int dl$ is equal to πR, the length of wire in the semicircle.) We have shown that the two straight sections make zero contribution to $\vec{B}$, so $B_1 = \mu_0 I_1 / 4R$ and is directed into the page.

For current in the direction shown in Figure 28.31e, a similar analysis gives $B_2 = \mu_0 I_2 / 4R$, out of the paper

Figure 28.31e

$\vec{B}_1$ and $\vec{B}_2$ are in opposite directions, so the magnitude of the net field at P is $B = \left| B_1 - B_2 \right| = \dfrac{\mu_0 \left| I_1 - I_2 \right|}{4R}$.

EVALUATE: When $I_1 = I_2$, $B = 0$.

28.33. **IDENTIFY:** Apply Eq.(28.16).

SET UP: At the center of the coil, $x = 0$. a is the radius of the coil, 0.020 m.

EXECUTE: (a) $B_{\text{center}} = \dfrac{\mu_0 NI}{2a} = \dfrac{\mu_0 (600)\,(0.500\ \text{A})}{2(0.020\ \text{m})} = 9.42 \times 10^{-3}\ \text{T}$.

(b) $B(x) = \dfrac{\mu_0 NI a^2}{2(x^2 + a^2)^{3/2}}$. $B(0.08\ \text{m}) = \dfrac{\mu_0 (600)(0.500\ \text{A})(0.020\ \text{m})^2}{2((0.080\ \text{m})^2 + (0.020\ \text{m})^2)^{3/2}} = 1.34 \times 10^{-4}\ \text{T}$.

EVALUATE: As shown in Figure 28.41 in the textbook, the field has its largest magnitude at the center of the coil and decreases with distance along the axis from the center.

28.35. **IDENTIFY:** Apply Ampere's law.

SET UP: $\mu_0 = 4\pi \times 10^{-7}\ \text{T} \cdot \text{m/A}$

EXECUTE: (a) $\oint \vec{B} \cdot d\vec{l} = \mu_0 I_{\text{encl}} = 3.83 \times 10^{-4}\ \text{T} \cdot \text{m}$ and $I_{\text{encl}} = 305\ \text{A}$.

(b) $-3.83 \times 10^{-4}\ \text{T} \cdot \text{m}$ since at each point on the curve the direction of $d\vec{l}$ is reversed.

EVALUATE: The line integral $\oint \vec{B} \cdot d\vec{l}$ around a closed path is proportional to the net current that is enclosed by the path.

28.37. **IDENTIFY:** Apply Ampere's law.

SET UP: To calculate the magnetic field at a distance r from the center of the cable, apply Ampere's law to a circular path of radius r. By symmetry, $\oint \vec{B} \cdot d\vec{l} = B(2\pi r)$ for such a path.

EXECUTE: (a) For $a < r < b$, $I_{\text{encl}} = I \Rightarrow \oint \vec{B} \cdot d\vec{l} = \mu_0 I \Rightarrow B 2\pi r = \mu_0 I \Rightarrow B = \dfrac{\mu_0 I}{2\pi r}$.

(b) For $r > c$, the enclosed current is zero, so the magnetic field is also zero.

EVALUATE: A useful property of coaxial cables for many applications is that the current carried by the cable doesn't produce a magnetic field outside the cable.

28.41.   **(a) IDENTIFY** and **SET UP:**   The magnetic field near the center of a long solenoid is given by Eq.(28.23), $B = \mu_0 nI$.

EXECUTE: Turns per unit length $n = \dfrac{B}{\mu_0 I} = \dfrac{0.0270 \text{ T}}{(4\pi \times 10^{-7} \text{ T} \cdot \text{m/A})(12.0 \text{ A})} = 1790$ turns/m

(b) $N = nL = (1790 \text{ turns/m})(0.400 \text{ m}) = 716$ turns

Each turn of radius R has a length $2\pi R$ of wire. The total length of wire required is
$N(2\pi R) = (716)(2\pi)(1.40 \times 10^{-2} \text{ m}) = 63.0$ m.

EVALUATE: A large length of wire is required. Due to the length of wire the solenoid will have appreciable resistance.

28.45.   **IDENTIFY:**   Example 28.10 shows that inside a toroidal solenoid, $B = \dfrac{\mu_0 NI}{2\pi r}$.

SET UP: $r = 0.070$ m

EXECUTE: $B = \dfrac{\mu_0 NI}{2\pi r} = \dfrac{\mu_0 (600)(0.650 \text{ A})}{2\pi (0.070 \text{ m})} = 1.11 \times 10^{-3}$ T.

EVALUATE: If the radial thickness of the torus is small compared to its mean diameter, B is approximately uniform inside its windings.

28.47.   **IDENTIFY** and **SET UP:** $B = \dfrac{K_m \mu_0 NI}{2\pi r}$ (Eq.28.24, with μ_0 replaced by $K_m \mu_0$)

EXECUTE: **(a)** $K_m = 1400$

$I = \dfrac{2\pi r B}{\mu_0 K_m N} = \dfrac{(2.90 \times 10^{-2} \text{ m})(0.350 \text{ T})}{(2 \times 10^{-7} \text{ T} \cdot \text{m/A})(1400)(500)} = 0.0725$ A

(b) $K_m = 5200$

$I = \dfrac{2\pi r B}{\mu_0 K_m N} = \dfrac{(2.90 \times 10^{-2} \text{ m})(0.350 \text{ T})}{(2 \times 10^{-7} \text{ T} \cdot \text{m/A})(5200)(500)} = 0.0195$ A

EVALUATE: If the solenoid were air-filled instead, a much larger current would be required to produce the same magnetic field.

28.49.   **IDENTIFY:**   The magnetic field from the solenoid alone is $B_0 = \mu_0 nI$. The total magnetic field is $B = K_m B_0$. M is given by Eq.(28.29).

SET UP: $n = 6000$ turns/m

EXECUTE: **(a) (i)** $B_0 = \mu_0 nI = \mu_0 (6000 \text{ m}^{-1})(0.15 \text{ A}) = 1.13 \times 10^{-3}$ T.

(ii) $M = \dfrac{K_m - 1}{\mu_0} B_0 = \dfrac{5199}{\mu_0}(1.13 \times 10^{-3} \text{ T}) = 4.68 \times 10^6$ A/m.

(iii) $B = K_m B_0 = (5200)(1.13 \times 10^{-3} \text{ T}) = 5.88$ T.

(b) The directions of $\vec{B}$, $\vec{B}_0$ and $\vec{M}$ are shown in Figure 28.49. Silicon steel is paramagnetic and $\vec{B}_0$ and $\vec{M}$ are in the same direction.

EVALUATE: The total magnetic field is much larger than the field due to the solenoid current alone.

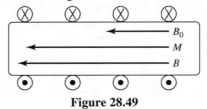

Figure 28.49

28.51.   **IDENTIFY:**   Moving charges create magnetic fields. The net field is the vector sum of the two fields. A charge moving in an external magnetic field feels a force.

(a) SET UP: The magnetic field due to a moving charge is $B = \dfrac{\mu_0}{4\pi} \dfrac{qv \sin\phi}{r^2}$. Both fields are into the paper, so their magnitudes add, giving $B_{net} = B + B' = \dfrac{\mu_0}{4\pi}\left(\dfrac{qv \sin\phi}{r^2} + \dfrac{q'v' \sin\phi'}{r'^2} \right)$.

EXECUTE: Substituting numbers gives
$B_{net} = \dfrac{\mu_0}{4\pi}\left[\dfrac{(8.00 \ \mu\text{C})(9.00 \times 10^4 \text{ m/s})\sin 90°}{(0.300 \text{ m})^2} + \dfrac{(5.00 \ \mu\text{C})(6.50 \times 10^4 \text{ m/s})\sin 90°}{(0.400 \text{ m})^2} \right]$

$B_{net} = 1.00 \times 10^{-6}$ T $= 1.00 \mu$T, into the paper.

(b) SET UP: The magnetic force on a moving charge is $\vec{F} = q\vec{v} \times \vec{B}$, and the magnetic field of charge q' at the location of charge q is into the page. The force on q is

$$\vec{F} = q\vec{v} \times \vec{B}' = (qv)\hat{i} \times \frac{\mu_0}{4\pi}\frac{q\vec{v}' \times \hat{r}}{r^2} = (qv)\hat{i} \times \left(\frac{\mu_0}{4\pi}\frac{qv'\sin\phi}{r^2}\right)(-\hat{k}) = \left(\frac{\mu_0}{4\pi}\frac{qq'vv'\sin\phi}{r^2}\right)\hat{j}$$

where ϕ is the angle between $\vec{v}'$ and $\hat{r}'$.
EXECUTE: Substituting numbers gives

$$\vec{F} = \frac{\mu_0}{4\pi}\left[\frac{(8.00\times10^{-6}\,\text{C})(5.00\times10^{-6}\,\text{C})(9.00\times10^{-6}\,\text{m/s})(6.50\times10^{-6}\,\text{m/s})}{(0.500\,\text{m})^2}\left(\frac{0.400}{0.500}\right)\right]\hat{j}$$

$$\vec{F} = (7.49\times10^{-8}\,\text{N})\hat{j}.$$

EVALUATE: These are small fields and small forces, but if the charge has small mass, the force can affect its motion.

28.53.   **IDENTIFY:**   Find the force that the magnetic field of the wire exerts on the electron.
SET UP: The force on a moving charge has magnitude $F = |q|vB\sin\phi$ and direction given by the right-hand rule.

For a long straight wire, $B = \dfrac{\mu_0 I}{2\pi r}$ and the direction of $\vec{B}$ is given by the right-hand rule.

EXECUTE: **(a)** $a = \dfrac{F}{m} = \dfrac{|q|vB\sin\phi}{m} = \dfrac{ev}{m}\left(\dfrac{\mu_0 I}{2\pi r}\right)$

$$a = \frac{(1.6\times10^{-17}\,\text{C})(2.50\times10^{5}\,\text{m/s})(4\pi\times10^{-7}\,\text{T}\cdot\text{m/A})(25.0\,\text{A})}{(9.11\times10^{-31}\,\text{kg})(2\pi)(0.0200\,\text{m})} = 1.1\times10^{13}\,\text{m/s}^2,$$

away from the wire.
(b) The electric force must balance the magnetic force. $eE = evB$, and

$E = vB = v\dfrac{\mu_0 I}{2\pi r} = \dfrac{(250,000\,\text{m/s})(4\pi\times10^{-7}\,\text{T}\cdot\text{m/A})(25.0\,\text{A})}{2\pi(0.0200\,\text{m})} = 62.5\,\text{N/C}$. The magnetic force is directed away from the wire so the force from the electric field must be toward the wire. Since the charge of the electron is negative, the electric field must be directed away from the wire to produce a force in the desired direction.
EVALUATE: **(c)** $mg = (9.11\times10^{-31}\,\text{kg})(9.8\,\text{m/s}^2) \approx 10^{-29}\,\text{N}$. $F_{el} = eE = (1.6\times10^{-19}\,\text{C})(62.5\,\text{N/C}) \approx 10^{-17}\,\text{N}$.

$F_{el} \approx 10^{12}\,F_{grav}$, so we can neglect gravity.

28.55.   **IDENTIFY:**   Find the net magnetic field due to the two loops at the location of the proton and then find the force these fields exert on the proton.

SET UP: For a circular loop, the field on the axis, a distance x from the center of the loop is $B = \dfrac{\mu_0 I R^2}{2(R^2 + x^2)^{3/2}}$.

$R = 0.200$ m and $x = 0.125$ m.

EXECUTE: The fields add, so $B = B_1 + B_2 = 2B_1 = 2\left[\dfrac{\mu_0 I R^2}{2(R^2 + x^2)^{3/2}}\right]$.

$$B = \frac{(4\pi\times10^{-7}\,\text{T}\cdot\text{m/A})(1.50\,\text{A})(0.200\,\text{m})^2}{[(0.200\,\text{m})^2 + (0.125\,\text{m})^2]^{3/2}} = 5.75\times10^{-6}\,\text{T}.$$

$F = |q|vB\sin\phi = (1.6\times10^{-19}\,\text{C})(2400\,\text{m/s})(5.75\times10^{-6}\,\text{T})\sin 90° = 2.21\times10^{-21}\,\text{N}$, perpendicular to the line ab and to the velocity.
EVALUATE: The weight of a proton is $w = mg = 1.6\times10^{-24}\,\text{N}$, so the force from the loops is much greater than the gravity force on the proton.

28.59.   **IDENTIFY:**   Use Eq.(28.9) and the right-hand rule to calculate the magnitude and direction of the magnetic field at P produced by each wire. Add these two field vectors to find the net field.

(a) SET UP: The directions of the fields at point P due to the two wires are sketched in Figure 28.59a.

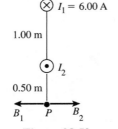

EXECUTE: $\vec{B}_1$ and $\vec{B}_2$ must be equal and opposite for the resultant field at P to be zero. $\vec{B}_2$ is to the right so I_2 is out of the page.

Figure 28.59a

$$B_1 = \frac{\mu_0 I_1}{2\pi r_1} = \frac{\mu_0}{2\pi}\left(\frac{6.00\ \text{A}}{1.50\ \text{m}}\right) \qquad B_2 = \frac{\mu_0 I_2}{2\pi r_2} = \frac{\mu_0}{2\pi}\left(\frac{I_2}{0.50\ \text{m}}\right)$$

$$B_1 = B_2 \text{ says } \frac{\mu_0}{2\pi}\left(\frac{6.00\ \text{A}}{1.50\ \text{m}}\right) = \frac{\mu_0}{2\pi}\left(\frac{I_2}{0.50\ \text{m}}\right)$$

$$I_2 = \left(\frac{0.50\ \text{m}}{1.50\ \text{m}}\right)(6.00\ \text{A}) = 2.00\ \text{A}$$

(b) SET UP: The directions of the fields at point Q are sketched in Figure 28.59b.

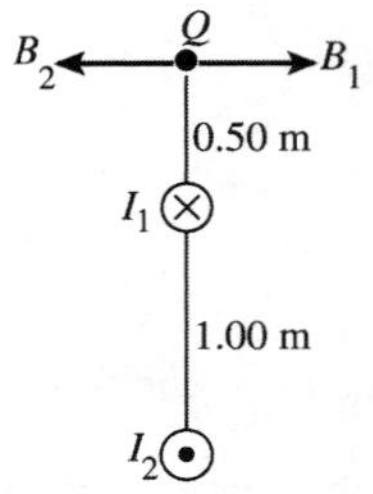

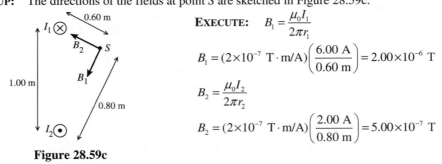

EXECUTE: $B_1 = \frac{\mu_0 I_1}{2\pi r_1}$

$$B_1 = (2\times10^{-7}\ \text{T}\cdot\text{m/A})\left(\frac{6.00\ \text{A}}{0.50\ \text{m}}\right) = 2.40\times10^{-6}\ \text{T}$$

$$B_2 = \frac{\mu_0 I_2}{2\pi r_2}$$

$$B_2 = (2\times10^{-7}\ \text{T}\cdot\text{m/A})\left(\frac{2.00\ \text{A}}{1.50\ \text{m}}\right) = 2.67\times10^{-7}\ \text{T}$$

Figure 28.59b

$\vec{B}_1$ and $\vec{B}_2$ are in opposite directions and $B_1 > B_2$ so

$B = B_1 - B_2 = 2.40\times10^{-6}\ \text{T} - 2.67\times10^{-7}\ \text{T} = 2.13\times10^{-6}\ \text{T}$, and $\vec{B}$ is to the right.

(c) SET UP: The directions of the fields at point S are sketched in Figure 28.59c.

EXECUTE: $B_1 = \frac{\mu_0 I_1}{2\pi r_1}$

$$B_1 = (2\times10^{-7}\ \text{T}\cdot\text{m/A})\left(\frac{6.00\ \text{A}}{0.60\ \text{m}}\right) = 2.00\times10^{-6}\ \text{T}$$

$$B_2 = \frac{\mu_0 I_2}{2\pi r_2}$$

$$B_2 = (2\times10^{-7}\ \text{T}\cdot\text{m/A})\left(\frac{2.00\ \text{A}}{0.80\ \text{m}}\right) = 5.00\times10^{-7}\ \text{T}$$

Figure 28.59c

$\vec{B}_1$ and $\vec{B}_2$ are right angles to each other, so the magnitude of their resultant is given by

$$B = \sqrt{B_1^2 + B_2^2} = \sqrt{(2.00\times10^{-6}\ \text{T})^2 + (5.00\times10^{-7}\ \text{T})^2} = 2.06\times10^{-6}\ \text{T}$$

EVALUATE: The magnetic field lines for a long, straight wire are concentric circles with the wire at the center. The magnetic field at each point is tangent to the field line, so $\vec{B}$ is perpendicular to the line from the wire to the point where the field is calculated.

28.63.   **IDENTIFY:**   Apply $\sum \vec{F} = 0$ to one of the wires. The force one wire exerts on the other depends on I so $\sum \vec{F} = 0$ gives two equations for the two unknowns T and I.

SET UP: The force diagram for one of the wires is given in Figure 28.63.

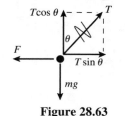

The force one wire exerts on the other is $F = \left(\dfrac{\mu_0 I^2}{2\pi r} \right) L$,

where $r = 2(0.040 \text{ m}) \sin \theta = 8.362 \times 10^{-3}$ m is the distance between the two wires.

Figure 28.63

EXECUTE: $\sum F_y = 0$ gives $T \cos \theta = mg$ and $T = mg / \cos \theta$

$\sum F_x = 0$ gives $F = T \sin \theta = (mg / \cos \theta) \sin \theta = mg \tan \theta$

And $m = \lambda L$, so $F = \lambda L g \tan \theta$

$\left(\dfrac{\mu_0 I^2}{2\pi r} \right) L = \lambda L g \tan \theta$

$I = \sqrt{\dfrac{\lambda g r \tan \theta}{(\mu_0 / 2\pi)}}$

$I = \sqrt{\dfrac{(0.0125 \text{ kg/m})(9.80 \text{ m/s})^2 (\tan\ 6.00^\circ)(8.362 \times 10^{-3} \text{ m})}{2 \times 10^{-7}\,\text{T} \cdot \text{m/A}}} = 23.2 \text{ A}$

EVALUATE: Since the currents are in opposite directions the wires repel. When I is increased, the angle θ from the vertical increases; a large current is required even for the small displacement specified in this problem.

28.69.   **IDENTIFY:**   Apply $d\vec{B} = \dfrac{\mu_0}{4\pi} \dfrac{I d\vec{l} \times \hat{r}}{r^2}$.

SET UP: The contribution from the straight segments is zero since $d\vec{l} \times \vec{r} = 0$. The magnetic field from the curved wire is just one quarter of a full loop.

EXECUTE: $B = \tfrac{1}{4} \left(\dfrac{\mu_0 I}{2R} \right) = \dfrac{\mu_0 I}{8R}$ and is directed out of the page.

EVALUATE: It is very simple to calculate B at point P but it would be much more difficult to calculate B at other points.

28.71.   **(a) IDENTIFY:**   Consider current density J for a small concentric ring and integrate to find the total current in terms of α and R.

SET UP: We can't say $I = JA = J\pi R^2$, since J varies across the cross section.

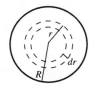

To integrate J over the cross section of the wire divide the wire cross section up into thin concentric rings of radius r and width dr, as shown in Figure 28.71.

Figure 28.71

EXECUTE: The area of such a ring is dA, and the current through it is $dI = J\, dA$; $dA = 2\pi r\, dr$ and

$dI = J\, dA = \alpha r (2\pi r\, dr) = 2\pi \alpha r^2 dr$

$I = \int dI = 2\pi \alpha \int_0^R r^2 dr = 2\pi \alpha (R^3 / 3)$ so $\alpha = \dfrac{3I}{2\pi R^3}$

(b) IDENTIFY and **SET UP:**   (i) $r \leq R$

Apply Ampere's law to a circle of radius $r < R$. Use the method of part (a) to find the current enclosed by the Ampere's law path.

EXECUTE: $\oint \vec{B} \cdot d\vec{l} = \oint B\, dl = B \oint dl = B(2\pi r)$, by the symmetry and direction of $\vec{B}$. The current passing through

the path is $I_{\text{encl}} = \int dl$, where the integration is from 0 to r. $I_{\text{encl}} = 2\pi \alpha \int_0^r r^2 dr = \dfrac{2\pi \alpha r^3}{3} = \dfrac{2\pi}{3} \left(\dfrac{3I}{2\pi R^3} \right) r^3 = \dfrac{Ir^3}{R^3}$. Thus

$\oint \vec{B} \cdot d\vec{l} = \mu_0 I_{\text{encl}}$ gives $B(2\pi r) = \mu_0 \left(\dfrac{Ir^3}{R^3} \right)$ and $B = \dfrac{\mu_0 I r^2}{2\pi R^3}$

(ii) IDENTIFY and **SET UP:** $r \geq R$

Apply Ampere's law to a circle of radius $r > R$.

EXECUTE: $\oint \vec{B} \cdot d\vec{l} = \oint B \, dl = B \oint dl = B(2\pi r)$

$I_{\text{encl}} = I$; all the current in the wire passes through this path. Thus $\oint \vec{B} \cdot d\vec{l} = \mu_0 I_{\text{encl}}$ gives $B(2\pi r) = \mu_0 I$ and $B = \dfrac{\mu_0 I}{2\pi r}$

EVALUATE: Note that at $r = R$ the expression in (i) (for $r \le R$) gives $B = \dfrac{\mu_0 I}{2\pi R}$. At $r = R$ the expression in (ii) (for $r \ge R$) gives $B = \dfrac{\mu_0 I}{2\pi R}$, which is the same.

28.75. **IDENTIFY:** Use Ampere's law to find the magnetic field at $r = 2a$ from the axis. The analysis of Example 28.9 shows that the field outside the cylinder is the same as for a long, straight wire along the axis of the cylinder.
SET UP:

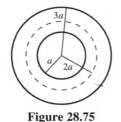

EXECUTE: Apply Ampere's law to a circular path of radius $2a$, as shown in Figure 28.75.
$B(2\pi) = \mu_0 I_{\text{encl}}$

$I_{\text{encl}} = I\left(\dfrac{(2a)^2 - a^2}{(3a)^2 - a^2}\right) = 3I/8$

Figure 28.75

$B = \dfrac{3}{16}\dfrac{\mu_0 I}{2\pi a}$; this is the magnetic field inside the metal at a distance of $2a$ from the cylinder axis. Outside the cylinder, $B = \dfrac{\mu_0 I}{2\pi r}$. The value of r where these two fields are equal is given by $1/r = 3/(16a)$ and $r = 16a/3$.

EVALUATE: For $r < 3a$, as r increases the magnetic field increases from zero at $r = 0$ to $\mu_0 I /(2\pi(3a))$ at $r = 3a$. For $r > 3a$ the field decreases as r increases so it is reasonable for there to be a $r > 3a$ where the field is the same as at $r = 2a$.

28.77. **IDENTIFY:** Use the current density J to find dI through a concentric ring and integrate over the appropriate cross section to find the current through that cross section. Then use Ampere's law to find $\vec{B}$ at the specified distance from the center of the wire.
(a) SET UP:

Divide the cross section of the cylinder into thin concentric rings of radius r and width dr, as shown in Figure 28.77a. The current through each ring is $dI = J \, dA = J \, 2\pi r \, dr$.

Figure 28.77a

EXECUTE: $dI = \dfrac{2I_0}{\pi a^2}\left[1 - (r/a)^2\right] 2\pi r \, dr = \dfrac{4I_0}{a^2}\left[1 - (r/a)^2\right] r \, dr$. The total current I is obtained by integrating dI

over the cross section $I = \int_0^a dI = \left(\dfrac{4I_0}{a^2}\right)\int_0^a (1 - r^2/a^2) r \, dr = \left(\dfrac{4I_0}{a^2}\right)\left[\dfrac{1}{2}r^2 - \dfrac{1}{4}r^4/a^2\right]_0^a = I_0$, as was to be shown.

(b) SET UP: Apply Ampere's law to a path that is a circle of radius $r > a$, as shown in Figure 28.77b.

$\oint \vec{B} \cdot d\vec{l} = B(2\pi r)$

$I_{\text{encl}} = I_0$ (the path encloses the entire cylinder)

Figure 28.77b

EXECUTE: $\oint \vec{B} \cdot d\vec{l} = \mu_0 I_{\text{encl}}$ says $B(2\pi r) = \mu_0 I_0$ and $B = \dfrac{\mu_0 I_0}{2\pi r}$.

(c) SET UP:

Figure 28.77c

Divide the cross section of the cylinder into concentric rings of radius r' and width dr', as was done in part (a). See Figure 28.77c. The current dI through each ring is $dI = \dfrac{4I_0}{a^2}\left[1-\left(\dfrac{r'}{a}\right)^2\right]r'\,dr'$

EXECUTE: The current I is obtained by integrating dI from $r'=0$ to $r'=r$:

$$I = \int dI = \frac{4I_0}{a^2}\int_0^r \left[1-\left(\frac{r'}{a}\right)^2\right]r'\,dr' = \frac{4I_0}{a^2}\left[\tfrac{1}{2}(r')^2 - \tfrac{1}{4}(r')^4/a^2\right]_0^r$$

$$I = \frac{4I_0}{a^2}(r^2/2 - r^4/4a^2) = \frac{I_0 r^2}{a^2}\left(2-\frac{r^2}{a^2}\right)$$

(d) SET UP: Apply Ampere's law to a path that is a circle of radius $r<a$, as shown in Figure 28.77d.

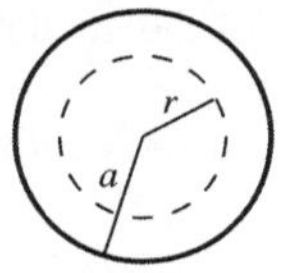

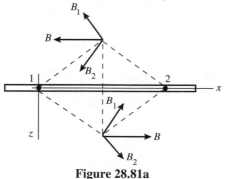

Figure 28.77d

$$\oint \vec{B}\cdot d\vec{l} = B(2\pi r)$$

$$I_{\text{encl}} = \frac{I_0 r^2}{a^2}\left(2-\frac{r^2}{a^2}\right) \text{ (from part (c))}$$

EXECUTE: $\oint \vec{B}\cdot d\vec{l} = \mu_0 I_{\text{encl}}$ says $B(2\pi r) = \mu_0 \dfrac{I_0 r^2}{a^2}(2-r^2/a^2)$ and $B = \dfrac{\mu_0 I_0}{2\pi}\dfrac{r}{a^2}(2-r^2/a^2)$

EVALUATE: Result in part (b) evaluated at $r=a$: $B = \dfrac{\mu_0 I_0}{2\pi a}$. Result in part (d) evaluated at

$r=a$: $B = \dfrac{\mu_0 I_0}{2\pi}\dfrac{a}{a^2}(2-a^2/a^2) = \dfrac{\mu_0 I_0}{2\pi a}$. The two results, one for $r>a$ and the other for $r<a$, agree at $r=a$.

28.81.   **IDENTIFY:**   Use what we know about the magnetic field of a long, straight conductor to deduce the symmetry of the magnetic field. Then apply Ampere's law to calculate the magnetic field at a distance a above and below the current sheet.

SET UP: Do parts (a) and (b) together.

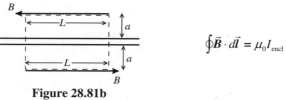

Figure 28.81a

Consider the individual currents in pairs, where the currents in each pair are equidistant on either side of the point where $\vec{B}$ is being calculated. Figure 28.81a shows that for each pair the z-components cancel, and that above the sheet the field is in the $-x$-direction and that below the sheet it is in the $+x$-direction.

Also, by symmetry the magnitude of $\vec{B}$ a distance a above the sheet must equal the magnitude of $\vec{B}$ a distance a below the sheet. Now that we have deduced the symmetry of $\vec{B}$, apply Ampere's law. Use a path that is a rectangle, as shown in Figure 28.81b.

$$\oint \vec{B}\cdot d\vec{l} = \mu_0 I_{\text{encl}}$$

Figure 28.81b

I is directed out of the page, so for I to be positive the integral around the path is taken in the counterclockwise direction.

EXECUTE: Since $\vec{B}$ is parallel to the sheet, on the sides of the rectangle that have length $2a$, $\oint \vec{B} \cdot d\vec{l} = 0$. On the long sides of length L, $\vec{B}$ is parallel to the side, in the direction we are integrating around the path, and has the same magnitude, B, on each side. Thus $\oint \vec{B} \cdot d\vec{l} = 2BL$. n conductors per unit length and current I out of the page in each conductor gives $I_{encl} = InL$. Ampere's law then gives $2BL = \mu_0 InL$ and $B = \frac{1}{2}\mu_0 In$.

EVALUATE: Note that B is independent of the distance a from the sheet. Compare this result to the electric field due to an infinite sheet of charge (Example 22.7).

28.83.   **IDENTIFY** and **SET UP:**   Use Eq.(28.28) to calculate the total magnetic moment of a volume V of the iron. Use the density and atomic mass of iron to find the number of atoms in this volume and use that to find the magnetic dipole moment per atom.

EXECUTE: $M = \dfrac{\mu_{total}}{V}$, so $\mu_{total} = MV$ The average magnetic moment per atom is $\mu_{atom} = \mu_{total}/N = MV/N$, where N is the number of atoms in volume V. The mass of volume V is $m = \rho V$, where ρ is the density. ($\rho_{iron} = 7.8 \times 10^3$ kg/m^3). The number of moles of iron in volume V is

$n = \dfrac{m}{55.847 \times 10^{-3} \text{ kg/mol}} = \dfrac{\rho V}{55.847 \times 10^{-3} \text{ kg/mol}}$, where 55.847×10^{-3} kg/mol is the atomic mass

of iron from appendix D. $N = nN_A$, where $N_A = 6.022 \times 10^{23}$ atoms/mol is Avogadro's number. Thus

$N = nN_A = \dfrac{\rho V N_A}{55.847 \times 10^{-3} \text{ kg/mol}}$.

$\mu_{atom} = \dfrac{MV}{N} = MV\left(\dfrac{55.847 \times 10^{-3} \text{ kg/mol}}{\rho V N_A}\right) = \dfrac{M(55.847 \times 10^{-3} \text{ kg/mol})}{\rho N_A}$.

$\mu_{atom} = \dfrac{(6.50 \times 10^4 \text{ A/m})(55.847 \times 10^{-3} \text{ kg/mol})}{(7.8 \times 10^3 \text{ kg/m}^3)(6.022 \times 10^{23} \text{ atoms/mol})}$

$\mu_{atom} = 7.73 \times 10^{-25} \text{ A} \cdot \text{m}^2 = 7.73 \times 10^{-25} \text{ J/T}$

$\mu_B = 9.274 \times 10^{-24} \text{ A} \cdot \text{m}^2$, so $\mu_{atom} = 0.0834\mu_B$.

EVALUATE: The magnetic moment per atom is much less than one Bohr magneton. The magnetic moments of each electron in the iron must be in different directions and mostly cancel each other.

ELECTROMAGNETIC INDUCTION

29.3. **IDENTIFY** and **SET UP:** Use Faraday's law to calculate the average induced emf and apply Ohm's law to the coil to calculate the average induced current and charge that flows.

(a) EXECUTE: The magnitude of the average emf induced in the coil is $|\mathcal{E}_{av}| = N \left| \dfrac{\Delta \Phi I_B}{\Delta t} \right|$. Initially,

$\Phi_{Bi} = BA\cos\phi = BA$. The final flux is zero, so $|\mathcal{E}_{av}| = N\dfrac{|\Phi_{Bf} - \Phi_{Bi}|}{\Delta t} = \dfrac{NBA}{\Delta t}$. The average induced current is

$I = \dfrac{|\mathcal{E}_{av}|}{R} = \dfrac{NBA}{R\Delta t}$. The total charge that flows through the coil is $Q = I\Delta t = \left(\dfrac{NBA}{R\Delta t}\right)\Delta t = \dfrac{NBA}{R}$.

EVALUATE: The charge that flows is proportional to the magnetic field but does not depend on the time Δt.

(b) The magnetic stripe consists of a pattern of magnetic fields. The pattern of charges that flow in the reader coil tell the card reader the magnetic field pattern and hence the digital information coded onto the card.

(c) According to the result in part (a) the charge that flows depends only on the change in the magnetic flux and it does not depend on the rate at which this flux changes.

29.5. **IDENTIFY:** Apply Faraday's law.

SET UP: Let $+z$ be the positive direction for $\vec{A}$. Therefore, the initial flux is positive and the final flux is zero.

EXECUTE: **(a)** and **(b)** $\mathcal{E} = -\dfrac{\Delta \Phi_B}{\Delta t} = -\dfrac{0 - (1.5\text{ T})\pi(0.120\text{ m})^2}{2.0 \times 10^{-3}\text{ s}} = +34$ V. Since $\mathcal{E}$ is positive and $\vec{A}$ is toward us,

the induced current is counterclockwise.

EVALUATE: The shorter the removal time, the larger the average induced emf.

29.7. **IDENTIFY:** Calculate the flux through the loop and apply Faraday's law.

SET UP: To find the total flux integrate $d\Phi_B$ over the width of the loop. The magnetic field of a long straight

wire, at distance r from the wire, is $B = \dfrac{\mu_0 I}{2\pi r}$. The direction of $\vec{B}$ is given by the right-hand rule.

EXECUTE: **(a)** When $B = \dfrac{\mu_0 i}{2\pi r}$, into the page.

(b) $d\Phi_B = BdA = \dfrac{\mu_0 i}{2\pi r} L\,dr$.

(c) $\Phi_B = \displaystyle\int_a^b d\Phi_B = \dfrac{\mu_0 iL}{2\pi} \int_a^b \dfrac{dr}{r} = \dfrac{\mu_0 iL}{2\pi}\ln(b/a)$.

(d) $\mathcal{E} = \dfrac{d\Phi_B}{dt} = \dfrac{\mu_0 L}{2\pi}\ln(b/a)\dfrac{di}{dt}$.

(e) $\mathcal{E} = \dfrac{\mu_0(0.240\text{ m})}{2\pi}\ln(0.360/0.120)(9.60\text{ A/s}) = 5.06 \times 10^{-7}$ V.

EVALUATE: The induced emf is proportional to the rate at which the current in the long straight wire is changing

29.9. **IDENTIFY** and **SET UP:** Use Faraday's law to calculate the emf (magnitude and direction). The direction of the induced current is the same as the direction of the emf. The flux changes because the area of the loop is changing; relate dA/dt to dc/dt, where c is the circumference of the loop.

(a) EXECUTE: $c = 2\pi r$ and $A = \pi r^2$ so $A = c^2/4\pi$

$\Phi_B = BA = (B/4\pi)c^2$

$|\mathcal{E}| = \left|\dfrac{d\Phi_B}{dt}\right| = \left(\dfrac{B}{2\pi}\right)c\left|\dfrac{dc}{dt}\right|$

At $t = 9.0$ s, $c = 1.650\text{ m} - (9.0\text{ s})(0.120\text{ m/s}) = 0.570$ m

$|\mathcal{E}| = (0.500\text{ T})(1/2\pi)(0.570\text{ m})(0.120\text{ m/s}) = 5.44$ mV

(b) SET UP: The loop and magnetic field are sketched in Figure 29.9.

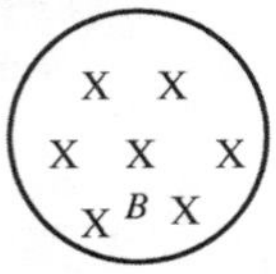

Take into the page to be the positive direction for $\vec{A}$. Then the magnetic flux is positive.

Figure 29.9

EXECUTE: The positive flux is decreasing in magnitude; $d\Phi_B/dt$ is negative and $\mathcal{E}$ is positive. By the right-hand rule, for $\vec{A}$ into the page, positive $\mathcal{E}$ is clockwise.

EVALUATE: Even though the circumference is changing at a constant rate, dA/dt is not constant and $|\mathcal{E}|$ is not constant. Flux $\otimes$ is decreasing so the flux of the induced current is $\otimes$ and this means that I is clockwise, which checks.

29.11.   **IDENTIFY:**   A change in magnetic flux through a coil induces an emf in the coil.
SET UP: The flux through a coil is $\Phi = NBA \cos \phi$ and the induced emf is $\mathcal{E} = d\Phi/dt$.
EXECUTE: (a) $\mathcal{E} = d\Phi/dt = d[A(B_0 + bx)]/dt = bA\, dx/dt = bAv$
(b) clockwise
(c) Same answers except the current is counterclockwise.
EVALUATE: Even though the coil remains within the magnetic field, the flux through it increases because the strength of the field is increasing.

29.15.   **IDENTIFY** and **SET UP:**   The field of the induced current is directed to oppose the change in flux.
EXECUTE: (a) The field is into the page and is increasing so the flux is increasing. The field of the induced current is out of the page. To produce field out of the page the induced current is counterclockwise.
(b) The field is into the page and is decreasing so the flux is decreasing. The field of the induced current is into the page. To produce field into the page the induced current is clockwise.
(c) The field is constant so the flux is constant and there is no induced emf and no induced current.
EVALUATE: The direction of the induced current depends on the direction of the external magnetic field and whether the flux due to this field is increasing or decreasing.

29.17.   **IDENTIFY** and **SET UP:**   Apply Lenz's law, in the form that states that the flux of the induced current tends to oppose the change in flux.
EXECUTE: (a) With the switch closed the magnetic field of coil A is to the right at the location of coil B. When the switch is opened the magnetic field of coil A goes away. Hence by Lenz's law the field of the current induced in coil B is to the right, to oppose the decrease in the flux in this direction. To produce magnetic field that is to the right the current in the circuit with coil B must flow through the resistor in the direction a to b.
(b) With the switch closed the magnetic field of coil A is to the right at the location of coil B. This field is stronger at points closer to coil A so when coil B is brought closer the flux through coil B increases. By Lenz's law the field of the induced current in coil B is to the left, to oppose the increase in flux to the right. To produce magnetic field that is to the left the current in the circuit with coil B must flow through the resistor in the direction b to a.
(c) With the switch closed the magnetic field of coil A is to the right at the location of coil B. The current in the circuit that includes coil A increases when R is decreased and the magnetic field of coil A increases when the current through the coil increases. By Lenz's law the field of the induced current in coil B is to the left, to oppose the increase in flux to the right. To produce magnetic field that is to the left the current in the circuit with coil B must flow through the resistor in the direction b to a.
EVALUATE: In parts (b) and (c) the change in the circuit causes the flux through circuit B to increase and in part (a) it causes the flux to decrease. Therefore, the direction of the induced current is the same in parts (b) and (c) and opposite in part (a).

29.19.   **IDENTIFY** and **SET UP:**   Lenz's law requires that the flux of the induced current opposes the change in flux.
EXECUTE: (a) Φ_B is $\odot$ and increasing so the flux Φ_{ind} of the induced current is $\otimes$ and the induced current is clockwise.
(b) The current reaches a constant value so Φ_B is constant. $d\Phi_B/dt = 0$ and there is no induced current.
(c) Φ_B is $\odot$ and decreasing, so Φ_{ind} is $\odot$ and current is counterclockwise.
EVALUATE: Only a change in flux produces an induced current. The induced current is in one direction when the current in the outer ring is increasing and is in the opposite direction when that current is decreasing.

29.25.   **IDENTIFY** and **SET UP:** $\mathcal{E} = vBL$. Use Lenz's law to determine the direction of the induced current. The force F_{ext} required to maintain constant speed is equal and opposite to the force F_I that the magnetic field exerts on the rod because of the current in the rod.
EXECUTE: (a) $\mathcal{E} = vBL = (7.50 \text{ m/s})(0.800 \text{ T})(0.500 \text{ m}) = 3.00 \text{ V}$
(b) $\vec{B}$ is into the page. The flux increases as the bar moves to the right, so the magnetic field of the induced current is out of the page inside the circuit. To produce magnetic field in this direction the induced current must be counterclockwise, so from b to a in the rod.

(c) $I = \dfrac{\mathcal{E}}{R} = \dfrac{3.00 \text{ V}}{1.50 \ \Omega} = 2.00 \text{ A}$. $F_I = ILB\sin\phi = (2.00 \text{ A})(0.500 \text{ m})(0.800 \text{ T})\sin 90° = 0.800 \text{ N}$. $\vec{F}_I$ is to the left. To keep the bar moving to the right at constant speed an external force with magnitude $F_{ext} = 0.800 \text{ N}$ and directed to the right must be applied to the bar.

(d) The rate at which work is done by the force F_{ext} is $F_{ext}v = (0.800 \text{ N})(7.50 \text{ m/s}) = 6.00 \text{ W}$. The rate at which thermal energy is developed in the circuit is $I^2R = (2.00 \text{ A})(1.50 \ \Omega) = 6.00 \text{ W}$. These two rates are equal, as is required by conservation of energy.

EVALUATE: The force on the rod due to the induced current is directed to oppose the motion of the rod. This agrees with Lenz's law.

29.27. **IDENTIFY:** A bar moving in a magnetic field has an emf induced across its ends.

SET UP: The induced potential is $\mathcal{E} = vBL \sin\phi$.

EXECUTE: Note that $\phi = 90°$ in all these cases because the bar moved perpendicular to the magnetic field. But the effective length of the bar, $L \sin\theta$, is different in each case.

(a) $\mathcal{E} = vBL \sin\theta = (2.50 \text{ m/s})(1.20 \text{ T})(1.41 \text{ m}) \sin(37.0°) = 2.55 \text{ V}$, with a at the higher potential because positive charges are pushed toward that end.

(b) Same as (a) except $\theta = 53.0°$, giving 3.38 V, with a at the higher potential.

(c) Zero, since the velocity is parallel to the magnetic field.

(d) The bar must move perpendicular to its length, for which the emf is 4.23 V. For $V_b > V_a$, it must move upward and to the left (toward the second quadrant) perpendicular to its length.

EVALUATE: The orientation of the bar affects the potential induced across its ends.

29.29. **IDENTIFY:** Apply Eqs.(29.9) and (29.10).

SET UP: Evaluate the integral if Eq.(29.10) for a path which is a circle of radius r and concentric with the solenoid. The magnetic field of the solenoid is confined to the region inside the solenoid, so $B(r) = 0$ for $r > R$

EXECUTE: **(a)** $\dfrac{d\Phi_B}{dt} = A\dfrac{dB}{dt} = \pi r_1^2 \dfrac{dB}{dt}$.

(b) $E = \dfrac{1}{2\pi r_1}\dfrac{d\Phi_B}{dt} = \dfrac{\pi r_1^2}{2\pi r_1}\dfrac{dB}{dt} = \dfrac{r_1}{2}\dfrac{dB}{dt}$. The direction of $\vec{E}$ is shown in Figure 29.29a.

(c) All the flux is within $r < R$, so outside the solenoid $E = \dfrac{1}{2\pi r_2}\dfrac{d\Phi_B}{dt} = \dfrac{\pi R^2}{2\pi r_2}\dfrac{dB}{dt} = \dfrac{R^2}{2r_2}\dfrac{dB}{dt}$.

(d) The graph is sketched in Figure 29.29b.

(e) At $r = R/2$, $\mathcal{E} = \dfrac{d\Phi_B}{dt} = \pi(R/2)^2 \dfrac{dB}{dt} = \dfrac{\pi R^2}{4}\dfrac{dB}{dt}$.

(f) At $r = R$, $\mathcal{E} = \dfrac{d\Phi_B}{dt} = \pi R^2 \dfrac{dB}{dt}$.

(g) At $r = 2R$, $\mathcal{E} = \dfrac{d\Phi_B}{dt} = \pi R^2 \dfrac{dB}{dt}$.

EVALUATE: The emf is independent of the distance from the center of the cylinder at all points outside it. Even though the magnetic field is zero for $r > R$, the induced electric field is nonzero outside the solenoid and a nonzero emf is induced in a circular turn that has $r > R$.

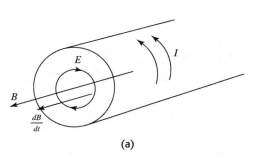

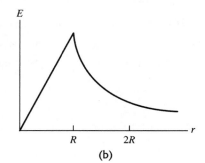

(a)   (b)

Figure 29.29

29.31. **IDENTIFY:** Apply Eq.(29.1) with $\Phi_B = \mu_0 niA$.

SET UP: $A = \pi r^2$, where $r = 0.0110 \text{ m}$. In Eq.(29.11), $r = 0.0350 \text{ m}$.

EXECUTE: $|\mathcal{E}| = \left|\dfrac{d\Phi_B}{dt}\right| = \left|\dfrac{d}{dt}(BA)\right| = \left|\dfrac{d}{dt}(\mu_0 niA)\right| = \mu_0 nA\left|\dfrac{di}{dt}\right|$ and $|\mathcal{E}| = E(2\pi r)$. Therefore, $\left|\dfrac{di}{dt}\right| = \dfrac{E2\pi r}{\mu_0 nA}$.

$\left|\dfrac{di}{dt}\right| = \dfrac{(8.00\times10^{-6}\ \text{V/m})2\pi(0.0350\ \text{m})}{\mu_0(400\ \text{m}^{-1})\pi(0.0110\ \text{m})^2} = 9.21\ \text{A/s}.$

EVALUATE: Outside the solenoid the induced electric field decreases with increasing distance from the axis of the solenoid.

29.35.   **IDENTIFY:**   Apply Eq.(29.14), where $\epsilon = K\epsilon_0$.

      SET UP: $d\Phi_E/dt = 4(8.76\times10^3\ \text{V}\cdot\text{m/s}^4)t^3$. $\epsilon_0 = 8.854\times10^{-12}$ F/m.

      EXECUTE: $\epsilon = \dfrac{i_D}{(d\Phi_E/dt)} = \dfrac{12.9\times10^{-12}\ \text{A}}{4(8.76\times10^3\ \text{V}\cdot\text{m/s}^4)(26.1\times10^{-3}\ \text{s})^3} = 2.07\times10^{-11}$ F/m. The dielectric constant is

$K = \dfrac{\epsilon}{\epsilon_0} = 2.34.$

      EVALUATE: The larger the dielectric constant, the larger is the displacement current for a given $d\Phi_E/dt$.

29.37.   **IDENTIFY:** $q = CV$. For a parallel-plate capacitor, $C = \dfrac{\epsilon A}{d}$, where $\epsilon = K\epsilon_0$. $i_C = dq/dt$. $j_D = \epsilon\dfrac{E}{dt}$.

      SET UP: $E = q/\epsilon A$ so $dE/dt = i_C/\epsilon A$.

      EXECUTE: **(a)** $q = CV = \left(\dfrac{\epsilon A}{d}\right)V = \dfrac{(4.70)\epsilon_0(3.00\times10^{-4}\ \text{m}^2)(120\ \text{V})}{2.50\times10^{-3}\ \text{m}} = 5.99\times10^{-10}$ C.

      (b) $\dfrac{dq}{dt} = i_C = 6.00\times10^{-3}$ A.

      (c) $j_D = \epsilon\dfrac{dE}{dt} = K\epsilon_0\dfrac{i_C}{K\epsilon_0 A} = \dfrac{i_C}{A} = j_C$, so $i_D = i_C = 6.00\times10^{-3}$ A.

      EVALUATE: $i_D = i_C$, so Kirchhoff's junction rule is satisfied where the wire connects to each capacitor plate.

29.43.   **IDENTIFY:**   Apply $\vec{B} = \vec{B}_0 + \mu_0\vec{M}$.

      SET UP: When the magnetic flux is expelled from the material the magnetic field $\vec{B}$ in the material is zero. When the material is completely normal, the magnetization is close to zero.

      EXECUTE: **(a)** When $\vec{B}_0$ is just under $\vec{B}_{c1}$ (threshold of superconducting phase), the magnetic field in the

material must be zero, and $\vec{M} = -\dfrac{\vec{B}_{c1}}{\mu_0} = -\dfrac{(55\times10^{-3}\ \text{T})\hat{i}}{\mu_0} = -(4.38\times10^4\ \text{A/m})\hat{i}$.

      (b) When $\vec{B}_0$ is just over $\vec{B}_{c2}$ (threshold of normal phase), there is zero magnetization, and $\vec{B} = \vec{B}_{c2} = (15.0\ \text{T})\hat{i}$.

      EVALUATE: Between B_{c1} and B_{c2} there are filaments of normal phase material and there is magnetic field along these filaments.

29.45.   **IDENTIFY:**   Apply Faraday's law and Lenz's law.

      SET UP: For a discharging RC circuit, $i(t) = \dfrac{V_0}{R}e^{-t/RC}$, where V_0 is the initial voltage across the capacitor. The resistance of the small loop is $(25)(0.600\ \text{m})(1.0\ \Omega/\text{m}) = 15.0\ \Omega$.

      EXECUTE: **(a)** The large circuit is an RC circuit with a time constant of $\tau = RC = (10\ \Omega)(20\times10^{-6}\ \text{F}) = 200\ \mu\text{s}$. Thus, the current as a function of time is $i = ((100\ \text{V})/(10\ \Omega))e^{-t/200\ \mu s}$. At $t = 200\ \mu\text{s}$, we obtain $i = (10\ \text{A})(e^{-1}) = 3.7$ A.

      (b) Assuming that only the long wire nearest the small loop produces an appreciable magnetic flux through the small loop and referring to the solution of Exercise 29.7 we obtain $\Phi_B = \displaystyle\int_c^{c+a}\dfrac{\mu_0 ib}{2\pi r}dr = \dfrac{\mu_0 ib}{2\pi}\ln\left(1+\dfrac{a}{c}\right)$. Therefore,

the emf induced in the small loop at $t = 200\mu$s is $\mathcal{E} = -\dfrac{d\Phi}{dt} = -\dfrac{\mu_0 b}{2\pi}\ln\left(1+\dfrac{a}{c}\right)\dfrac{di}{dt}$.

$\mathcal{E} = -\dfrac{(4\pi\times10^{-7}\ \text{Wb/A}\cdot\text{m}^2)(0.200\ \text{m})}{2\pi}\ln(3.0)\left(-\dfrac{3.7\ \text{A}}{200\times10^{-6}\text{s}}\right) = +0.81$ mV. Thus, the induced current in the small

loop is $i' = \dfrac{\mathcal{E}}{R} = \dfrac{0.81\ \text{mV}}{15.0\ \Omega} = 54\mu\text{A}.$

      (c) The magnetic field from the large loop is directed out of the page within the small loop. The induced current will act to oppose the decrease in flux from the large loop. Thus, the induced current flows counterclockwise.

EVALUATE: **(d)** Three of the wires in the large loop are too far away to make a significant contribution to the flux in the small loop–as can be seen by comparing the distance c to the dimensions of the large loop.

29.49.   **(a) IDENTIFY:**   (i) $\left|\mathcal{E}\right| = \left|\dfrac{d\Phi_B}{dt}\right|$. The flux is changing because the magnitude of the magnetic field of the wire decreases

with distance from the wire. Find the flux through a narrow strip of area and integrate over the loop to find the total flux.
SET UP:

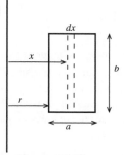

Consider a narrow strip of width dx and a distance x from the long wire, as shown in Figure 29.49a. The magnetic field of the wire at the strip is $B = \mu_0 I / 2\pi x$. The flux through the strip is $d\Phi_B = Bb\ dx = (\mu_0 Ib/2\pi)(dx/x)$

Figure 29.49a

EXECUTE: The total flux through the loop is $\Phi_B = \int d\Phi_B = \left(\dfrac{\mu_0 Ib}{2\pi}\right)\displaystyle\int_r^{r+a} \dfrac{dx}{x}$

$$\Phi_B = \left(\frac{\mu_0 Ib}{2\pi}\right)\ln\left(\frac{r+a}{r}\right)$$

$$\frac{d\Phi_B}{dt} = \frac{d\Phi_B}{dt}\frac{dr}{dt} = \frac{\mu_0 Ib}{2\pi}\left(-\frac{a}{r(r+a)}\right)v$$

$$\left|\mathcal{E}\right| = \frac{\mu_0 Iabv}{2\pi r(r+a)}$$

(ii) **IDENTIFY:** $\mathcal{E} = Bvl$ for a bar of length l moving at speed v perpendicular to a magnetic field B. Calculate the induced emf in each side of the loop, and combine the emfs according to their polarity.
SET UP: The four segments of the loop are shown in Figure 29.49b.

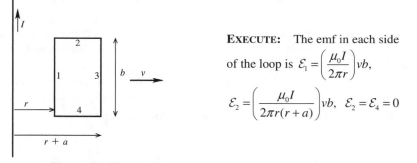

EXECUTE: The emf in each side of the loop is $\mathcal{E}_1 = \left(\dfrac{\mu_0 I}{2\pi r}\right)vb,$

$$\mathcal{E}_2 = \left(\frac{\mu_0 I}{2\pi r(r+a)}\right)vb, \quad \mathcal{E}_2 = \mathcal{E}_4 = 0$$

Figure 29.49b

Both emfs $\mathcal{E}_1$ and $\mathcal{E}_2$ are directed toward the top of the loop so oppose each other. The net emf is

$$\mathcal{E} = \mathcal{E}_1 - \mathcal{E}_2 = \frac{\mu_0 Ivb}{2\pi}\left(\frac{1}{r} - \frac{1}{r+a}\right) = \frac{\mu_0 Iabv}{2\pi r(r+a)}$$

This expression agrees with what was obtained in (i) using Faraday's law.
(b) (i) **IDENTIFY and SET UP:**   The flux of the induced current opposes the change in flux.
EXECUTE: $\vec{B}$ is $\otimes$. Φ_B is $\otimes$ and decreasing, so the flux Φ_{ind} of the induced current is $\otimes$ and the current is clockwise.

(ii) **IDENTIFY** and **SET UP:** Use the right-hand rule to find the force on the positive charges in each side of the loop. The forces on positive charges in segments 1 and 2 of the loop are shown in Figure 29.49c.

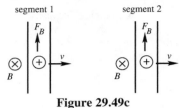

Figure 29.49c

EXECUTE: B is larger at segment 1 since it is closer to the long wire, so F_B is larger in segment 1 and the induced current in the loop is clockwise. This agrees with the direction deduced in (i) using Lenz's law.

(c) **EVALUATE:** When $v = 0$ the induced emf should be zero; the expression in part (a) gives this. When $a \to 0$ the flux goes to zero and the emf should approach zero; the expression in part (a) gives this. When $r \to \infty$ the magnetic field through the loop goes to zero and the emf should go to zero; the expression in part (a) gives this.

29.51. **IDENTIFY:** Apply the results of Example 29.4, so $\mathcal{E}_{max} = N\omega BA$ for N loops.

SET UP: For the minimum ω, let the rotating loop have an area equal to the area of the uniform magnetic field, so $A = (0.100 \text{ m})^2$.

EXECUTE: $N = 400$, $B = 1.5 \text{ T}$, $A = (0.100 \text{ m})^2$ and $\mathcal{E}_{max} = 120 \text{ V}$ gives
$\omega = \mathcal{E}_{max}/NBA = (20 \text{ rad/s})(1 \text{ rev}/2\pi \text{ rad})(60 \text{ s}/1 \text{ min}) = 190 \text{ rpm}$.

EVALUATE: In $\mathcal{E}_{max} = \omega BA$, ω is in rad/s.

29.53. **IDENTIFY:** Apply Faraday's law in the form $\mathcal{E}_{av} = -N\dfrac{\Delta\Phi_B}{\Delta t}$ to calculate the average emf. Apply Lenz's law to calculate the direction of the induced current.

SET UP: $\Phi_B = BA$. The flux changes because the area of the loop changes.

EXECUTE: (a) $\mathcal{E}_{av} = \left|\dfrac{\Delta\Phi_B}{\Delta t}\right| = B\left|\dfrac{\Delta A}{\Delta t}\right| = B\dfrac{\pi r^2}{\Delta t} = (0.950 \text{ T})\dfrac{\pi(0.0650/2 \text{ m})^2}{0.250 \text{ s}} = 0.0126 \text{ V}$.

(b) Since the magnetic field is directed into the page and the magnitude of the flux through the loop is decreasing, the induced current must produce a field that goes into the page. Therefore the current flows from point a through the resistor to point b.

EVALUATE: Faraday's law can be used to find the direction of the induced current. Let $\vec{A}$ be into the page. Then Φ_B is positive and decreasing in magnitude, so $d\Phi_B/dt < 0$. Therefore $\mathcal{E} > 0$ and the induced current is clockwise around the loop.

29.55. **IDENTIFY:** Use Faraday's law to calculate the induced emf and Ohm's law to find the induced current. Use Eq.(27.19) to calculate the magnetic force F_I on the induced current. Use the net force $F - F_I$ in Newton's 2nd law to calculate the acceleration of the rod and use that to describe its motion.

(a) **SET UP:** The forces in the rod are shown in Figure 29.55a.

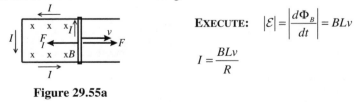

EXECUTE: $|\mathcal{E}| = \left|\dfrac{d\Phi_B}{dt}\right| = BLv$

$I = \dfrac{BLv}{R}$

Figure 29.55a

Use $\mathcal{E} = -\dfrac{d\Phi_B}{dt}$ to find the direction of I: Let $\vec{A}$ be into the page. Then $\Phi_B > 0$. The area of the circuit is increasing, so $\dfrac{d\Phi_B}{dt} > 0$. Then $\mathcal{E} < 0$ and with our direction for $\vec{A}$ this means that $\mathcal{E}$ and I are counterclockwise, as shown in the sketch. The force F_I on the rod due to the induced current is given by $\vec{F}_I = I\vec{l} \times \vec{B}$. This gives $\vec{F}_I$ to the left with magnitude $F_I = ILB = (BLv/R)LB = B^2L^2v/R$. Note that $\vec{F}_I$ is directed to oppose the motion of the rod, as required by Lenz's law.

EVALUATE: The net force on the rod is $F - F_I$, so its acceleration is $a = (F - F_I)/m = (F - B^2L^2v/R)/m$. The rod starts with $v = 0$ and $a = F/m$. As the speed v increases the acceleration a decreases. When $a = 0$ the rod has reached its terminal speed v_t. The graph of v versus t is sketched in Figure 29.55b.

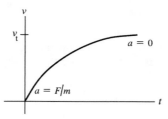

(Recall that a is the slope of the tangent to the v versus t curve.)

Figure 29.55b

(b) EXECUTE: $v = v_t$ when $a = 0$ so $\dfrac{F - B^2L^2v_t/R}{m} = 0$ and $v_t = \dfrac{RF}{B^2L^2}$.

EVALUATE: A large F produces a large v_t. If B is larger, or R is smaller, the induced current is larger at a given v so F_I is larger and the terminal speed is less.

29.59.   **IDENTIFY:**   Find the magnetic field at a distance r from the center of the wire. Divide the rectangle into narrow strips of width dr, find the flux through each strip and integrate to find the total flux.
SET UP: Example 28.8 uses Ampere's law to show that the magnetic field inside the wire, a distance r from the axis, is $B(r) = \mu_0 Ir/2\pi R^2$.

EXECUTE: Consider a small strip of length W and width dr that is a distance r from the axis of the wire, as shown in Figure 29.59. The flux through the strip is $d\Phi_B = B(r)W\,dr = \dfrac{\mu_0 IW}{2\pi R^2} r\,dr$. The total flux through the rectangle is

$$\Phi_B = \int d\Phi_B = \left(\frac{\mu_0 IW}{2\pi R^2}\right)\int_0^R r\,dr = \frac{\mu_0 IW}{4\pi}.$$

EVALUATE: Note that the result is independent of the radius R of the wire.

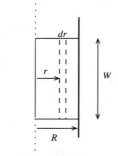

Figure 29.59

29.61.   **(a) and (b) IDENTIFY and Set Up:**

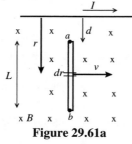

Figure 29.61a

The magnetic field of the wire is given by $B = \dfrac{\mu_0 I}{2\pi r}$ and varies along the length of the bar. At every point along the bar $\vec{B}$ has direction into the page. Divide the bar up into thin slices, as shown in Figure 29.61a.

EXECUTE: The emf $d\mathcal{E}$ induced in each slice is given by $d\mathcal{E} = \vec{v} \times \vec{B} \cdot d\vec{l}$. $\vec{v} \times \vec{B}$ is directed toward the wire, so $d\mathcal{E} = -vB\,dr = -v\left(\dfrac{\mu_0 I}{2\pi r}\right)dr$. The total emf induced in the bar is

$$V_{ba} = \int_a^b d\mathcal{E} = -\int_d^{d+L}\left(\frac{\mu_0 Iv}{2\pi r}\right)dr = -\frac{\mu_0 Iv}{2\pi}\int_d^{d+L}\frac{dr}{r} = -\frac{\mu_0 Iv}{2\pi}\big[\ln(r)\big]_d^{d+L}$$

$$V_{ba} = -\frac{\mu_0 Iv}{2\pi}(\ln(d+L) - \ln(d)) = -\frac{\mu_0 Iv}{2\pi}\ln(1 + L/d)$$

EVALUATE: The minus sign means that V_{ba} is negative, point a is at higher potential than point b. (The force $\vec{F} = q\vec{v} \times \vec{B}$ on positive charge carriers in the bar is towards a, so a is at higher potential.) The potential difference increases when I or v increase, or d decreases.

(c) IDENTIFY: Use Faraday's law to calculate the induced emf.

SET UP: The wire and loop are sketched in Figure 29.61b.

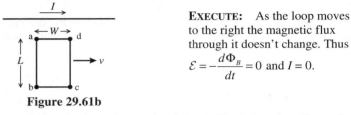

EXECUTE: As the loop moves to the right the magnetic flux through it doesn't change. Thus

$$\mathcal{E} = -\frac{d\Phi_B}{dt} = 0 \text{ and } I = 0.$$

Figure 29.61b

EVALUATE: This result can also be understood as follows. The induced emf in section ab puts point a at higher potential; the induced emf in section dc puts point d at higher potential. If you travel around the loop then these two induced emf's sum to zero. There is no emf in the loop and hence no current.

29.63.   **(a) IDENTIFY:**   Use the expression for motional emf to calculate the emf induced in the rod.

SET UP: The rotating rod is shown in Figure 29.63a.

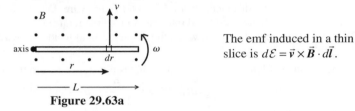

The emf induced in a thin slice is $d\mathcal{E} = \vec{v} \times \vec{B} \cdot d\vec{l}$.

Figure 29.63a

EXECUTE: Assume that $\vec{B}$ is directed out of the page. Then $\vec{v} \times \vec{B}$ is directed radially outward and $dl = dr$, so $\vec{v} \times \vec{B} \cdot d\vec{l} = vB \, dr$

$v = r\omega$ so $d\mathcal{E} = \omega Br \, dr$.

The $d\mathcal{E}$ for all the thin slices that make up the rod are in series so they add:

$$\mathcal{E} = \int d\mathcal{E} = \int_0^L \omega Br \, dr = \tfrac{1}{2}\omega BL^2 = \tfrac{1}{2}(8.80 \text{ rad/s})(0.650 \text{ T})(0.240 \text{ m})^2 = 0.165 \text{ V}$$

EVALUATE: $\mathcal{E}$ increases with ω, B or L^2.

(b) No current flows so there is no IR drop in potential. Thus the potential difference between the ends equals the emf of 0.165 V calculated in part (a).

(c) SET UP: The rotating rod is shown in Figure 29.63b.

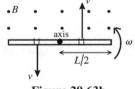

Figure 29.63b

EXECUTE: The emf between the center of the rod and each end is $\mathcal{E} = \tfrac{1}{2}\omega B(L/2)^2 = \tfrac{1}{4}(0.165 \text{ V}) = 0.0412 \text{ V}$, with the direction of the emf from the center of the rod toward each end. The emfs in each half of the rod thus oppose each other and there is no net emf between the ends of the rod.

EVALUATE: ω and B are the same as in part (a) but L of each half is $\tfrac{1}{2}L$ for the whole rod. $\mathcal{E}$ is proportional to L^2, so is smaller by a factor of $\tfrac{1}{4}$.

29.65.   **(a) IDENTIFY:**   Use Faraday's law to calculate the induced emf, Ohm's law to calculate I, and Eq.(27.19) to calculate the force on the rod due to the induced current.

SET UP: The force on the wire is shown in Figure 29.65.

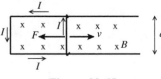

EXECUTE: When the wire has speed v the induced emf is $\mathcal{E} = Bva$ and the induced current is $I = \mathcal{E}/R = \dfrac{Bva}{R}$

Figure 29.65

The induced current flows upward in the wire as shown, so the force $\vec{F} = I\vec{l} \times \vec{B}$ exerted by the magnetic field on the induced current is to the left. $\vec{F}$ opposes the motion of the wire, as it must by Lenz's law. The magnitude of the force is $F = IaB = B^2a^2v/R$.

(b) Apply $\sum \vec{F} = m\vec{a}$ to the wire. Take $+x$ to be toward the right and let the origin be at the location of the wire at $t = 0$, so $x_0 = 0$.

$\sum F_x = ma_x$ says $-F = ma_x$

$a_x = -\dfrac{F}{m} = -\dfrac{B^2a^2v}{mR}$

Use this expression to solve for $v(t)$:

$a_x = \dfrac{dv}{dt} = -\dfrac{B^2a^2v}{mR}$ and $\dfrac{dv}{v} = -\dfrac{B^2a^2}{mR}dt$

$\displaystyle\int_{v_0}^{v} \dfrac{dv'}{v'} = -\dfrac{B^2a^2}{mR}\int_0^t dt'$

$\ln(v) - \ln(v_0) = -\dfrac{B^2a^2t}{mR}$

$\ln\left(\dfrac{v}{v_0}\right) = -\dfrac{B^2a^2t}{mR}$ and $v = v_0 e^{-B^2a^2t/mR}$

Note: At $t = 0$, $v = v_0$ and $v \to 0$ when $t \to \infty$

Now solve for $x(t)$:

$v = \dfrac{dx}{dt} = v_0 e^{-B^2a^2t/mR}$ so $dx = v_0 e^{-B^2a^2t/mR}dt$

$\displaystyle\int_0^x dx' = \int_0^t v_0 e^{-B^2a^2t/mR}dt'$

$x = v_0\left(-\dfrac{mR}{B^2a^2}\right)\left[e^{-B^2a^2t'/mR}\right]_0^t = \dfrac{mRv_0}{B^2a^2}\left(1 - e^{-B^2a^2t/mR}\right)$

Comes to rest implies $v = 0$. This happens when $t \to \infty$.

$t \to \infty$ gives $x = \dfrac{mRv_0}{B^2a^2}$. Thus this is the distance the wire travels before coming to rest.

EVALUATE: The motion of the slide wire causes an induced emf and current. The magnetic force on the induced current opposes the motion of the wire and eventually brings it to rest. The force and acceleration depend on v and are constant. If the acceleration were constant, not changing from its initial value of $a_x = -B^2a^2v_0/mR$, then the stopping distance would be $x = -v_0^2/2a_x = mRv_0/2B^2a^2$. The actual stopping distance is twice this.

29.67.   **IDENTIFY:**   Use Eq.(29.10) to calculate the induced electric field at each point and then use $\vec{F} = q\vec{E}$.

SET UP:

Figure 29.67a

Apply $\oint \vec{E} \cdot d\vec{l} = -\dfrac{d\Phi_B}{dt}$ to a concentric circle of radius r, as shown in Figure 29.67a. Take $\vec{A}$ to be into the page, in the direction of $\vec{B}$.

EXECUTE: B increasing then gives $\dfrac{d\Phi_B}{dt} > 0$, so $\oint \vec{E} \cdot d\vec{l}$ is negative. This means that E is tangent to the circle in the counterclockwise direction, as shown in Figure 29.67b.

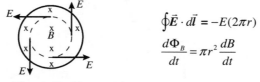

Figure 29.67b

$\oint \vec{E} \cdot d\vec{l} = -E(2\pi r)$

$\dfrac{d\Phi_B}{dt} = \pi r^2 \dfrac{dB}{dt}$

$-E(2\pi r) = -\pi r^2 \dfrac{dB}{dt}$ so $E = \tfrac{1}{2}r\dfrac{dB}{dt}$

point _a_   The induced electric field and the force on q are shown in Figure 29.67c.

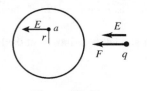

$$F = qE = \tfrac{1}{2}qr\frac{dB}{dt}$$

$\vec{F}$ is to the left

($\vec{F}$ is in the same direction as $\vec{E}$ since

q is positive.)

Figure 29.67c

point _b_   The induced electric field and the force on q are shown in Figure 29.67d.

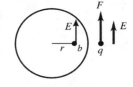

$$F = qE = \tfrac{1}{2}qr\frac{dB}{dt}$$

$\vec{F}$ is toward the top of the page.

Figure 29.67d

point _c_ $r = 0$ here, so $E = 0$ and $F = 0$.

EVALUATE: If there were a concentric conducting ring of radius r in the magnetic field region, Lenz's law tells us that the increasing magnetic field would induce a counterclockwise current in the ring. This agrees with the direction of the force we calculated for the individual positive point charges.

29.73.   **IDENTIFY:**   The conduction current density is related to the electric field by Ohm's law. The displacement current density is related to the rate of change of the electric field by Eq.(29.16).

SET UP: $dE/dt = \omega E_0 \cos \omega t$

EXECUTE: **(a)** $j_C(\text{max}) = \dfrac{E_0}{\rho} = \dfrac{0.450 \text{ V/m}}{2300 \ \Omega \cdot \text{m}} = 1.96 \times 10^{-4} \text{ A/m}^2$

(b) $j_D(\text{max}) = \epsilon_0 \left(\dfrac{dE}{dt} \right)_{\text{max}} = \epsilon_0 \omega E_0 = 2\pi \epsilon_0 f E_0 = 2\pi \epsilon_0 (120 \text{ Hz})(0.450 \text{ V/m}) = 3.00 \times 10^{-9} \text{ A/m}^2$

(c) If $j_C = j_D$ then $\dfrac{E_0}{\rho} = \omega \epsilon_0 E_0$ and $\omega = \dfrac{1}{\rho \epsilon_0} = 4.91 \times 10^7 \text{ rad/s}$

$$f = \frac{\omega}{2\pi} = \frac{4.91 \times 10^7 \text{ rad/s}}{2\pi} = 7.82 \times 10^6 \text{ Hz}.$$

EVALUATE: **(d)** The two current densities are out of phase by $90°$ because one has a sine function and the other has a cosine, so the displacement current leads the conduction current by $90°$.

30

INDUCTANCE

30.1. **IDENTIFY** and **SET UP:** Apply Eq.(30.4).

EXECUTE: **(a)** $|\mathcal{E}_2| = M \left| \dfrac{di_1}{dt} \right| = (3.25 \times 10^{-4} \text{ H})(830 \text{ A/s}) = 0.270 \text{ V}$; yes, it is constant.

(b) $|\mathcal{E}_1| = M \left| \dfrac{di_2}{dt} \right|$; M is a property of the pair of coils so is the same as in part (a). Thus $|\mathcal{E}_1| = 0.270$ V.

EVALUATE: The induced emf is the same in either case. A constant di/dt produces a constant emf.

30.5. **IDENTIFY** and **SET UP:** Apply Eq.(30.5).

EXECUTE: **(a)** $M = \dfrac{N_2 \Phi_{B2}}{i_1} = \dfrac{400(0.0320 \text{ Wb})}{6.52 \text{ A}} = 1.96$ H

(b) $M = \dfrac{N_1 \Phi_{B1}}{i_2}$ so $\Phi_{B1} = \dfrac{M i_2}{N_1} = \dfrac{(1.96 \text{ H})(2.54 \text{ A})}{700} = 7.11 \times 10^{-3}$ Wb

EVALUATE: M relates the current in one coil to the flux through the other coil. Eq.(30.5) shows that M is the same for a pair of coils, no matter which one has the current and which one has the flux.

30.9. **IDENTIFY** and **SET UP:** Apply $|\mathcal{E}| = L|di/dt|$. Apply Lenz's law to determine the direction of the induced emf in the coil.

EXECUTE: **(a)** $|\mathcal{E}| = L(di/dt) = (0.260 \text{ H})(0.0180 \text{ A/s}) = 4.68 \times 10^{-3}$ V

(b) Terminal a is at a higher potential since the coil pushes current through from b to a and if replaced by a battery it would have the $+$ terminal at a.

EVALUATE: The induced emf is directed so as to oppose the decrease in the current.

30.11. **IDENTIFY** and **SET UP:** Use Eq.(30.6) to relate L to the flux through each turn of the solenoid. Use Eq.(28.23) for the magnetic field through the solenoid.

EXECUTE: $L = \dfrac{N\Phi_B}{i}$. If the magnetic field is uniform inside the solenoid $\Phi_B = BA$. From Eq.(28.23),

$B = \mu_0 n i = \mu_0 \left(\dfrac{N}{l} \right) i$ so $\Phi_B = \dfrac{\mu_0 N i A}{l}$. Then $L = \dfrac{N}{i} \left(\dfrac{\mu_0 N i A}{l} \right) = \dfrac{\mu_0 N^2 A}{l}$.

EVALUATE: Our result is the same as L for a torodial solenoid calculated in Example 30.3, except that the average circumference $2\pi r$ of the toroid is replaced by the length l of the straight solenoid.

30.13. **IDENTIFY** and **SET UP:** Use Eq.(30.9) to relate the energy stored to the inductance. Example 30.3 gives the inductance of a toroidal solenoid to be $L = \dfrac{\mu_0 N^2 A}{2\pi r}$, so once we know L we can solve for N.

EXECUTE: $U = \frac{1}{2}LI^2$ so $L = \dfrac{2U}{I^2} = \dfrac{2(0.390 \text{ J})}{(12.0 \text{ A})^2} = 5.417 \times 10^{-3}$ H

$N = \sqrt{\dfrac{2\pi r L}{\mu_0 A}} = \sqrt{\dfrac{2\pi (0.150 \text{ m})(5.417 \times 10^{-3} \text{ H})}{(4\pi \times 10^{-7} \text{ T} \cdot \text{m/A})(5.00 \times 10^{-4} \text{ m}^2)}} = 2850.$

EVALUATE: L and hence U increase according to the square of N.

30.15. **IDENTIFY:** A current-carrying inductor has a magnetic field inside of itself and hence stores magnetic energy.

(a) SET UP: The magnetic field inside a solenoid is $B = \mu_0 n I$.

EXECUTE: $B = \dfrac{(4\pi \times 10^{-7} \text{ T} \cdot \text{m/A})(400)(80.0 \text{ A})}{0.250 \text{ m}} = 0.161$ T

(b) SET UP: The energy density in a magnetic field is $u = \dfrac{B^2}{2\mu_0}$.

EXECUTE: $u = \dfrac{(0.161 \text{ T})^2}{2(4\pi \times 10^{-7} \text{ T} \cdot \text{m/A})} = 1.03 \times 10^4 \text{ J/m}^3$

(c) SET UP: The total stored energy is $U = uV$.

EXECUTE: $U = uV = u(lA) = (1.03 \times 10^4 \text{ J/m}^3)(0.250 \text{ m})(0.500 \times 10^{-4} \text{ m}^2) = 0.129 \text{ J}$

(d) SET UP: The energy stored in an inductor is $U = \frac{1}{2} L I^2$.

EXECUTE: Solving for L and putting in the numbers gives

$$L = \frac{2U}{I^2} = \frac{2(0.129 \text{ J})}{(80.0 \text{ A})^2} = 4.02 \times 10^{-5} \text{ H}$$

EVALUATE: An inductor stores its energy in the magnetic field inside of it.

30.19.   **IDENTIFY:**   Apply Kirchhoff's loop rule to the circuit. $i(t)$ is given by Eq.(30.14).

SET UP: The circuit is sketched in Figure 30.19.

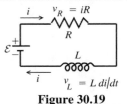

$\dfrac{di}{dt}$ is positive as the current

increases from its initial value of zero.

Figure 30.19

EXECUTE: $\mathcal{E} - v_R - v_L = 0$

$\mathcal{E} - iR - L\dfrac{di}{dt} = 0$ so $i = \dfrac{\mathcal{E}}{R}\left(1 - e^{-(R/L)t}\right)$

(a) Initially ($t = 0$), $i = 0$ so $\mathcal{E} - L\dfrac{di}{dt} = 0$

$\dfrac{di}{dt} = \dfrac{\mathcal{E}}{L} = \dfrac{6.00 \text{ V}}{2.50 \text{ H}} = 2.40 \text{ A/s}$

(b) $\mathcal{E} - iR - L\dfrac{di}{dt} = 0$ (Use this equation rather than Eq.(30.15) since i rather than t is given.)

Thus $\dfrac{di}{dt} = \dfrac{\mathcal{E} - iR}{L} = \dfrac{6.00 \text{ V} - (0.500 \text{ A})(8.00 \text{ }\Omega)}{2.50 \text{ H}} = 0.800 \text{ A/s}$

(c) $i = \dfrac{\mathcal{E}}{R}\left(1 - e^{-(R/L)t}\right) = \left(\dfrac{6.00 \text{ V}}{8.00 \text{ }\Omega}\right)\left(1 - e^{-(8.00 \text{ }\Omega/2.50 \text{ H})(0.250 \text{ s})}\right) = 0.750 \text{ A}(1 - e^{-0.800}) = 0.413 \text{ A}$

(d) Final steady state means $t \to \infty$ and $\dfrac{di}{dt} \to 0$, so $\mathcal{E} - iR = 0$.

$i = \dfrac{\mathcal{E}}{R} = \dfrac{6.00 \text{ V}}{8.00 \text{ }\Omega} = 0.750 \text{ A}$

EVALUATE: Our results agree with Fig.30.12 in the textbook. The current is initially zero and increases to its final value of $\mathcal{E}/R$. The slope of the current in the figure, which is di/dt, decreases with t.

30.21.   **IDENTIFY:** $i = \mathcal{E}/R(1 - e^{-t/\tau})$, with $\tau = L/R$. The energy stored in the inductor is $U = \frac{1}{2}Li^2$.

SET UP: The maximum current occurs after a long time and is equal to $\mathcal{E}/R$.

EXECUTE: **(a)** $i_{max} = \mathcal{E}/R$ so $i = i_{max}/2$ when $(1 - e^{-t/\tau}) = \frac{1}{2}$ and $e^{-t/\tau} = \frac{1}{2}$. $-t/\tau = \ln\left(\frac{1}{2}\right)$.

$t = \dfrac{L\ln 2}{R} = \dfrac{(\ln 2)(1.25 \times 10^{-3} \text{ H})}{50.0 \text{ }\Omega} = 17.3 \text{ }\mu s$

(b) $U = \frac{1}{2}U_{max}$ when $i = i_{max}/\sqrt{2}$. $1 - e^{-t/\tau} = 1/\sqrt{2}$, so $e^{-t/\tau} = 1 - 1/\sqrt{2} = 0.2929$. $t = -L\ln(0.2929)/R = 30.7 \text{ }\mu s$.

EVALUATE: $\tau = L/R = 2.50 \times 10^{-5} \text{ s} = 25.0 \text{ }\mu s$. The time in part (a) is 0.692τ and the time in part (b) is 1.23τ.

30.25.   **IDENTIFY:**   Apply the concepts of current decay in an R-L circuit. Apply the loop rule to the circuit. $i(t)$ is given by Eq.(30.18). The voltage across the resistor depends on i and the voltage across the inductor depends on di/dt.

SET UP: The circuit with S_1 closed and S_2 open is sketched in Figure 30.25a.

$\mathcal{E} - iR - L\dfrac{di}{dt} = 0$

Figure 30.25a

Constant current established means $\frac{di}{dt} = 0$.

EXECUTE: $i = \frac{\mathcal{E}}{R} = \frac{60.0 \text{ V}}{240 \text{ }\Omega} = 0.250 \text{ A}$

(a) SET UP: The circuit with S_2 closed and S_1 open is shown in Figure 30.25b.

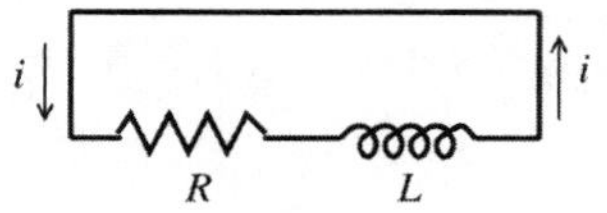

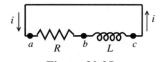

$i = I_0 e^{-(R/L)t}$

At $t = 0$, $i = I_0 = 0.250$ A

Figure 30.25b

The inductor prevents an instantaneous change in the current; the current in the inductor just after S_2 is closed and S_1 is opened equals the current in the inductor just before this is done.

(b) EXECUTE: $i = I_0 e^{-(R/L)t} = (0.250 \text{ A})e^{-(240 \text{ }\Omega/0.160 \text{ H})(4.00\times10^{-4} \text{ s})} = (0.250 \text{ A})e^{-0.600} = 0.137 \text{ A}$

(c) SET UP: See Figure 30.25c.

Figure 30.25c

EXECUTE: If we trace around the loop in the direction of the current the potential falls as we travel through the resistor so it must rise as we pass through the inductor: $v_{ab} > 0$ and $v_{bc} < 0$. So point c is at higher potential than point b.

$v_{ab} + v_{bc} = 0$ and $v_{bc} = -v_{ab}$

Or, $v_{cb} = v_{ab} = iR = (0.137 \text{ A})(240 \text{ }\Omega) = 32.9 \text{ V}$

(d) $i = I_0 e^{-(R/L)t}$

$i = \frac{1}{2}I_0$ says $\frac{1}{2}I_0 = I_0 e^{-(R/L)t}$ and $\frac{1}{2} = e^{-(R/L)t}$

Taking natural logs of both sides of this equation gives $\ln(\frac{1}{2}) = -Rt/L$

$t = \left(\frac{0.160 \text{ H}}{240 \text{ }\Omega}\right)\ln 2 = 4.62\times10^{-4} \text{ s}$

EVALUATE: The current decays, as shown in Fig. 30.13 in the textbook. The time constant is $\tau = L/R = 6.67\times10^{-4}$ s. The values of t in the problem are less than one time constant. At any instant the potential drop across the resistor (in the direction of the current) equals the potential rise across the inductor.

30.27. **IDENTIFY:** $i(t)$ is given by Eq.(30.14).

SET UP: The power input from the battery is $\mathcal{E}i$. The rate of dissipation of energy in the resistance is i^2R. The voltage across the inductor has magnitude Ldi/dt, so the rate at which energy is being stored in the inductor is $iLdi/dt$.

EXECUTE: **(a)** $P = \mathcal{E}i = \mathcal{E}I_0(1 - e^{-(R/L)t}) = \frac{\mathcal{E}^2}{R}(1 - e^{-(R/L)t}) = \frac{(6.00 \text{ V})^2}{8.00 \text{ }\Omega}(1 - e^{-(8.00 \text{ }\Omega/2.50 \text{ H})t})$.

$P = (4.50 \text{ W})(1 - e^{-(3.20 \text{ s}^{-1})t})$.

(b) $P_R = i^2R = \frac{\mathcal{E}^2}{R}(1 - e^{-(R/L)t})^2 = \frac{(6.00 \text{ V})^2}{8.00 \text{ }\Omega}(1 - e^{-(8.00 \text{ }\Omega/2.50 \text{ H})t})^2 = (4.50 \text{ W})(1 - e^{-(3.20 \text{ s}^{-1})t})^2$

(c) $P_L = iL\frac{di}{dt} = \frac{\mathcal{E}}{R}(1 - e^{-(R/L)t})L\left(\frac{\mathcal{E}}{L}e^{-(R/L)t}\right) = \frac{\mathcal{E}^2}{R}(e^{-(R/L)t} - e^{-2(R/L)t})$

$P_L = (4.50 \text{ W})(e^{-(3.20 \text{ s}^{-1})t} - e^{-(6.40 \text{ s}^{-1})t})$.

EVALUATE: **(d)** Note that if we expand the square in part (b), then parts (b) and (c) add to give part (a), and the total power delivered is dissipated in the resistor and inductor. Conservation of energy requires that this be so.

30.29. **IDENTIFY:** The energy moves back and forth between the inductor and capacitor.

(a) SET UP: The period is $T = \frac{1}{f} = \frac{1}{\omega/2\pi} = \frac{2\pi}{\omega} = 2\pi\sqrt{LC}$.

EXECUTE: Solving for L gives

$$L = \frac{T^2}{4\pi^2 C} = \frac{(8.60\times10^{-5} \text{ s})^2}{4\pi^2 (7.50\times10^{-9} \text{ C})} = 2.50\times10^{-2} \text{ H} = 25.0 \text{ mH}$$

(b) SET UP: The charge on a capacitor is $Q = CV$.
EXECUTE: $Q = CV = (7.50 \times 10^{-9} \text{ F})(12.0 \text{ V}) = 9.00 \times 10^{-8} \text{ C}$
(c) SET UP: The stored energy is $U = Q^2/2C$.

EXECUTE: $U = \dfrac{\left(9.00 \times 10^{-8} \text{ C}\right)^2}{2\left(7.50 \times 10^{-9} \text{ F}\right)} = 5.40 \times 10^{-7} \text{ J}$

(d) SET UP: The maximum current occurs when the capacitor is discharged, so the inductor has all the initial energy. $U_L + U_C = U_{\text{Total}}.$ $\frac{1}{2}LI^2 + 0 = U_{\text{Total}}.$
EXECUTE: Solve for the current:

$$I = \sqrt{\frac{2U_{\text{Total}}}{L}} = \sqrt{\frac{2\left(5.40 \times 10^{-7} \text{ J}\right)}{2.50 \times 10^{-2} \text{ H}}} = 6.58 \times 10^{-3} \text{ A} = 6.58 \text{ mA}$$

EVALUATE: The energy oscillates back and forth forever. However if there is any resistance in the circuit, no matter how small, all this energy will eventually be dissipated as heat in the resistor.

30.31. **IDENTIFY** and **SET UP:**   The angular frequency is given by Eq.(30.22). $q(t)$ and $i(t)$ are given by Eqs.(30.21) and (30.23). The energy stored in the capacitor is $U_C = \frac{1}{2}CV^2 = q^2/2C$. The energy stored in the inductor is $U_L = \frac{1}{2}Li^2$.

EXECUTE: **(a)** $\omega = \dfrac{1}{\sqrt{LC}} = \dfrac{1}{\sqrt{(1.50 \text{ H})(6.00 \times 10^{-5} \text{ F})}} = 105.4 \text{ rad/s}$, which rounds to 105 rad/s. The period is

given by $T = \dfrac{2\pi}{\omega} = \dfrac{2\pi}{105.4 \text{ rad/s}} = 0.0596 \text{ s}$

(b) The circuit containing the battery and capacitor is sketched in Figure 30.31.

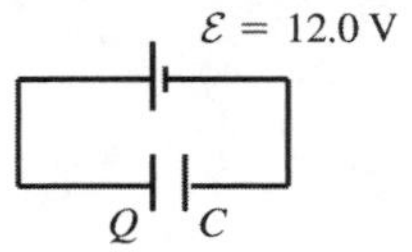

$\mathcal{E} - \dfrac{Q}{C} = 0$

$Q = \mathcal{E}C = (12.0 \text{ V})(6.00 \times 10^{-5} \text{ F}) = 7.20 \times 10^{-4} \text{ C}$

Figure 30.31

(c) $U = \frac{1}{2}CV^2 = \frac{1}{2}(6.00 \times 10^{-5} \text{ F})(12.0 \text{ V})^2 = 4.32 \times 10^{-3} \text{ J}$
(d) $q = Q\cos(\omega t + \phi)$ (Eq.30.21)
$q = Q$ at $t = 0$ so $\phi = 0$
$q = Q\cos\omega t = (7.20 \times 10^{-4} \text{ C})\cos([105.4 \text{ rad/s}][0.0230 \text{ s}]) = -5.42 \times 10^{-4} \text{ C}$
The minus sign means that the capacitor has discharged fully and then partially charged again by the current maintained by the inductor; the plate that initially had positive charge now has negative charge and the plate that initially had negative charge now has positive charge.
(e) $i = -\omega Q\sin(\omega t + \phi)$ (Eq.30.23)
$i = -(105 \text{ rad/s})(7.20 \times 10^{-4} \text{ C})\sin([105.4 \text{ rad/s}][0.0230 \text{ s}]) = -0.050 \text{ A}$
The negative sign means the current is counterclockwise in Figure 30.15 in the textbook.
or

$\frac{1}{2}Li^2 + \dfrac{q^2}{2C} = \dfrac{Q^2}{2C}$ gives $i = \pm\sqrt{\dfrac{1}{LC}}\sqrt{Q^2 - q^2}$ (Eq.30.26)

$i = \pm(105 \text{ rad/s})\sqrt{(7.20 \times 10^{-4} \text{ C})^2 - (-5.42 \times 10^{-4} \text{ C})^2} = \pm 0.050 \text{ A}$, which checks.

(f) $U_C = \dfrac{q^2}{2C} = \dfrac{(-5.42 \times 10^{-4} \text{ C})^2}{2(6.00 \times 10^{-5} \text{ F})} = 2.45 \times 10^{-3} \text{ J}$

$U_L = \frac{1}{2}Li^2 = \frac{1}{2}(1.50 \text{ H})(0.050 \text{ A})^2 = 1.87 \times 10^{-3} \text{ J}$

EVALUATE: Note that $U_C + U_L = 2.45 \times 10^{-3} \text{ J} + 1.87 \times 10^{-3} \text{ J} = 4.32 \times 10^{-3} \text{ J}.$

This agrees with the total energy initially stored in the capacitor, $U = \dfrac{Q^2}{2C} = \dfrac{(7.20 \times 10^{-4} \text{ C})^2}{2(6.00 \times 10^{-5} \text{ F})} = 4.32 \times 10^{-3} \text{ J}.$

Energy is conserved. At some times there is energy stored in both the capacitor and the inductor. When $i = 0$ all the energy is stored in the capacitor and when $q = 0$ all the energy is stored in the inductor. But at all times the total energy stored is the same.

30.33. **IDENTIFY:**   Apply energy conservation and Eqs. (30.22) and (30.23).

SET UP: If I is the maximum current, $\frac{1}{2}LI^2 = \dfrac{Q^2}{2C}$. For the inductor, $U_L = \frac{1}{2}Li^2$.

EXECUTE: (a) $\frac{1}{2}LI^2 = \frac{Q^2}{2C}$ gives $Q = i\sqrt{LC} = (0.750 \text{ A})\sqrt{(0.0800 \text{ H})(1.25\times10^{-9} \text{ F})} = 7.50\times10^{-6} \text{ C}$.

(b) $\omega = \frac{1}{\sqrt{LC}} = \frac{1}{\sqrt{(0.0800 \text{ H})(1.25\times10^{-9} \text{ F})}} = 1.00\times10^5 \text{ rad/s}$. $f = \frac{\omega}{2\pi} = 1.59\times10^4 \text{ Hz}$.

(c) $q = Q$ at $t = 0$ means $\phi = 0$. $i = -\omega Q \sin(\omega t)$, so

$i = -(1.00\times10^5 \text{ rad/s})(7.50\times10^{-6} \text{ C})\sin([1.00\times10^5 \text{ rad/s}][2.50\times10^{-3} \text{ s}]) = -0.7279 \text{ A}$.

$U_L = \frac{1}{2}Li^2 = \frac{1}{2}(0.0800 \text{ H})(-0.7279 \text{ A})^2 = 0.0212 \text{ J}$.

EVALUATE: The total energy of the system is $\frac{1}{2}LI^2 = 0.0225 \text{ J}$. At $t = 2.50$ ms, the current is close to its maximum value and most of the system's energy is stored in the inductor.

30.41. **IDENTIFY:** Evaluate Eq.(30.29).

SET UP: The angular frequency of the circuit is ω'.

EXECUTE: (a) When $R = 0$, $\omega_0 = \frac{1}{\sqrt{LC}} = \frac{1}{\sqrt{(0.450 \text{ H})(2.50\times10^{-5} \text{ F})}} = 298 \text{ rad/s}$.

(b) We want $\frac{\omega}{\omega_0} = 0.95$, so $\frac{(1/LC - R^2/4L^2)}{1/LC} = 1 - \frac{R^2C}{4L} = (0.95)^2$. This gives

$R = \sqrt{\frac{4L}{C}(1-(0.95)^2)} = \sqrt{\frac{4(0.450 \text{ H})(0.0975)}{(2.50\times10^{-5} \text{ F})}} = 83.8 \text{ }\Omega$.

EVALUATE: When R increases, the angular frequency decreases and approaches zero as $R \rightarrow 2\sqrt{L/C}$.

30.43. **IDENTIFY:** The emf $\mathcal{E}_2$ in solenoid 2 produced by changing current i_1 in solenoid 1 is given by $\mathcal{E}_2 = M\left|\frac{\Delta i_1}{\Delta t}\right|$. The mutual inductance of two solenoids is derived in Example 30.1. For the two solenoids in this problem $M = \frac{\mu_0 A N_1 N_2}{l}$, where A is the cross-sectional area of the inner solenoid and l is the length of the outer solenoid.

SET UP: $\mu_0 = 4\pi\times10^{-7} \text{ T}\cdot\text{m/A}$. Let the outer solenoid be solenoid 1.

EXECUTE: (a) $M = \frac{(4\pi\times10^{-7} \text{ T}\cdot\text{m/A})\pi(6.00\times10^{-4} \text{ m})^2(6750)(15)}{0.500 \text{ m}} = 2.88\times10^{-7} \text{ H} = 0.288 \text{ }\mu\text{H}$

(b) $\mathcal{E}_2 = \left|\frac{\Delta i_1}{\Delta t}\right| = (2.88\times10^{-7} \text{ H})(37.5 \text{ A/s}) = 1.08\times10^{-5} \text{ V}$

EVALUATE: If current in the inner solenoid changed at 37.5 A/s, the emf induced in the outer solenoid would be 1.08×10^{-5} V.

30.47. **IDENTIFY:** Apply $\mathcal{E} = -L\frac{di}{dt}$ to the series and parallel combinations.

SET UP: In series, $i_1 = i_2$ and the voltages add. In parallel the voltages are the same and the currents add.

EXECUTE: (a) Series: $L_1\frac{di_1}{dt} + L_2\frac{di_2}{dt} = L_{eq}\frac{di}{dt}$, but $i_1 = i_2 = i$ for series components so $\frac{di_1}{dt} = \frac{di_2}{dt} = \frac{di}{dt}$ and $L_1 + L_2 = L_{eq}$.

(b) Parallel: Now $L_1\frac{di_1}{dt} = L_2\frac{di_2}{dt} = L_{eq}\frac{di}{dt}$, where $i = i_1 + i_2$. Therefore, $\frac{di}{dt} = \frac{di_1}{dt} + \frac{di_2}{dt}$. But $\frac{di_1}{dt} = \frac{L_{eq}}{L_1}\frac{di}{dt}$ and

$\frac{di_2}{dt} = \frac{L_{eq}}{L_2}\frac{di}{dt}$. $\frac{di}{dt} = \frac{L_{eq}}{L_1}\frac{di}{dt} + \frac{L_{eq}}{L_2}\frac{di}{dt}$ and $L_{eq} = \left(\frac{1}{L_1} + \frac{1}{L_2}\right)^{-1}$.

EVALUATE: Inductors in series and parallel combine in the same way as resistors.

30.49. (a) **IDENTIFY and SET UP:** An end view is shown in Figure 30.49.

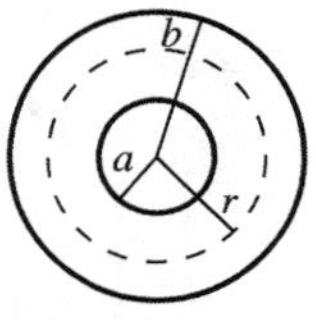

Apply Ampere's law to a circular path of radius r.

$\oint \vec{B} \cdot d\vec{l} = \mu_0 I_{encl}$

Figure 30.49

EXECUTE: $\oint \vec{B} \cdot d\vec{l} = B(2\pi r)$

$I_{encl} = i$, the current in the inner conductor

Thus $B(2\pi r) = \mu_0 i$ and $B = \dfrac{\mu_0 i}{2\pi r}$.

(b) IDENTIFY and **SET UP:**   Follow the procedure specified in the problem.

EXECUTE: $u = \dfrac{B^2}{2\mu_0}$

$dU = u\, dV$, where $dV = 2\pi r l\, dr$

$dU = \dfrac{1}{2\mu_0}\left(\dfrac{\mu_0 i}{2\pi r}\right)^2 (2\pi r l)\, dr = \dfrac{\mu_0 i^2 l}{4\pi r}\, dr$

(c) $U = \int dU = \dfrac{\mu_0 i^2 l}{4\pi}\displaystyle\int_a^b \dfrac{dr}{r} = \dfrac{\mu_0 i^2 l}{4\pi}[\ln r]_a^b$

$U = \dfrac{\mu_0 i^2 l}{4\pi}(\ln b - \ln a) = \dfrac{\mu_0 i^2 l}{4\pi}\ln\left(\dfrac{b}{a}\right)$

(d) Eq.(30.9): $U = \tfrac{1}{2}Li^2$

Part (c): $U = \dfrac{\mu_0 i^2 l}{4\pi}\ln\left(\dfrac{b}{a}\right)$

$\tfrac{1}{2}Li^2 = \dfrac{\mu_0 i^2 l}{4\pi}\ln\left(\dfrac{b}{a}\right)$

$L = \dfrac{\mu_0 l}{2\pi}\ln\left(\dfrac{b}{a}\right).$

EVALUATE: The value of L we obtain from these energy considerations agrees with L calculated in part (d) of Problem 30.48 by considering flux and Eq.(30.6)

30.53.   **IDENTIFY** and **SET UP:**   Follow the procedure specified in the problem. $L = 2.50$ H, $R = 8.00\ \Omega$,

$\mathcal{E} = 6.00$ V. $i = (\mathcal{E}/R)(1 - e^{-t/\tau})$, $\tau = L/R$

EXECUTE: (a) Eq.(30.9): $U_L = \tfrac{1}{2}Li^2$

$t = \tau$ so $i = (\mathcal{E}/R)(1 - e^{-1}) = (6.00\ \text{V}/8.00\ \Omega)(1 - e^{-1}) = 0.474$ A

Then $U_L = \tfrac{1}{2}Li^2 = \tfrac{1}{2}(2.50\ \text{H})(0.474\ \text{A})^2 = 0.281$ J

Exercise 30.27 (c): $P_L = \dfrac{dU_L}{dt} = Li\dfrac{di}{dt}$

$i = \left(\dfrac{\mathcal{E}}{R}\right)(1 - e^{-t/\tau}); \quad \dfrac{di}{dt} = \left(\dfrac{\mathcal{E}}{L}\right)e^{-(R/L)t} = \dfrac{\mathcal{E}}{L}e^{-t/\tau}$

$P_L = L\left(\dfrac{\mathcal{E}}{R}(1 - e^{-t/\tau})\right)\left(\dfrac{\mathcal{E}}{L}e^{-t/\tau}\right) = \dfrac{\mathcal{E}^2}{R}(e^{-t/\tau} - e^{-2\pm t/\tau})$

$U_L = \displaystyle\int_0^\tau P_L dt = \dfrac{\mathcal{E}^2}{R}\int_0^\tau (e^{-t/\tau} - e^{-2t/\tau})dt = \dfrac{\mathcal{E}^2}{R}\left[-\tau e^{-t/\tau} + \dfrac{\tau}{2}e^{-2t/\tau}\right]_0^\tau$

$U_L = -\dfrac{\mathcal{E}^2}{R}\tau\left[e^{-t/\tau} - \tfrac{1}{2}e^{-2t/\tau}\right]_0^\tau = \dfrac{\mathcal{E}^2}{R}\tau\left[1 - \tfrac{1}{2} - e^{-1} + \tfrac{1}{2}e^{-2}\right]$

$U_L = \left(\dfrac{\mathcal{E}^2}{2R}\right)\left(\dfrac{L}{R}\right)(1 - 2e^{-1} + e^{-2}) = \tfrac{1}{2}\left(\dfrac{\mathcal{E}}{R}\right)^2 L(1 - 2e^{-1} + e^{-2})$

$U_L = \tfrac{1}{2}\left(\dfrac{6.00\ \text{V}}{8.00\ \Omega}\right)^2 (2.50\ \text{H})(0.3996) = 0.281$ J, which checks.

(b) Exercise 30.27(a): The rate at which the battery supplies energy is $P_\mathcal{E} = \mathcal{E}i = \mathcal{E}\left(\dfrac{\mathcal{E}}{R}(1 - e^{-t/\tau})\right) = \dfrac{\mathcal{E}^2}{R}(1 - e^{-t/\tau})$

$U_\mathcal{E} = \displaystyle\int_0^\tau P_\mathcal{E} dt = \dfrac{\mathcal{E}^2}{R}\int_0^\tau (1 - e^{-t/\tau})dt = \dfrac{\mathcal{E}^2}{R}\left[t + \tau e^{-t/\tau}\right]_0^\tau = \left(\dfrac{\mathcal{E}^2}{R}\right)(\tau + \tau e^{-1} - \tau)$

$U_\mathcal{E} = \left(\dfrac{\mathcal{E}^2}{R}\right)\tau e^{-1} = \left(\dfrac{\mathcal{E}^2}{R}\right)\left(\dfrac{L}{R}\right)e^{-1} = \left(\dfrac{\mathcal{E}}{R}\right)^2 Le^{-1}$

$U_\mathcal{E} = \left(\dfrac{6.00\ \text{V}}{8.00\ \Omega}\right)^2 (2.50\ \text{H})(0.3679) = 0.517$ J

(c) $P_R = i^2 R = \left(\dfrac{\mathcal{E}^2}{R}\right)(1 - e^{-t/\tau})^2 = \dfrac{\mathcal{E}^2}{R}(1 - 2e^{-t/\tau} + e^{-2t/\tau})$

$U_R = \displaystyle\int_0^\tau P_R dt = \dfrac{\mathcal{E}^2}{R}\int_0^\tau (1 - 2e^{-t/\tau} + e^{-2t/\tau})dt = \dfrac{\mathcal{E}^2}{R}\left[t + 2\tau e^{-t/\tau} - \dfrac{\tau}{2}e^{-2t/\tau}\right]_0^\tau$

$U_R = \dfrac{\mathcal{E}^2}{R}\left[\tau + 2\tau e^{-1} - \dfrac{\tau}{2}e^{-2} - 2\tau + \dfrac{\tau}{2}\right] = \dfrac{\mathcal{E}^2}{R}\left[-\dfrac{\tau}{2} + 2\tau e^{-1} - \dfrac{\tau}{2}e^{-2}\right]$

$U_R = \left(\dfrac{\mathcal{E}^2}{2R}\right)\left(\dfrac{L}{R}\right)\left[-1 + 4e^{-1} - e^{-2}\right]$

$U_R = \left(\dfrac{\mathcal{E}}{R}\right)^2 (\tfrac{1}{2}L)\left[-1 + 4e^{-1} - e^{-2}\right] = \left(\dfrac{6.00\text{ V}}{8.00\ \Omega}\right)^2 \tfrac{1}{2}(2.50\text{ H})(0.3362) = 0.236\text{ J}$

(d) EVALUATE: $U_\mathcal{E} = U_R + U_L.\ (0.517\text{ J} = 0.236\text{ J} + 0.281\text{ J})$

The energy supplied by the battery equals the sum of the energy stored in the magnetic field of the inductor and the energy dissipated in the resistance of the inductor.

30.55.   **IDENTIFY** and **SET UP:**   Follow the procedure specified in the problem. $\tfrac{1}{2}Li^2$ is the energy stored in the inductor

and $q^2/2C$ is the energy stored in the capacitor. The equation is $-iR - L\dfrac{di}{dt} - \dfrac{q}{C} = 0$.

EXECUTE: Multiplying by $-i$ gives $i^2 R + Li\dfrac{di}{dt} + \dfrac{qi}{C} = 0$. $\dfrac{d}{dt}U_L = \dfrac{d}{dt}\left(\tfrac{1}{2}Li^2\right) = \tfrac{1}{2}L\dfrac{d}{dt}(i^2) = \tfrac{1}{2}L\left(2i\dfrac{di}{dt}\right) = Li\dfrac{di}{dt}$, the

second term. $\dfrac{d}{dt}U_C = \dfrac{d}{dt}\left(\dfrac{q^2}{2C}\right) = \dfrac{1}{2C}\dfrac{d}{dt}(q^2) = \dfrac{1}{2C}(2q)\dfrac{dq}{dt} = \dfrac{qi}{C}$, the third term. $i^2 R = P_R$, the rate at which

electrical energy is dissipated in the resistance. $\dfrac{d}{dt}U_L = P_L$, the rate at which the amount of energy stored in the

inductor is changing. $\dfrac{d}{dt}U_C = P_C$, the rate at which the amount of energy stored in the capacitor is changing.

EVALUATE: The equation says that $P_R + P_L + P_C = 0$; the net rate of change of energy in the circuit is zero. Note that at any given time one of P_C or P_L is negative. If the current and U_L are increasing the charge on the capacitor and U_C are decreasing, and vice versa.

30.57.   **IDENTIFY** and **SET UP:**   Use $U_C = \tfrac{1}{2}CV_C^2$ (energy stored in a capacitor) to solve for C. Then use Eq.(30.22) and $\omega = 2\pi f$ to solve for the L that gives the desired current oscillation frequency.

EXECUTE: $V_C = 12.0\text{ V};\ U_C = \tfrac{1}{2}CV_C^2$ so $C = 2U_C/V_C^2 = 2(0.0160\text{ J})/(12.0\text{ V})^2 = 222\ \mu\text{F}$

$f = \dfrac{1}{2\pi\sqrt{LC}}$ so $L = \dfrac{1}{(2\pi f)^2 C}$

$f = 3500\text{ Hz}$ gives $L = 9.31\ \mu\text{H}$

EVALUATE: f is in Hz and ω is in rad/s; we must be careful not to confuse the two.

30.61.   **IDENTIFY** and **SET UP:**   The current grows in the circuit as given by Eq.(30.14). In an R-L circuit the full emf initially is across the inductance and after a long time is totally across the resistance. A solenoid in a circuit is represented as a resistance in series with an inductance. Apply the loop rule to the circuit; the voltage across a resistance is given by Ohm's law.

EXECUTE: (a) In the R-L circuit the voltage across the resistor starts at zero and increases to the battery voltage. The voltage across the solenoid (inductor) starts at the battery voltage and decreases to zero. In the graph, the voltage drops, so the oscilloscope is across the solenoid.

(b) At $t \to \infty$ the current in the circuit approaches its final, constant value. The voltage doesn't go to zero because the solenoid has some resistance R_L. The final voltage across the solenoid is IR_L, where I is the final current in the circuit.

(c) The emf of the battery is the initial voltage across the inductor, 50 V. Just after the switch is closed, the current is zero and there is no voltage drop across any of the resistance in the circuit.

(d) As $t \to \infty$, $\mathcal{E} - IR - IR_L = 0$

$\mathcal{E} = 50\text{ V}$ and from the graph $IR_L = 15\text{ V}$ (the final voltage across the inductor), so

$IR = 35\text{ V}$ and $I = (35\text{ V})/R = 3.5\text{ A}$

(e) $IR_L = 15$ V, so $R_L = (15$ V$)/(3.5$ A$) = 4.3\ \Omega$

$\mathcal{E} - V_L - iR = 0,$ where V_L includes the voltage across the resistance of the solenoid.

$$V_L = \mathcal{E} - iR,\ i = \frac{\mathcal{E}}{R_{tot}}\left(1 - e^{-t/\tau}\right),\ \text{so}\ V_L = \mathcal{E}\left[1 - \frac{R}{R_{tot}}(1 - e^{-t/\tau})\right]$$

$\mathcal{E} = 50$ V, $R = 10\ \Omega$, $R_{tot} = 14.3\ \Omega$, so when $t = \tau$, $V_L = 27.9$ V. From the graph, V_L has this value when $t = 3.0$ ms (read approximately from the graph), so $\tau = L/R_{tot} = 3.0$ ms. Then $L = (3.0\ \text{ms})(14.3\ \Omega) = 43$ mH.

EVALUATE: At $t = 0$ there is no current and the 50 V measured by the oscilloscope is the induced emf due to the inductance of the solenoid. As the current grows, there are voltage drops across the two resistances in the circuit. We derived an equation for V_L, the voltage across the solenoid. At $t = 0$ it gives $V_L = \mathcal{E}$ and at $t \to \infty$ it gives $V_L = \mathcal{E}R/R_{tot} = iR$.

30.63. **IDENTIFY and SET UP:** Just after the switch is closed, the current in each branch containing an inductor is zero and the voltage across any capacitor is zero. The inductors can be treated as breaks in the circuit and the capacitors can be replaced by wires. After a long time there is no voltage across each inductor and no current in any branch containing a capacitor. The inductors can be replaced by wires and the capacitors by breaks in the circuit.

EXECUTE: (a) Just after the switch is closed the voltage V_5 across the capacitor is zero and there is also no current through the inductor, so $V_3 = 0$. $V_2 + V_3 = V_4 = V_5$, and since $V_5 = 0$ and $V_3 = 0$, V_4 and V_2 are also zero.

$V_4 = 0$ means V_3 reads zero. V_1 then must equal 40.0 V, and this means the current read by A_1 is $(40.0\ \text{V})/(50.0\ \Omega) = 0.800$ A. $A_2 + A_3 + A_4 = A_1$, but $A_2 = A_3 = 0$ so $A_4 = A_1 = 0.800$ A. $A_1 = A_4 = 0.800$ A; all other ammeters read zero. $V_1 = 40.0$ V and all other voltmeters read zero.

(b) After a long time the capacitor is fully charged so $A_4 = 0$. The current through the inductor isn't changing, so $V_2 = 0$. The currents can be calculated from the equivalent circuit that replaces the inductor by a short circuit, as shown in Figure 30.63a.

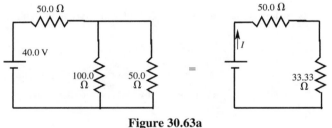

Figure 30.63a

$I = (40.0\ \text{V})/(83.33\ \Omega) = 0.480$ A; A_1 reads 0.480 A

$V_1 = I(50.0\ \Omega) = 24.0$ V

The voltage across each parallel branch is 40.0 V – 24.0 V = 16.0 V

$V_2 = 0$, $V_3 = V_4 = V_5 = 16.0$ V

$V_3 = 16.0$ V means A_2 reads 0.160 A. $V_4 = 16.0$ V means A_3 reads 0.320 A. A_4 reads zero. Note that $A_2 + A_3 = A_1$.

(c) $V_5 = 16.0$ V so $Q = CV = (12.0\ \mu\text{F})(16.0\ \text{V}) = 192\ \mu\text{C}$

(d) At $t = 0$ and $t \to \infty$, $V_2 = 0$. As the current in this branch increases from zero to 0.160 A the voltage V_2 reflects the rate of change of the current. The graph is sketched in Figure 30.63b.

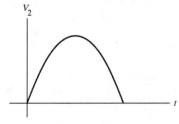

Figure 30.63b

EVALUATE: This reduction of the circuit to resistor networks only apply at $t = 0$ and $t \to \infty$. At intermediate times the analysis is complicated.

30.67. **IDENTIFY:** Apply the loop rule to each parallel branch. The voltage across a resistor is given by iR and the voltage across an inductor is given by $L|di/dt|$. The rate of change of current through the inductor is limited.

SET UP: With S closed the circuit is sketched in Figure 30.67a.

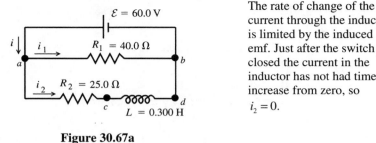

The rate of change of the current through the inductor is limited by the induced emf. Just after the switch is closed the current in the inductor has not had time to increase from zero, so

$i_2 = 0.$

Figure 30.67a

EXECUTE: (a) $\mathcal{E} - v_{ab} = 0$, so $v_{ab} = 60.0$ V

(b) The voltage drops across R, as we travel through the resistor in the direction of the current, so point a is at higher potential.

(c) $i_2 = 0$ so $v_{R_2} = i_2 R_2 = 0$

$\mathcal{E} - v_{R_2} - v_L = 0$ so $v_L = \mathcal{E} = 60.0$ V

(d) The voltage rises when we go from b to a through the emf, so it must drop when we go from a to b through the inductor. Point c must be at higher potential than point d.

(e) After the switch has been closed a long time, $\dfrac{di_2}{dt} \to 0$ so $v_L = 0$. Then $\mathcal{E} - v_{R_2} = 0$ and $i_2 R_2 = \mathcal{E}$

so $i_2 = \dfrac{\mathcal{E}}{R_2} = \dfrac{60.0 \text{ V}}{25.0 \ \Omega} = 2.40$ A.

SET UP: The rate of change of the current through the inductor is limited by the induced emf. Just after the switch is opened again the current through the inductor hasn't had time to change and is still $i_2 = 2.40$ A. The circuit is sketched in Figure 30.67b.

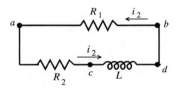

EXECUTE: The current through R_1 is $i_2 = 2.40$ A, in the direction b to a. Thus

$v_{ab} = -i_2 R_1 = -(2.40 \text{ A})(40.0 \ \Omega)$

$v_{ab} = -96.0$ V

Figure 30.67b

(f) Point where current enters resistor is at higher potential; point b is at higher potential.

(g) $v_L - v_{R_1} - v_{R_2} = 0$

$v_L = v_{R_1} + v_{R_2}$

$v_{R_1} = -v_{ab} = 96.0$ V; $v_{R_2} = i_2 R_2 = (2.40 \text{ A})(25.0 \ \Omega) = 60.0$ V

Then $v_L = v_{R_1} + v_{R_2} = 96.0 \text{ V} + 60.0 \text{ V} = 156$ V.

As you travel counterclockwise around the circuit in the direction of the current, the voltage drops across each resistor, so it must rise across the inductor and point d is at higher potential than point c. The current is decreasing, so the induced emf in the inductor is directed in the direction of the current. Thus, $v_{cd} = -156$ V.

(h) Point d is at higher potential.

EVALUATE: The voltage across R_1 is constant once the switch is closed. In the branch containing R_2, just after S is closed the voltage drop is all across L and after a long time it is all across R_2. Just after S is opened the same current flows in the single loop as had been flowing through the inductor and the sum of the voltage across the resistors equals the voltage across the inductor. This voltage dies away, as the energy stored in the inductor is dissipated in the resistors.

30.69. **IDENTIFY** and **SET UP:** The circuit is sketched in Figure 30.69a. Apply the loop rule. Just after S_1 is closed, $i = 0$.
After a long time i has reached its final value and $di/dt = 0$. The voltage across a resistor depends on i and the voltage across an inductor depends on di/dt.

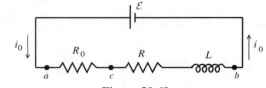

Figure 30.69a

EXECUTE: (a) At time $t = 0$, $i_0 = 0$ so $v_{ac} = i_0 R_0 = 0$. By the loop rule $\mathcal{E} - v_{ac} - v_{cb} = 0$, so $v_{cb} = \mathcal{E} - v_{ac} = \mathcal{E} = 36.0$ V.
($i_0 R = 0$ so this potential difference of 36.0 V is across the inductor and is an induced emf produced by the changing current.)

(b) After a long time $\dfrac{di_0}{dt} \to 0$ so the potential $-L\dfrac{di_0}{dt}$ across the inductor becomes zero. The loop rule gives
$\mathcal{E} - i_0 (R_0 + R) = 0$.

$i_0 = \dfrac{\mathcal{E}}{R_0 + R} = \dfrac{36.0 \text{ V}}{50.0 \ \Omega + 150 \ \Omega} = 0.180$ A

$v_{ac} = i_0 R_0 = (0.180 \text{ A})(50.0 \ \Omega) = 9.0$ V

Thus $v_{cb} = i_0 R + L\dfrac{di_0}{dt} = (0.180 \text{ A})(150 \ \Omega) + 0 = 27.0$ V (Note that $v_{ac} + v_{cb} = \mathcal{E}$.)

(c) $\mathcal{E} - v_{ac} - v_{cb} = 0$

$\mathcal{E} - iR_0 - iR - L\dfrac{di}{dt} = 0$

$L\dfrac{di}{dt} = \mathcal{E} - i(R_0 + R)$ and $\left(\dfrac{L}{R + R_0}\right)\dfrac{di}{dt} = -i + \dfrac{\mathcal{E}}{R + R_0}$

$\dfrac{di}{-i + \mathcal{E}/(R + R_0)} = \left(\dfrac{R + R_0}{L}\right)dt$

Integrate from $t = 0$, when $i = 0$, to t, when $i = i_0$: $\displaystyle\int_0^{i_0} \dfrac{di}{-i + \mathcal{E}/(R + R_0)} = \dfrac{R + R_0}{L}\int_0^t dt = -\ln\left[-i + \dfrac{\mathcal{E}}{R + R_0}\right]_0^{i_0} = \left(\dfrac{R + R_0}{L}\right)t$,

so $\ln\left(-i_0 + \dfrac{\mathcal{E}}{R + R_0}\right) - \ln\left(\dfrac{\mathcal{E}}{R + R_0}\right) = -\left(\dfrac{R + R_0}{L}\right)t$

$\ln\left(\dfrac{-i_0 + \mathcal{E}/(R + R_0)}{\mathcal{E}/(R + R_0)}\right) = -\left(\dfrac{R + R_0}{L}\right)t$

Taking exponentials of both sides gives $\dfrac{-i_0 + \mathcal{E}/(R + R_0)}{\mathcal{E}/(R + R_0)} = e^{-(R + R_0)t/L}$ and $i_0 = \dfrac{\mathcal{E}}{R + R_0}\left(1 - e^{-(R + R_0)t/L}\right)$

Substituting in the numerical values gives $i_0 = \dfrac{36.0 \text{ V}}{50 \ \Omega + 150 \ \Omega}\left(1 - e^{-(200 \ \Omega/4.00 \text{ H})t}\right) = (0.180 \text{ A})\left(1 - e^{-t/0.020 \text{ s}}\right)$

At $t \to 0$, $i_0 = (0.180 \text{ A})(1 - 1) = 0$ (agrees with part (a)). At $t \to \infty$, $i_0 = (0.180 \text{ A})(1 - 0) = 0.180$ A (agrees with part (b)).

$v_{ac} = i_0 R_0 = \dfrac{\mathcal{E} R_0}{R + R_0}\left(1 - e^{-(R + R_0)t/L}\right) = 9.0 \text{ V}\left(1 - e^{-t/0.020 \text{ s}}\right)$

$v_{cb} = \mathcal{E} - v_{ac} = 36.0 \text{ V} - 9.0 \text{ V}(1 - e^{-t/0.020 \text{ s}}) = 9.0 \text{ V}(3.00 + e^{-t/0.020 \text{ s}})$

At $t \to 0$, $v_{ac} = 0$, $v_{cb} = 36.0$ V (agrees with part (a)). At $t \to \infty$, $v_{ac} = 9.0$ V, $v_{cb} = 27.0$ V (agrees with part (b)).
The graphs are given in Figure 30.69b.

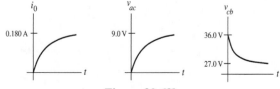

Figure 30.69b

EVALUATE: The expression for $i(t)$ we derived becomes Eq.(30.14) if the two resistors R_0 and R in series are replaced by a single equivalent resistance $R_0 + R$.

30.71.   **IDENTIFY:**   The current through an inductor doesn't change abruptly. After a long time the current isn't changing and the voltage across each inductor is zero.

SET UP: Problem 30.47 shows how to find the equivalent inductance of inductors in series and parallel.

EXECUTE: **(a)** Just after the switch is closed there is no current in the inductors. There is no current in the resistors so there is no voltage drop across either resistor. A reads zero and V reads 20.0 V.

(b) After a long time the currents are no longer changing, there is no voltage across the inductors, and the inductors can be replaced by short-circuits. The circuit becomes equivalent to the circuit shown in Figure 30.71a.

$I = (20.0\text{ V})/(75.0\ \Omega) = 0.267\text{ A}$. The voltage between points a and b is zero, so the voltmeter reads zero.

(c) Use the results of Problem 30.49 to combine the inductor network into its equivalent, as shown in Figure 30.71b. $R = 75.0\ \Omega$ is the equivalent resistance. Eq.(30.14) says $i = (\mathcal{E}/R)(1 - e^{-t/\tau})$ with

$\tau = L/R = (10.8\text{ mH})/(75.0\ \Omega) = 0.144\text{ ms}$. $\mathcal{E} = 20.0\text{ V}$, $R = 75.0\ \Omega$, $t = 0.115\text{ ms}$ so $i = 0.147\text{ A}$.

$V_R = iR = (0.147\text{ A})(75.0\ \Omega) = 11.0\text{ V}$. $20.0\text{ V} - V_R - V_L = 0$ and $V_L = 20.0\text{ V} - V_R = 9.0\text{ V}$. The ammeter reads 0.147 A and the voltmeter reads 9.0 V.

EVALUATE: The current through the battery increases from zero to a final value of 0.267 A. The voltage across the inductor network drops from 20.0 V to zero.

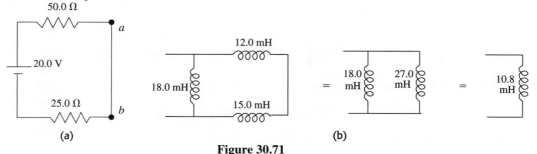

(a)   **Figure 30.71**   (b)

30.75.   **(a) IDENTIFY** and **SET UP:**   With switch S closed the circuit is shown in Figure 30.75a.

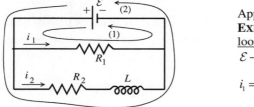

Figure 30.75a

Apply the loop rule to loops 1 and 2.
EXECUTE:
loop 1
$\mathcal{E} - i_1 R_1 = 0$

$i_1 = \dfrac{\mathcal{E}}{R_1}$ (independent of t)

loop (2)

$\mathcal{E} - i_2 R_2 - L\dfrac{di_2}{dt} = 0$

This is in the form of equation (30.12), so the solution is analogous to Eq.(30.14): $i_2 = \dfrac{\mathcal{E}}{R_2}\left(1 - e^{-R_2 t/L}\right)$

(b) EVALUATE: The expressions derived in part (a) give that as $t \to \infty$, $i_1 = \dfrac{\mathcal{E}}{R_1}$ and $i_2 = \dfrac{\mathcal{E}}{R_2}$. Since $\dfrac{di_2}{dt} \to 0$ at steady-state, the inductance then has no effect on the circuit. The current in R_1 is constant; the current in R_2 starts at zero and rises to $\mathcal{E}/R_2$.

(c) IDENTIFY and **SET UP:**   The circuit now is as shown in Figure 30.75b.

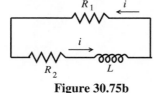

Figure 30.75b

Let $t = 0$ now be when S is opened.
At $t = 0$, $i = \dfrac{\mathcal{E}}{R_2}$.

Apply the loop rule to the single current loop.

EXECUTE: $-i(R_1 + R_2) - L\dfrac{di}{dt} = 0.$ (Now $\dfrac{di}{dt}$ is negative.)

$L\dfrac{di}{dt} = -i(R_1 + R_2)$ gives $\dfrac{di}{i} = -\left(\dfrac{R_1 + R_2}{L}\right)dt$

Integrate from $t = 0$, when $i = I_0 = \mathcal{E}/R_2$, to t.

$\displaystyle\int_{I_0}^{i} \dfrac{di}{i} = -\left(\dfrac{R_1 + R_2}{L}\right)\int_0^t dt$ and $\ln\left(\dfrac{i}{I_0}\right) = -\left(\dfrac{R_1 + R_2}{L}\right)t$

Taking exponentials of both sides of this equation gives $i = I_0 e^{-(R_1 + R_2)t/L} = \dfrac{\mathcal{E}}{R_2}e^{-(R_1 + R_2)t/L}$

(d) IDENTIFY and SET UP: Use the equation derived in part (c) and solve for R_2 and $\mathcal{E}$.

EXECUTE: $L = 22.0$ H

$R_{R_1} = \dfrac{V^2}{R_1} = 40.0$ W gives $R_1 = \dfrac{V^2}{P_{R_1}} = \dfrac{(120 \text{ V})^2}{40.0 \text{ W}} = 360\ \Omega.$

We are asked to find R_2 and $\mathcal{E}$. Use the expression derived in part (c).

$I_0 = 0.600$ A so $\mathcal{E}/R_2 = 0.600$ A

$i = 0.150$ A when $t = 0.080$ s, so $i = \dfrac{\mathcal{E}}{R_2}e^{-(R_1 + R_2)t/L}$ gives $0.150 \text{ A} = (0.600 \text{ A})e^{-(R_1 + R_2)t/L}$

$\frac{1}{4} = e^{-(R_1 + R_2)t/L}$ so $\ln 4 = (R_1 + R_2)t/L$

$R_2 = \dfrac{L\ln 4}{t} - R_1 = \dfrac{(22.0 \text{ H})\ln 4}{0.080 \text{ s}} - 360\ \Omega = 381.2\ \Omega - 360\ \Omega = 21.2\ \Omega$

Then $\mathcal{E} = (0.600 \text{ A})R_2 = (0.600 \text{ A})(21.2\ \Omega) = 12.7$ V.

(e) IDENTIFY and SET UP: Use the expressions derived in part (a).

EXECUTE: The current through the light bulb before the switch is opened is $i_1 = \dfrac{\mathcal{E}}{R_1} = \dfrac{12.7 \text{ V}}{360\ \Omega} = 0.0353$ A

EVALUATE: When the switch is opened the current through the light bulb jumps from 0.0353 A to 0.600 A. Since the electrical power dissipated in the bulb (brightness) depend on i^2, the bulb suddenly becomes much brighter.

ALTERNATING CURRENT

31.9. **IDENTIFY and SET UP:** Use Eqs.(31.12) and (31.18).
EXECUTE: (a) $X_L = \omega L = 2\pi f L = 2\pi(80.0 \text{ Hz})(3.00 \text{ H}) = 1510 \ \Omega$

(b) $X_L = 2\pi f L$ gives $L = \dfrac{X_L}{2\pi f} = \dfrac{120 \ \Omega}{2\pi(80.0 \text{ Hz})} = 0.239 \text{ H}$

(c) $X_C = \dfrac{1}{\omega C} = \dfrac{1}{2\pi f C} = \dfrac{1}{2\pi(80.0 \text{ Hz})(4.00\times10^{-6} \text{ F})} = 497 \ \Omega$

(d) $X_C = \dfrac{1}{2\pi f C}$ gives $C = \dfrac{1}{2\pi f X_C} = \dfrac{1}{2\pi(80.0 \text{ Hz})(120 \ \Omega)} = 1.66\times10^{-5} \text{ F}$

EVALUATE: X_L increases when L increases; X_C decreases when C increases.

31.11. **IDENTIFY and SET UP:** Apply Eqs.(31.18) and (31.19).

EXECUTE: $V = IX_C$ so $X_C = \dfrac{V}{I} = \dfrac{170 \text{ V}}{0.850 \text{ A}} = 200 \ \Omega$

$X_C = \dfrac{1}{\omega C}$ gives $C = \dfrac{1}{2\pi f X_C} = \dfrac{1}{2\pi(60.0 \text{ Hz})(200 \ \Omega)} = 1.33\times10^{-5} \text{ F} = 13.3 \ \mu\text{F}$

EVALUATE: The reactance relates the voltage amplitude to the current amplitude and is similar to Ohm's law.

31.13. **IDENTIFY and SET UP:** The voltage and current for a resistor are related by $v_R = iR$. Deduce the frequency of the voltage and use this in Eq.(31.12) to calculate the inductive reactance. Eq.(31.10) gives the voltage across the inductor.

EXECUTE: (a) $v_R = (3.80 \text{ V})\cos[(720 \text{ rad/s})t]$

$v_R = iR$, so $i = \dfrac{v_R}{R} = \left(\dfrac{3.80 \text{ V}}{150 \ \Omega}\right)\cos[(720 \text{ rad/s})t] = (0.0253 \text{ A})\cos[(720 \text{ rad/s})t]$

(b) $X_L = \omega L$

$\omega = 720 \text{ rad/s}$, $L = 0.250 \text{ H}$, so $X_L = \omega L = (720 \text{ rad/s})(0.250 \text{ H}) = 180 \ \Omega$

(c) If $i = I\cos\omega t$ then $v_L = V_L \cos(\omega t + 90°)$ (from Eq.31.10). $V_L = I\omega L = IX_L = (0.02533 \text{ A})(180 \ \Omega) = 4.56 \text{ V}$

$v_L = (4.56 \text{ V})\cos[(720 \text{ rad/s})t + 90°]$

But $\cos(a + 90°) = -\sin a$ (Appendix B), so $v_L = -(4.56 \text{ V})\sin[(720 \text{ rad/s})t]$.

EVALUATE: The current is the same in the resistor and inductor and the voltages are $90°$ out of phase, with the voltage across the inductor leading.

31.15. **IDENTIFY:** $v_R(t)$ is given by Eq.(31.8). $v_L(t)$ is given by Eq.(31.10).

SET UP: From Exercise 31.14, $V = 30.0 \text{ V}$, $V_R = 26.8 \text{ V}$, $V_L = 13.4 \text{ V}$ and $\phi = 26.6°$.

EXECUTE: (a) The graph is given in Figure 31.15.

(b) The different voltages are $v = (30.0 \text{ V})\cos(250t + 26.6°)$, $v_R = (26.8 \text{ V})\cos(250t)$,

$v_L = (13.4 \text{ V})\cos(250t + 90°)$. At $t = 20$ ms: $v = 20.5 \text{ V}$, $v_R = 7.60 \text{ V}$, $v_L = 12.85 \text{ V}$. Note that $v_R + v_L = v$.

(c) At $t = 40$ ms: $v = -15.2 \text{ V}$, $v_R = -22.49 \text{ V}$, $v_L = 7.29 \text{ V}$. Note that $v_R + v_L = v$.

EVALUATE: It is important to be careful with radians versus degrees in above expressions!

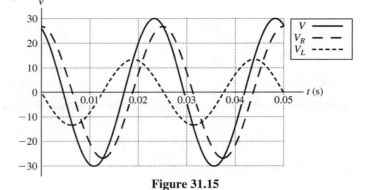

Figure 31.15

31.17.   **IDENTIFY and SET UP:**   Calculate the impendance of the circuit and use Eq.(31.22) to find the current amplitude. The voltage amplitudes across each circuit element are given by Eqs.(31.7), (31.13), and (31.19). The phase angle is calculated using Eq.(31.24). The circuit is shown in Figure 31.17a.

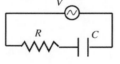

No inductor means $X_L = 0$

$R = 200\ \Omega,\ C = 6.00 \times 10^{-6}$ F,
$V = 30.0$ V, $\omega = 250$ rad/s

Figure 31.17a

EXECUTE: (a) $X_C = \dfrac{1}{\omega C} = \dfrac{1}{(250\ \text{rad/s})(6.00 \times 10^{-6}\ \text{F})} = 666.7\ \Omega$

$Z = \sqrt{R^2 + (X_L - X_C)^2} = \sqrt{(200\ \Omega)^2 + (666.7\ \Omega)^2} = 696\ \Omega$

(b) $I = \dfrac{V}{Z} = \dfrac{30.0\ \text{V}}{696\ \Omega} = 0.0431\ \text{A} = 43.1\ \text{mA}$

(c) Voltage amplitude across the resistor: $V_R = IR = (0.0431\ \text{A})(200\ \Omega) = 8.62$ V

Voltage amplitude across the capacitor: $V_C = IX_C = (0.0431\ \text{A})(666.7\ \Omega) = 28.7$ V

(d) $\tan\phi = \dfrac{X_L - X_C}{R} = \dfrac{0 - 666.7\ \Omega}{200\ \Omega} = -3.333$ so $\phi = -73.3°$

The phase angle is negative, so the source voltage lags behind the current.
(e) The phasor diagram is sketched qualitatively in Figure 31.17b.

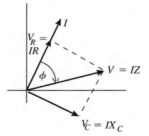

Figure 31.17b

EVALUATE: The voltage across the resistor is in phase with the current and the capacitor voltage lags the current by 90°. The presence of the capacitor causes the source voltage to lag behind the current. Note that $V_R + V_C > V$.

The instantaneous voltages in the circuit obey the loop rule at all times but because of the phase differences the voltage amplitudes do not.

31.19.   **IDENTIFY:**   Apply the equations in Section 31.3.
SET UP: $\omega = 250$ rad/s, $R = 200\ \Omega$, $L = 0.400$ H, $C = 6.00\ \mu$F and $V = 30.0$ V.

EXECUTE: (a) $Z = \sqrt{R^2 + (\omega L - 1/\omega C)^2}$.

$Z = \sqrt{(200\ \Omega)^2 + ((250\ \text{rad/s})(0.0400\ \text{H}) - 1/((250\ \text{rad/s})(6.00 \times 10^{-6}\ \text{F})))^2} = 601\ \Omega$

(b) $I = \dfrac{V}{Z} = \dfrac{30\ \text{V}}{601\ \Omega} = 0.0499$ A.

(c) $\phi = \arctan\left(\dfrac{\omega L - 1/\omega C}{R}\right) = \arctan\left(\dfrac{100\,\Omega - 667\,\Omega}{200\,\Omega}\right) = -70.6°$, and the voltage lags the current.

(d) $V_R = IR = (0.0499\text{ A})(200\,\Omega) = 9.98$ V;

$V_L = I\omega L = (0.0499\text{ A})(250\text{ rad/s})(0.400\text{ H}) = 4.99$ V; $V_C = \dfrac{I}{\omega C} = \dfrac{(0.0499\text{ A})}{(250\text{ rad/s})(6.00\times10^{-6}\text{ F})} = 33.3$ V.

EVALUATE: **(e)** At any instant, $v = v_R + v_C + v_L$. But v_C and v_L are 180° out of phase, so v_C can be larger than v at a value of t, if $v_L + v_R$ is negative at that t.

31.21. **IDENTIFY and SET UP:** The current is largest at the resonance frequency. At resonance, $X_L = X_C$ and $Z = R$. For part (b), calculate Z and use $I = V/Z$.

EXECUTE: **(a)** $f_0 = \dfrac{1}{2\pi\sqrt{LC}} = 113$ Hz. $I = V/R = 15.0$ mA.

(b) $X_C = 1/\omega C = 500\,\Omega$. $X_L = \omega L = 160\,\Omega$. $Z = \sqrt{R^2 + (X_L - X_C)^2} = \sqrt{(200\,\Omega)^2 + (160\,\Omega - 500\,\Omega)^2} = 394.5\,\Omega$.
$I = V/Z = 7.61$ mA. $X_C > X_L$ so the source voltage lags the current.

EVALUATE: $\omega_0 = 2\pi f_0 = 710$ rad/s. $\omega = 400$ rad/s and is less than ω_0. When $\omega < \omega_0$, $X_C > X_L$. Note that I in part (b) is less than I in part (a).

31.23. **IDENTIFY and SET UP:** Use the equation that preceeds Eq.(31.20): $V^2 = V_R^2 + (V_L - V_C)^2$

EXECUTE: $V = \sqrt{(30.0\text{ V})^2 + (50.0\text{ V} - 90.0\text{ V})^2} = 50.0$ V

EVALUATE: The equation follows directly from the phasor diagrams of Fig.31.13 (b or c). Note that the voltage amplitudes do not simply add to give 170.0 V for the source voltage.

31.27. **IDENTIFY:** The power factor is $\cos\phi$, where ϕ is the phase angle in Fig.31.13. The average power is given by Eq.(31.31). Use the result of part (a) to rewrite this expression.
(a) SET UP: The phasor diagram is sketched in Figure 31.27.

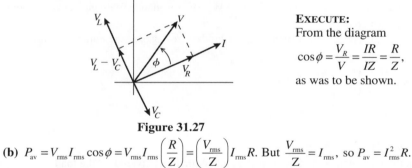

EXECUTE:
From the diagram
$$\cos\phi = \frac{V_R}{V} = \frac{IR}{IZ} = \frac{R}{Z},$$
as was to be shown.

Figure 31.27

(b) $P_{av} = V_{rms}I_{rms}\cos\phi = V_{rms}I_{rms}\left(\dfrac{R}{Z}\right) = \left(\dfrac{V_{rms}}{Z}\right)I_{rms}R$. But $\dfrac{V_{rms}}{Z} = I_{rms}$, so $P_{av} = I_{rms}^2 R$.

EVALUATE: In an L-R-C circuit, electrical energy is stored and released in the inductor and capacitor but none is dissipated in either of these circuit elements. The power delivered by the source equals the power dissipated in the resistor.

31.29. **IDENTIFY and SET UP:** Use the equations of Section 31.3 to calculate ϕ, Z and V_{rms}. The average power delivered by the source is given by Eq.(31.31) and the average power dissipated in the resistor is $I_{rms}^2 R$

EXECUTE: **(a)** $X_L = \omega L = 2\pi f L = 2\pi(400\text{ Hz})(0.120\text{ H}) = 301.6\,\Omega$

$X_C = \dfrac{1}{\omega C} = \dfrac{1}{2\pi f C} = \dfrac{1}{2\pi(400\text{ Hz})(7.3\times10^{-6}\text{ Hz})} = 54.51\,\Omega$

$\tan\phi = \dfrac{X_L - X_C}{R} = \dfrac{301.6\,\Omega - 54.41\,\Omega}{240\,\Omega}$, so $\phi = +45.8°$. The power factor is $\cos\phi = +0.697$.

(b) $Z = \sqrt{R^2 + (X_L - X_C)^2} = \sqrt{(240\,\Omega)^2 + (301.6\,\Omega - 54.51\,\Omega)^2} = 344\,\Omega$

(c) $V_{rms} = I_{rms}Z = (0.450\text{ A})(344\,\Omega) = 155$ V

(d) $P_{av} = I_{rms}V_{rms}\cos\phi = (0.450\text{ A})(155\text{ V})(0.697) = 48.6$ W

(e) $P_{av} = I_{rms}^2 R = (0.450\text{ A})^2(240\,\Omega) = 48.6$ W

EVALUATE: The average electrical power delivered by the source equals the average electrical power consumed in the resistor.
(f) All the energy stored in the capacitor during one cycle of the current is released back to the circuit in another part of the cycle. There is no net dissipation of energy in the capacitor.

(g) The answer is the same as for the capacitor. Energy is repeatedly being stored and released in the inductor, but no net energy is dissipated there.

31.31.   **IDENTIFY** and **SET UP:**   At the resonance frequency, $Z = R$. Use that $V = IZ$, $V_R = IR$, $V_L = IX_L$ and $V_C = IX_C$. P_{av} is given by Eq.(31.31).

(a) EXECUTE: $V = IZ = IR = (0.500 \text{ A})(300 \text{ }\Omega) = 150 \text{ V}$

(b) $V_R = IR = 150 \text{ V}$

$X_L = \omega L = L(1/\sqrt{LC}) = \sqrt{L/C} = 2582 \text{ }\Omega$; $V_L = IX_L = 1290 \text{ V}$

$X_C = 1/(\omega C) = \sqrt{L/C} = 2582 \text{ }\Omega$; $V_C = IX_C = 1290 \text{ V}$

(c) $P_{av} = \frac{1}{2}VI\cos\phi = \frac{1}{2}I^2 R$, since $V = IR$ and $\cos\phi = 1$ at resonance.

$P_{av} = \frac{1}{2}(0.500 \text{ A})^2 (300 \text{ }\Omega) = 37.5 \text{ W}$

EVALUATE: At resonance $V_L = V_C$. Note that $V_L + V_C > V$. However, at any instant $v_L + v_C = 0$.

31.33.   **IDENTIFY** and **SET UP:**   At resonance $X_L = X_C$, $\phi = 0$ and $Z = R$. $R = 150 \text{ }\Omega$, $L = 0.750 \text{ H}$, $C = 0.0180 \text{ }\mu\text{F}$, $V = 150 \text{ V}$

EXECUTE: **(a)** At the resonance frequency $X_L = X_C$ and from $\tan\phi = \dfrac{X_L - X_C}{R}$ we have that $\phi = 0°$ and the power factor is $\cos\phi = 1.00$.

(b) $P_{av} = \frac{1}{2}VI\cos\phi$ (Eq.31.31)

At the resonance frequency $Z = R$, so $I = \dfrac{V}{Z} = \dfrac{V}{R}$

$P_{av} = \frac{1}{2}V\left(\dfrac{V}{R}\right)\cos\phi = \frac{1}{2}\dfrac{V^2}{R} = \frac{1}{2}\dfrac{(150 \text{ V})^2}{150 \text{ }\Omega} = 75.0 \text{ W}$

(c) EVALUATE: When C and f are changed but the circuit is kept on resonance, nothing changes in $P_{av} = V^2/(2R)$, so the average power is unchanged: $P_{av} = 75.0 \text{ W}$. The resonance frequency changes but since $Z = R$ at resonance the current doesn't change.

31.37.   **IDENTIFY** and **SET UP:**   Eq.(31.35) relates the primary and secondary voltages to the number of turns in each. $I = V/R$ and the power consumed in the resistive load is $I_{rms}^2 = V_{rms}^2/R$.

EXECUTE: **(a)** $\dfrac{V_2}{V_1} = \dfrac{N_2}{N_1}$ so $\dfrac{N_1}{N_2} = \dfrac{V_1}{V_2} = \dfrac{120 \text{ V}}{12.0 \text{ V}} = 10$

(b) $I_2 = \dfrac{V_2}{R} = \dfrac{12.0 \text{ V}}{5.00 \text{ }\Omega} = 2.40 \text{ A}$

(c) $P_{av} = I_2^2 R = (2.40 \text{ A})^2 (5.00 \text{ }\Omega) = 28.8 \text{ W}$

(d) The power drawn from the line by the transformer is the 28.8 W that is delivered by the load.

$$P_{av} = \dfrac{V^2}{R} \text{ so } R = \dfrac{V^2}{P_{av}} = \dfrac{(120 \text{ V})^2}{28.8 \text{ W}} = 500 \text{ }\Omega$$

And $\left(\dfrac{N_1}{N_2}\right)^2 (5.00 \text{ }\Omega) = (10)^2 (5.00 \text{ }\Omega) = 500 \text{ }\Omega$, as was to be shown.

EVALUATE: The resistance is "transformed". A load of resistance R connected to the secondary draws the same power as a resistance $(N_1/N_2)^2 R$ connected directly to the supply line, without using the transformer.

31.41.   **IDENTIFY** and **SET UP:**   Use Eq.(31.24) to relate L and R to ϕ. The voltage across the coil leads the current in it by 52.3°, so $\phi = +52.3°$.

EXECUTE: $\tan\phi = \dfrac{X_L - X_C}{R}$. But there is no capacitance in the circuit so $X_C = 0$. Thus $\tan\phi = \dfrac{X_L}{R}$ and $X_L = R\tan\phi = (48.0 \text{ }\Omega)\tan 52.3° = 62.1 \text{ }\Omega$. $X_L = \omega L = 2\pi f L$ so $L = \dfrac{X_L}{2\pi f} = \dfrac{62.1 \text{ }\Omega}{2\pi(80.0 \text{ Hz})} = 0.124 \text{ H}$.

EVALUATE: $\phi > 45°$ when $(X_L - X_C) > R$, which is the case here.

31.43.   **IDENTIFY** and **SET UP:**   The rectified current equals the absolute value of the current i. Evaluate the integral as specified in the problem.

EXECUTE: **(a)** From Fig.31.3b, the rectified current is zero at the same values of t for which the sinusoidal current is zero. At these t, $\cos\omega t = 0$ and $\omega t = \pm\pi/2$, $\pm 3\pi/2,\ldots$. The two smallest positive times are $t_1 = \pi/2\omega$, $t_2 = 3\pi/2\omega$.

(b) $A = \left| \int_{t_1}^{t_2} i\, dt \right| = -\int_{t_1}^{t_2} I\cos\omega t\, dt = -I\left[\frac{1}{\omega}\sin\omega t\right]_{t_1}^{t_2} = -\frac{I}{\omega}(\sin\omega t_2 - \sin\omega t_1)$

$\sin\omega t_1 = \sin[\omega(\pi/2\omega)] = \sin(\pi/2) = 1$

$\sin\omega t_2 = \sin[\omega(3\pi/2\omega)] = \sin(3\pi/2) = -1$

$A = \left(\frac{I}{\omega}\right)(1-(-1)) = \frac{2I}{\omega}$

(c) $I_{\text{rav}}(t_2 - t_1) = 2I/\omega$

$I_{\text{rav}} = \dfrac{2I}{\omega(t_2 - t_1)} = \dfrac{2I}{\omega(3\pi/2\omega - \pi/2\omega)} = \dfrac{2I}{\pi}$, which is Eq.(31.3).

EVALUATE: We have shown that Eq.(31.3) is correct. The average rectified current is less than the current amplitude I, since the rectified current varies between 0 and I. The average of the current is zero, since it has both positive and negative values.

31.45. **(a) IDENTIFY** and **SET UP:** Source voltage lags current so it must be that $X_C > X_L$ and we must add an inductor in series with the circuit. When $X_C = X_L$ the power factor has its maximum value of unity, so calculate the additional L needed to raise X_L to equal X_C.

(b) EXECUTE: power factor $\cos\phi$ equals 1 so $\phi = 0$ and $X_C = X_L$. Calculate the present value of $X_C - X_L$ to see how much more X_L is needed: $R = Z\cos\phi = (60.0\ \Omega)(0.720) = 43.2\ \Omega$

$\tan\phi = \dfrac{X_L - X_C}{R}$ so $X_L - X_C = R\tan\phi$

$\cos\phi = 0.720$ gives $\phi = -43.95°$ (ϕ is negative since the voltage lags the current)

Then $X_L - X_C = R\tan\phi = (43.2\ \Omega)\tan(-43.95°) = -41.64\ \Omega$.

Therefore need to add $41.64\ \Omega$ of X_L.

$X_L = \omega L = 2\pi f L$ and $L = \dfrac{X_L}{2\pi f} = \dfrac{41.64\ \Omega}{2\pi(50.0\ \text{Hz})} = 0.133\ \text{H}$, amount of inductance to add.

EVALUATE: From the information given we can't calculate the original value of L in the circuit, just how much to add. When this L is added the current in the circuit will increase.

31.49. **IDENTIFY** and **SET UP:** Express Z and I in terms of ω, L, C and R. The voltages across the resistor and the inductor are 90° out of phase, so $V_{\text{out}} = \sqrt{V_R^2 + V_L^2}$.

EXECUTE: The circuit is sketched in Figure 31.49.

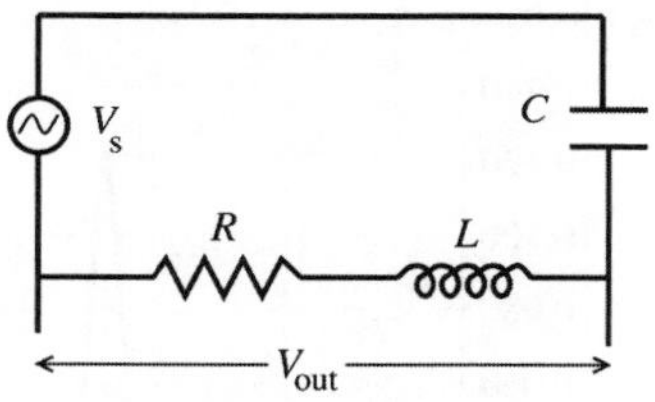

$X_L = \omega L,\ X_C = \dfrac{1}{\omega C}$

$Z = \sqrt{R^2 + \left(\omega L - \dfrac{1}{\omega C}\right)^2}$

$I = \dfrac{V_s}{Z} = \dfrac{V_s}{\sqrt{R^2 + \left(\omega L - \dfrac{1}{\omega C}\right)^2}}$

Figure 31.49

$V_{\text{out}} = I\sqrt{R^2 + X_L^2} = I\sqrt{R^2 + \omega^2 L^2} = V_s\sqrt{\dfrac{R^2 + \omega^2 L^2}{R^2 + \left(\omega L - \dfrac{1}{\omega C}\right)^2}}$

$\dfrac{V_{\text{out}}}{V_s} = \sqrt{\dfrac{R^2 + \omega^2 L^2}{R^2 + \left(\omega L - \dfrac{1}{\omega C}\right)^2}}$

ω small

As ω gets small, $R^2 + \left(\omega L - \dfrac{1}{\omega C}\right)^2 \to \dfrac{1}{\omega^2 C^2}, R^2 + \omega^2 L^2 \to R^2$

Therefore $\dfrac{V_{\text{out}}}{V_s} \to \sqrt{\dfrac{R^2}{(1/\omega^2 C^2)}} = \omega RC$ as ω becomes small.

ω large

As ω gets large, $R^2 + \left(\omega L - \dfrac{1}{\omega C}\right)^2 \rightarrow R^2 + \omega^2 L^2 \rightarrow \omega^2 L^2$, $R^2 + \omega^2 L^2 \rightarrow \omega^2 L^2$

Therefore, $\dfrac{V_{\text{out}}}{V_{\text{s}}} \rightarrow \sqrt{\dfrac{\omega^2 L^2}{\omega^2 L^2}} = 1$ as ω becomes large.

EVALUATE: $V_{\text{out}}/V_{\text{s}} \rightarrow 0$ as ω becomes small, so there is V_{out} only when the frequency ω of V_{s} is large. If the source voltage contains a number of frequency components, only the high frequency ones are passed by this filter.

31.53.   **IDENTIFY:** $U_B = \tfrac{1}{2}Li^2$. $U_E = \tfrac{1}{2}Cv^2$.

SET UP: Let $\langle x \rangle$ denote the average value of the quantity x. $\langle i^2 \rangle = \tfrac{1}{2}I^2$ and $\langle v_C^2 \rangle = \tfrac{1}{2}V_C^2$. Problem 31.51 shows

that $I = \dfrac{V}{\sqrt{R^2 + (\omega L - 1/[\omega C])^2}}$. Problem 31.52 shows that $V_C = \dfrac{V}{\omega C \sqrt{R^2 + (\omega L - 1/[\omega C])^2}}$.

EXECUTE: (a) $U_B = \tfrac{1}{2}Li^2 \Rightarrow \langle U_B \rangle = \tfrac{1}{2}L\langle i^2 \rangle = \tfrac{1}{2}LI_{\text{rms}}^2 = \tfrac{1}{2}L\left(\dfrac{I}{\sqrt{2}}\right)^2 = \tfrac{1}{4}LI^2$.

$U_E = \tfrac{1}{2}Cv_C^2 \Rightarrow \langle U_E \rangle = \dfrac{1}{2}C\langle v_C^2 \rangle = \tfrac{1}{2}CV_{C,\text{rms}}^2 = \tfrac{1}{2}C\left(\dfrac{V_C}{\sqrt{2}}\right)^2 = \tfrac{1}{4}CV_C^2$

(b) Using Problem 31.51a

$\langle U_B \rangle = \dfrac{1}{4}LI^2 = \dfrac{1}{4}L\left(\dfrac{V^2}{\sqrt{R^2 + (\omega L - 1/\omega C)^2}}\right)^2 = \dfrac{LV^2}{4\left(R^2 + (\omega L - 1/\omega C)^2\right)}$.

Using Problem (31.47b): $\langle U_E \rangle = \dfrac{1}{4}CV_C^2 = \dfrac{1}{4}C\dfrac{V^2}{\omega^2 C^2\left(R^2 + (\omega L - 1/\omega C)^2\right)} = \dfrac{V^2}{4\omega^2 C\left(R^2 + (\omega L - 1/\omega C)^2\right)}$.

(c) The graphs of the magnetic and electric energies are given in Figure 31.53.

EVALUATE: **(d)** When the angular frequency is zero, the magnetic energy stored in the inductor is zero, while the electric energy in the capacitor is $U_E = CV^2/4$. As the frequency goes to infinity, the energy noted in both

inductor and capacitor go to zero. The energies equal each other at the resonant frequency where $\omega_0 = \dfrac{1}{\sqrt{LC}}$ and

$U_B = U_E = \dfrac{LV^2}{4R^2}$.

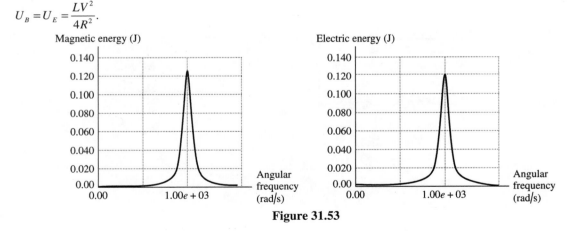

Figure 31.53

31.55.   **IDENTIFY:**   Apply the expression for I from problem 31.54 when $\omega_0 = 1/\sqrt{LC}$.

SET UP: From Problem 31.54, $I = V\sqrt{\dfrac{1}{R^2} + \left(\omega C - \dfrac{1}{\omega L}\right)^2}$

EXECUTE: (a) At resonance, $\omega_0 = \dfrac{1}{\sqrt{LC}} \Rightarrow \omega_0 C = \dfrac{1}{\omega_0 L} \Rightarrow I_C = V\omega_0 C = \dfrac{V}{\omega_0 L} = I_L$ so $I = I_R$ and I is a minimum.

(b) $P_{\text{av}} = \dfrac{V_{\text{rms}}^2}{Z}\cos\phi = \dfrac{V^2}{R}$ at resonance where $R < Z$ so power is a maximum.

(c) At $\omega = \omega_0$, I and V are in phase, so the phase angle is zero, which is the same as a series resonance.

EVALUATE: **(d)** The parallel circuit is sketched in Figure 31.55. At resonance, $\left|i_C\right|=\left|i_L\right|$ and at any instant of time these two currents are in opposite directions. Therefore, the net current between a and b is always zero.
(e) If the inductor and capacitor each have some resistance, and these resistances aren't the same, then it is no longer true that $i_C+i_L=0$ and the statement in part (d) isn't valid.

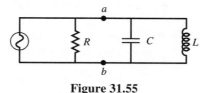

Figure 31.55

31.57.   **IDENTIFY** and **SET UP:**   Refer to the results and the phasor diagram in Problem 31.54. The source voltage is applied across each parallel branch.

EXECUTE: **(a)** $I_R=\dfrac{V}{R};\ I_C=V\omega C;\ I_L=\dfrac{V}{\omega L}.$

(b) The graph of each current versus ω is given in Figure 31.57a.

(c) $\omega\to 0:I_C\to 0;\ I_L\to\infty.$ $\omega\to\infty:I_C\to\infty;\ I_L\to 0.$

At low frequencies, the current is not changing much so the inductor's back-emf doesn't "resist." This allows the current to pass fairly freely. However, the current in the capacitor goes to zero because it tends to "fill up" over the slow period, making it less effective at passing charge. At high frequency, the induced emf in the inductor resists the violent changes and passes little current. The capacitor never gets a chance to fill up so passes charge freely.

(d) $\omega=\dfrac{1}{\sqrt{LC}}=\dfrac{1}{\sqrt{(2.0\text{ H})(0.50\times 10^{-6}\text{ F})}}=1000\text{ rad}/\sec$ and $f=159$ Hz. The phasor diagram is sketched in Figure 31.57b.

(e) $I=\sqrt{\left(\dfrac{V}{R}\right)^2+\left(V\omega C-\dfrac{V}{\omega L}\right)^2}.$

$I=\sqrt{\left(\dfrac{100\text{ V}}{200\ \Omega}\right)^2+\left((100\text{ V})(1000\text{ s}^{-1})(0.50\times 10^{-6}\text{ F})-\dfrac{100\text{ V}}{(1000\text{ s}^{-1})(2.0\text{ H})}\right)^2}=0.50\text{ A}$

(f) At resonance $I_L=I_C=V\omega C=(100\text{ V})(1000\text{ s}^{-1})(0.50\times 10^{-6}\text{ F})=0.0500$ A and $I_R=\dfrac{V}{R}=\dfrac{100\text{ V}}{200\ \Omega}=0.50$ A.

EVALUATE: At resonance $i_C=i_L=0$ at all times and the current through the source equals the current through the resistor.

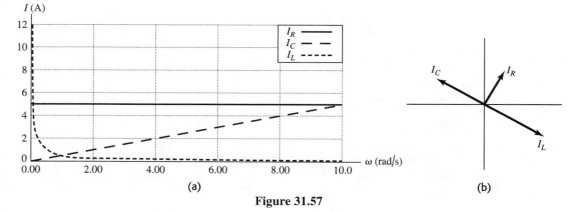

(a)   (b)

Figure 31.57

31.59.   **IDENTIFY:**   We know R, X_C and ϕ so Eq.(31.24) tells us X_L. Use $P_{av}=I_{rms}^2R$ from Exercise 31.27 to calculate I_{rms}. Then calculate Z and use Eq.(31.26) to calculate V_{rms} for the source.

SET UP: Source voltage lags current so $\phi=-54.0°$. $X_C=350\ \Omega$, $R=180\ \Omega$, $P_{av}=140$ W

EXECUTE: **(a)** $\tan\phi=\dfrac{X_L-X_C}{R}$

$X_L=R\tan\phi+X_C=(180\ \Omega)\tan(-54.0°)+350\ \Omega=-248\ \Omega+350\ \Omega=102\ \Omega$

(b) $P_{av}=V_{rms}I_{rms}\cos\phi=I_{rms}^2R$ (Exercise 31.27). $I_{rms}=\sqrt{\dfrac{P_{av}}{R}}=\sqrt{\dfrac{140\text{ W}}{180\ \Omega}}=0.882$ A

(c) $Z = \sqrt{R^2 + (X_L - X_C)^2} = \sqrt{(180\ \Omega)^2 + (102\ \Omega - 350\ \Omega)^2} = 306\ \Omega$

$V_{rms} = I_{rms}Z = (0.882\ A)(306\ \Omega) = 270\ V.$

EVALUATE: We could also use Eq.(31.31): $P_{av} = V_{rms}I_{rms}\cos\phi$

$V_{rms} = \dfrac{P_{av}}{I_{rms}\cos\phi} = \dfrac{140\ W}{(0.882\ A)\cos(-54.0°)} = 270\ V$, which agrees. The source voltage lags the current when

$X_C > X_L$, and this agrees with what we found.

31.61.   **IDENTIFY** and **SET UP:**   Eq.(31.19) allows us to calculate I and then Eq.(31.22) gives Z. Solve Eq.(31.21) for L.

EXECUTE: **(a)** $V_C = IX_C$ so $I = \dfrac{V_C}{X_C} = \dfrac{360\ V}{480\ \Omega} = 0.750\ A$

(b) $V = IZ$ so $Z = \dfrac{V}{I} = \dfrac{120\ V}{0.750\ A} = 160\ \Omega$

(c) $Z^2 = R^2 + (X_L - X_C)^2$

$X_L - X_C = \pm\sqrt{Z^2 - R^2}$, so

$X_L = X_C \pm \sqrt{Z^2 - R^2} = 480\ \Omega \pm \sqrt{(160\ \Omega)^2 - (80.0\ \Omega)^2} = 480\ \Omega \pm 139\ \Omega$

$X_L = 619\ \Omega$ or $341\ \Omega$

(d) EVALUATE: $X_C = \dfrac{1}{\omega C}$ and $X_L = \omega L$. At resonance, $X_C = X_L$. As the frequency is lowered below the

resonance frequency X_C increases and X_L decreases. Therefore, for $\omega < \omega_0, X_L < X_C$. So for $X_L = 341\ \Omega$ the

angular frequency is less than the resonance angular frequency. ω is greater than ω_0 when $X_L = 619\ \Omega$. But at

these two values of X_L, the magnitude of $X_L - X_C$ is the same so Z and I are the same. In one case $(X_L = 691\ \Omega)$

the source voltage leads the current and in the other $(X_L = 341\ \Omega)$ the source voltage lags the current.

31.63.   **IDENTIFY** and **SET UP:**   Consider the cycle of the repeating current that lies between $t_1 = \tau/2$ and $t_2 = 3\tau/2$. In

this interval $i = \dfrac{2I_0}{\tau}(t - \tau)$. $I_{av} = \dfrac{1}{t_2 - t_1}\displaystyle\int_{t_1}^{t_2} i\,dt$ and $I_{rms}^2 = \dfrac{1}{t_2 - t_1}\displaystyle\int_{t_1}^{t_2} i^2\,dt$

EXECUTE: $I_{av} = \dfrac{1}{t_2 - t_1}\displaystyle\int_{t_1}^{t_2} i\,dt = \dfrac{1}{\tau}\displaystyle\int_{\tau/2}^{3\tau/2}\dfrac{2I_0}{\tau}(t - \tau)\,dt = \dfrac{2I_0}{\tau^2}\left[\dfrac{1}{2}t^2 - \tau t\right]_{\tau/2}^{3\tau/2}$

$I_{av} = \left(\dfrac{2I_0}{\tau^2}\right)\left(\dfrac{9\tau^2}{8} - \dfrac{3\tau 2}{2} - \dfrac{\tau^2}{8} + \dfrac{\tau^2}{2}\right) = (2I_0)\tfrac{1}{8}(9 - 12 - 1 + 4) = \dfrac{I_0}{4}(13 - 13) = 0.$

$I_{rms}^2 = (I^2)_{av} = \dfrac{1}{t_2 - t_1}\displaystyle\int_{t_1}^{t_2} i^2\,dt = \dfrac{1}{\tau}\displaystyle\int_{\tau/2}^{3\tau/2}\dfrac{4I_0^2}{\tau^2}(t - \tau)^2\,dt$

$I_{rms}^2 = \dfrac{4I_0^2}{\tau^3}\displaystyle\int_{\tau/2}^{3\tau/2}(t - \tau)^2\,dt = \dfrac{4I_0^2}{\tau^3}\left[\tfrac{1}{3}(t - \tau)^3\right]_{\tau/2}^{3\tau/2} = \dfrac{4I_0^2}{3\tau^3}\left[\left(\dfrac{\tau}{2}\right)^3 - \left(-\dfrac{\tau}{2}\right)^3\right]$

$I_{rms}^2 = \dfrac{I_0^2}{6}[1 + 1] = \tfrac{1}{3}I_0^2$

$I_{rms} = \sqrt{I_{rms}^2} = \dfrac{I_0}{\sqrt{3}}.$

EVALUATE: In each cycle the current has as much negative value as positive value and its average is zero. i^2 is

always positive and its average is not zero. The relation between I_{rms} and the current amplitude for this current is

different from that for a sinusoidal current (Eq.31.4).

31.69.   **IDENTIFY:** $I = V/R$. $V_R = IR$, $V_C = IX_C$ and $V_L = IX_L$. $U_E = \tfrac{1}{2}CV_C^2$ and $U_L = \tfrac{1}{2}LI^2$.

SET UP: The amplitudes of each time dependent quantity correspond to the maximum values of those quantities.

EXECUTE: $\omega = \dfrac{\omega_0}{2}.$

(a) $I = \dfrac{V}{Z} = \dfrac{V}{\sqrt{R^2 + \left(\dfrac{\omega_0 L}{2} - 2/\omega_0 C\right)^2}} = \dfrac{V}{\sqrt{R^2 + \dfrac{9}{4}\dfrac{L}{C}}}.$

(b) $V_C = IX_C = \dfrac{2}{\omega_0 C} \dfrac{V}{\sqrt{R^2 + \dfrac{9}{4}\dfrac{L}{C}}} = \sqrt{\dfrac{L}{C}} \dfrac{2V}{\sqrt{R^2 + \dfrac{9}{4}\dfrac{L}{C}}}.$

(c) $V_L = IX_L = \dfrac{\omega_0 L}{2} \dfrac{V}{\sqrt{R^2 + \dfrac{9}{4}\dfrac{L}{C}}} = \sqrt{\dfrac{L}{C}} \dfrac{V/2}{\sqrt{R^2 + \dfrac{9}{4}\dfrac{L}{C}}}.$

(d) $U_C = \dfrac{1}{2}CV_C^2 = \dfrac{2LV^2}{R^2 + \dfrac{9}{4}\dfrac{L}{C}}.$

(e) $U_L = \dfrac{1}{2}LI^2 = \dfrac{1}{2}\dfrac{LV^2}{R^2 + \dfrac{9}{4}\dfrac{L}{C}}.$

EVALUATE: For $\omega < \omega_0$, $V_C > V_L$ and the maximum energy stored in the capacitor is greater than the maximum energy stored in the inductor.

31.73. **IDENTIFY:** $p_R = i^2 R.$ $p_L = iL\dfrac{di}{dt}.$ $p_C = \dfrac{q}{C}i.$

SET UP: $i = I\cos\omega t$

EXECUTE: **(a)** $p_R = i^2 R = I^2 \cos^2(\omega t)R = V_R I\cos^2(\omega t) = \dfrac{1}{2}V_R I(1 + \cos(2\omega t)).$

$P_{av}(R) = \dfrac{1}{T}\int_0^T p_R dt = \dfrac{V_R I}{2T}\int_0^T (1 + \cos(2\omega t))dt = \dfrac{V_R I}{2T}[t]_0^T = \tfrac{1}{2}V_R I.$

(b) $p_L = Li\dfrac{di}{dt} = -\omega LI^2 \cos(\omega t)\sin(\omega t) = -\tfrac{1}{2}V_L I\sin(2\omega t).$ But $\int_0^T \sin(2\omega t)dt = 0 \Rightarrow P_{av}(L) = 0.$

(c) $p_C = \dfrac{q}{C}i = v_C i = V_C I\sin(\omega t)\cos(\omega t) = \tfrac{1}{2}V_C I\sin(2\omega t).$ But $\int_0^T \sin(2\omega t)dt = 0 \Rightarrow P_{av}(C) = 0.$

(d) $p = p_R + p_L + p_c = V_R I\cos^2(\omega t) - \tfrac{1}{2}V_L I\sin(2\omega t) + \tfrac{1}{2}V_C I\sin(2\omega t)$ and

$p = I\cos(\omega t)(V_R \cos(\omega t) - V_L \sin(\omega t) + V_C \sin(\omega t)).$ But $\cos\phi = \dfrac{V_R}{V}$ and $\sin\phi = \dfrac{V_L - V_C}{V}$, so

$p = VI\cos(\omega t)(\cos\phi\cos(\omega t) - \sin\phi\sin(\omega t)),$ at any instant of time.

EVALUATE: At an instant of time the energy stored in the capacitor and inductor can be changing, but there is no net consumption of electrical energy in these components.

ELECTROMAGNETIC WAVES

32.7. **IDENTIFY** and **SET UP:** The equations are of the form of Eqs.(32.17), with x replaced by z. $\vec{B}$ is along the y-axis; deduce the direction of $\vec{E}$.

EXECUTE: $\omega = 2\pi f = 2\pi(6.10 \times 10^{14} \text{ Hz}) = 3.83 \times 10^{15} \text{ rad/s}$

$k = \dfrac{2\pi}{\lambda} = \dfrac{2\pi f}{c} = \dfrac{\omega}{c} = \dfrac{3.83 \times 10^{15} \text{ rad/s}}{3.00 \times 10^8 \text{ m/s}} = 1.28 \times 10^7 \text{ rad/m}$

$B_{\text{max}} = 5.80 \times 10^{-4} \text{ T}$

$E_{\text{max}} = cB_{\text{max}} = (3.00 \times 10^8 \text{ m/s})(5.80 \times 10^{-4} \text{ T}) = 1.74 \times 10^5 \text{ V/m}$

$\vec{B}$ is along the y-axis. $\vec{E} \times \vec{B}$ is in the direction of propagation (the $+z$-direction). From this we can deduce the direction of $\vec{E}$, as shown in Figure 32.7.

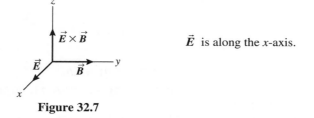

$\vec{E}$ is along the x-axis.

Figure 32.7

$\vec{E} = E_{\text{max}} \hat{i} \cos(kz - \omega t) = (1.74 \times 10^5 \text{ V/m}) \hat{i} \cos[(1.28 \times 10^7 \text{ rad/m})z - (3.83 \times 10^{15} \text{ rad/s})t]$

$\vec{B} = B_{\text{max}} \hat{j} \cos(kz - \omega t) = \left(5.80 \times 10^{-4} \text{ T}\right) \hat{j} \cos[(1.28 \times 10^7 \text{ rad/m})z - (3.83 \times 10^{15} \text{ rad/s})t]$

EVALUATE: $\vec{E}$ and $\vec{B}$ are perpendicular and oscillate in phase.

32.9. **IDENTIFY** and **SET UP:** Compare the $\vec{E}(y, t)$ given in the problem to the general form given by Eq.(32.17). Use the direction of propagation and of $\vec{E}$ to find the direction of $\vec{B}$.

(a) EXECUTE: The equation for the electric field contains the factor $\sin(ky - \omega t)$ so the wave is traveling in the $+y$-direction. The equation for $\vec{E}(y, t)$ is in terms of $\sin(ky - \omega t)$ rather than $\cos(ky - \omega t)$; the wave is shifted in phase by $90°$ relative to one with a $\cos(ky - \omega t)$ factor.

(b) $\vec{E}(y,t) = -(3.10 \times 10^5 \text{ V/m}) \hat{k} \sin[ky - (2.65 \times 10^{12} \text{ rad/s})t]$

Comparing to Eq.(32.17) gives $\omega = 2.65 \times 10^{12} \text{ rad/s}$

$\omega = 2\pi f = \dfrac{2\pi c}{\lambda}$ so $\lambda = \dfrac{2\pi c}{\omega} = \dfrac{2\pi(2.998 \times 10^8 \text{ m/s})}{(2.65 \times 10^{12} \text{ rad/s})} = 7.11 \times 10^{-4} \text{ m}$

(c)

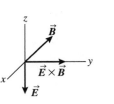

$\vec{E} \times \vec{B}$ must be in the $+y$-direction (the direction in which the wave is traveling). When $\vec{E}$ is in the $-z$-direction then $\vec{B}$ must be in the $-x$-direction, as shown in Figure 32.9.

Figure 32.9

$$k = \frac{2\pi}{\lambda} = \frac{\omega}{c} = \frac{2.65 \times 10^{12} \text{ rad/s}}{2.998 \times 10^8 \text{ m/s}} = 8.84 \times 10^3 \text{ rad/m}$$

$E_{max} = 3.10 \times 10^5$ V/m

Then $B_{max} = \dfrac{E_{max}}{c} = \dfrac{3.10 \times 10^5 \text{ V/m}}{2.998 \times 10^8 \text{ m/s}} = 1.03 \times 10^{-3}$ T

Using Eq.(32.17) and the fact that $\vec{B}$ is in the $-\hat{i}$ direction when $\vec{E}$ is in the $-\hat{k}$ direction,

$\vec{B} = -(1.03 \times 10^{-3} \text{ T})\hat{i} \sin[(8.84 \times 10^3 \text{ rad/m})y - (2.65 \times 10^{12} \text{ rad/s})t]$

EVALUATE: $\vec{E}$ and $\vec{B}$ are perpendicular and oscillate in phase.

32.11. **IDENTIFY** and **SET UP:** $c = f\lambda$ allows calculation of λ. $k = 2\pi / \lambda$ and $\omega = 2\pi f$. Eq.(32.18) relates the electric and magnetic field amplitudes.

EXECUTE: **(a)** $c = f\lambda$ so $\lambda = \dfrac{c}{f} = \dfrac{2.998 \times 10^8 \text{ m/s}}{830 \times 10^3 \text{ Hz}} = 361$ m

(b) $k = \dfrac{2\pi}{\lambda} = \dfrac{2\pi \text{ rad}}{361 \text{ m}} = 0.0174$ rad/m

(c) $\omega = 2\pi f = (2\pi)(830 \times 10^3 \text{ Hz}) = 5.22 \times 10^6$ rad/s

(d) Eq.(32.18): $E_{max} = cB_{max} = (2.998 \times 10^8 \text{ m/s})(4.82 \times 10^{-11} \text{ T}) = 0.0144$ V/m

EVALUATE: This wave has a very long wavelength; its frequency is in the AM radio braodcast band. The electric and magnetic fields in the wave are very weak.

32.13. **IDENTIFY** and **SET UP:** $v = f\lambda$ relates frequency and wavelength to the speed of the wave. Use Eq.(32.22) to calculate n and K.

EXECUTE: **(a)** $\lambda = \dfrac{v}{f} = \dfrac{2.17 \times 10^8 \text{ m/s}}{5.70 \times 10^{14} \text{ Hz}} = 3.81 \times 10^{-7}$ m

(b) $\lambda = \dfrac{c}{f} = \dfrac{2.998 \times 10^8 \text{ m/s}}{5.70 \times 10^{14} \text{ Hz}} = 5.26 \times 10^{-7}$ m

(c) $n = \dfrac{c}{v} = \dfrac{2.998 \times 10^8 \text{ m/s}}{2.17 \times 10^8 \text{ m/s}} = 1.38$

(d) $n = \sqrt{KK_m} \approx \sqrt{K}$ so $K = n^2 = (1.38)^2 = 1.90$

EVALUATE: In the material $v < c$ and f is the same, so λ is less in the material than in air. $v < c$ always, so n is always greater than unity.

32.17. **IDENTIFY:** $E_{max} = cB_{max}$. $\vec{E} \times \vec{B}$ is in the direction of propagation.

SET UP: $c = 3.00 \times 10^8$ m/s. $E_{max} = 4.00$ V/m.

EXECUTE: $B_{max} = E_{max}/c = 1.33 \times 10^{-8}$ T. For $\vec{E}$ in the $+x$-direction, $\vec{E} \times \vec{B}$ is in the $+z$-direction when $\vec{B}$ is in the $+y$-direction.

EVALUATE: $\vec{E}$, $\vec{B}$ and the direction of propagation are all mutually perpendicular.

32.19. **IDENTIFY** and **SET UP:** Use Eq.(32.29) to calculate I, Eq.(32.18) to calculate B_{max}, and use $I = P_{av}/4\pi r^2$ to calculate P_{av}.

(a) EXECUTE: $I = \frac{1}{2}\epsilon_0 E_{max}^2$; $E_{max} = 0.090$ V/m, so $I = 1.1 \times 10^{-5}$ W/m^2

(b) $E_{max} = cB_{max}$ so $B_{max} = E_{max}/c = 3.0 \times 10^{-10}$ T

(c) $P_{av} = I(4\pi r^2) = (1.075 \times 10^{-5} \text{ W/m}^2)(4\pi)(2.5 \times 10^3 \text{ m})^2 = 840$ W

(d) EVALUATE: The calculation in part (c) assumes that the transmitter emits uniformly in all directions.

32.23. **IDENTIFY:** $P_{av} = IA$ and $I = \frac{1}{2}\epsilon_0 c E_{max}^2$

SET UP: The surface area of a sphere is $A = 4\pi r^2$.

EXECUTE: $P_{av} = S_{av}A = \left(\dfrac{E_{max}^2}{2c\mu_0}\right)(4\pi r^2)$. $E_{max} = \sqrt{\dfrac{P_{av}c\mu_0}{2\pi r^2}} = \sqrt{\dfrac{(60.0 \text{ W})(3.00\times10^8 \text{ m/s})\mu_0}{2\pi(5.00 \text{ m})^2}} = 12.0 \text{ V/m}.$

$B_{max} = \dfrac{E_{max}}{c} = \dfrac{12.0 \text{ V/m}}{3.00\times10^8 \text{ m/s}} = 4.00\times10^{-8} \text{ T}.$

EVALUATE: E_{max} and B_{max} are both inversely proportional to the distance from the source.

32.25. **IDENTIFY:**   Use the radiation pressure to find the intensity, and then $P_{av} = I(4\pi r^2)$.

SET UP: For a perfectly absorbing surface, $p_{rad} = \dfrac{I}{c}$

EXECUTE: $p_{rad} = I/c$ so $I = cp_{rad} = 2.70\times10^3 \text{ W/m}^2$. Then

$P_{av} = I(4\pi r^2) = (2.70\times10^3 \text{ W/m}^2)(4\pi)(5.0 \text{ m})^2 = 8.5\times10^5 \text{ W}.$

EVALUATE: Even though the source is very intense the radiation pressure 5.0 m from the surface is very small.

32.27. **IDENTIFY and SET UP:**   Use Eqs.(32.30) and (32.31).

EXECUTE: **(a)** By Eq.(32.30) the average momentum density is $\dfrac{dp}{dV} = \dfrac{S_{av}}{c^2} = \dfrac{I}{c^2}$

$\dfrac{dp}{dV} = \dfrac{0.78\times10^3 \text{ W/m}^2}{(2.998\times10^8 \text{ m/s})^2} = 8.7\times10^{-15} \text{ kg/m}^2\cdot\text{s}$

(b) By Eq.(32.31) the average momentum flow rate per unit area is $\dfrac{S_{av}}{c} = \dfrac{I}{c} = \dfrac{0.78\times10^3 \text{ W/m}^2}{2.998\times10^8 \text{ m/s}} = 2.6\times10^{-6} \text{ Pa}$

EVALUATE: The radiation pressure that the sunlight would exert on an absorbing or reflecting surface is very small.

32.31. **IDENTIFY and SET UP:**   Apply Eqs.(32.36) and (32.37).

EXECUTE: **(a)** By Eq.(32.37) we see that the nodal planes of the $\vec{B}$ field are a distance $\lambda/2$ apart, so $\lambda/2 = 3.55$ mm and $\lambda = 7.10$ mm.

(b) By Eq.(32.36) we see that the nodal planes of the $\vec{E}$ field are also a distance $\lambda/2 = 3.55$ mm apart.

(c) $v = f\lambda = (2.20\times10^{10} \text{ Hz})(7.10\times10^{-3} \text{ m}) = 1.56\times10^8 \text{ m/s}.$

EVALUATE: The spacing between the nodes of $\vec{E}$ is the same as the spacing between the nodes of $\vec{B}$. Note that $v < c$, as it must.

32.33. **(a) IDENTIFY and SET UP:**   The distance between adjacent nodal planes of $\vec{B}$ is $\lambda/2$. There is an antinodal plane of $\vec{B}$ midway between any two adjacent nodal planes, so the distance between a nodal plane and an adjacent antinodal plane is $\lambda/4$. Use $v = f\lambda$ to calculate λ.

EXECUTE: $\lambda = \dfrac{v}{f} = \dfrac{2.10\times10^8 \text{ m/s}}{1.20\times10^{10} \text{ Hz}} = 0.0175 \text{ m}$

$\dfrac{\lambda}{4} = \dfrac{0.0175 \text{ m}}{4} = 4.38\times10^{-3} \text{ m} = 4.38 \text{ mm}$

(b) IDENTIFY and SET UP: The nodal planes of $\vec{E}$ are at $x = 0$, $\lambda/2$, λ, $3\lambda/2, \ldots$, so the antinodal planes of $\vec{E}$ are at $x = \lambda/4$, $3\lambda/4$, $5\lambda/4, \ldots$. The nodal planes of $\vec{B}$ are at $x = \lambda/4$, $3\lambda/4$, $5\lambda/4, \ldots$, so the antinodal planes of $\vec{B}$ are at $\lambda/2$, λ, $3\lambda/2, \ldots$.

EXECUTE: The distance between adjacent antinodal planes of $\vec{E}$ and antinodal planes of $\vec{B}$ is therefore $\lambda/4 = 4.38$ mm.

(c) From Eqs.(32.36) and (32.37) the distance between adjacent nodal planes of $\vec{E}$ and $\vec{B}$ is $\lambda/4 = 4.38$ mm.

EVALUATE: The nodes of $\vec{E}$ coincide with the antinodes of $\vec{B}$, and conversely. The nodes of $\vec{B}$ and the nodes of $\vec{E}$ are equally spaced.

32.37. **IDENTIFY and SET UP:**   Take partial derivatives of Eqs.(32.12) and (32.14), as specified in the problem.

EXECUTE: Eq.(32.12): $\dfrac{\partial E_y}{\partial x} = -\dfrac{\partial B_z}{\partial t}$

Taking $\frac{\partial}{\partial t}$ of both sides of this equation gives $\frac{\partial^2 E_y}{\partial x \partial t} = -\frac{\partial^2 B_z}{\partial t^2}$. Eq.(32.14) says $-\frac{\partial B_z}{\partial x} = \epsilon_0\mu_0\frac{\partial E_y}{\partial t}$. Taking $\frac{\partial}{\partial x}$ of

both sides of this equation gives $-\frac{\partial^2 B_z}{\partial x^2} = \epsilon_0\mu_0\frac{\partial^2 E_y}{\partial t \partial x}$, so $\frac{\partial^2 E_y}{\partial t \partial x} = -\frac{1}{\epsilon_0\mu_0}\frac{\partial^2 B_z}{\partial x^2}$. But $\frac{\partial^2 E_y}{\partial x \partial t} = \frac{\partial^2 E_y}{\partial t \partial x}$ (The order in

which the partial derivatives are taken doesn't change the result.) So $-\frac{\partial^2 B_z}{\partial t^2} = -\frac{1}{\epsilon_0\mu_0}\frac{\partial^2 B_z}{\partial x^2}$ and $\frac{\partial^2 B_z}{\partial x^2} = \epsilon_0\mu_0\frac{\partial^2 B_z}{\partial t^2}$,

as was to be shown.

EVALUATE: Both fields, electric and magnetic, satisfy the wave equation, Eq.(32.10). We have also shown that both fields propagate with the same speed $v = 1/\sqrt{\epsilon_0\mu_0}$.

32.39. **IDENTIFY:** The intensity of an electromagnetic wave depends on the amplitude of the electric and magnetic fields. Such a wave exerts a force because it carries energy.

SET UP: The intensity of the wave is $I = P_{av}/A = \frac{1}{2}\epsilon_0 c E_{max}^2$, and the force is $F = P_{av}A$ where $P_{av} = I/c$.

EXECUTE: **(a)** $I = P_{av}/A = (25{,}000\text{ W})/[4\pi(575\text{ m})^2] = 0.00602\text{ W/m}^2$

(b) $I = \frac{1}{2}\epsilon_0 c E_{max}^2$, so $E_{max} = \sqrt{\frac{2I}{\epsilon_0 c}} = \sqrt{\frac{2(0.00602\text{ W/m}^2)}{(8.85\times10^{-12}\text{ C}^2/\text{N}\cdot\text{m}^2)(3.00\times10^8\text{ m/s})}} = 2.13\text{ N/C}.$

$$B_{max} = E_{max}/c = (2.13\text{ N/C})/(3.00\times10^8\text{ m/s}) = 7.10\times10^{-9}\text{ T}$$

(c) $F = P_{av}A = (I/c)A = (0.00602\text{ W/m}^2)(0.150\text{ m})(0.400\text{ m})/(3.00\times10^8\text{ m/s}) = 1.20\times10^{-12}\text{ N}$

EVALUATE: The fields are very weak compared to ordinary laboratory fields, and the force is hardly worth worrying about!

32.41. **(a) IDENTIFY and SET UP:** Calculate I and then use Eq.(32.29) to calculate E_{max} and Eq.(32.18) to calculate B_{max}.

EXECUTE: The intensity is power per unit area: $I = \frac{P}{A} = \frac{3.20\times10^{-3}\text{ W}}{\pi(1.25\times10^{-3}\text{ m})^2} = 652\text{ W/m}^2.$

$I = \frac{E_{max}^2}{2\mu_0 c}$, so $E_{max} = \sqrt{2\mu_0 c I}$

$E_{max} = \sqrt{2(4\pi\times10^{-7}\text{ T}\cdot\text{m/A})(2.998\times10^8\text{ m/s})(652\text{ W/m}^2)} = 701\text{ V/m}$

$B_{max} = \frac{E_{max}}{c} = \frac{701\text{ V/m}}{2.998\times10^8\text{ m/s}} = 2.34\times10^{-6}\text{ T}$

EVALUATE: The magnetic field amplitude is quite small.

(b) IDENTIFY and SET UP: Eqs.(24.11) and (30.10) give the energy density in terms of the electric and magnetic field values at any time. For sinusoidal fields average over E^2 and B^2 to get the average energy densities.

EXECUTE: The energy density in the electric field is $u_E = \frac{1}{2}\epsilon_0 E^2$. $E = E_{max}\cos(kx - \omega t)$ and the average value of $\cos^2(kx - \omega t)$ is $\frac{1}{2}$. The average energy density in the electric field then is

$u_{E,av} = \frac{1}{4}\epsilon_0 E_{max}^2 = \frac{1}{4}(8.854\times10^{-12}\text{ C}^2/\text{N}\cdot\text{m}^2)(701\text{ V/m})^2 = 1.09\times10^{-6}\text{ J/m}^3$. The energy density in the magnetic field

is $u_B = \frac{B^2}{2\mu_0}$. The average value is $u_{B,av} = \frac{B_{max}^2}{4\mu_0} = \frac{(2.34\times10^{-6}\text{ T})^2}{4(4\pi\times10^{-7}\text{ T}\cdot\text{m/A})} = 1.09\times10^{-6}\text{ J/m}^3.$

EVALUATE: Our result agrees with the statement in Section 32.4 that the average energy density for the electric field is the same as the average energy density for the magnetic field.

(c) IDENTIFY and SET UP: The total energy in this length of beam is the total energy density

$u_{av} = u_{E,av} + u_{B,av} = 2.18\times10^{-6}\text{ J/m}^3$ times the volume of this part of the beam.

EXECUTE: $U = u_{av}LA = (2.18\times10^{-6}\text{ J/m}^3)(1.00\text{ m})\pi(1.25\times10^{-3}\text{ m})^2 = 1.07\times10^{-11}\text{ J}.$

EVALUATE: This quantity can also be calculated as the power output times the time it takes the light to travel $L = 1.00$ m: $U = P\left(\frac{L}{c}\right) = (3.20\times10^{-3}\text{ W})\left(\frac{1.00\text{ m}}{2.998\times10^8\text{ m/s}}\right) = 1.07\times10^{-11}\text{ J}$, which checks.

32.45. **IDENTIFY:** The same intensity light falls on both reflectors, but the force on the reflecting surface will be twice as great as the force on the absorbing surface. Therefore there will be a net torque about the rotation axis.

SET UP: For a totally absorbing surface, $F = P_{av}A = (I/c)A$, while for a totally reflecting surface the force will be twice as great. The intensity of the wave is $I = \frac{1}{2}\epsilon_0 c E_{max}^2$. Once we have the torque, we can use the rotational form of Newton's second law, $\tau_{net} = I\alpha$, to find the angular acceleration.

EXECUTE: The force on the absorbing reflector is $F_{Abs} = p_{av}A = (I/c)A = \dfrac{\frac{1}{2}\epsilon_0 cE^2_{max}A}{c} = \frac{1}{2}\epsilon_0 AE^2_{max}$

For a totally reflecting surface, the force will be twice as great, which is $\epsilon_0 cE^2_{max}$. The net torque is therefore

$\tau_{net} = F_{Refl}(L/2) - F_{Abs}(L/2) = \epsilon_0 AE^2_{max}L/4$

Newton's 2nd law for rotation gives $\tau_{net} = I\alpha$. $\epsilon_0 AE^2_{max}L/4 = 2m(L/2)^2\alpha$

Solving for α gives $\alpha = \epsilon_0 AE^2_{max}/(2mL) = \dfrac{(8.85\times 10^{-12}\ \text{C}^2/\text{N}\cdot\text{m}^2)(0.0150\ \text{m})^2(1.25\ \text{N/C})^2}{(2)(0.00400\ \text{kg})(1.00\ \text{m})} = 3.89\times 10^{-13}\ \text{rad/s}^2$

EVALUATE: This is an extremely small angular acceleration. To achieve a larger value, we would have to greatly increase the intensity of the light wave or decrease the mass of the reflectors.

32.47.   **IDENTIFY** and **SET UP:**   In the wire the electric field is related to the current density by Eq.(25.7). Use Ampere's law to calculate $\vec{B}$. The Poynting vector is given by Eq.(32.28) and the equation that follows it relates the energy flow through a surface to $\vec{S}$.

EXECUTE: **(a)** The direction of $\vec{E}$ is parallel to the axis of the cylinder, in the direction of the current. From Eq.(25.7), $E = \rho J = \rho I/\pi a^2$. ($E$ is uniform across the cross section of the conductor.)

(b) A cross-sectional view of the conductor is given in Figure 32.47a; take the current to be coming out of the page.

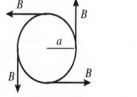

Apply Ampere's law to a circle of radius a.

$\oint \vec{B} \cdot d\vec{l} = B(2\pi a)$

$I_{encl} = I$

Figure 32.47a

$\oint \vec{B} \cdot d\vec{l} = \mu_0 I_{encl}$ gives $B(2\pi a) = \mu_0 I$ and $B = \dfrac{\mu_0 I}{2\pi a}$

The direction of $\vec{B}$ is counterclockwise around the circle.

(c) The directions of $\vec{E}$ and $\vec{B}$ are shown in Figure 32.47b.

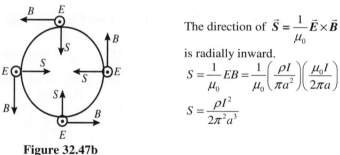

The direction of $\vec{S} = \dfrac{1}{\mu_0}\vec{E} \times \vec{B}$

is radially inward.

$S = \dfrac{1}{\mu_0}EB = \dfrac{1}{\mu_0}\left(\dfrac{\rho I}{\pi a^2}\right)\left(\dfrac{\mu_0 I}{2\pi a}\right)$

$S = \dfrac{\rho I^2}{2\pi^2 a^3}$

Figure 32.47b

(d) EVALUATE: Since S is constant over the surface of the conductor, the rate of energy flow P is given by S times the surface of a length l of the conductor: $P = SA = S(2\pi al) = \dfrac{\rho I^2}{2\pi^2 a^3}(2\pi al) = \dfrac{\rho lI^2}{\pi a^2}$. But $R = \dfrac{\rho l}{\pi a^2}$, so the result from the Poynting vector is $P = RI^2$. This agrees with $P_R = I^2R$, the rate at which electrical energy is being dissipated by the resistance of the wire. Since $\vec{S}$ is radially inward at the surface of the wire and has magnitude equal to the rate at which electrical energy is being dissipated in the wire, this energy can be thought of as entering through the cylindrical sides of the conductor.

32.49.   **IDENTIFY** and **SET UP:**   The magnitude of the induced emf is given by Faraday's law: $|\mathcal{E}| = \left|\dfrac{d\Phi_B}{dt}\right|$. To calculate $d\Phi_B/dt$ we need dB/dt at the antenna. Use the total power output to calculate I and then combine Eq.(32.29) and (32.18) to calculate B_{max}. The time dependence of B is given by Eq.(32.17).

EXECUTE: $\Phi_B = B\pi R^2$, where $R = 0.0900$ m is the radius of the loop. (This assumes that the magnetic field is uniform across the loop, an excellent approximation.) $|\mathcal{E}| = \pi R^2 \left|\dfrac{dB}{dt}\right|$

$B = B_{max}\cos(kx - \omega t)$ so $\left|\dfrac{dB}{dt}\right| = B_{max}\omega\sin(kx - \omega t)$

The maximum value of $\left|\dfrac{dB}{dt}\right|$ is $B_{max}\omega$, so $\left|\mathcal{E}\right|_{max} = \pi R^2 B_{max}\omega$.

$R = 0.0900$ m, $\omega = 2\pi f = 2\pi(95.0\times10^6 \text{ Hz}) = 5.97\times10^8$ rad/s

Calculate the intensity I at this distance from the source, and from that the magnetic field amplitude B_{max}:

$$I = \frac{P}{4\pi r^2} = \frac{55.0\times10^3 \text{ W}}{4\pi(2.50\times10^3 \text{ m})^2} = 7.00\times10^{-4} \text{ W/m}^2. \quad I = \frac{E_{max}^2}{2\mu_0 c} = \frac{(cB_{max})^2}{2\mu_0 c} = \frac{c}{2\mu_0}B_{max}^2$$

Thus $B_{max} = \sqrt{\dfrac{2\mu_0 I}{c}} = \sqrt{\dfrac{2(4\pi\times10^{-7} \text{ T}\cdot\text{m/A})(7.00\times10^{-4} \text{ W/m}^2)}{2.998\times10^8 \text{ m/s}}} = 2.42\times10^{-9}$ T. Then

$\left|\mathcal{E}\right|_{max} = \pi R^2 B_{max}\omega = \pi(0.0900 \text{ m})^2(2.42\times10^{-9} \text{ T})(5.97\times10^8 \text{ rad/s}) = 0.0368$ V.

EVALUATE: An induced emf of this magnitude is easily detected.

32.51. **IDENTIFY** and **SET UP:** Find the force on you due to the momentum carried off by the light. Express this force in terms of the radiated power of the flashlight. Use this force to calculate your acceleration and use a constant acceleration equation to find the time.

(a) **EXECUTE:** $p_{rad} = I/c$ and $F = p_{rad}A$ gives $F = IA/c = P_{av}/c$

$a_x = F/m = P_{av}/(mc) = (200 \text{ W})/[(150 \text{ kg})(3.00\times10^8 \text{ m/s})] = 4.44\times10^{-9}$ m/s^2

Then $x - x_0 = v_{0x}t + \frac{1}{2}a_x t^2$ gives $t = \sqrt{2(x - x_0)/a_x} = \sqrt{2(16.0 \text{ m})/(4.44\times10^{-9} \text{ m/s}^2)} = 8.49\times10^4$ s $= 23.6$ h

EVALUATE: The radiation force is very small. In the calculation we have ignored any other forces on you.

(b) You could throw the flashlight in the direction away from the ship. By conservation of linear momentum you would move toward the ship with the same magnitude of momentum as you gave the flashlight.

32.55. **IDENTIFY** and **SET UP:** The gravitational force is given by Eq.(12.2). Express the mass of the particle in terms of its density and volume. The radiation pressure is given by Eq.(32.32); relate the power output L of the sun to the intensity at a distance r. The radiation force is the pressure times the cross sectional area of the particle.

EXECUTE: (a) The gravitational force is $F_g = G\dfrac{mM}{r^2}$. The mass of the dust particle is $m = \rho V = \rho\frac{4}{3}\pi R^3$. Thus

$$F_g = \frac{4\rho G\pi MR^3}{3r^2}.$$

(b) For a totally absorbing surface $p_{rad} = \dfrac{I}{c}$. If L is the power output of the sun, the intensity of the solar radiation

a distance r from the sun is $I = \dfrac{L}{4\pi r^2}$. Thus $p_{rad} = \dfrac{L}{4\pi c r^2}$. The force F_{rad} that corresponds to p_{rad} is in the

direction of propagation of the radiation, so $F_{rad} = p_{rad}A_\perp$, where $A_\perp = \pi R^2$ is the component of area of the particle

perpendicular to the radiation direction. Thus $F_{rad} = \left(\dfrac{L}{4\pi c r^2}\right)(\pi R^2) = \dfrac{LR^2}{4cr^2}$.

(c) $F_g = F_{rad}$

$$\frac{4\rho G\pi MR^3}{3r^2} = \frac{LR^2}{4cr^2}$$

$$\left(\frac{4\rho G\pi M}{3}\right)R = \frac{L}{4c} \text{ and } R = \frac{3L}{16c\rho G\pi M}$$

$$R = \frac{3(3.9\times10^{26} \text{ W})}{16(2.998\times10^8 \text{ m/s})(3000 \text{ kg/m}^3)(6.673\times10^{-11} \text{ N}\cdot\text{m}^2/\text{kg}^2)\pi(1.99\times10^{30} \text{ kg})}$$

$R = 1.9\times10^{-7}$ m $= 0.19$ μm.

EVALUATE: The gravitational force and the radiation force both have a r^{-2} dependence on the distance from the sun, so this distance divides out in the calculation of R.

(d) $\dfrac{F_{rad}}{F_g} = \left(\dfrac{LR^2}{4cr^2}\right)\left(\dfrac{3r^2}{4\rho G\pi mR^3}\right) = \dfrac{3L}{16c\rho G\pi MR}$. F_{rad} is proportional to R^2 and F_g is proportional to R^3, so this

ratio is proportional to $1/R$. If $R < 0.20$ μm then $F_{rad} > F_g$ and the radiation force will drive the particles out of the solar system.

THE NATURE AND PROPAGATION OF LIGHT

33.1. **IDENTIFY:** For reflection, $\theta_r = \theta_a$.

SET UP: The desired path of the ray is sketched in Figure 33.1.

EXECUTE: $\tan\phi = \dfrac{14.0 \text{ cm}}{11.5 \text{ cm}}$, so $\phi = 50.6°$. $\theta_r = 90° - \phi = 39.4°$ and $\theta_r = \theta_a = 39.4°$.

EVALUATE: The angle of incidence is measured from the normal to the surface.

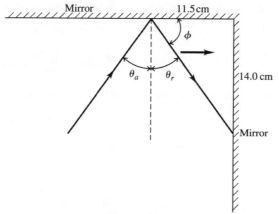

Figure 33.1

33.3. **IDENTIFY and SET UP:** Use Eqs.(33.1) and (33.5) to calculate v and λ.

EXECUTE: **(a)** $n = \dfrac{c}{v}$ so $v = \dfrac{c}{n} = \dfrac{2.998 \times 10^8 \text{ m/s}}{1.47} = 2.04 \times 10^8 \text{ m/s}$

(b) $\lambda = \dfrac{\lambda_0}{n} = \dfrac{650 \text{ nm}}{1.47} = 442 \text{ nm}$

EVALUATE: Light is slower in the liquid than in vacuum. By $v = f\lambda$, when v is smaller, λ is smaller.

33.7. **IDENTIFY:** Apply Eqs.(33.2) and (33.4) to calculate θ_r and θ_b. The angles in these equations are measured with respect to the normal, not the surface.

(a) SET UP: The incident, reflected and refracted rays are shown in Figure 33.7.

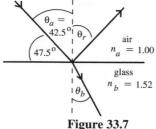

EXECUTE: $\theta_r = \theta_a = 42.5°$
The reflected ray makes an angle of $90.0° - \theta_r = 47.5°$
with the surface of the glass.

Figure 33.7

(b) $n_a \sin\theta_a = n_b \sin\theta_b$, where the angles are measured from the normal to the interface.

$\sin\theta_b = \dfrac{n_a \sin\theta_a}{n_b} = \dfrac{(1.00)(\sin 42.5°)}{1.66} = 0.4070$

$\theta_b = 24.0°$

The refracted ray makes an angle of $90.0° - \theta_b = 66.0°$ with the surface of the glass.

EVALUATE: The light is bent toward the normal when the light enters the material of larger refractive index.

33.9.   **IDENTIFY** and **SET UP:**   Use Snell's law to find the index of refraction of the plastic and then use Eq.(33.1) to calculate the speed v of light in the plastic.

EXECUTE: $n_a \sin\theta_a = n_b \sin\theta_b$

$$n_b = n_a \left(\frac{\sin\theta_a}{\sin\theta_b} \right) = 1.00 \left(\frac{\sin 62.7°}{\sin 48.1°} \right) = 1.194$$

$$n = \frac{c}{v} \text{ so } v = \frac{c}{n} = (3.00 \times 10^8 \text{ m/s})/1.194 = 2.51 \times 10^8 \text{ m/s}$$

EVALUATE: Light is slower in plastic than in air. When the light goes from air into the plastic it is bent toward the normal.

33.15.   **IDENTIFY:**   Apply $n_a \sin\theta_a = n_b \sin\theta_b$.

SET UP: The light refracts from the liquid into the glass, so $n_a = 1.70$, $\theta_a = 62.0°$. $n_b = 1.58$.

EXECUTE: $\sin\theta_b = \left(\frac{n_a}{n_b} \right) \sin\theta_a = \left(\frac{1.70}{1.58} \right) \sin 62.0° = 0.950$ and $\theta_b = 71.8°$.

EVALUATE: The ray refracts into a material of smaller n, so it is bent away from the normal.

33.17.   **IDENTIFY:**   The critical angle for total internal reflection is θ_a that gives $\theta_b = 90°$ in Snell's law.

SET UP: In Figure 33.17 the angle of incidence θ_a is related to angle θ by $\theta_a + \theta = 90°$.

EXECUTE: **(a)** Calculate θ_a that gives $\theta_b = 90°$. $n_a = 1.60$, $n_b = 1.00$ so $n_a \sin\theta_a = n_b \sin\theta_b$ gives

$(1.60)\sin\theta_a = (1.00)\sin 90°$. $\sin\theta_a = \frac{1.00}{1.60}$ and $\theta_a = 38.7°$. $\theta = 90° - \theta_a = 51.3°$.

(b) $n_a = 1.60$, $n_b = 1.333$. $(1.60)\sin\theta_a = (1.333)\sin 90°$. $\sin\theta_a = \frac{1.333}{1.60}$ and $\theta_a = 56.4°$. $\theta = 90° - \theta_a = 33.6°$.

EVALUATE: The critical angle increases when the ratio $\frac{n_a}{n_b}$ increases.

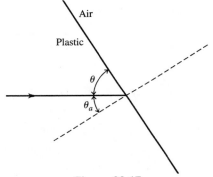

Figure 33.17

33.19.   **IDENTIFY:**   Use the critical angle to find the index of refraction of the liquid.

SET UP: Total internal reflection requires that the light be incident on the material with the larger n, in this case the liquid. Apply $n_a \sin\theta_a = n_b \sin\theta_b$ with a = liquid and b = air, so $n_a = n_{\text{liq}}$ and $n_b = 1.0$.

EXECUTE: $\theta_a = \theta_{\text{crit}}$ when $\theta_b = 90°$, so $n_{\text{liq}} \sin\theta_{\text{crit}} = (1.0)\sin 90°$

$$n_{\text{liq}} = \frac{1}{\sin\theta_{\text{crit}}} = \frac{1}{\sin 42.5°} = 1.48.$$

(a) $n_a \sin\theta_a = n_b \sin\theta_b$ (a = liquid, b = air)

$$\sin\theta_b = \frac{n_a \sin\theta_a}{n_b} = \frac{(1.48)\sin 35.0°}{1.0} = 0.8489 \text{ and } \theta_b = 58.1°$$

(b) Now $n_a \sin\theta_a = n_b \sin\theta_b$ with a = air, b = liquid

$$\sin\theta_b = \frac{n_a \sin\theta_a}{n_b} = \frac{(1.0)\sin 35.0°}{1.48} = 0.3876 \text{ and } \theta_b = 22.8°$$

EVALUATE: For light traveling liquid $\rightarrow$ air the light is bent away from the normal. For light traveling air $\rightarrow$ liquid the light is bent toward the normal.

33.21.   **IDENTIFY** and **SET UP:**   For glass $\rightarrow$ water, $\theta_{\text{crit}} = 48.7°$. Apply Snell's law with $\theta_a = \theta_{\text{crit}}$ to calculate the index of refraction n_a of the glass.

EXECUTE: $n_a \sin\theta_{\text{crit}} = n_b \sin 90°$, so $n_a = \frac{n_b}{\sin\theta_{\text{crit}}} = \frac{1.333}{\sin 48.7°} = 1.77$

EVALUATE: For total internal reflection to occur the light must be incident in the material of larger refractive index. Our results give $n_{glass} > n_{water}$, in agreement with this.

33.25.   **IDENTIFY:**   When unpolarized light passes through a polarizer the intensity is reduced by a factor of $\frac{1}{2}$ and the transmitted light is polarized along the axis of the polarizer. When polarized light of intensity I_{max} is incident on a polarizer, the transmitted intensity is $I = I_{max} \cos^2 \phi$, where ϕ is the angle between the polarization direction of the incident light and the axis of the filter.

SET UP: For the second polarizer $\phi = 60°$. For the third polarizer, $\phi = 90° - 60° = 30°$.

EXECUTE: (a) At point A the intensity is $I_0/2$ and the light is polarized along the vertical direction. At point B the intensity is $(I_0/2)(\cos 60°)^2 = 0.125 I_0$, and the light is polarized along the axis of the second polarizer. At point C the intensity is $(0.125 I_0)(\cos 30°)^2 = 0.0938 I_0$.

(b) Now for the last filter $\phi = 90°$ and $I = 0$.

EVALUATE: Adding the middle filter increases the transmitted intensity.

33.27.   **IDENTIFY** and **SET UP:**   Reflected beam completely linearly polarized implies that the angle of incidence equals the polarizing angle, so $\theta_p = 54.5°$. Use Eq.(33.8) to calculate the refractive index of the glass. Then use Snell's law to calculate the angle of refraction.

EXECUTE: (a) $\tan \theta_p = \dfrac{n_b}{n_a}$ gives $n_{glass} = n_{air} \tan \theta_p = (1.00) \tan 54.5° = 1.40$.

(b) $n_a \sin \theta_a = n_b \sin \theta_b$

$\sin \theta_b = \dfrac{n_a \sin \theta_a}{n_b} = \dfrac{(1.00) \sin 54.5°}{1.40} = 0.5815$ and $\theta_b = 35.5°$

EVALUATE:

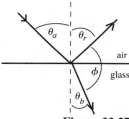

Note: $\phi = 180.0° - \theta_r - \theta_b$ and $\theta_r = \theta_a$. Thus $\phi = 180.0° - 54.5° - 35.5° = 90.0°$; the reflected ray and the refracted ray are perpendicular to each other. This agrees with Fig.33.28.

Figure 33.27

33.33.   **IDENTIFY** and **SET UP:**   Apply Eq.(33.7) to polarizers #2 and #3. The light incident on the first polarizer is unpolarized, so the transmitted light has half the intensity of the incident light, and the transmitted light is polarized.

(a) **EXECUTE:**   The axes of the three filters are shown in Figure 33.33a.

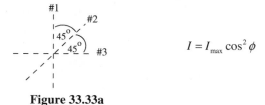

$I = I_{max} \cos^2 \phi$

Figure 33.33a

After the first filter the intensity is $I_1 = \frac{1}{2} I_0$ and the light is linearly polarized along the axis of the first polarizer. After the second filter the intensity is $I_2 = I_1 \cos^2 \phi = (\frac{1}{2} I_0)(\cos 45.0°)^2 = 0.250 I_0$ and the light is linearly polarized along the axis of the second polarizer. After the third filter the intensity is $I_3 = I_2 \cos^2 \phi = 0.250 I_0 (\cos 45.0°)^2 = 0.125 I_0$ and the light is linearly polarized along the axis of the third polarizer.

(b) The axes of the remaining two filters are shown in Figure 33.33b.

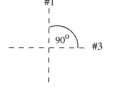

After the first filter the intensity is $I_1 = \frac{1}{2} I_0$ and the light is linearly polarized along the axis of the first polarizer.

Figure 33.33b

After the next filter the intensity is $I_3 = I_1 \cos^2 \phi = (\frac{1}{2} I_0)(\cos 90.0°)^2 = 0$. No light is passed.

EVALUATE: Light is transmitted through all three filters, but no light is transmitted if the middle polarizer is removed.

33.35.   **IDENTIFY:**   The shorter the wavelength of light, the more it is scattered. The intensity is inversely proportional to the fourth power of the wavelength.

SET UP: The intensity of the scattered light is proportional to $1/\lambda^4$, we can write it as $I = $ (constant)$/\lambda^4$.

EXECUTE: **(a)** Since I is proportional to $1/\lambda^4$, we have $I = $ (constant)$/\lambda^4$. Taking the ratio of the intensity of the red light to that of the green light gives $\dfrac{I_R}{I} = \dfrac{(\text{constant})/\lambda_R^4}{(\text{constant})/\lambda_G^4} = \left(\dfrac{\lambda_G}{\lambda_R}\right)^4 = \left(\dfrac{520\,\text{nm}}{665\,\text{nm}}\right)^4 = 0.374$, so $I_R = 0.374I$.

(b) Following the same procedure as in part (a) gives $\dfrac{I_V}{I} = \left(\dfrac{\lambda_G}{\lambda_V}\right)^4 = \left(\dfrac{520\,\text{nm}}{420\,\text{nm}}\right)^4 = 2.35$, so $I_V = 2.35I$.

EVALUATE: In the scattered light, the intensity of the short-wavelength violet light is about 7 times as great as that of the red light, so this scattered light will have a blue-violet color.

33.41.   **IDENTIFY:**   The angle of incidence at A is to be the critical angle. Apply Snell's law at the air to glass refraction at the top of the block.

SET UP: The ray is sketched in Figure 33.41.

EXECUTE: For glass $\rightarrow$ air at point A, Snell's law gives $(1.38)\sin\theta_{crit} = (1.00)\sin 90°$ and $\theta_{crit} = 46.4°$.

$\theta_b = 90° - \theta_{crit} = 43.6°$. Snell's law applied to the refraction from air to glass at the top of the block gives $(1.00)\sin\theta_a = (1.38)\sin(43.6°)$ and $\theta_a = 72.1°$.

EVALUATE: If θ_a is larger than $72.1°$ then the angle of incidence at point A is less than the initial critical angle and total internal reflection doesn't occur.

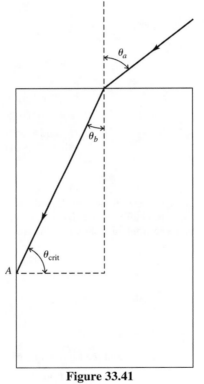

Figure 33.41

33.45.   **IDENTIFY:**   Find the critical angle for glass $\rightarrow$ air. Light incident at this critical angle is reflected back to the edge of the halo.

SET UP: The ray incident at the critical angle is sketched in Figure 33.45.

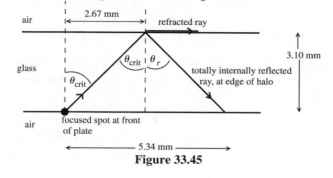

Figure 33.45

EXECUTE: From the distances given in the sketch, $\tan\theta_{crit} = \dfrac{2.67 \text{ mm}}{3.10 \text{ mm}} = 0.8613;\ \theta_{crit} = 40.7°.$

Apply Snell's law to the total internal reflection to find the refractive index of the glass: $n_a \sin\theta_a = n_b \sin\theta_b$

$n_{glass} \sin\theta_{crit} = 1.00 \sin 90°$

$n_{glass} = \dfrac{1}{\sin\theta_{crit}} = \dfrac{1}{\sin 40.7°} = 1.53$

EVALUATE: Light incident on the back surface is also totally reflected if it is incident at angles greater than θ_{crit}.

If it is incident at less than θ_{crit} it refracts into the air and does not reflect back to the emulsion.

33.47.   **IDENTIFY:**   Use Snell's law to determine the effect of the liquid on the direction of travel of the light as it enters the liquid.

SET UP: Use geometry to find the angles of incidence and refraction. Before the liquid is poured in the ray along your line of sight has the path shown in Figure 33.47a.

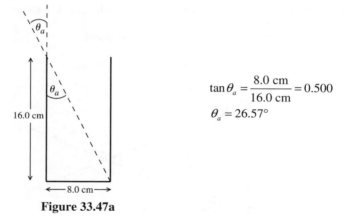

$\tan\theta_a = \dfrac{8.0 \text{ cm}}{16.0 \text{ cm}} = 0.500$

$\theta_a = 26.57°$

Figure 33.47a

After the liquid is poured in, θ_a is the same and the refracted ray passes through the center of the bottom of the glass, as shown in Figure 33.47b.

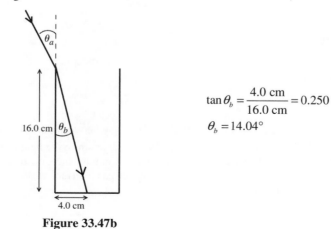

$\tan\theta_b = \dfrac{4.0 \text{ cm}}{16.0 \text{ cm}} = 0.250$

$\theta_b = 14.04°$

Figure 33.47b

EXECUTE: Use Snell's law to find n_b, the refractive index of the liquid:

$$n_a \sin\theta_a = n_b \sin\theta_b$$

$$n_b = \frac{n_a \sin\theta_a}{\sin\theta_b} = \frac{(1.00)(\sin 26.57°)}{\sin 14.04°} = 1.84$$

EVALUATE: When the light goes from air to liquid (larger refractive index) it is bent toward the normal.

33.49.   **IDENTIFY:**   Apply Snell's law to the water → ice and ice → air interfaces.

(a) **SET UP:**   Consider the ray shown in Figure 33.49.

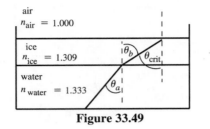

We want to find the incident angle θ_a at the water-ice interface that causes the incident angle at the ice-air interface to be the critical angle.

Figure 33.49

EXECUTE: ice-air interface: $n_{\text{ice}} \sin\theta_{\text{crit}} = 1.0 \sin 90°$

$$n_{\text{ice}} \sin\theta_{\text{crit}} = 1.0 \text{ so } \sin\theta_{\text{crit}} = \frac{1}{n_{\text{ice}}}$$

But from the diagram we see that $\theta_b = \theta_{\text{crit}}$, so $\sin\theta_b = \frac{1}{n_{\text{ice}}}$.

water-ice interface: $n_w \sin\theta_a = n_{\text{ice}} \sin\theta_b$

But $\sin\theta_b = \frac{1}{n_{\text{ice}}}$ so $n_w \sin\theta_a = 1.0$. $\sin\theta_a = \frac{1}{n_w} = \frac{1}{1.333} = 0.7502$ and $\theta_a = 48.6°$.

(b) **EVALUATE:**   The angle calculated in part (a) is the critical angle for a water-air interface; the answer would be the same if the ice layer wasn't there!

33.51.   **IDENTIFY:**   Apply Snell's law to the refraction of each ray as it emerges from the glass. The angle of incidence equals the angle $A = 25.0°$.

SET UP: The paths of the two rays are sketched in Figure 33.51.

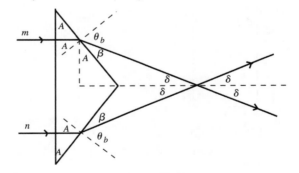

Figure 33.51

EXECUTE: $n_a \sin\theta_a = n_b \sin\theta_b$

$n_{\text{glas}} \sin 25.0° = 1.00 \sin\theta_b$

$\sin\theta_b = n_{\text{glass}} \sin 25.0°$

$\sin\theta_b = 1.66 \sin 25.0° = 0.7015$

$\theta_b = 44.55°$

$\beta = 90.0° - \theta_b = 45.45°$

Then $\delta = 90.0° - A - \beta = 90.0° - 25.0° - 45.45° = 19.55°$. The angle between the two rays is $2\delta = 39.1°$.

EVALUATE: The light is incident normally on the front face of the prism so the light is not bent as it enters the prism.

33.53.   **IDENTIFY:**   No light enters the gas because total internal reflection must have occurred at the water-gas interface.

SET UP: At the minimum value of S, the light strikes the water-gas interface at the critical angle. We apply Snell's law, $n_a \sin\theta_a = n_b \sin\theta_b$, at that surface.

EXECUTE: (a) In the water, $\theta = \dfrac{S}{R} = (1.09 \text{ m})/(1.10 \text{ m}) = 0.991 \text{ rad} = 56.77°$. This is the critical angle. So, using

the refractive index for water from Table 33.1, we get $n = (1.333) \sin 56.77° = 1.12$
(b) (i) The laser beam stays in the water all the time, so

$$t = 2R/v = 2R / \left(\dfrac{c}{n_{\text{water}}} \right) = \dfrac{Dn_{\text{water}}}{c} = (2.20 \text{ m})(1.333)/(3.00 \times 10^8 \text{ m/s}) = 9.78 \text{ ns}$$

(ii) The beam is in the water half the time and in the gas the other half of the time.

$$t_{\text{gas}} = \dfrac{Rn_{\text{gas}}}{c} = (1.10 \text{ m})(1.12)/(3.00 \times 10^8 \text{ m/s}) = 4.09 \text{ ns}$$

The total time is 4.09 ns + (9.78 ns)/2 = 8.98 ns
EVALUATE: The gas must be under considerable pressure to have a refractive index as high as 1.12.

33.55.   (a) IDENTIFY: Apply Snell's law to the refraction of the light as it enters the atmosphere.
SET UP: The path of a ray from the sun is sketched in Figure 33.55.

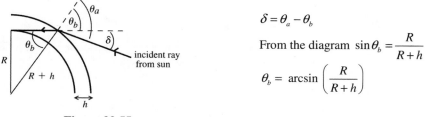

$\delta = \theta_a - \theta_b$

From the diagram $\sin\theta_b = \dfrac{R}{R+h}$

$\theta_b = \arcsin\left(\dfrac{R}{R+h}\right)$

Figure 33.55

EXECUTE: Apply Snell's law to the refraction that occurs at the top of the atmosphere: $n_a \sin\theta_a = n_b \sin\theta_b$
(a = vacuum of space, refractive, index 1.0; b = atmosphere, refractive index n)

$\sin\theta_a = n\sin\theta_b = n\left(\dfrac{R}{R+h}\right)$ so $\theta_a = \arcsin\left(\dfrac{nR}{R+h}\right)$

$\delta = \theta_a - \theta_b = \arcsin\left(\dfrac{nR}{R+h}\right) - \arcsin\left(\dfrac{R}{R+h}\right)$

(b) $\dfrac{R}{R+h} = \dfrac{6.38 \times 10^6 \text{ m}}{6.38 \times 10^6 \text{ m} + 20 \times 10^3 \text{ m}} = 0.99688$

$\dfrac{nR}{R+h} = 1.0003(0.99688) = 0.99718$

$\theta_b = \arcsin\left(\dfrac{R}{R+h}\right) = 85.47°$

$\theta_b = \arcsin\left(\dfrac{nR}{R+h}\right) 85.70°$

$\delta = \theta_a - \theta_b = 85.70° - 85.47° = 0.23°$

EVALUATE: The calculated δ is about the same as the angular radius of the sun.

33.57.   IDENTIFY and SET UP: Find the distance that the ray travels in each medium. The travel time in each medium is the distance divided by the speed in that medium.
(a) EXECUTE: The light travels a distance $\sqrt{h_1^2 + x^2}$ in traveling from point A to the interface. Along this path

the speed of the light is v_1, so the time it takes to travel this distance is $t_1 = \dfrac{\sqrt{h_1^2 + x^2}}{v_1}$. The light travels a

distance $\sqrt{h_2^2 + (l - x)^2}$ in traveling from the interface to point B. Along this path the speed of the light is v_2,

so the time it takes to travel this distance is $t_2 = \dfrac{\sqrt{h_2^2 + (l - x)^2}}{v_2}$. The total time to go from A to B is

$$t = t_1 + t_2 = \dfrac{\sqrt{h_1^2 + x^2}}{v_1} + \dfrac{\sqrt{h_2^2 + (l + x)^2}}{v_2}.$$

(b) $\dfrac{dt}{dx} = \dfrac{1}{v_1}\left(\dfrac{1}{2}\right)(h_1^2 + x^2)^{-1/2}(2x) + \dfrac{1}{v_2}\left(\dfrac{1}{2}\right)(h_2^2 + (l-x)^2)^{-1/2}2(l-x)(-1) = 0$

$\dfrac{x}{v_1\sqrt{h_1^2 + x^2}} = \dfrac{l-x}{v_2\sqrt{h_2^2 + (l-x)^2}}$

Multiplying both sides by c gives $\dfrac{c}{v_1}\dfrac{x}{\sqrt{h_1^2 + x^2}} = \dfrac{c}{v_2}\dfrac{l-x}{\sqrt{h_2^2 + (l-x)^2}}$

$\dfrac{c}{v_1} = n_1$ and $\dfrac{c}{v_2} = n_2$ (Eq.33.1)

From Fig.33.55 in the textbook, $\sin\theta_1 = \dfrac{x}{\sqrt{h_1^2 + x^2}}$ and $\sin\theta_2 = \dfrac{l-x}{\sqrt{h_2^2 + (l-x)^2}}$.

So $n_1\sin\theta_1 = n_2\sin\theta_2$, which is Snell's law.

EVALUATE: Snell's law is a result of a change in speed when light goes from one material to another.

33.61. **IDENTIFY:** Apply Snell's law to the two refractions of the ray.

SET UP: Refer to the figure that accompanies the problem.

EXECUTE: **(a)** $n_a\sin\theta_a = n_b\sin\theta_b$ gives $\sin\theta_a = n_b\sin\dfrac{A}{2}$. But $\theta_a = \dfrac{A}{2} + \alpha$, so $\sin\left(\dfrac{A}{2} + \alpha\right) = \sin\dfrac{A + 2\alpha}{2} = n\sin\dfrac{A}{2}$.

At each face of the prism the deviation is α, so $2\alpha = \delta$ and $\sin\dfrac{A+\delta}{2} = n\sin\dfrac{A}{2}$.

(b) From part (a), $\delta = 2\arcsin\left(n\sin\dfrac{A}{2}\right) - A$. $\delta = 2\arcsin\left((1.52)\sin\dfrac{60.0°}{2}\right) - 60.0° = 38.9°$.

(c) If two colors have different indices of refraction for the glass, then the deflection angles for them will differ:

$$\delta_{\text{red}} = 2\arcsin\left((1.61)\sin\dfrac{60.0°}{2}\right) - 60.0° = 47.2°$$

$$\delta_{\text{violet}} = 2\arcsin\left((1.66)\sin\dfrac{60.0°}{2}\right) - 60.0° = 52.2° \Rightarrow \Delta\delta = 52.2° - 47.2° = 5.0°$$

EVALUATE: The violet light has a greater refractive index and therefore the angle of deviation is greater for the violet light.

33.63. **IDENTIFY** and **SET UP:** The polarizer passes $\frac{1}{2}$ of the intensity of the unpolarized component, independent of ϕ.

Out of the intensity I_p of the polarized component the polarizer passes intensity $I_p\cos^2(\phi - \theta)$, where $\phi - \theta$ is the angle between the plane of polarization and the axis of the polarizer.

(a) Use the angle where the transmitted intensity is maximum or minimum to find θ. See Figure 33.63.

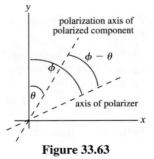

Figure 33.63

EXECUTE: The total transmitted intensity is $I = \frac{1}{2}I_0 + I_p\cos^2(\phi - \theta)$. This is maximum when $\theta = \phi$ and from the table of data this occurs for ϕ between 30° and 40°, say at 35° and $\theta = 35°$. Alternatively, the total transmitted intensity is minimum when $\phi - \theta = 90°$ and from the data this occurs for $\phi = 125°$. Thus, $\theta = \phi - 90° = 125° - 90° = 35°$, in agreement with the above.

(b) **IDENTIFY** and **SET UP:** $I = \frac{1}{2}I_0 + I_p\cos^2(\phi - \theta)$

Use data at two values of ϕ to determine the two constants I_0 and I_p. Use data where the I_p term is large ($\phi = 30°$) and where it is small ($\phi = 130°$) to have the greatest sensitivity to both I_0 and I_p:

EXECUTE: $\phi = 30°$ gives 24.8 W/m$^2 = \frac{1}{2}I_0 + I_p \cos^2(30° - 35°)$

24.8 W/m$^2 = 0.500I_0 + 0.9924I_p$

$\phi = 130°$ gives 5.2 W/m$^2 = \frac{1}{2}I_0 + I_p \cos^2(130° - 35°)$

5.2 W/m$^2 = 0.500I_0 + 0.0076I_p$

Subtracting the second equation from the first gives 19.6 W/m$^2 = 0.9848I_p$ and $I_p = 19.9$ W/m^2. And then

$I_0 = 2(5.2$ W/m$^2 - 0.0076(19.9$ W/m$^2)) = 10.1$ W/m^2.

EVALUATE: Now that we have I_0, I_p and θ we can verify that $I = \frac{1}{2}I_0 + I_p \cos^2(\phi - \theta)$ describes that data in the table.

GEOMETRIC OPTICS

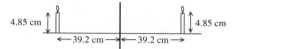

34.1. **IDENTIFY** and **SET UP:** Plane mirror: $s = -s'$ (Eq.34.1) and $m = y'/y = -s'/s = +1$ (Eq.34.2). We are given s and y and are asked to find s' and y'.

EXECUTE: The object and image are shown in Figure 34.1.

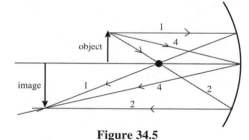

$$s' = -s = -39.2 \text{ cm}$$
$$|y'| = |m||y| = (+1)(4.85 \text{ cm})$$
$$|y'| = 4.85 \text{ cm}$$

Figure 34.1

The image is 39.2 cm to the right of the mirror and is 4.85 cm tall.

EVALUATE: For a plane mirror the image is always the same distance behind the mirror as the object is in front of the mirror. The image always has the same height as the object.

34.5. **IDENTIFY** and **SET UP:** Use Eq.(34.6) to calculate s' and use Eq.(34.7) to calculate y'. The image is real if s' is positive and is erect if $m > 0$. Concave means R and f are positive, $R = +22.0$ cm; $f = R/2 = +11.0$ cm.

EXECUTE: **(a)**

Three principal rays, numbered as in Sect. 34.2, are shown in Figure 34.5. The principal ray diagram shows that the image is real, inverted, and enlarged.

Figure 34.5

(b) $\dfrac{1}{s} + \dfrac{1}{s'} = \dfrac{1}{f}$

$\dfrac{1}{s'} = \dfrac{1}{f} - \dfrac{1}{s} = \dfrac{s-f}{sf}$ so $s' = \dfrac{sf}{s-f} = \dfrac{(16.5 \text{ cm})(11.0 \text{ cm})}{16.5 \text{ cm} - 11.0 \text{ cm}} = +33.0$ cm

$s' > 0$ so real image, 33.0 cm to left of mirror vertex

$m = -\dfrac{s'}{s} = -\dfrac{33.0 \text{ cm}}{16.5 \text{ cm}} = -2.00$ ($m < 0$ means inverted image) $|y'| = |m||y| = 2.00(0.600 \text{ cm}) = 1.20$ cm

EVALUATE: The image is 33.0 cm to the left of the mirror vertex. It is real, inverted, and is 1.20 cm tall (enlarged). The calculation agrees with the image characterization from the principal ray diagram. A concave mirror used alone always forms a real, inverted image if $s > f$ and the image is enlarged if $f < s < 2f$.

34.9. **IDENTIFY:** The shell behaves as a spherical mirror.

SET UP: The equation relating the object and image distances to the focal length of a spherical mirror is $\dfrac{1}{s} + \dfrac{1}{s'} = \dfrac{1}{f}$, and its magnification is given by $m = -\dfrac{s'}{s}$.

EXECUTE: $\dfrac{1}{s} + \dfrac{1}{s'} = \dfrac{1}{f} \Rightarrow \dfrac{1}{s} = \dfrac{2}{-18.0 \text{ cm}} - \dfrac{1}{-6.00 \text{ cm}} \Rightarrow s = 18.0$ cm from the vertex.

$m = -\dfrac{s'}{s} = -\dfrac{-6.00 \text{ cm}}{18.0 \text{ cm}} = \dfrac{1}{3} \Rightarrow y' = \dfrac{1}{3}(1.5 \text{ cm}) = 0.50$ cm. The image is 0.50 cm tall, erect, and virtual.

EVALUATE: Since the magnification is less than one, the image is smaller than the object.

34.13.   **IDENTIFY:** $\dfrac{1}{s}+\dfrac{1}{s'}=\dfrac{1}{f}$ and $m=\dfrac{y'}{y}=-\dfrac{s'}{s}$.

SET UP: $m=+2.00$ and $s=1.25$ cm. An erect image must be virtual.

EXECUTE: (a) $s'=\dfrac{sf}{s-f}$ and $m=-\dfrac{f}{s-f}$. For a concave mirror, m can be larger than 1.00. For a convex mirror,

$|f|=-f$ so $m=+\dfrac{|f|}{s+|f|}$ and m is always less than 1.00. The mirror must be concave ($f>0$).

(b) $\dfrac{1}{f}=\dfrac{s'+s}{ss'}$. $f=\dfrac{ss'}{s+s'}$. $m=-\dfrac{s'}{s}=+2.00$ and $s'=-2.00s$. $f=\dfrac{s(-2.00s)}{s-2.00s}=+2.00s=+2.50$ cm.

$R=2f=+5.00$ cm.

(c) The principal ray diagram is drawn in Figure 34.13.

EVALUATE: The principal-ray diagram agrees with the description from the equations.

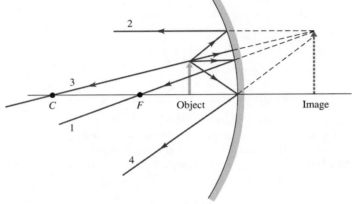

Figure 34.13

34.15.   **IDENTIFY:**   Apply Eq.(34.11), with $R\to\infty$. $|s'|$ is the apparent depth.

SET UP   The image and object are shown in Figure 34.15.

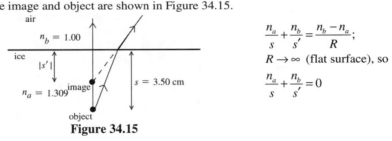

$\dfrac{n_a}{s}+\dfrac{n_b}{s'}=\dfrac{n_b-n_a}{R}$;

$R\to\infty$ (flat surface), so

$\dfrac{n_a}{s}+\dfrac{n_b}{s'}=0$

Figure 34.15

EXECUTE: $s'=-\dfrac{n_b s}{n_a}=-\dfrac{(1.00)(3.50\text{ cm})}{1.309}=-2.67$ cm

The apparent depth is 2.67 cm.

EVALUATE: When the light goes from ice to air (larger to smaller n), it is bent away from the normal and the virtual image is closer to the surface than the object is.

34.17.   **IDENTIFY:** $\dfrac{n_a}{s}+\dfrac{n_b}{s'}=\dfrac{n_b-n_a}{R}$. $m=-\dfrac{n_a s'}{n_b s}$. Light comes from the fish to the person's eye.

SET UP: $R=-14.0$ cm. $s=+14.0$ cm. $n_a=1.333$ (water). $n_b=1.00$ (air). Figure 34.17 shows the object and the refracting surface.

EXECUTE: (a) $\dfrac{1.333}{14.0\text{ cm}}+\dfrac{1.00}{s'}=\dfrac{1.00-1.333}{-14.0\text{ cm}}$. $s'=-14.0$ cm. $m=-\dfrac{(1.333)(-14.0\text{ cm})}{(1.00)(14.0\text{ cm})}=+1.33$.

The fish's image is 14.0 cm to the left of the bowl surface so is at the center of the bowl and the magnification is 1.33.

(b) The focal point is at the image location when $s\to\infty$. $\dfrac{n_b}{s'}=\dfrac{n_b-n_a}{R}$. $n_a=1.00$. $n_b=1.333$. $R=+14.0$ cm.

$\dfrac{1.333}{s'}=\dfrac{1.333-1.00}{14.0\text{ cm}}$. $s'=+56.0$ cm. s' is greater than the diameter of the bowl, so the surface facing the sunlight does not focus the sunlight to a point inside the bowl. The focal point is outside the bowl and there is no danger to the fish.

EVALUATE: In part (b) the rays refract when they exit the bowl back into the air so the image we calculated is not the final image.

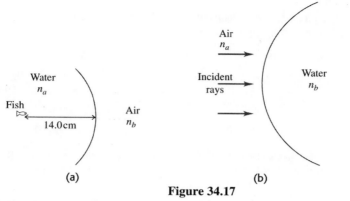

Figure 34.17

34.19.   **IDENTIFY:**   The hemispherical glass surface forms an image by refraction. The location of this image depends on the curvature of the surface and the indices of refraction of the glass and oil.

   SET UP: The image and object distances are related to the indices of refraction and the radius of curvature by the equation $\dfrac{n_a}{s} + \dfrac{n_b}{s'} = \dfrac{n_b - n_a}{R}$.

   EXECUTE: $\dfrac{n_a}{s} + \dfrac{n_b}{s'} = \dfrac{n_b - n_a}{R} \Rightarrow \dfrac{1.45}{s} + \dfrac{1.60}{1.20 \text{ m}} = \dfrac{0.15}{0.0300 \text{ m}} \Rightarrow s = 0.395 \text{ cm}$

   EVALUATE: The presence of the oil changes the location of the image.

34.21.   **IDENTIFY:**   Apply Eqs.(34.11) and (34.12). Calculate s' and y'. The image is erect if $m > 0$.

   SET UP: The object and refracting surface are shown in Figure 34.21.

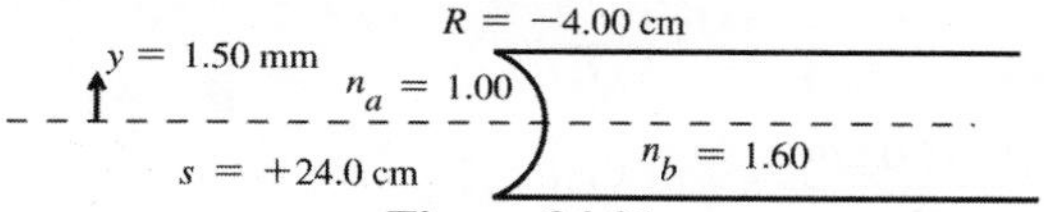

$$R = -4.00 \text{ cm}$$
$$y = 1.50 \text{ mm} \qquad n_a = 1.00$$
$$s = +24.0 \text{ cm} \qquad n_b = 1.60$$

Figure 34.21

   EXECUTE: $\dfrac{n_a}{s} + \dfrac{n_b}{s'} = \dfrac{n_b - n_a}{R}$

$$\dfrac{1.00}{24.0 \text{ cm}} + \dfrac{1.60}{s'} = \dfrac{1.60 - 1.00}{-4.00 \text{ cm}}$$

   Multiplying by 24.0 cm gives $1.00 + \dfrac{38.4}{s'} = -3.60$

   $\dfrac{38.4 \text{ cm}}{s'} = -4.60$ and $s' = -\dfrac{38.4 \text{ cm}}{4.60} = -8.35 \text{ cm}$

   Eq.(34.12): $m = -\dfrac{n_a s'}{n_b s} = -\dfrac{(1.00)(-8.35 \text{ cm})}{(1.60)(+24.0 \text{ cm})} = +0.217$

   $|y'| = |m||y| = (0.217)(1.50 \text{ mm}) = 0.326 \text{ mm}$

   EVALUATE: The image is virtual $(s' < 0)$ and is 8.35 cm to the left of the vertex. The image is erect $(m > 0)$ and is 0.326 mm tall. R is negative since the center of curvature of the surface is on the incoming side.

34.23.   **IDENTIFY:**   Use $\dfrac{1}{f} = (n-1)\left(\dfrac{1}{R_1} - \dfrac{1}{R_2}\right)$ to calculate f. The apply $\dfrac{1}{s} + \dfrac{1}{s'} = \dfrac{1}{f}$ and $m = \dfrac{y'}{y} = -\dfrac{s'}{s}$.

   SET UP: $R_1 \to \infty$. $R_2 = -13.0 \text{ cm}$. If the lens is reversed, $R_1 = +13.0 \text{ cm}$ and $R_2 \to \infty$.

   EXECUTE: **(a)** $\dfrac{1}{f} = (0.70)\left(\dfrac{1}{\infty} - \dfrac{1}{-13.0 \text{ cm}}\right) = \dfrac{0.70}{13.0 \text{ cm}}$ and $f = 18.6 \text{ cm}$. $\dfrac{1}{s'} = \dfrac{1}{f} - \dfrac{1}{s} = \dfrac{s-f}{sf}$.

   $s' = \dfrac{sf}{s-f} = \dfrac{(22.5 \text{ cm})(18.6 \text{ cm})}{22.5 \text{ cm} - 18.6 \text{ cm}} = 107 \text{ cm}$. $m = -\dfrac{s'}{s} = -\dfrac{107 \text{ cm}}{22.5 \text{ cm}} = -4.76$.

   $y' = my = (-4.76)(3.75 \text{ mm}) = -17.8 \text{ mm}$. The image is 107 cm to the right of the lens and is 17.8 mm tall. The image is real and inverted.

(b) $\frac{1}{f} = (n-1)\left(\frac{1}{13.0 \text{ cm}} - \frac{1}{\infty}\right)$ and $f = 18.6 \text{ cm}$. The image is the same as in part (a).

EVALUATE: Reversing a lens does not change the focal length of the lens.

34.27.   **IDENTIFY:** The thin-lens equation applies in this case.

SET UP: The thin-lens equation is $\frac{1}{s} + \frac{1}{s'} = \frac{1}{f}$, and the magnification is $m = -\frac{s'}{s} = \frac{y'}{y}$.

EXECUTE: $m = \frac{y'}{y} = \frac{34.0 \text{ mm}}{8.00 \text{ mm}} = 4.25 = -\frac{s'}{s} = -\frac{-12.0 \text{ cm}}{s} \Rightarrow s = 2.82 \text{ cm}$. The thin-lens equation gives

$\frac{1}{s} + \frac{1}{s'} = \frac{1}{f} \Rightarrow f = 3.69 \text{ cm}$.

EVALUATE: Since the focal length is positive, this is a converging lens. The image distance is negative because the object is inside the focal point of the lens.

34.29.   **IDENTIFY:** Apply $\frac{1}{f} = (n-1)\left(\frac{1}{R_1} - \frac{1}{R_2}\right)$.

SET UP: For a distant object the image is at the focal point of the lens. Therefore, $f = 1.87 \text{ cm}$. For the double-convex lens, $R_1 = +R$ and $R_2 = -R$, where $R = 2.50 \text{ cm}$.

EXECUTE: $\frac{1}{f} = (n-1)\left(\frac{1}{R} - \frac{1}{-R}\right) = \frac{2(n-1)}{R}$. $n = \frac{R}{2f} + 1 = \frac{2.50 \text{ cm}}{2(1.87 \text{ cm})} + 1 = 1.67$.

EVALUATE: $f > 0$ and the lens is converging. A double-convex lens is always converging.

34.33.   **IDENTIFY:** Use Eq.(34.16) to calculate the object distance s. m calculated from Eq.(34.17) determines the size and orientation of the image.
SET UP: $f = -48.0 \text{ cm}$. Virtual image 17.0 cm from lens so $s' = -17.0 \text{ cm}$.

EXECUTE: $\frac{1}{s} + \frac{1}{s'} = \frac{1}{f}$, so $\frac{1}{s'} = \frac{1}{f} - \frac{1}{s} = \frac{s-f}{sf}$

$s = \frac{s'f}{s'-f} = \frac{(-17.0 \text{ cm})(-48.0 \text{ cm})}{-17.0 \text{ cm} - (-48.0 \text{ cm})} = +26.3 \text{ cm}$

$m = -\frac{s'}{s} = -\frac{-17.0 \text{ cm}}{+26.3 \text{ cm}} = +0.646$

$m = \frac{y'}{y}$ so $|y| = \frac{|y'|}{|m|} = \frac{8.00 \text{ mm}}{0.646} = 12.4 \text{ mm}$

The principal-ray diagram is sketched in Figure 34.33.
EVALUATE: Virtual image, real object ($s > 0$) so image and object are on same side of lens. $m > 0$ so image is erect with respect to the object. The height of the object is 12.4 mm.

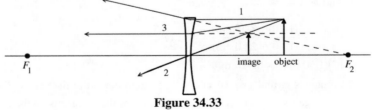

Figure 34.33

34.39.   **IDENTIFY and SET UP:** Find the lateral magnification that results in this desired image size. Use Eq.(34.17) to relate m and s' and Eq.(34.16) to relate s and s' to f.

EXECUTE: **(a)** We need $m = -\frac{24 \times 10^{-3} \text{ m}}{160 \text{ m}} = -1.5 \times 10^{-4}$. Alternatively, $m = -\frac{36 \times 10^{-3} \text{ m}}{240 \text{ m}} = -1.5 \times 10^{-4}$.

$s \gg f$ so $s' \approx f$

Then $m = -\frac{s'}{s} = -\frac{f}{s} = -1.5 \times 10^{-4}$ and $f = (1.5 \times 10^{-4})(600 \text{ m}) = 0.090 \text{ m} = 90 \text{ mm}$.

A smaller f means a smaller s' and a smaller m, so with $f = 85 \text{ mm}$ the object's image nearly fills the picture area.

(b) We need $m = -\frac{36 \times 10^{-3} \text{ m}}{9.6 \text{ m}} = -3.75 \times 10^{-3}$. Then, as in part (a), $\frac{f}{s} = 3.75 \times 10^{-3}$ and

$f = (40.0 \text{ m})(3.75 \times 10^{-3}) = 0.15 \text{ m} = 150 \text{ mm}$. Therefore use the 135 mm lens.

EVALUATE: When $s \gg f$ and $s' \approx f$, $y' = -f(y/s)$. For the mobile home y/s is smaller so a larger f is needed. Note that m is very small; the image is much smaller than the object.

34.41.   **IDENTIFY:**   The f-number of a lens is the ratio of its focal length to its diameter. To maintain the same exposure, the amount of light passing through the lens during the exposure must remain the same.

SET UP: The f-number is f/D.

EXECUTE: (a) $f\text{-number} = \dfrac{f}{D} \Rightarrow f\text{-number} = \dfrac{180.0 \text{ mm}}{16.36 \text{ mm}} \Rightarrow f\text{-number} = f/11$. (The f-number is an integer.)

(b) $f/11$ to $f/2.8$ is four steps of 2 in intensity, so one needs $1/16^{\text{th}}$ the exposure. The exposure should be $1/480$ s = 2.1×10^{-3} s = 2.1 ms.

EVALUATE: When opening the lens from $f/11$ to $f/2.8$, the area increases by a factor of 16, so 16 times as much light is allowed in. Therefore the exposure time must be decreased by a factor of 1/16 to maintain the same exposure on the film or light receptors of a digital camera.

34.43.   **IDENTIFY and SET UP:**   Apply $\dfrac{1}{s} + \dfrac{1}{s'} = \dfrac{1}{f}$ to calculate s' .

EXECUTE: (a) A real image is formed at the film, so the lens must be convex.

(b) $\dfrac{1}{s} + \dfrac{1}{s'} = \dfrac{1}{f}$ so $\dfrac{1}{s'} = \dfrac{s-f}{sf}$ and $s' = \dfrac{sf}{s-f}$, with $f = +50.0.0$ mm . For $s = 45$ cm $= 450$ mm, $s' = 56$ mm. For $s = \infty$, $s' = f = 50$ mm.   The range of distances between the lens and film is 50 mm to 56 mm.

EVALUATE: The lens is closer to the film when photographing more distant objects.

34.45.   **(a) IDENTIFY:**   The purpose of the corrective lens is to take an object 25 cm from the eye and form a virtual image at the eye's near point. Use Eq.(34.16) to solve for the image distance when the object distance is 25 cm.

SET UP: $\dfrac{1}{f} = +2.75$ diopters means $f = +\dfrac{1}{2.75}$ m $= +0.3636$ m (converging lens)

$f = 36.36$ cm; $s = 25$ cm; $s' = ?$

EXECUTE: $\dfrac{1}{s} + \dfrac{1}{s'} = \dfrac{1}{f}$ so

$s' = \dfrac{sf}{s-f} = \dfrac{(25 \text{ cm})(36.36 \text{ cm})}{25 \text{ cm} - 36.36 \text{ cm}} = -80.0$ cm

The eye's near point is 80.0 cm from the eye.

(b) IDENTIFY: The purpose of the corrective lens is to take an object at infinity and form a virtual image of it at the eye's far point. Use Eq.(34.16) to solve for the image distance when the object is at infinity.

SET UP: $\dfrac{1}{f} = -1.30$ diopters means $f = -\dfrac{1}{1.30}$ m $= -0.7692$ m (diverging lens)

$f = -76.92$ cm; $s = \infty$; $s' = ?$

EXECUTE: $\dfrac{1}{s} + \dfrac{1}{s'} = \dfrac{1}{f}$ and $s = \infty$ says $\dfrac{1}{s'} = \dfrac{1}{f}$ and $s' = f = -76.9$ cm. The eye's far point is 76.9 cm from the eye.

EVALUATE: In each case a virtual image is formed by the lens. The eye views this virtual image instead of the object. The object is at a distance where the eye can't focus on it, but the virtual image is at a distance where the eye can focus.

34.49.   **IDENTIFY:**   Use Eqs.(34.16) and (34.17) to calculate s and y'.

(a) SET UP: $f = 8.00$ cm; $s' = -25.0$ cm; $s = ?$

$\dfrac{1}{s} + \dfrac{1}{s'} = \dfrac{1}{f}$, so $\dfrac{1}{s} = \dfrac{1}{f} - \dfrac{1}{s'} = \dfrac{s'-f}{s'f}$

EXECUTE: $s = \dfrac{s'f}{s'-f} = \dfrac{(-25.0 \text{ cm})(+8.00 \text{ cm})}{-25.0 \text{ cm} - 8.00 \text{ cm}} = +6.06$ cm

(b) $m = -\dfrac{s'}{s} = -\dfrac{-25.0 \text{ cm}}{6.06 \text{ cm}} = +4.125$

$|m| = \dfrac{|y'|}{|y|}$ so $|y'| = |m||y| = (4.125)(1.00 \text{ mm}) = 4.12$ mm

EVALUATE: The lens allows the object to be much closer to the eye than the near point. The lense allows the eye to view an image at the near point rather than the object.

34.53.   **(a) IDENTIFY and SET UP:**

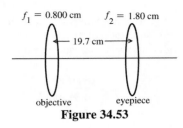

Figure 34.53

Final image is at ∞ so the object for the eyepiece is at its focal point. But the object for the eyepiece is the image of the objective so the image formed by the objective is 19.7 cm – 1.80 cm = 17.9 cm to the right of the lens. Apply Eq.(34.16) to the image formation by the objective, solve for the object distance s.
$f = 0.800$ cm; $s' = 17.9$ cm; $s = ?$

$$\frac{1}{s}+\frac{1}{s'}=\frac{1}{f}, \text{ so } \frac{1}{s}=\frac{1}{f}-\frac{1}{s'}=\frac{s'-f}{s'f}$$

EXECUTE: $s = \dfrac{s'f}{s'-f}=\dfrac{(17.9 \text{ cm})(+0.800 \text{ cm})}{17.9 \text{ cm}-0.800 \text{ cm}}=+8.37$ mm

(b) SET UP: Use Eq.(34.17).

EXECUTE: $m_1 = -\dfrac{s'}{s}=-\dfrac{17.9 \text{ cm}}{0.837 \text{ cm}}=-21.4$

The linear magnification of the objective is 21.4.
(c) SET UP: Use Eq.(34.23): $M = m_1 M_2$

EXECUTE: $M_2 = \dfrac{25 \text{ cm}}{f_2}=\dfrac{25 \text{ cm}}{1.80 \text{ cm}}=13.9$

$M = m_1 M_2 = (-21.4)(13.9) = -297$

EVALUATE: M is not accurately given by $(25 \text{ cm})s_1'/f_1 f_2 = 311$, because the object is not quite at the focal point of the objective ($s_1 = 0.837$ cm and $f_1 = 0.800$ cm).

34.57.   **(a) IDENTIFY and SET UP:**   Use Eq.(34.24), with $f_1 = 95.0$ cm (objective) and $f_2 = 15.0$ cm (eyepiece).

EXECUTE: $M = -\dfrac{f_1}{f_2}=-\dfrac{95.0 \text{ cm}}{15.0 \text{ cm}}=-6.33$

(b) IDENTIFY and SET UP: Use Eq.(34.17) to calculate y'.

SET UP: $s = 3.00\times10^3$ m
$s' = f_1 = 95.0$ cm (since s is very large, $s' \approx f$)

EXECUTE: $m = -\dfrac{s'}{s}=-\dfrac{0.950 \text{ m}}{3.00\times10^3 \text{ m}}=-3.167\times10^{-4}$

$|y'| = |m||y| = (3.167\times10^{-4})(60.0 \text{ m}) = 0.0190 \text{ m} = 1.90$ cm

(c) IDENTIFY: Use Eq.(34.21) and the angular magnification M obtained in part (a) to calculate θ'. The angular size θ of the image formed by the objective (object for the eyepiece) is its height divided by its distance from the objective.

EXECUTE: The angular size of the object for the eyepiece is $\theta = \dfrac{0.0190 \text{ m}}{0.950 \text{ m}}=0.0200$ rad.

(Note that this is also the angular size of the object for the objective: $\theta = \dfrac{60.0 \text{ m}}{3.00\times10^3 \text{ m}}=0.0200$ rad. For a thin lens the object and image have the same angular size and the image of the objective is the object for the eyepiece.)

$M = \dfrac{\theta'}{\theta}$ (Eq.34.21) so the angular size of the image is $\theta' = M\theta = -(6.33)(0.0200 \text{ rad}) = -0.127$ rad (The minus sign shows that the final image is inverted.)

EVALUATE: The lateral magnification of the objective is small; the image it forms is much smaller than the object. But the total angular magnification is larger than 1.00; the angular size of the final image viewed by the eye is 6.33 times larger than the angular size of the original object, as viewed by the unaided eye.

34.59.   **IDENTIFY:** $f = R/2$ and $M = -\dfrac{f_1}{f_2}$.

SET UP: For object and image both at infinity, $f_1 + f_2$ equals the distance d between the two mirrors.
$f_2 = 1.10$ cm . $R_1 = 1.30$ m .

EXECUTE: (a) $f_1 = \dfrac{R_1}{2} = 0.650 \text{ m} \Rightarrow d = f_1 + f_2 = 0.661 \text{ m}.$

(b) $|M| = \dfrac{f_1}{f_2} = \dfrac{0.650 \text{ m}}{0.011 \text{ m}} = 59.1.$

EVALUATE: For a telescope, $f_1 \gg f_2$.

34.61.   **IDENTIFY** and **SET UP:**   For a plane mirror $s' = -s$. $v = \dfrac{ds}{dt}$ and $v' = \dfrac{ds'}{dt}$, so $v' = -v$.

EXECUTE: The velocities of the object and image relative to the mirror are equal in magnitude and opposite in direction. Thus both you and your image are receding from the mirror surface at 2.40 m/s, in opposite directions. Your image is therefore moving at 4.80 m/s relative to you.

EVALUATE: The result derives from the fact that for a plane mirror the image is the same distance behind the mirror as the object is in front of the mirror.

34.65.   **IDENTIFY:**   We are given the image distance, the image height, and the object height. Use Eq.(34.7) to calculate the object distance s. Then use Eq.(34.4) to calculate R.

(a) **SET UP:**   Image is to be formed on screen so is real image; $s' > 0$. Mirror to screen distance is 8.00 m, so $s' = +800$ cm. $m = -\dfrac{s'}{s} < 0$ since both s and s' are positive.

EXECUTE: $|m| = \dfrac{|y'|}{|y|} = \dfrac{36.0 \text{ m}}{0.600 \text{ cm}} = 60.0$ and $m = -60.0$. Then $m = -\dfrac{s'}{s}$ gives $s = -\dfrac{s'}{m} = -\dfrac{800 \text{ cm}}{-60.0} = +13.3$ cm.

(b) $\dfrac{1}{s} + \dfrac{1}{s'} = \dfrac{2}{R}$, so $\dfrac{2}{R} = \dfrac{s + s'}{ss'}$

$R = 2\left(\dfrac{ss'}{s + s'} \right) = 2\left(\dfrac{(13.3 \text{ cm})(800 \text{ cm})}{800 \text{ cm} + 13.3 \text{ cm}} \right) = 26.2$ cm

EVALUATE: R is calculated to be positive, which is correct for a concave mirror. Also, in part (a) s is calculated to be positive, as it should be for a real object.

34.69.   **IDENTIFY** and **SET UP:**   Apply Eqs.(34.6) and (34.7). For a virtual object $s < 0$. The image is real if $s' > 0$.

EXECUTE: (a) Convex implies $R < 0$; $R = -24.0$ cm; $f = R/2 = -12.0$ cm

$\dfrac{1}{s} + \dfrac{1}{s'} = \dfrac{1}{f}$, so $\dfrac{1}{s'} = \dfrac{1}{f} - \dfrac{1}{s} = \dfrac{s - f}{sf}$

$s' = \dfrac{sf}{s - f} = \dfrac{(-12.0 \text{ cm})s}{s + 12.0 \text{ cm}}$

s is negative, so write as $s = -|s|$; $s' = +\dfrac{(12.0 \text{ cm})|s|}{12.0 \text{ cm} - |s|}$. Thus $s' > 0$ (real image) for $|s| < 12.0$ cm. Since s is negative this means $-12.0 \text{ cm} < s < 0$. A real image is formed if the virtual object is closer to the mirror than the focus.

(b) $m = -\dfrac{s'}{s}$; real image implies $s' > 0$; virtual object implies $s < 0$. Thus $m > 0$ and the image is erect.

(c) The principal-ray diagram is given in Figure 34.69.

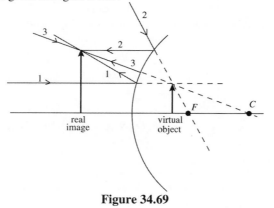

Figure 34.69

EVALUATE: For a real object, only virtual images are formed by a convex mirror. The virtual object considered in this problem must have been produced by some other optical element, by another lens or mirror in addition to the convex one we considered.

34.73. **IDENTIFY:** Since the truck is moving toward the mirror, its image will also be moving toward the mirror.

SET UP: The equation relating the object and image distances to the focal length of a spherical mirror is

$\frac{1}{s} + \frac{1}{s'} = \frac{1}{f}$, where $f = R/2$.

EXECUTE: Since the mirror is convex, $f = R/2 = (-1.50 \text{ m})/2 = -0.75 \text{ m}$. Applying the equation for a spherical

mirror gives $\frac{1}{s} + \frac{1}{s'} = \frac{1}{f} \Rightarrow s' = \frac{fs}{s-f}$.

Using the chain rule from calculus and the fact that $v = ds/dt$, we have $v' = \frac{ds'}{dt} = \frac{ds'}{ds}\frac{ds}{dt} = v \frac{f^2}{(s-f)^2}$

Solving for v gives $v = v'\left(\frac{s-f}{f}\right)^2 = (1.5 \text{ m/s})\left[\frac{2.0 \text{ m} - (-0.75 \text{ m})}{-0.75 \text{ m}}\right]^2 = 20.2 \text{ m/s}$.

This is the velocity of the truck relative to the mirror, so the truck is approaching the mirror at 20.2 m/s. You are traveling at 25 m/s, so the truck must be traveling at 25 m/s + 20.2 m/s = 45 m/s relative to the highway.

EVALUATE: Even though the truck and car are moving at constant speed, the image of the truck is *not* moving at constant speed because its location depends on the distance from the mirror to the truck.

34.79. **IDENTIFY:** Apply Eqs.(34.11) and (34.12) to the refraction as the light enters the rod and as it leaves the rod. The image formed by the first surface serves as the object for the second surface. The total magnification is $m_{\text{tot}} = m_1 m_2$, where m_1 and m_2 are the magnifications for each surface.

SET UP: The object and rod are shown in Figure 34.79.

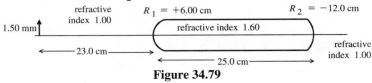

Figure 34.79

(a) image formed by refraction at first surface (left end of rod):

$s = +23.0 \text{ cm}; n_a = 1.00; n_b = 1.60; R = +6.00 \text{ cm}$

$\frac{n_a}{s} + \frac{n_b}{s'} = \frac{n_b - n_a}{R}$

EXECUTE: $\frac{1}{23.0 \text{ cm}} + \frac{1.60}{s'} = \frac{1.60 - 1.00}{6.00 \text{ cm}}$

$\frac{1.60}{s'} = \frac{1}{10.0 \text{ cm}} - \frac{1}{23.0 \text{ cm}} = \frac{23 - 10}{230 \text{ cm}} = \frac{13}{230 \text{ cm}}$

$s' = 1.60\left(\frac{230 \text{ cm}}{13}\right) = +28.3 \text{ cm};$ image is 28.3 cm to right of first vertex.

This image serves as the object for the refraction at the second surface (right-hand end of rod). It is 28.3 cm − 25.0 cm = 3.3 cm to the right of the second vertex. For the second surface $s = -3.3 \text{ cm}$ (virtual object).

(b) EVALUATE: Object is on side of outgoing light, so is a virtual object.

(c) SET UP: Image formed by refraction at second surface (right end of rod):

$s = -3.3 \text{ cm}; n_a = 1.60; n_b = 1.00; R = -12.0 \text{ cm}$

$\frac{n_a}{s} + \frac{n_b}{s'} = \frac{n_b - n_a}{R}$

EXECUTE: $\frac{1.60}{-3.3 \text{ cm}} + \frac{1.00}{s'} = \frac{1.00 - 1.60}{-12.0 \text{ cm}}$

$s' = +1.9 \text{ cm}; s' > 0$ so image is 1.9 cm to right of vertex at right-hand end of rod.

(d) $s' > 0$ so final image is real.

Magnification for first surface:

$m = -\frac{n_a s'}{n_b s} = -\frac{(1.60)(+28.3 \text{ cm})}{(1.00)(+23.0 \text{ cm})} = -0.769$

Magnification for second surface:

$m = -\frac{n_a s'}{n_b s} = -\frac{(1.60)(+1.9 \text{ cm})}{(1.00)(-3.3 \text{ cm})} = +0.92$

The overall magnification is $m_{\text{tot}} = m_1 m_2 = (-0.769)(+0.92) = -0.71$ $m_{\text{tot}} < 0$ so final image is inverted with respect to the original object.

(e) $y' = m_{tot} y = (-0.71)(1.50 \text{ mm}) = -1.06 \text{ mm}$

The final image has a height of 1.06 mm.

EVALUATE: The two refracting surfaces are not close together and Eq.(34.18) does not apply.

34.81.   **IDENTIFY:** $\frac{1}{s} + \frac{1}{s'} = \frac{1}{f}$. The type of lens determines the sign of f. $m = \frac{y'}{y} = -\frac{s'}{s}$. The sign of s' depends on whether the image is real or virtual. $s = 16.0 \text{ cm}$.

SET UP: $s' = -22.0 \text{ cm}$; s' is negative because the image is on the same side of the lens as the object.

EXECUTE: **(a)** $\frac{1}{f} = \frac{s + s'}{ss'}$ and $f = \frac{ss'}{s + s'} = \frac{(16.0 \text{ cm})(-22.0 \text{ cm})}{16.0 \text{ cm} - 22.0 \text{ cm}} = +58.7 \text{ cm}$. f is positive so the lens is converging.

(b) $m = -\frac{s'}{s} = -\frac{-22.0 \text{ cm}}{16.0 \text{ cm}} = 1.38$. $y' = my = (1.38)(3.25 \text{ mm}) = 4.48 \text{ mm}$. $s' < 0$ and the image is virtual.

EVALUATE: A converging lens forms a virtual image when the object is closer to the lens than the focal point.

34.83.   **(a) IDENTIFY:**   Apply Snell's law to the refraction of a ray at each side of the beam to find where these rays strike the table.

SET UP: The path of a ray is sketched in Figure 34.83.

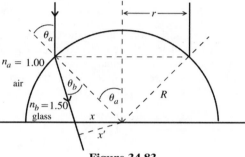

Figure 34.83

The width of the incident beam is exaggerated in the sketch, to make it easier to draw. Since the diameter of the beam is much less than the radius of the hemisphere, angles θ_a and θ_b are small. The diameter of the circle of light formed on the table is $2x$. Note the two right triangles containing the angles θ_a and θ_b.

$r = 0.190 \text{ cm}$ is the radius of the incident beam.

$R = 12.0 \text{ cm}$ is the radius of the glass hemisphere.

EXECUTE: θ_a and θ_b small imply $x \approx x'$; $\sin\theta_a = \frac{r}{R}$, $\sin\theta_b = \frac{x'}{R} \approx \frac{x}{R}$

Snell's law: $n_a \sin\theta_a = n_b \sin\theta_b$

Using the above expressions for $\sin\theta_a$ and $\sin\theta_b$ gives $n_a \frac{r}{R} = n_b \frac{x}{R}$

$n_a r = n_b x$ so $x = \frac{n_a r}{n_b} = \frac{1.00(0.190 \text{ cm})}{1.50} = 0.1267 \text{ cm}$

The diameter of the circle on the table is $2x = 2(0.1267 \text{ cm}) = 0.253 \text{ cm}$.

(b) EVALUATE: R divides out of the expression; the result for the diameter of the spot is independent of the radius R of the hemisphere. It depends only on the diameter of the incident beam and the index of refraction of the glass.

34.87.   **IDENTIFY:**   Apply Eq.(34.11) to the image formed by refraction at the front surface of the sphere.

SET UP: Let n_g be the index of refraction of the glass. The image formation is shown in Figure 34.87.

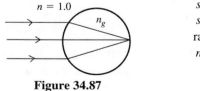

Figure 34.87

$s = \infty$

$s' = +2r$, where r is the radius of the sphere

$n_a = 1.00$, $n_b = n_g$, $R = +r$

$$\frac{n_a}{s} + \frac{n_b}{s'} = \frac{n_b - n_a}{R}$$

EXECUTE: $\dfrac{1}{\infty}+\dfrac{n_g}{2r}=\dfrac{n_g-1.00}{r}$

$\dfrac{n_g}{2r}=\dfrac{n_g}{r}-\dfrac{1}{r}$; $\dfrac{n_g}{2r}=\dfrac{1}{r}$ and $n_g=2.00$

EVALUATE: The required refractive index of the glass does not depend on the radius of the sphere.

34.89.   **IDENTIFY:**   The first lens forms an image which then acts as the object for the second lens.

SET UP: The thin-lens equation is $\dfrac{1}{s}+\dfrac{1}{s'}=\dfrac{1}{f}$ and the magnification is $m=-\dfrac{s'}{s}$.

EXECUTE: **(a)** For the first lens: $\dfrac{1}{s}+\dfrac{1}{s'}=\dfrac{1}{f}\Rightarrow\dfrac{1}{5.00\text{ cm}}+\dfrac{1}{s'}=\dfrac{1}{-15.0\text{ cm}}\Rightarrow s'=-3.75\text{ cm}$, to the left of the lens

(virtual image).

(b) For the second lens, $s=12.0\text{ cm}+3.75\text{ cm}=15.75\text{ cm}$.

$\dfrac{1}{s}+\dfrac{1}{s'}=\dfrac{1}{f}\Rightarrow\dfrac{1}{15.75\text{ cm}}+\dfrac{1}{s'}=\dfrac{1}{15.0\text{ cm}}\Rightarrow s'=315\text{ cm}$, or 332 cm from the object.

(c) The final image is real.

(d) $m=-\dfrac{s'}{s}$, $m_1=0.750$, $m_2=-20.0$, $m_{\text{total}}=-15.0\Rightarrow y'=-6.00$ cm, inverted.

EVALUATE: Note that the total magnification is the product of the individual magnifications.

34.91.   **IDENTIFY** and **SET UP:**   Apply Eq.(34.16) for each lens position. The lens to screen distance in each case is the image distance. There are two unknowns, the original object distance x and the focal length f of the lens. But each lens position gives an equation, so there are two equations for these two unknowns. The object, lens and screen before and after the lens is moved are shown in Figure 34.91.

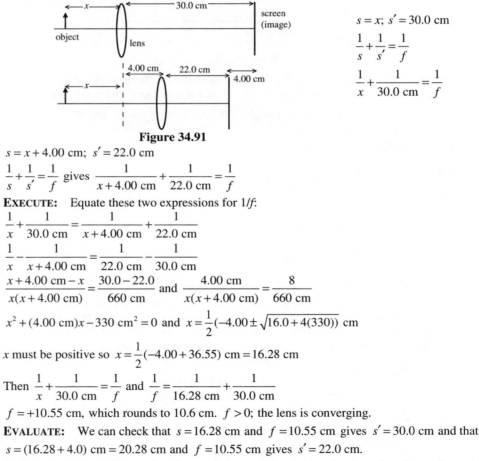

$s=x$; $s'=30.0$ cm

$\dfrac{1}{s}+\dfrac{1}{s'}=\dfrac{1}{f}$

$\dfrac{1}{x}+\dfrac{1}{30.0\text{ cm}}=\dfrac{1}{f}$

Figure 34.91

$s=x+4.00$ cm; $s'=22.0$ cm

$\dfrac{1}{s}+\dfrac{1}{s'}=\dfrac{1}{f}$ gives $\dfrac{1}{x+4.00\text{ cm}}+\dfrac{1}{22.0\text{ cm}}=\dfrac{1}{f}$

EXECUTE: Equate these two expressions for $1/f$:

$\dfrac{1}{x}+\dfrac{1}{30.0\text{ cm}}=\dfrac{1}{x+4.00\text{ cm}}+\dfrac{1}{22.0\text{ cm}}$

$\dfrac{1}{x}-\dfrac{1}{x+4.00\text{ cm}}=\dfrac{1}{22.0\text{ cm}}-\dfrac{1}{30.0\text{ cm}}$

$\dfrac{x+4.00\text{ cm}-x}{x(x+4.00\text{ cm})}=\dfrac{30.0-22.0}{660\text{ cm}}$ and $\dfrac{4.00\text{ cm}}{x(x+4.00\text{ cm})}=\dfrac{8}{660\text{ cm}}$

$x^2+(4.00\text{ cm})x-330\text{ cm}^2=0$ and $x=\dfrac{1}{2}(-4.00\pm\sqrt{16.0+4(330)})$ cm

x must be positive so $x=\dfrac{1}{2}(-4.00+36.55)\text{ cm}=16.28\text{ cm}$

Then $\dfrac{1}{x}+\dfrac{1}{30.0\text{ cm}}=\dfrac{1}{f}$ and $\dfrac{1}{f}=\dfrac{1}{16.28\text{ cm}}+\dfrac{1}{30.0\text{ cm}}$

$f=+10.55$ cm, which rounds to 10.6 cm. $f>0$; the lens is converging.

EVALUATE: We can check that $s=16.28$ cm and $f=10.55$ cm gives $s'=30.0$ cm and that

$s=(16.28+4.0)\text{ cm}=20.28\text{ cm}$ and $f=10.55$ cm gives $s'=22.0$ cm.

34.93.   **(a) IDENTIFY:**   Use Eq.(34.6) to locate the image formed by each mirror. The image formed by the first mirror serves as the object for the 2nd mirror.

SET UP: The positions of the object and the two mirrors are shown in Figure 34.93a.

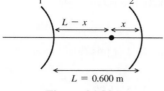

$|R| = 0.360$ m

$|f| = |R|/2 = 0.180$ m

Figure 34.93a

EXECUTE: <u>Image formed by convex mirror (mirror #1):</u>

convex means $f_1 = -0.180$ m; $s_1 = L - x$

$$s_1' = \frac{s_1 f_1}{s_1 - f_1} = \frac{(L-x)(-0.180 \text{ m})}{L - x + 0.180 \text{ m}} = -(0.180 \text{ m})\left(\frac{0.600 \text{ m} - x}{0.780 \text{ m} - x}\right) < 0$$

The image is $(0.180 \text{ m})\left(\dfrac{0.600 \text{ m} - x}{0.780 \text{ m} - x}\right)$ to the left of mirror #1 so is

$$0.600 \text{ m} + (0.180 \text{ m})\left(\frac{0.600 \text{ m} - x}{0.780 \text{ m} - x}\right) = \frac{0.576 \text{ m}^2 - (0.780 \text{ m})x}{0.780 \text{ m} - x} \quad \text{to the left of mirror #2.}$$

<u>Image formed by concave mirror (mirror #2);</u>

concave implies $f_2 = +0.180$ m

$$s_2 = \frac{0.576 \text{ m}^2 - (0.780 \text{ m})x}{0.780 \text{ m} - x}$$

Rays return to the source implies $s_2' = x$. Using these expressions in $s_2 = \dfrac{s_2' f_2}{s_2' - f_2}$ gives

$$\frac{0.576 \text{ m}^2 - (0.780 \text{ m})x}{0.780 \text{ m} - x} = \frac{(0.180 \text{ m})x}{x - 0.180 \text{ m}}$$

$$0.600 x^2 - (0.576 \text{ m})x + 0.10368 \text{ m}^2 = 0$$

$$x = \frac{1}{1.20}(0.576 \pm \sqrt{(0.576)^2 - 4(0.600)(0.10368)} \,) \text{ m} = \frac{1}{1.20}(0.576 \pm 0.288) \text{ m}$$

$x = 0.72$ m (impossible; can't have $x > L = 0.600$ m) or $x = 0.24$ m.

(b) SET UP: Which mirror is #1 and which is #2 is now reversed form part (a). This is shown in Figure 34.93b.

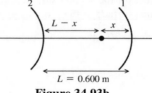

Figure 34.93b

EXECUTE: <u>Image formed by concave mirror (mirror #1):</u>

concave means $f_1 = +0.180$ m; $s_1 = x$

$$s_1' = \frac{s_1 f_1}{s_1 - f_1} = \frac{(0.180 \text{ m})x}{x - 0.180 \text{ m}}$$

The image is $\dfrac{(0.180 \text{ m})x}{x - 0.180 \text{ m}}$ to the left of mirror #1, so $s_2 = 0.600 \text{ m} - \dfrac{(0.180 \text{ m})x}{x - 0.180 \text{ m}} = \dfrac{(0.420 \text{ m})x - 0.180 \text{ m}^2}{x - 0.180 \text{ m}}$

<u>Image formed by convex mirror (mirror #2):</u>

convex means $f_2 = -0.180$ m

rays return to the source means $s_2' = L - x = 0.600 \text{ m} - x$

$$\frac{1}{s} + \frac{1}{s'} = \frac{1}{f} \text{ gives}$$

$$\frac{x - 0.180 \text{ m}}{(0.420 \text{ m})x - 0.180 \text{ m}^2} + \frac{1}{0.600 \text{ m} - x} = -\frac{1}{0.180 \text{ m}}$$

$$\frac{x - 0.180 \text{ m}}{(0.420 \text{ m})x - 0.180 \text{ m}^2} = -\left(\frac{0.780 \text{ m} - x}{0.180 \text{ m}^2 - (0.180 \text{ m})x}\right)$$

$$0.600 x^2 - (0.576 \text{ m})x + 0.1036 \text{ m}^2 = 0$$

This is the same quadratic equation as obtained in part (a), so again $x = 0.24$ m.

EVALUATE: For $x = 0.24$ m the image is at the location of the source, both for rays that initially travel from the source toward the left and for rays that travel from the source toward the right.

34.95.   **IDENTIFY:**   Apply $\dfrac{n_a}{s} + \dfrac{n_b}{s'} = \dfrac{n_b - n_a}{R}$ to each case.

SET UP: $s = 20.0$ cm. $R > 0$. Use $s' = +9.12$ cm to find R. For this calculation, $n_a = 1.00$ and $n_b = 1.55$. Then repeat the calculation with $n_a = 1.33$.

EXECUTE: $\dfrac{n_a}{s} + \dfrac{n_b}{s'} = \dfrac{n_b - n_a}{R}$ gives $\dfrac{1.00}{20.0 \text{ cm}} + \dfrac{1.55}{9.12 \text{ cm}} = \dfrac{1.55 - 1.00}{R}$. $R = 2.50$ cm.

Then $\dfrac{1.33}{20.0 \text{ cm}} + \dfrac{1.55}{s'} = \dfrac{1.55 - 1.33}{2.50 \text{ cm}}$ gives $s' = -72.1$ cm. The image is 72.1 cm to the left of the surface vertex.

EVALUATE: With the rod in air the image is real and with the rod in water the image is virtual.

34.97.   **IDENTIFY:**   Apply Eq.(34.11) with $R \to \infty$ to the refraction at each surface. For refraction at the first surface the point P serves as a virtual object. The image formed by the first refraction serves as the object for the second refraction.

SET UP: The glass plate and the two points are shown in Figure 37.97.

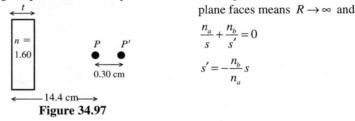

plane faces means $R \to \infty$ and

$$\frac{n_a}{s} + \frac{n_b}{s'} = 0$$

$$s' = -\frac{n_b}{n_a} s$$

Figure 34.97

EXECUTE: refraction at the first (left-hand) surface of the piece of glass:
The rays converging toward point P constitute a virtual object for this surface, so
$s = -14.4$ cm.
$n_a = 1.00$, $n_b = 1.60$.

$$s' = -\frac{1.60}{1.00}(-14.4 \text{ cm}) = +23.0 \text{ cm}$$

This image is 23.0 cm to the right of the first surface so is a distance $23.0 \text{ cm} - t$ to the right of the second surface. This image serves as a virtual object for the second surface.
refraction at the second (right-hand) surface of the piece of glass:

The image is at P' so $s' = 14.4 \text{ cm} + 0.30 \text{ cm} - t = 14.7 \text{ cm} - t$. $s = -(23.0 \text{ cm} - t)$; $n_a = 1.60$; $n_b = 1.00$ $s' = -\dfrac{n_b}{n_a} s$

gives $14.7 \text{ cm} - t = -\left(\dfrac{1.00}{1.60}\right)(-[23.0 \text{ cm} - t])$. $14.7 \text{ cm} - t = +14.4 \text{ cm} - 0.625t$.

$0.375t = 0.30$ cm and $t = 0.80$ cm

EVALUATE: The overall effect of the piece of glass is to diverge the rays and move their convergence point to the right. For a real object, refraction at a plane surface always produces a virtual image, but with a virtual object the image can be real.

34.99.   **IDENTIFY:**   Apply $\dfrac{1}{s} + \dfrac{1}{s'} = \dfrac{1}{f}$.

SET UP: The image formed by the converging lens is 30.0 cm from the converging lens, and becomes a virtual object for the diverging lens at a position 15.0 cm to the right of the diverging lens. The final image is projected $15 \text{ cm} + 19.2 \text{ cm} = 34.2 \text{ cm}$ from the diverging lens.

EXECUTE: $\dfrac{1}{s} + \dfrac{1}{s'} = \dfrac{1}{f} \Rightarrow \dfrac{1}{-15.0 \text{ cm}} + \dfrac{1}{34.2 \text{ cm}} = \dfrac{1}{f} \Rightarrow f = -26.7$ cm.

EVALUATE: Our calculation yields a negative value of f, which should be the case for a diverging lens.

34.101.   **IDENTIFY:**   In the sketch in Figure 34.101 the light travels upward from the object. Apply Eq.(34.11) with $R \to \infty$ to the refraction at each surface. The image formed by the first surface serves as the object for the second surface.

SET UP: The locations of the object and the glass plate are shown in Figure 34.101.

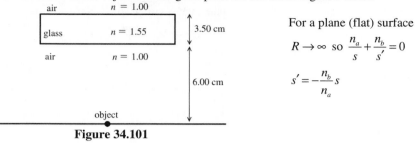

For a plane (flat) surface

$$R \rightarrow \infty \quad \text{so} \quad \frac{n_a}{s} + \frac{n_b}{s'} = 0$$

$$s' = -\frac{n_b}{n_a} s$$

Figure 34.101

EXECUTE: First refraction (air → glass):

$n_a = 1.00; \; n_b = 1.55; \; s = 6.00 \text{ cm}$

$$s' = -\frac{n_b}{n_a} s = -\frac{1.55}{1.00}(6.00 \text{ cm}) = -9.30 \text{ cm}$$

The image is 9.30 cm below the lower surface of the glass, so is $9.30 \text{ cm} + 3.50 \text{ cm} = 12.8 \text{ cm}$ below the upper surface.

Second refraction (glass → air):

$n_a = 1.55; \; n_b = 1.00; \; s = +12.8 \text{ cm}$

$$s' = -\frac{n_b}{n_a} s = -\frac{1.00}{1.55}(12.8 \text{ cm}) = -8.26 \text{ cm}$$

The image of the page is 8.26 cm below the top surface of the glass plate and therefore $9.50 \text{ cm} - 8.26 \text{ cm} = 1.24 \text{ cm}$ above the page.

EVALUATE: The image is virtual. If you view the object by looking down from above the plate, the image of the page that you see is closer to your eye than the page is.

34.105.   **IDENTIFY:**   Apply Eq.(34.16) to calculate the image distance for each lens. The image formed by the 1st lens serves as the object for the 2nd lens, and the image formed by the 2nd lens serves as the object for the 3rd lens.
   SET UP: The positions of the object and lenses are shown in Figure 34.105.

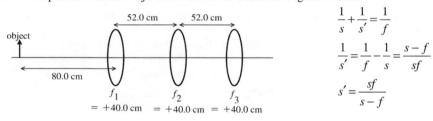

$$\frac{1}{s} + \frac{1}{s'} = \frac{1}{f}$$

$$\frac{1}{s'} = \frac{1}{f} - \frac{1}{s} = \frac{s - f}{sf}$$

$$s' = \frac{sf}{s - f}$$

Figure 34.105

EXECUTE: lens #1

$s = +80.0 \text{ cm}; f = +40.0 \text{ cm}$

$$s' = \frac{sf}{s - f} = \frac{(+80.0 \text{ cm})(+40.0 \text{ cm})}{+80.0 \text{ cm} - 40.0 \text{ cm}} = +80.0 \text{ cm}$$

The image formed by the first lens is 80.0 cm to the right of the first lens, so it is $80.0 \text{ cm} - 52.0 \text{ cm} = 28.0 \text{ cm}$ to the right of the second lens.

lens #2

$s = -28.0 \text{ cm}; f = +40.0 \text{ cm}$

$$s' = \frac{sf}{s - f} = \frac{(-28.0 \text{ cm})(+40.0 \text{ cm})}{-28.0 \text{ cm} - 40.0 \text{ cm}} = +16.47 \text{ cm}$$

The image formed by the second lens is 16.47 cm to the right of the second lens, so it is $52.0 \text{ cm} - 16.47 \text{ cm} = 35.53 \text{ cm}$ to the left of the third lens.

lens #3

$s = +35.53 \text{ cm}; f = +40.0 \text{ cm}$

$$s' = \frac{sf}{s - f} = \frac{(+35.53 \text{ cm})(+40.0 \text{ cm})}{+35.53 \text{ cm} - 40.0 \text{ cm}} = -318 \text{ cm}$$

The final image is 318 cm to the left of the third lens, so it is $318 \text{ cm} - 52 \text{ cm} - 52 \text{ cm} - 80 \text{ cm} = 134 \text{ cm}$ to the left of the object.

EVALUATE: We used the separation between the lenses and the sign conventions for s and s' to determine the object distances for the 2nd and 3rd lenses. The final image is virtual since the final s' is negative.

34.107. **IDENTIFY** and **SET UP:** The generalization of Eq.(34.22) is $M = \dfrac{\text{near point}}{f}$, so $f = \dfrac{\text{near point}}{M}$.

EXECUTE: (a) age 10, near point $= 7$ cm

$$f = \frac{7 \text{ cm}}{2.0} = 3.5 \text{ cm}$$

(b) age 30, near point $= 14$ cm

$$f = \frac{14 \text{ cm}}{2.0} = 7.0 \text{ cm}$$

(c) age 60, near point $= 200$ cm

$$f = \frac{200 \text{ cm}}{2.0} = 100 \text{ cm}$$

(d) $f = 3.5$ cm (from part (a)) and near point $= 200$ cm (for 60-year-old)

$$M = \frac{200 \text{ cm}}{3.5 \text{ cm}} = 57$$

(e) **EVALUATE:** No. The reason $f = 3.5$ cm gives a larger M for a 60-year-old than for a 10-year-old is that the eye of the older person can't focus on as close an object as the younger person can. The unaided eye of the 60-year-old must view a much smaller angular size, and that is why the same f gives a much larger M. The angular size of the image depends only on f and is the same for the two ages.

34.109. **IDENTIFY:** Apply $\dfrac{1}{s} + \dfrac{1}{s'} = \dfrac{1}{f}$. The near point is at infinity, so that is where the image must be formed for any objects that are close.

SET UP: The power in diopters equals $\dfrac{1}{f}$, with f in meters.

EXECUTE: $\dfrac{1}{f} = \dfrac{1}{s} + \dfrac{1}{s'} = \dfrac{1}{24 \text{ cm}} + \dfrac{1}{-\infty} = \dfrac{1}{0.24 \text{ m}} = 4.17$ diopters.

EVALUATE: To focus on closer objects, the power must be increased.

34.111. **IDENTIFY:** Use similar triangles in Figure 34.63 in the textbook and Eq.(34.16) to derive the expressions called for in the problem.

(a) **SET UP:** The effect of the converging lens on the ray bundle is sketched in Figure 34.111.

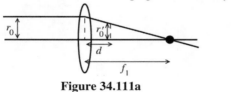

EXECUTE: From similar triangles in Figure 34.111a,

$$\frac{r_0}{f_1} = \frac{r_0'}{f_1 - d}$$

Figure 34.111a

Thus $r_0' = \left(\dfrac{f_1 - d}{f_1} \right) r_0$, as was to be shown.

(b) **SET UP:** The image at the focal point of the first lens, a distance f_1 to the right of the first lens, serves as the object for the second lens. The image is a distance $f_1 - d$ to the right of the second lens, so $s_2 = -(f_1 - d) = d - f_1$.

EXECUTE: $s_2' = \dfrac{s_2 f_2}{s_2 - f_2} = \dfrac{(d - f_1) f_2}{d - f_1 - f_2}$

$f_2 < 0$ so $|f_2| = -f_2$ and $s_2' = \dfrac{(f_1 - d)|f_2|}{|f_2| - f_1 + d}$, as was to be shown.

(c) **SET UP:** The effect of the diverging lens on the ray bundle is sketched in Figure 34.111b.

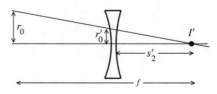

EXECUTE: From similar triangles in the sketch, $\dfrac{r_0}{f} = \dfrac{r_0'}{s_2'}$

Thus $\dfrac{r_0}{r_0'} = \dfrac{f}{s_2'}$

Figure 34.111b

From the results of part (a), $\dfrac{r_0}{r_0'} = \dfrac{f_1}{f_1 - d}$. Combining the two results gives $\dfrac{f_1}{f_1 - d} = \dfrac{f}{s_2'}$

$$f = s_2'\left(\frac{f_1}{f_1 - d}\right) = \frac{(f_1 - d)|f_2|f_1}{(|f_2| - f_1 + d)(f_1 - d)} = \frac{f_1|f_2|}{|f_2| - f_1 + d}, \text{ as was to be shown.}$$

(d) SET UP: Put the numerical values into the expression derived in part (c).

EXECUTE: $f = \dfrac{f_1|f_2|}{|f_2| - f_1 + d}$

$f_1 = 12.0$ cm, $|f_2| = 18.0$ cm, so $f = \dfrac{216 \text{ cm}^2}{6.0 \text{ cm} + d}$

$d = 0$ gives $f = 36.0$ cm; maximum f

$d = 4.0$ cm gives $f = 21.6$ cm; minimum f

$f = 30.0$ cm says 30.0 cm $= \dfrac{216 \text{ cm}^2}{6.0 \text{ cm} + d}$

$6.0 \text{ cm} + d = 7.2$ cm and $d = 1.2$ cm

EVALUATE: Changing d produces a range of effective focal lengths. The effective focal length can be both smaller and larger than $f_1 + |f_2|$.

34.113. IDENTIFY and SET UP: The image formed by the objective is the object for the eyepiece. The total lateral magnification is $m_{\text{tot}} = m_1 m_2$. $f_1 = 8.00$ mm (objective); $f_2 = 7.50$ cm (eyepiece)

(a) The locations of the object, lenses and screen are shown in Figure 34.113.

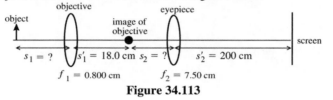

Figure 34.113

EXECUTE: Find the object distance s_1 for the objective:

$s_1' = +18.0$ cm, $f_1 = 0.800$ cm, $s_1 = ?$

$\dfrac{1}{s_1} + \dfrac{1}{s_1'} = \dfrac{1}{f_1}$, so $\dfrac{1}{s_1} = \dfrac{1}{f_1} - \dfrac{1}{s_1'} = \dfrac{s_1' - f_1}{s_1' f_1}$

$s_1 = \dfrac{s_1' f_1}{s_1' - f_1} = \dfrac{(18.0 \text{ cm})(0.800 \text{ cm})}{18.0 \text{ cm} - 0.800 \text{ cm}} = 0.8372$ cm

Find the object distance s_2 for the eyepiece:

$s_2' = +200$ cm, $f_2 = 7.50$ cm, $s_2 = ?$

$\dfrac{1}{s_2} + \dfrac{1}{s_2'} = \dfrac{1}{f_2}$

$s_2 = \dfrac{s_2' f_2}{s_2' - f_2} = \dfrac{(200 \text{ cm})(7.50 \text{ cm})}{200 \text{ cm} - 7.50 \text{ cm}} = 7.792$ cm

Now we calculate the magnification for each lens:

$m_1 = -\dfrac{s_1'}{s_1} = -\dfrac{18.0 \text{ cm}}{0.8372 \text{ cm}} = -21.50$

$m_2 = -\dfrac{s_2'}{s_2} = -\dfrac{200 \text{ cm}}{7.792 \text{ cm}} = -25.67$

$m_{\text{tot}} = m_1 m_2 = (-21.50)(-25.67) = 552.$

(b) From the sketch we can see that the distance between the two lenses is $s_1' + s_2 = 18.0 \text{ cm} + 7.792 \text{ cm} = 25.8$ cm.

EVALUATE: The microscope is not being used in the conventional way; it merely serves as a two-lens system. In particular, the final image formed by the eyepiece in the problem is real, not virtual as is the case normally for a microscope. Eq.(34.23) does not apply here, and in any event gives the angular not the lateral magnification.

INTERFERENCE

35.1. **IDENTIFY:** Compare the path difference to the wavelength.
SET UP: The separation between sources is 5.00 m, so for points between the sources the largest possible path difference is 5.00 m.
EXECUTE: **(a)** For constructive interference the path difference is $m\lambda$, $m = 0, \pm 1, \pm 2, \ldots$ Thus only the path difference of zero is possible. This occurs midway between the two sources, 2.50 m from A.
(b) For destructive interference the path difference is $(m + \frac{1}{2})\lambda$, $m = 0, \pm 1, \pm 2, \ldots$
A path difference of $\pm\lambda/2 = 3.00$ m is possible but a path difference as large as $3\lambda/2 = 9.00$ m is not possible. For a point a distance x from A and $5.00 - x$ from B the path difference is
$x - (5.00 \text{ m} - x)$. $x - (5.00 \text{ m} - x) = +3.00$ m gives $x = 4.00$ m. $x - (5.00 \text{ m} - x) = -3.00$ m gives $x = 1.00$ m.
EVALUATE: The point of constructive interference is midway between the points of destructive interference.

35.3. **IDENTIFY:** Use $c = f\lambda$ to calculate the wavelength of the transmitted waves. Compare the difference in the distance from A to P and from B to P. For constructive interference this path difference is an integer multiple of the wavelength.
SET UP: Consider Figure 35.3

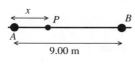

The distance of point P from each coherent source is $r_A = x$ and $r_B = 9.00$ m $- x$.

Figure 35.3

EXECUTE: The path difference is $r_B - r_A = 9.00$ m $- 2x$.
$r_B - r_A = m\lambda$, $m = 0, \pm 1, \pm 2, \ldots$
$$\lambda = \frac{c}{f} = \frac{2.998 \times 10^8 \text{ m/s}}{120 \times 10^6 \text{ Hz}} = 2.50 \text{ m}$$
Thus 9.00 m $- 2x = m(2.50$ m$)$ and $x = \dfrac{9.00 \text{ m} - m(2.50 \text{ m})}{2} = 4.50$ m $- (1.25$ m$)m$. x must lie in the range 0 to 9.00 m since P is said to be between the two antennas.
$m = 0$ gives $x = 4.50$ m
$m = +1$ gives $x = 4.50$ m $- 1.25$ m $= 3.25$ m
$m = +2$ gives $x = 4.50$ m $- 2.50$ m $= 2.00$ m
$m = +3$ gives $x = 4.50$ m $- 3.75$ m $= 0.75$ m
$m = -1$ gives $x = 4.50$ m $+ 1.25$ m $= 5.75$ m
$m = -2$ gives $x = 4.50$ m $+ 2.50$ m $= 7.00$ m
$m = -3$ gives $x = 4.50$ m $+ 3.75$ m $= 8.25$ m
All other values of m give values of x out of the allowed range. Constructive interference will occur for $x = 0.75$ m, 2.00 m, 3.25 m, 4.50 m, 5.75 m, 7.00 m, and 8.25 m.
EVALUATE: Constructive interference occurs at the midpoint between the two sources since that point is the same distance from each source. The other points of constructive interference are symmetrically placed relative to this point.

35.9.   **IDENTIFY** and **SET UP:**   The dark lines correspond to destructive interference and hence are located by Eq.(35.5):

$$d\sin\theta = \left(m + \frac{1}{2}\right)\lambda \text{ so } \sin\theta = \frac{\left(m + \frac{1}{2}\right)\lambda}{d}, \ m = 0, \pm 1, \pm 2, \ldots$$

Solve for θ that locates the second and third dark lines. Use $y = R\tan\theta$ to find the distance of each of the dark lines from the center of the screen.

EXECUTE: 1st dark line is for $m = 0$

2nd dark line is for $m = 1$ and $\sin\theta_1 = \frac{3\lambda}{2d} = \frac{3(500 \times 10^{-9} \text{ m})}{2(0.450 \times 10^{-3}\text{m})} = 1.667 \times 10^{-3}$ and $\theta_1 = 1.667 \times 10^{-3}$ rad

3rd dark line is for $m = 2$ and $\sin\theta_2 = \frac{5\lambda}{2d} = \frac{5(500 \times 10^{-9} \text{ m})}{2(0.450 \times 10^{-3}\text{m})} = 2.778 \times 10^{-3}$ and $\theta_2 = 2.778 \times 10^{-3}$ rad

(Note that θ_1 and θ_2 are small so that the approximation $\theta \approx \sin\theta \approx \tan\theta$ is valid.) The distance of each dark line from the center of the central bright band is given by $y_m = R\tan\theta$, where $R = 0.850$ m is the distance to the screen.

$\tan\theta \approx \theta$ so $y_m = R\theta_m$

$y_1 = R\theta_1 = (0.750 \text{ m})(1.667 \times 10^{-3} \text{ rad}) = 1.25 \times 10^{-3}$ m

$y_2 = R\theta_2 = (0.750 \text{ m})(2.778 \times 10^{-3} \text{ rad}) = 2.08 \times 10^{-3}$ m

$\Delta y = y_2 - y_1 = 2.08 \times 10^{-3} \text{ m} - 1.25 \times 10^{-3} \text{ m} = 0.83$ mm

EVALUATE: Since θ_1 and θ_2 are very small we could have used Eq.(35.6), generalized to destructive interference: $y_m = R\left(m + \frac{1}{2}\right)\lambda/d$.

35.11.   **IDENTIFY** and **SET UP:**   The positions of the bright fringes are given by Eq.(35.6): $y_m = R(m\lambda/d)$. For each fringe the adjacent fringe is located at $y_{m+1} = R(m+1)\lambda/d$. Solve for λ.

EXECUTE: The separation between adjacent fringes is $\Delta y = y_{m+1} - y_m = R\lambda/d$.

$\lambda = \frac{d\Delta y}{R} = \frac{(0.460 \times 10^{-3} \text{ m})(2.82 \times 10^{-3} \text{ m})}{2.20 \text{ m}} = 5.90 \times 10^{-7}$ m $= 590$ nm

EVALUATE: Eq.(35.6) requires that the angular position on the screen be small. The angular position of bright fringes is given by $\sin\theta = m\lambda/d$. The slit separation is much larger than the wavelength ($\lambda/d = 1.3 \times 10^{-3}$), so θ is small so long as m is not extremely large.

35.13.   **IDENTIFY** and **SET UP:**   The dark lines are located by $d\sin\theta = \left(m + \frac{1}{2}\right)\lambda$. The distance of each line from the center of the screen is given by $y = R\tan\theta$.

EXECUTE: First dark line is for $m = 0$ and $d\sin\theta_1 = \lambda/2$.

$\sin\theta_1 = \frac{\lambda}{2d} = \frac{550 \times 10^{-9} \text{ m}}{2(1.80 \times 10^{-6} \text{ m})} = 0.1528$ and $\theta_1 = 8.789°$. Second dark line is for $m = 1$ and $d\sin\theta_2 = 3\lambda/2$.

$\sin\theta_2 = \frac{3\lambda}{2d} = 3\left(\frac{550 \times 10^{-9} \text{ m}}{2(1.80 \times 10^{-6} \text{ m})}\right) = 0.4583$ and $\theta_2 = 27.28°$.

$y_1 = R\tan\theta_1 = (0.350 \text{ m})\tan 8.789° = 0.0541$ m

$y_2 = R\tan\theta_2 = (0.350 \text{ m})\tan 27.28° = 0.1805$ m

The distance between the lines is $\Delta y = y_2 - y_1 = 0.1805 \text{ m} - 0.0541 \text{ m} = 0.126 \text{ m} = 12.6$ cm.

EVALUATE: $\sin\theta_1 = 0.1528$ and $\tan\theta_1 = 0.1546$. $\sin\theta_2 = 0.4583$ and $\tan\theta_2 = 0.5157$. As the angle increases, $\sin\theta \approx \tan\theta$ becomes a poorer approximation.

35.15.   **IDENTIFY** and **SET UP:**   Use the information given about the bright fringe to find the distance d between the two slits. Then use Eq.(35.5) and $y = R\tan\theta$ to calculate λ for which there is a first-order dark fringe at this same place on the screen.

EXECUTE: $y_1 = \dfrac{R\lambda_1}{d}$, so $d = \dfrac{R\lambda_1}{y_1} = \dfrac{(3.00 \text{ m})(600\times10^{-9} \text{ m})}{4.84\times10^{-3} \text{ m}} = 3.72\times10^{-4}$ m. (R is much greater than d, so Eq.35.6

is valid.) The dark fringes are located by $d\sin\theta = \left(m + \dfrac{1}{2}\right)\lambda$, $m = 0, \pm 1, \pm 2, \ldots$ The first order dark fringe is located

by $\sin\theta = \lambda_2/2d$, where λ_2 is the wavelength we are seeking.

$$y = R\tan\theta \approx R\sin\theta = \dfrac{\lambda_2 R}{2d}$$

We want λ_2 such that $y = y_1$. This gives $\dfrac{R\lambda_1}{d} = \dfrac{R\lambda_2}{2d}$ and $\lambda_2 = 2\lambda_1 = 1200$ nm.

EVALUATE: For $\lambda = 600$ nm the path difference from the two slits to this point on the screen is 600 nm. For this same path difference (point on the screen) the path difference is $\lambda/2$ when $\lambda = 1200$ nm.

35.21. **IDENTIFY** and **SET UP:** The phase difference ϕ is given by $\phi = (2\pi d/\lambda)\sin\theta$ (Eq.35.13.)

EXECUTE: $\phi = [2\pi(0.340\times10^{-3} \text{ m})/(500\times10^{-9} \text{ m})]\sin 23.0° = 1670$ rad

EVALUATE: The mth bright fringe occurs when $\phi = 2\pi m$, so there are a large number of bright fringes within $23.0°$ from the centerline. Note that Eq.(35.13) gives ϕ in radians.

35.23. **(a) IDENTIFY** and **SET UP:** The minima are located at angles θ given by $d\sin\theta = \left(m + \dfrac{1}{2}\right)\lambda$. The first minimum

corresponds to $m = 0$. Solve for θ. Then the distance on the screen is $y = R\tan\theta$.

EXECUTE: $\sin\theta = \dfrac{\lambda}{2d} = \dfrac{660\times10^{-9} \text{ m}}{2(0.260\times10^{-3} \text{ m})} = 1.27\times10^{-3}$ and $\theta = 1.27\times10^{-3}$ rad

$y = (0.700 \text{ m})\tan(1.27\times10^{-3} \text{ rad}) = 0.889$ mm.

(b) IDENTIFY and **SET UP:** Eq.(35.15) given the intensity I as a function of the position y on the screen:

$I = I_0\cos^2\left(\dfrac{\pi dy}{\lambda R}\right)$. Set $I = I_0/2$ and solve for y.

EXECUTE: $I = \dfrac{1}{2}I_0$ says $\cos^2\left(\dfrac{\pi dy}{\lambda R}\right) = \dfrac{1}{2}$

$\cos\left(\dfrac{\pi dy}{\lambda R}\right) = \dfrac{1}{\sqrt{2}}$ so $\dfrac{\pi dy}{\lambda R} = \dfrac{\pi}{4}$ rad

$y = \dfrac{\lambda R}{4d} = \dfrac{(660\times10^{-9} \text{ m})(0.700 \text{ m})}{4(0.260\times10^{-3} \text{ m})} = 0.444$ mm

EVALUATE: $I = I_0/2$ at a point on the screen midway between where $I = I_0$ and $I = 0$.

35.25. **IDENTIFY:** The intensity decreases as we move away from the central maximum.

SET UP: The intensity is given by $I = I_0\cos^2\left(\dfrac{\pi dy}{\lambda R}\right)$.

EXECUTE: First find the wavelength: $\lambda = c/f = (3.00\times10^8 \text{ m/s})/(12.5 \text{ MHz}) = 24.00$ m
At the farthest the receiver can be placed, $I = I_0/4$, which gives

$$\dfrac{I_0}{4} = I_0\cos^2\left(\dfrac{\pi dy}{\lambda R}\right) \Rightarrow \cos^2\left(\dfrac{\pi dy}{\lambda R}\right) = \dfrac{1}{4} \Rightarrow \cos\left(\dfrac{\pi dy}{\lambda R}\right) = \pm\dfrac{1}{2}$$

The solutions are $\pi dy/\lambda R = \pi/3$ and $2\pi/3$. Using $\pi/3$, we get
$$y = \lambda R/3d = (24.00 \text{ m})(500 \text{ m})/[3(56.0 \text{ m})] = 71.4 \text{ m}$$
It must remain within 71.4 m of point C.

EVALUATE: Using $\pi dy/\lambda R = 2\pi/3$ gives $y = 142.8$ m. But to reach this point, the receiver would have to go beyond 71.4 m from C, where the signal would be too weak, so this second point is not possible.

35.27. **IDENTIFY:** Consider interference between rays reflected at the upper and lower surfaces of the film. Consider phase difference due to the path difference of $2t$ and any phase differences due to phase changes upon reflection.

SET UP: Consider Figure 35.27.

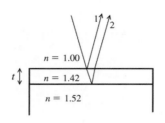

Figure 35.27

Both rays (1) and (2) undergo a 180° phase change on reflection, so these is no net phase difference introduced and the condition for destructive interference is

$$2t = \left(m + \frac{1}{2} \right) \lambda.$$

EXECUTE: $t = \dfrac{\left(m + \frac{1}{2} \right) \lambda}{2}$; thinnest film says $m = 0$ so $t = \dfrac{\lambda}{4}$

$\lambda = \dfrac{\lambda_0}{1.42}$ and $t = \dfrac{\lambda_0}{4(1.42)} = \dfrac{650 \times 10^{-9} \text{ m}}{4(1.42)} = 1.14 \times 10^{-7}$ m $= 114$ nm

EVALUATE: We compared the path difference to the wavelength in the film, since that is where the path difference occurs.

35.29.   IDENTIFY: The fringes are produced by interference between light reflected from the top and bottom surfaces of the air wedge. The refractive index of glass is greater than that of air, so the waves reflected from the top surface of the air wedge have no reflection phase shift and the waves reflected from the bottom surface of the air wedge do have a half-cycle reflection phase shift. The condition for constructive interference (bright fringes) is therefore $2t = (m + \frac{1}{2}) \lambda$.

SET UP: The geometry of the air wedge is sketched in Figure 35.29. At a distance x from the point of contact of the two plates, the thickness of the air wedge is t.

EXECUTE: $\tan \theta = \dfrac{t}{x}$ so $t = x \tan \theta$. $t_m = (m + \frac{1}{2}) \dfrac{\lambda}{2}$. $x_m = (m + \frac{1}{2}) \dfrac{\lambda}{2 \tan \theta}$ and $x_{m+1} = (m + \frac{3}{2}) \dfrac{\lambda}{2 \tan \theta}$. The

distance along the plate between adjacent fringes is $\Delta x = x_{m+1} - x_m = \dfrac{\lambda}{2 \tan \theta}$. 15.0 fringes/cm $= \dfrac{1.00}{\Delta x}$ and

$\Delta x = \dfrac{1.00}{15.0 \text{ fringes/cm}} = 0.0667$ cm . $\tan \theta = \dfrac{\lambda}{2 \Delta x} = \dfrac{546 \times 10^{-9} \text{ m}}{2(0.0667 \times 10^{-2} \text{ m})} = 4.09 \times 10^{-4}$. The angle of the wedge is

4.09×10^{-4} rad $= 0.0234°$.

EVALUATE: The fringes are equally spaced; Δx is independent of m.

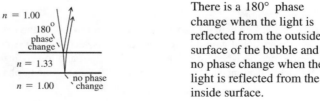

Figure 35.29

35.33.   IDENTIFY: Consider the interference between rays reflected from the two surfaces of the soap film. Strongly reflected means constructive interference. Consider phase difference due to the path difference of $2t$ and any phase difference due to phase changes upon reflection.

(a) SET UP: Consider Figure 35.33.

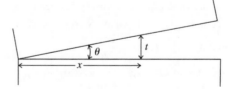

Figure 35.33

There is a 180° phase change when the light is reflected from the outside surface of the bubble and no phase change when the light is reflected from the inside surface.

EXECUTE: The reflections produce a net 180° phase difference and for there to be constructive interference the path difference $2t$ must correspond to a half-integer number of wavelengths to compensate for the $\lambda/2$ shift due to the reflections. Hence the condition for constructive interference is $2t = \left(m + \frac{1}{2}\right)(\lambda_0/n)$, $m = 0,1,2,\ldots$ Here λ_0 is the wavelength in air and (λ_0/n) is the wavelength in the bubble, where the path difference occurs.

$$\lambda_0 = \frac{2tn}{m+\frac{1}{2}} = \frac{2(290 \text{ nm})(1.33)}{m+\frac{1}{2}} = \frac{771.4 \text{ nm}}{m+\frac{1}{2}}$$

for $m = 0$, $\lambda = 1543$ nm; for $m = 1$, $\lambda = 514$ nm; for $m = 2$, $\lambda = 308$ nm;... Only 514 nm is in the visible region; the color for this wavelength is green.

(b) $\lambda_0 = \dfrac{2tn}{m+\frac{1}{2}} = \dfrac{2(340 \text{ nm})(1.33)}{m+\frac{1}{2}} = \dfrac{904.4 \text{ nm}}{m+\frac{1}{2}}$

for $m = 0$, $\lambda = 1809$ nm; for $m = 1$, $\lambda = 603$ nm; for $m = 2$, $\lambda = 362$ nm;... Only 603 nm is in the visible region; the color for this wavelength is orange.

EVALUATE: The dominant color of the reflected light depends on the thickness of the film. If the bubble has varying thickness at different points, these points will appear to be different colors when the light reflected from the bubble is viewed.

35.37.   **IDENTIFY and SET UP:**   Apply Eq.(35.19) and calculate y for $m = 1800$.

   EXECUTE: Eq.(35.19): $y = m(\lambda/2) = 1800(633 \times 10^{-9} \text{ m})/2 = 5.70 \times 10^{-4}$ m $= 0.570$ mm

   EVALUATE: A small displacement of the mirror corresponds to many wavelengths and a large number of fringes cross the line.

35.39.   **IDENTIFY:**   Consider the interference between light reflected from the top and bottom surfaces of the air film between the lens and the glass plate.

   SET UP: For maximum intensity, with a net half-cycle phase shift due to reflections, $2t = \left(m + \frac{1}{2}\right)\lambda$.

$t = R - \sqrt{R^2 - r^2}$.

   EXECUTE: $\dfrac{(2m+1)\lambda}{4} = R - \sqrt{R^2 - r^2} \Rightarrow \sqrt{R^2 - r^2} = R - \dfrac{(2m+1)\lambda}{4}$

$\Rightarrow R^2 - r^2 = R^2 + \left[\dfrac{(2m+1)\lambda}{4}\right]^2 - \dfrac{(2m+1)\lambda R}{2} \Rightarrow r = \sqrt{\dfrac{(2m+1)\lambda R}{2} - \left[\dfrac{(2m+1)\lambda}{4}\right]^2}$

$\Rightarrow r \approx \sqrt{\dfrac{(2m+1)\lambda R}{2}}$, for $R \gg \lambda$.

The second bright ring is when $m = 1$:

$$r \approx \sqrt{\frac{(2(1)+1)\,(5.80 \times 10^{-7}\text{ m})\,(0.952\text{ m})}{2}} = 9.10 \times 10^{-4}\text{ m} = 0.910\text{ mm}.$$

So the diameter of the second bright ring is 1.82 mm.

   EVALUATE: The diameter of the mth ring is proportional to $\sqrt{2m+1}$, so the rings get closer together as m increases. This agrees with Figure 35.17b in the textbook.

35.41.   **IDENTIFY:**   The liquid alters the wavelength of the light and that affects the locations of the interference minima.

   SET UP: The interference minima are located by $d\sin\theta = (m + \frac{1}{2})\lambda$. For a liquid with refractive index n,

$\lambda_{\text{liq}} = \dfrac{\lambda_{\text{air}}}{n}$.

   EXECUTE: $\dfrac{\sin\theta}{\lambda} = \dfrac{(m + \frac{1}{2})}{d} = $ constant , so $\dfrac{\sin\theta_{\text{air}}}{\lambda_{\text{air}}} = \dfrac{\sin\theta_{\text{liq}}}{\lambda_{\text{liq}}}$. $\dfrac{\sin\theta_{\text{air}}}{\lambda_{\text{air}}} = \dfrac{\sin\theta_{\text{liq}}}{\lambda_{\text{air}}/n}$ and $n = \dfrac{\sin\theta_{\text{air}}}{\sin\theta_{\text{liq}}} = \dfrac{\sin 35.20°}{\sin 19.46°} = 1.730$.

   EVALUATE: In the liquid the wavelength is shorter and $\sin\theta = (m + \frac{1}{2})\dfrac{\lambda}{d}$ gives a smaller θ than in air, for the same m.

35.43.   **IDENTIFY:**   *Both* frequencies will interfere constructively when the path difference from both of them is an integral number of wavelengths.

   SET UP: Constructive interference occurs when $\sin\theta = m\lambda/d$.

EXECUTE: First find the two wavelengths.

$$\lambda_1 = v/f_1 = (344 \text{ m/s})/(900 \text{ Hz}) = 0.3822 \text{ m}$$

$$\lambda_2 = v/f_2 = (344 \text{ m/s})/(1200 \text{ Hz}) = 0.2867 \text{ m}$$

To interfere constructively at the same angle, the angles must be the same, and hence the sines of the angles must be equal. Each sine is of the form $\sin \theta = m\lambda/d$, so we can equate the sines to get

$$m_1 \lambda_1/d = m_2 \lambda 2/d$$

$$m_1 (0.3822 \text{ m}) = m_2 (0.2867 \text{ m})$$

$$m_2 = 4/3 \ m_1$$

Since both m_1 and m_2 must be integers, the allowed pairs of values of m_1 and m_2 are

$$m_1 = m_2 = 0$$

$$m_1 = 3, \ m_2 = 4$$

$$m_1 = 6, \ m_2 = 8$$

$$m_1 = 9, \ m_2 = 12$$

etc.

For $m_1 = m_2 = 0$, we have $\theta = 0$.
For $m_1 = 3$, $m_2 = 4$, we have $\sin \theta_1 = (3)(0.3822 \text{ m})/(2.50 \text{ m})$, giving $\theta_1 = 27.3°$
For $m_1 = 6$, $m_2 = 8$, we have $\sin \theta_1 = (6)(0.3822 \text{ m})/(2.50 \text{ m})$, giving $\theta_1 = 66.5°$
For $m_1 = 9$, $m_2 = 12$, we have $\sin \theta_1 = (9)(0.3822 \text{ m})/(2.50 \text{ m}) = 1.38 > 1$, so no angle is possible.
EVALUATE: At certain other angles, one frequency will interfere constructively, but the other will not.

35.45.   **IDENTIFY:**   The two scratches are parallel slits, so the light that passes through them produces an interference pattern. However the light is traveling through a medium (plastic) that is different from air.
SET UP: The central bright fringe is bordered by a dark fringe on each side of it. At these dark fringes, $d \sin \theta = ½ \lambda/n$, where n is the refractive index of the plastic.
EXECUTE: First use geometry to find the angles at which the two dark fringes occur. At the first dark fringe $\tan \theta = [(5.82 \text{ mm})/2]/(3250 \text{ mm})$, giving $\theta = \pm 0.0513°$
For destructive interference, we have $d \sin \theta = ½ \lambda/n$ and

$$n = \lambda/(2d\sin \theta) = (632.8 \text{ nm})/[2(0.000225 \text{ m})(\sin 0.0513°)] = 1.57$$

EVALUATE: The wavelength of the light in the plastic is reduced compared to what it would be in air.

35.49.   **IDENTIFY:**   Follow the steps specified in the problem.
SET UP: The definition of hyperbola is the locus of points such that the difference between P to S_2 and P to S_1 is a constant.
EXECUTE: **(a)** $\Delta r = m\lambda$. $r_1 = \sqrt{x^2 + (y-d)^2}$ and $r_2 = \sqrt{x^2 + (y+d)^2}$.
$\Delta r = \sqrt{x^2 + (y+d)^2} - \sqrt{x^2 + (y-d)^2} = m\lambda$.
(b) For a given m and λ, Δr is a constant and we get a hyperbola. Or, in the case of all m for a given λ, a family of hyperbolas.
(c) $\sqrt{x^2 + (y+d)^2} - \sqrt{x^2 + (y-d)^2} = (m+\frac{1}{2})\lambda$.
EVALUATE: The hyperbolas approach straight lines at large distances from the source.

35.51.   **IDENTIFY and SET UP:**   Consider interference between rays reflected from the upper and lower surfaces of the film to relate the thickness of the film to the wavelengths for which there is destructive interference. The thermal expansion of the film changes the thickness of the film when the temperature changes.
EXECUTE: For this film on this glass, there is a net $\lambda/2$ phase change due to reflection and the condition for destructive interference is $2t = m(\lambda/n)$, where $n = 1.750$.

Smallest nonzero thickness is given by $t = \lambda/2n$.
At $20.0°C$, $t_0 = (582.4 \text{ nm})/[(2)(1.750)] = 166.4 \text{ nm}$.
At $170°C$, $t_0 = (588.5 \text{ nm})/[(2)(1.750)] = 168.1 \text{ nm}$.
$t = t_0(1 + \alpha\Delta T)$ so
$\alpha = (t - t_0)/(t_0\Delta T) = (1.7 \text{ nm})/[(166.4 \text{ nm})(150°C)] = 6.8 \times 10^{-5} (C°)^{-1}$

EVALUATE: When the film is heated its thickness increases, and it takes a larger wavelength in the film to equal $2t$. The value we calculated for α is the same order of magnitude as those given in Table 17.1.

35.55.   **IDENTIFY:**   Consider the phase difference due to the path difference and due to the reflection of one ray from the glass surface.
(a) **SET UP:**   Consider Figure 35.55

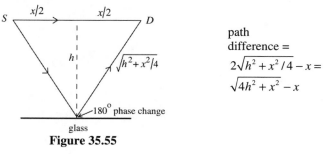

path difference =
$$2\sqrt{h^2 + x^2/4} - x =$$
$$\sqrt{4h^2 + x^2} - x$$

Figure 35.55

Since there is a 180° phase change for the reflected ray, the condition for constructive interference is path difference $= \left(m + \dfrac{1}{2}\right)\lambda$ and the condition for destructive interference is path difference $= m\lambda$.

(b) **EXECUTE:**   Constructive interference: $\left(m + \dfrac{1}{2}\right)\lambda = \sqrt{4h^2 + x^2} - x$ and $\lambda = \dfrac{\sqrt{4h^2 + x^2} - x}{m + \dfrac{1}{2}}$. Longest λ is for

$m = 0$ and then $\lambda = 2\left(\sqrt{4h^2 + x^2} - x\right) = 2\left(\sqrt{4(0.24 \text{ m})^2 + (0.14 \text{ m})^2} - 0.14 \text{ m}\right) = 0.72 \text{ m}$

EVALUATE: For $\lambda = 0.72$ m the path difference is $\lambda/2$.

35.59.   (a) **IDENTIFY:**   The wavelength in the glass is decreased by a factor of $1/n$, so for light through the upper slit a shorter path is needed to produce the same phase at the screen. Therefore, the interference pattern is shifted downward on the screen.
(b) **SET UP:**   Consider the total phase difference produced by the path length difference and also by the different wavelength in the glass.
EXECUTE: At a point on the screen located by the angle θ the difference in path length is $d \sin\theta$. This introduces a phase difference of $\phi = \left(\dfrac{2\pi}{\lambda_0}\right)(d \sin\theta)$, where λ_0 is the wavelength of the light in air or vacuum.

In the thickness L of glass the number of wavelengths is $\dfrac{L}{\lambda} = \dfrac{nL}{\lambda_0}$. A corresponding length L of the path of the ray through the lower slit, in air, contains L/λ_0 wavelengths. The phase difference this introduces is

$\phi = 2\pi\left(\dfrac{nL}{\lambda_0} - \dfrac{L}{\lambda_0}\right)$ and $\phi = 2\pi(n-1)(L/\lambda_0)$. The total phase difference is the sum of these two,

$\left(\dfrac{2\pi}{\lambda_0}\right)(d \sin\theta) + 2\pi(n-1)(L/\lambda_0) = (2\pi/\lambda_0)(d \sin\theta + L(n-1))$. Eq.(35.10) then gives

$I = I_0 \cos^2\left[\left(\dfrac{\pi}{\lambda_0}\right)(d \sin\theta + L(n-1))\right]$.

(c) Maxima means $\cos\phi/2 = \pm 1$ and $\phi/2 = m\pi$, $m = 0, \pm 1, \pm 2, \dots$ $(\pi/\lambda_0)(d \sin\theta + L(n-1)) = m\pi$

$d \sin\theta + L(n-1) = m\lambda_0$

$\sin\theta = \dfrac{m\lambda_0 - L(n-1)}{d}$

EVALUATE: When $L \to 0$ or $n \to 1$ the effect of the plate goes away and the maxima are located by Eq.(35.4).

36

DIFFRACTION

36.1. **IDENTIFY:** Use $y = x\tan\theta$ to calculate the angular position θ of the first minimum. The minima are located by Eq.(36.2): $\sin\theta = \dfrac{m\lambda}{a}$, $m = \pm 1, \pm 2, \ldots$ First minimum means $m = 1$ and $\sin\theta_1 = \lambda/a$ and $\lambda = a\sin\theta_1$. Use this equation to calculate λ.

SET UP: The central maximum is sketched in Figure 36.1.

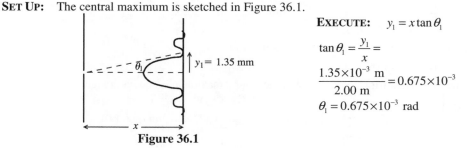

EXECUTE: $y_1 = x\tan\theta_1$

$$\tan\theta_1 = \frac{y_1}{x} =$$

$$\frac{1.35\times 10^{-3}\ \text{m}}{2.00\ \text{m}} = 0.675\times 10^{-3}$$

$$\theta_1 = 0.675\times 10^{-3}\ \text{rad}$$

$y_1 = 1.35$ mm

Figure 36.1

$\lambda = a\sin\theta_1 = (0.750\times 10^{-3}\ \text{m})\sin(0.675\times 10^{-3}\ \text{rad}) = 506\ \text{nm}$

EVALUATE: θ_1 is small so the approximation used to obtain Eq.(36.3) is valid and this equation could have been used.

36.3. **IDENTIFY:** The dark fringes are located at angles θ that satisfy $\sin\theta = \dfrac{m\lambda}{a}$, $m = \pm 1, \pm 2, \ldots$.

SET UP: The largest value of $\left|\sin\theta\right|$ is 1.00.

EXECUTE: **(a)** Solve for m that corresponds to $\sin\theta = 1$: $m = \dfrac{a}{\lambda} = \dfrac{0.0666\times 10^{-3}\ \text{m}}{585\times 10^{-9}\ \text{m}} = 113.8$. The largest value m can have is 113. $m = \pm 1, \pm 2, \ldots, \pm 113$ gives 226 dark fringes.

(b) For $m = \pm 113$, $\sin\theta = \pm 113\left(\dfrac{585\times 10^{-9}\ \text{m}}{0.0666\times 10^{-3}\ \text{m}}\right) = \pm 0.9926$ and $\theta = \pm 83.0°$.

EVALUATE: When the slit width a is decreased, there are fewer dark fringes. When $a < \lambda$ there are no dark fringes and the central maximum completely fills the screen.

36.7. **IDENTIFY:** We can model the hole in the concrete barrier as a single slit that will produce a single-slit diffraction pattern of the water waves on the shore.

SET UP: For single-slit diffraction, the angles at which destructive interference occurs are given by $\sin\theta_m = m\lambda/a$, where $m = 1, 2, 3, \ldots$.

EXECUTE: **(a)** The frequency of the water waves is $f = 75.0\ \text{min}^{-1} = 1.25\ \text{s}^{-1} = 1.25\ \text{Hz}$, so their wavelength is $\lambda = v/f = (15.0\ \text{cm/s})/(1.25\ \text{Hz}) = 12.0\ \text{cm}$.

At the first point for which destructive interference occurs, we have
$\tan\theta = (0.613\ \text{m})/(3.20\ \text{m}) \Rightarrow \theta = 10.84°$. $a\sin\theta = \lambda$ and

$$a = \lambda/\sin\theta = (12.0\ \text{cm})/(\sin 10.84°) = 63.8\ \text{cm}.$$

(b) First find the angles at which destructive interference occurs.

$$\sin\theta_2 = 2\lambda/a = 2(12.0\ \text{cm})/(63.8\ \text{cm}) \rightarrow \theta_2 = \pm 22.1°$$

$$\sin\theta_3 = 3\lambda/a = 3(12.0\ \text{cm})/(63.8\ \text{cm}) \rightarrow \theta_3 = \pm 34.3°$$

$$\sin\theta_4 = 4\lambda/a = 4(12.0\ \text{cm})/(63.8\ \text{cm}) \rightarrow \theta_4 = \pm 48.8°$$

$$\sin\theta_5 = 5\lambda/a = 5(12.0\ \text{cm})/(63.8\ \text{cm}) \rightarrow \theta_5 = \pm 70.1°$$

EVALUATE: These are large angles, so we cannot use the approximation that $\theta_m \approx m\lambda/a$.

36.9.   **IDENTIFY** and **SET UP:** $v = f\lambda$ gives λ. The person hears no sound at angles corresponding to diffraction minima. The diffraction minima are located by $\sin\theta = m\lambda/a$, $m = \pm1, \pm2,\ldots$ Solve for θ.

     EXECUTE: $\lambda = v/f = (344 \text{ m/s})/(1250 \text{ Hz}) = 0.2752$ m; $a = 1.00$ m. $m = \pm1$, $\theta = \pm16.0°$; $m = \pm2$, $\theta = \pm33.4°$; $m = \pm3$, $\theta = \pm55.6°$; no solution for larger m

     EVALUATE: $\lambda/a = 0.28$ so for the large wavelength sound waves diffraction by the doorway is a large effect. Diffraction would not be observable for visible light because its wavelength is much smaller and $\lambda/a \ll 1$.

36.13.   **IDENTIFY:**   Calculate the angular positions of the minima and use $y = x\tan\theta$ to calculate the distance on the screen between them.

     (a) SET UP: The central bright fringe is shown in Figure 36.13a.

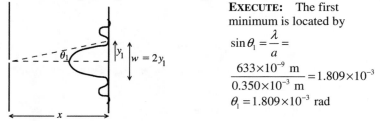

         EXECUTE: The first minimum is located by

$$\sin\theta_1 = \frac{\lambda}{a} = \frac{633\times10^{-9} \text{ m}}{0.350\times10^{-3} \text{ m}} = 1.809\times10^{-3}$$

$$\theta_1 = 1.809\times10^{-3} \text{ rad}$$

Figure 36.13a

$y_1 = x\tan\theta_1 = (3.00 \text{ m})\tan(1.809\times10^{-3} \text{ rad}) = 5.427\times10^{-3}$ m

$w = 2y_1 = 2(5.427\times10^{-3} \text{ m}) = 1.09\times10^{-2}$ m $= 10.9$ mm

     (b) SET UP: The first bright fringe on one side of the central maximum is shown in Figure 36.13b.

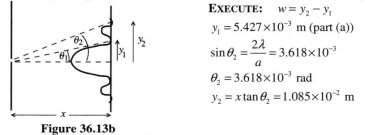

         EXECUTE: $w = y_2 - y_1$

$y_1 = 5.427\times10^{-3}$ m (part (a))

$\sin\theta_2 = \dfrac{2\lambda}{a} = 3.618\times10^{-3}$

$\theta_2 = 3.618\times10^{-3}$ rad

$y_2 = x\tan\theta_2 = 1.085\times10^{-2}$ m

Figure 36.13b

$w = y_2 - y_1 = 1.085\times10^{-2}$ m $- 5.427\times10^{-3}$ m $= 5.4$ mm

     EVALUATE: The central bright fringe is twice as wide as the other bright fringes.

36.15.   **(a) IDENTIFY:**   Use Eq.(36.2) with $m = 1$ to locate the angular position of the first minimum and then use $y = x\tan\theta$ to find its distance from the center of the screen.

     SET UP: The diffraction pattern is sketched in Figure 36.15.

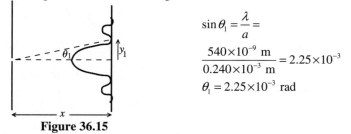

$$\sin\theta_1 = \frac{\lambda}{a} = \frac{540\times10^{-9} \text{ m}}{0.240\times10^{-3} \text{ m}} = 2.25\times10^{-3}$$

$$\theta_1 = 2.25\times10^{-3} \text{ rad}$$

Figure 36.15

$y_1 = x\tan\theta_1 = (3.00 \text{ m})\tan(2.25\times10^{-3} \text{ rad}) = 6.75\times10^{-3}$ m $= 6.75$ mm

     (b) IDENTIFY and **SET UP:**   Use Eqs.(36.5) and (36.6) to calculate the intensity at this point.

     EXECUTE: Midway between the center of the central maximum and the first minimum implies

$$y = \frac{1}{2}(6.75 \text{ mm}) = 3.375\times10^{-3} \text{ m}.$$

$$\tan\theta = \frac{y}{x} = \frac{3.375\times10^{-3} \text{ m}}{3.00 \text{ m}} = 1.125\times10^{-3}; \quad \theta = 1.125\times10^{-3} \text{ rad}$$

The phase angle β at this point on the screen is

$$\beta = \left(\frac{2\pi}{\lambda}\right)a\sin\theta = \frac{2\pi}{540\times10^{-9} \text{ m}}(0.240\times10^{-3} \text{ m})\sin(1.125\times10^{-3} \text{ rad}) = \pi.$$

Then $I = I_0 \left(\dfrac{\sin \beta/2}{\beta/2} \right)^2 = (6.00 \times 10^{-6} \text{ W/m}^2) \left(\dfrac{\sin \pi/2}{\pi/2} \right)^2$

$I = \left(\dfrac{4}{\pi^2} \right)(6.00 \times 10^{-6} \text{ W/m}^2) = 2.43 \times 10^{-6} \text{ W/m}^2.$

EVALUATE: The intensity at this point midway between the center of the central maximum and the first minimum is less than half the maximum intensity. Compare this result to the corresponding one for the two-slit pattern, Exercise 35.23.

36.17. **IDENTIFY** and **SET UP:** Use Eq.(36.6) to calculate λ and use Eq.(36.5) to calculate I. $\theta = 3.25°$, $\beta = 56.0$ rad, $a = 0.105 \times 10^{-3}$ m.

(a) EXECUTE: $\beta = \left(\dfrac{2\pi}{\lambda} \right) a \sin \theta$ so

$\lambda = \dfrac{2\pi a \sin \theta}{\beta} = \dfrac{2\pi (0.105 \times 10^{-3} \text{ m}) \sin 3.25°}{56.0 \text{ rad}} = 668$ nm

(b) $I = I_0 \left(\dfrac{\sin \beta/2}{\beta/2} \right)^2 = I_0 \left(\dfrac{4}{\beta^2} \right)(\sin(\beta/2))^2 = I_0 \dfrac{4}{(56.0 \text{ rad})^2}[\sin(28.0 \text{ rad})]^2 = 9.36 \times 10^{-5} I_0$

EVALUATE: At the first minimum $\beta = 2\pi$ rad and at the point considered in the problem $\beta = 17.8\pi$ rad, so the point is well outside the central maximum. Since β is close to $m\pi$ with $m = 18$, this point is near one of the minima. The intensity here is much less than I_0.

36.21. **(a) IDENTIFY** and **SET UP:** The interference fringes (maxima) are located by $d \sin \theta = m\lambda$, with

$m = 0, \pm 1, \pm 2, \dots$. The intensity I in the diffraction pattern is given by $I = I_0 \left(\dfrac{\sin \beta/2}{\beta/2} \right)^2$, with $\beta = \left(\dfrac{2\pi}{\lambda} \right) a \sin \theta$.

We want $m = \pm 3$ in the first equation to give θ that makes $I = 0$ in the second equation.

EXECUTE: $d \sin \theta = m\lambda$ gives $\beta = \left(\dfrac{2\pi}{\lambda} \right) a \left(\dfrac{3\lambda}{d} \right) = 2\pi(3a/d)$.

$I = 0$ says $\dfrac{\sin \beta/2}{\beta/2} = 0$ so $\beta = 2\pi$ and then $2\pi = 2\pi(3a/d)$ and $(d/a) = 3$.

(b) IDENTIFY and **SET UP:** Fringes $m = 0, \pm 1, \pm 2$ are within the central diffraction maximum and the $m = \pm 3$ fringes coincide with the first diffraction minimum. Find the value of m for the fringes that coincide with the second diffraction minimum.

EXECUTE: Second minimum implies $\beta = 4\pi$.

$\beta = \left(\dfrac{2\pi}{\lambda} \right) a \sin \theta = \left(\dfrac{2\pi}{\lambda} \right) a \left(\dfrac{m\lambda}{d} \right) = 2\pi m (a/d) = 2\pi(m/3)$

Then $\beta = 4\pi$ says $4\pi = 2\pi(m/3)$ and $m = 6$. Therefore the $m = \pm 4$ and $m = +5$ fringes are contained within the first diffraction maximum on one side of the central maximum; two fringes.

EVALUATE: The central maximum is twice as wide as the other maxima so it contains more fringes.

36.23. **(a) IDENTIFY** and **SET UP:** If the slits are very narrow then the central maximum of the diffraction pattern for each slit completely fills the screen and the intensity distribution is given solely by the two-slit interference. The maxima are given by
$d \sin \theta = m\lambda$ so $\sin \theta = m\lambda/d$. Solve for θ.

EXECUTE: 1st order maximum: $m = 1$, so $\sin \theta = \dfrac{\lambda}{d} = \dfrac{580 \times 10^{-9} \text{ m}}{0.530 \times 10^{-3} \text{ m}} = 1.094 \times 10^{-3}$; $\theta = 0.0627°$

2nd order maximum: $m = 2$, so $\sin \theta = \dfrac{2\lambda}{d} = 2.188 \times 10^{-3}$; $\theta = 0.125°$

(b) IDENTIFY and **SET UP:** The intensity is given by Eq.(36.12): $I = I_0 \cos^2(\phi/2) \left(\dfrac{\sin \beta/2}{\beta/2} \right)^2$. Calculate ϕ and β at each θ from part (a).

EXECUTE: $\phi = \left(\dfrac{2\pi d}{\lambda} \right) \sin \theta = \left(\dfrac{2\pi d}{\lambda} \right) \left(\dfrac{m\lambda}{d} \right) = 2\pi m$, so $\cos^2(\phi/2) = \cos^2(m\pi) = 1$

(Since the angular positions in part (a) correspond to interference maxima.)

$\beta = \left(\dfrac{2\pi a}{\lambda} \right) \sin \theta = \left(\dfrac{2\pi a}{\lambda} \right) \left(\dfrac{m\lambda}{d} \right) = 2\pi m (a/d) = m2\pi \left(\dfrac{0.320 \text{ mm}}{0.530 \text{ mm}} \right) = m(3.794 \text{ rad})$

1st order maximum: $m = 1$, so $I = I_0(1)\left(\dfrac{\sin(3.794/2)\text{ rad}}{(3.794/2)\text{ rad}}\right)^2 = 0.249I_0$

2nd order maximum: $m = 2$, so $I = I_0(1)\left(\dfrac{\sin 3.794\text{ rad}}{3.794\text{ rad}}\right)^2 = 0.0256I_0$

EVALUATE: The first diffraction minimum is at an angle θ given by $\sin\theta = \lambda/a$ so $\theta = 0.104°$. The first order fringe is within the central maximum and the second order fringe is inside the first diffraction maximum on one side of the central maximum. The intensity here at this second fringe is much less than I_0.

36.25.   **IDENTIFY** and **SET UP:**   The phasor diagrams are similar to those in Fig.36.14. An interference minimum occurs when the phasors add to zero.

EXECUTE: **(a)** The phasor diagram is given in Figure 36.25a

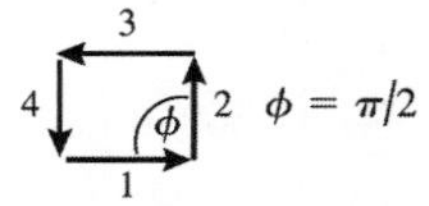

Figure 36.25a

There is destructive interference between the light through slits 1 and 3 and between 2 and 4.
(b) The phasor diagram is given in Figure 36.25b.

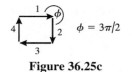

Figure 36.25b

There is destructive interference between the light through slits 1 and 2 and between 3 and 4.
(c) The phasor diagram is given in Figure 36.25c.

Figure 36.25c

There is destructive interference between light through slits 1 and 3 and between 2 and 4.
EVALUATE: Maxima occur when $\phi = 0$, 2π, 4π, etc. Our diagrams show that there are three minima between the maxima at $\phi = 0$ and $\phi = 2\pi$. This agrees with the general result that for N slits there are $N-1$ minima between each pair of principal maxima.

36.29.   **IDENTIFY** and **SET UP:**   The bright bands are at angles θ given by $d\sin\theta = m\lambda$. Solve for d and then solve for θ for the specified order.

EXECUTE: **(a)** $\theta = 78.4°$ for $m = 3$ and $\lambda = 681$ nm, so $d = m\lambda/\sin\theta = 2.086\times 10^{-4}$ cm
The number of slits per cm is $1/d = 4790$ slits/cm

(b) 1st order: $m = 1$, so $\sin\theta = \lambda/d = (681\times 10^{-9}\text{ m})/(2.086\times 10^{-6}\text{ m})$ and $\theta = 19.1°$

2nd order: $m = 2$, so $\sin\theta = 2\lambda/d$ and $\theta = 40.8°$
(c) For $m = 4$, $\sin\theta = 4\lambda/d$ is greater than 1.00, so there is no 4th-order bright band.
EVALUATE: The angular position of the bright bands for a particular wavelength increases as the order increases.

36.31.   **IDENTIFY** and **SET UP:**   Calculate d for the grating. Use Eq.(36.13) to calculate θ for the longest wavelength in the visible spectrum and verify that θ is small. Then use Eq.(36.3) to relate the linear separation of lines on the screen to the difference in wavelength.

EXECUTE: **(a)** $d = \left(\dfrac{1}{900}\right)$ cm $= 1.111\times 10^{-5}$ m

For $\lambda = 700$ nm, $\lambda/d = 6.3\times 10^{-2}$. The first-order lines are located at $\sin\theta = \lambda/d$; $\sin\theta$ is small enough for $\sin\theta \approx \theta$ to be an excellent approximation.
(b) $y = x\lambda/d$, where $x = 2.50$ m.

The distance on the screen between 1st order bright bands for two different wavelengths is $\Delta y = x(\Delta\lambda)/d$, so

$\Delta\lambda = d(\Delta y)/x = (1.111\times 10^{-5}\text{ m})(3.00\times 10^{-3}\text{ m})/(2.50\text{ m}) = 13.3$ nm

EVALUATE: The smaller d is (greater number of lines per cm) the smaller the $\Delta\lambda$ that can be measured.

36.37.   **IDENTIFY:**   The resolving power depends on the line density and the width of the grating.
SET UP: The resolving power is given by $R = Nm = = \lambda/\Delta\lambda$.
EXECUTE: **(a)** $R = Nm = (5000\text{ lines/cm})(3.50\text{ cm})(1) = 17{,}500$

(b) The resolving power needed to resolve the sodium doublet is

$$R = \lambda/\Delta\lambda = (589 \text{ nm})/(589.59 \text{ nm} - 589.00 \text{ nm}) = 998$$

so this grating can easily resolve the doublet.

(c) (i) $R = \lambda/\Delta\lambda$. Since $R = 17,500$ when $m = 1$, $R = 2 \times 17,500 = 35,000$ for $m = 2$. Therefore

$$\Delta\lambda = \lambda/R = (587.8 \text{ nm})/35,000 = 0.0168 \text{ nm}$$

$$\lambda_{\min} = \lambda + \Delta\lambda = 587.8002 \text{ nm} + 0.0168 \text{ nm} = 587.8170 \text{ nm}$$

(ii) $\lambda_{\max} = \lambda - \Delta\lambda = 587.8002 \text{ nm} - 0.0168 \text{ nm} = 587.7834 \text{ nm}$

EVALUATE: (iii) Therefore the range of resolvable wavelengths is $587.7834 \text{ nm} < \lambda < 587.8170 \text{ nm}$.

36.39.   **IDENTIFY** and **SET UP:**   The maxima occur at angles θ given by Eq.(36.16), $2d\sin\theta = m\lambda$, where d is the spacing between adjacent atomic planes. Solve for d.

EXECUTE: second order says $m = 2$.

$$d = \frac{m\lambda}{2\sin\theta} = \frac{2(0.0850 \times 10^{-9} \text{ m})}{2\sin 21.5°} = 2.32 \times 10^{-10} \text{ m} = 0.232 \text{ nm}$$

EVALUATE: Our result is similar to d calculated in Example 36.5.

36.43.   **IDENTIFY:**   Apply $\sin\theta = 1.22\dfrac{\lambda}{D}$.

SET UP: $\theta = \dfrac{W}{h}$, where $W = 28$ km and $h = 1200$ km. θ is small, so $\sin\theta \approx \theta$.

EXECUTE: $D = \dfrac{1.22\lambda}{\sin\theta} = 1.22\lambda\dfrac{h}{W} = 1.22(0.036 \text{ m})\dfrac{1.2 \times 10^6 \text{ m}}{2.8 \times 10^4 \text{ m}} = 1.88$ m

EVALUATE: D must be significantly larger than the wavelength, so a much larger diameter is needed for microwaves than for visible wavelengths.

36.45.   **IDENTIFY** and **SET UP:**   The angular size of the first dark ring is given by $\sin\theta_1 = 1.22\lambda/D$ (Eq.36.17). Calculate θ_1, and then the diameter of the ring on the screen is $2(4.5 \text{ m})\tan\theta_1$.

EXECUTE: $\sin\theta_1 = 1.22\left(\dfrac{620 \times 10^{-9} \text{ m}}{7.4 \times 10^{-6} \text{ m}}\right) = 0.1022;$ $\theta_1 = 0.1024$ rad

The radius of the Airy disk (central bright spot) is $r = (4.5 \text{ m})\tan\theta_1 = 0.462$ m. The diameter is $2r = 0.92 \text{ m} = 92$ cm.

EVALUATE: $\lambda/D = 0.084$. For this small D the central diffraction maximum is broad.

36.47.   **IDENTIFY** and **SET UP:**   Resolved by Rayleigh's criterion means angular separation θ of the objects equals $1.22\lambda/D$. The angular separation θ of the objects is their linear separation divided by their distance from the telescope.

EXECUTE: $\theta = \dfrac{250 \times 10^3 \text{ m}}{5.93 \times 10^{11} \text{ m}}$, where 5.93×10^{11} m is the distance from earth to Jupiter. Thus $\theta = 4.216 \times 10^{-7}$.

Then $\theta = 1.22\dfrac{\lambda}{D}$ and $D = \dfrac{1.22\lambda}{\theta} = \dfrac{1.22(500 \times 10^{-9} \text{ m})}{4.216 \times 10^{-7}} = 1.45$ m

EVALUATE: This is a very large telescope mirror. The greater the angular resolution the greater the diameter the lens or mirror must be.

36.51.   **IDENTIFY:**   Let y be the separation between the two points being resolved and let s be their distance from the telescope. The limit of resolution corresponds to $1.22\,\lambda/D = y/s$.

SET UP: $s = 4.28 \text{ ly} = 4.05 \times 10^{16}$ m. Assume visible light, with $\lambda = 400$ m.

EXECUTE: $y = 1.22\,\lambda s/D = 1.22(400 \times 10^{-9} \text{ m})(4.05 \times 10^{16} \text{ m}/(10.0 \text{ m}) = 2.0 \times 10^9$ m

EVALUATE: The diameter of Jupiter is 1.38×10^8 m, so the resolution is insufficient, by about one order of magnitude.

36.53.   (a) **IDENTIFY** and **SET UP:**   The intensity in the diffraction pattern is given by Eq.(36.5): $I = I_0\left(\dfrac{\sin\beta/2}{\beta/2}\right)^2$, where $\beta = \left(\dfrac{2\pi}{\lambda}\right)a\sin\theta$. Solve for θ that gives $I = \dfrac{1}{2}I_0$. The angles θ_+ and θ_- are shown in Figure 36.53.

EXECUTE: $I = \dfrac{1}{2}I_0$ so $\dfrac{\sin\beta/2}{\beta/2} = \dfrac{1}{\sqrt{2}}$

Let $x = \beta/2$; the equation for x is $\dfrac{\sin x}{x} = \dfrac{1}{\sqrt{2}} = 0.7071$.

Use trial and error to find the value of x that is a solution to this equation.

x	$(\sin x)/x$
1.0 rad	0.841
1.5 rad	0.665
1.2 rad	0.777
1.4 rad	0.7039
1.39 rad	0.7077;

thus $x = 1.39$ rad and $\beta = 2x = 2.78$ rad

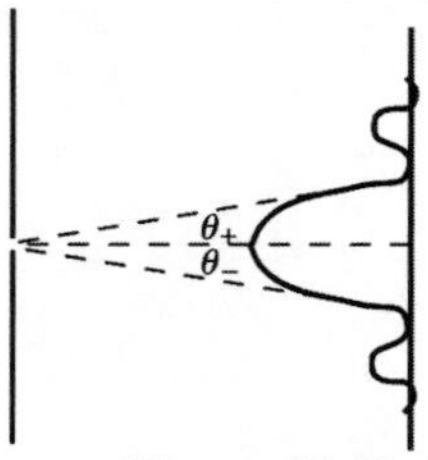

$$\Delta\theta = |\theta_+ - \theta_-| = 2\theta_+$$

$$\sin\theta_+ = \frac{\lambda\beta}{2\pi a} =$$

$$\frac{\lambda}{a}\left(\frac{2.78 \text{ rad}}{2\pi \text{ rad}}\right) = 0.4425\left(\frac{\lambda}{a}\right)$$

Figure 36.53

(i) For $\dfrac{a}{\lambda} = 2$, $\sin\theta_+ = 0.4425\left(\dfrac{1}{2}\right) = 0.2212$; $\theta_+ = 12.78°$; $\Delta\theta = 2\theta_+ = 25.6°$

(ii) For $\dfrac{a}{\lambda} = 5$, $\sin\theta_+ = 0.4425\left(\dfrac{1}{5}\right) = 0.0885$; $\theta_+ = 5.077°$; $\Delta\theta = 2\theta_+ = 10.2°$

(iii) For $\dfrac{a}{\lambda} = 10$, $\sin\theta_+ = 0.4425\left(\dfrac{1}{10}\right) = 0.04425$; $\theta_+ = 2.536°$; $\Delta\theta = 2\theta_+ = 5.1°$

(b) IDENTIFY and **SET UP:** $\sin\theta_0 = \dfrac{\lambda}{a}$ locates the first minimum. Solve for θ_0.

EXECUTE: (i) For $\dfrac{a}{\lambda} = 2$, $\sin\theta_0 = \dfrac{1}{2}$; $\theta_0 = 30.0°$; $2\theta_0 = 60.0°$

(ii) For $\dfrac{a}{\lambda} = 5$, $\sin\theta_0 = \dfrac{1}{5}$; $\theta_0 = 11.54°$; $2\theta_0 = 23.1°$

(iii) For $\dfrac{a}{\lambda} = 10$, $\sin\theta_0 = \left(\dfrac{1}{10}\right)$; $\theta_0 = 5.74°$; $2\theta_0 = 11.5°$

EVALUATE: Either definition of the width shows that the central maximum gets narrower as the slit gets wider.

36.55.   **IDENTIFY** and **SET UP:** $\sin\theta = \lambda/a$ locates the first dark band. In the liquid the wavelength changes and this changes the angular position of the first diffraction minimum.

EXECUTE: $\sin\theta_{air} = \dfrac{\lambda_{air}}{a}$; $\sin\theta_{liquid} = \dfrac{\lambda_{liquid}}{a}$

$$\lambda_{liquid} = \lambda_{air}\left(\frac{\sin\theta_{liquid}}{\sin\theta_{air}}\right) = 0.4836$$

$\lambda = \lambda_{air}/n$ (Eq.33.5), so $n = \lambda_{air}/\lambda_{liquid} = 1/0.4836 = 2.07$

EVALUATE: Light travels faster in air and n must be >1.00. The smaller λ in the liquid reduces θ that located the first dark band.

36.57.   **(a) IDENTIFY** and **SET UP:**   The angular position of the first minimum is given by $a\sin\theta = m\lambda$ (Eq.36.2), with $m = 1$. The distance of the minimum from the center of the pattern is given by $y = x\tan\theta$.

$$\sin\theta = \frac{\lambda}{a} = \frac{540\times10^{-9} \text{ m}}{0.360\times10^{-3} \text{ m}} = 1.50\times10^{-3}; \ \theta = 1.50\times10^{-3} \text{ rad}$$

$y_1 = x\tan\theta = (1.20 \text{ m})\tan(1.50\times10^{-3} \text{ rad}) = 1.80\times10^{-3} \text{ m} = 1.80 \text{ mm}$.

(Note that θ is small enough for $\theta \approx \sin\theta \approx \tan\theta$, and Eq.(36.3) applies.)

(b) IDENTIFY and **SET UP:**   Find the phase angle β where $I = I_0/2$. Then use Eq.(36.6) to solve for θ and $y = x\tan\theta$ to find the distance.

EXECUTE: From part (a) of Problem 36.53, $I = \frac{1}{2}I_0$ when $\beta = 2.78$ rad.

$\beta = \left(\dfrac{2\pi}{\lambda}\right) a \sin\theta$ (Eq.(36.6)), so $\sin\theta = \dfrac{\beta\lambda}{2\pi a}$.

$y = x\tan\theta \approx x\sin\theta \approx \dfrac{\beta\lambda x}{2\pi a} = \dfrac{(2.78 \text{ rad})(540\times10^{-9} \text{ m})(1.20 \text{ m})}{2\pi(0.360\times10^{-3} \text{ m})} = 7.96\times10^{-4} \text{ m} = 0.796 \text{ mm}$

EVALUATE: The point where $I = I_0/2$ is not midway between the center of the central maximum and the first minimum; see Exercise 36.15.

36.59.   **IDENTIFY** and **SET UP:**   Relate the phase difference between adjacent slits to the sum of the phasors for all slits. The phase difference between adjacent slits is $\phi = \dfrac{2\pi d}{\lambda}\sin\theta \approx \dfrac{2\pi d\theta}{\lambda}$ when θ is small and $\sin\theta \approx \theta$. Thus $\theta = \dfrac{\lambda\phi}{2\pi d}$.

EXECUTE: A principal maximum occurs when $\phi = \phi_{\text{max}} = m2\pi$, where m is an integer, since then all the phasors add. The first minima on either side of the mth principal maximum occur when $\phi = \phi_{\text{min}}^{\pm} = m2\pi \pm (2\pi/N)$ and the phasor diagram for N slits forms a closed loop and the resultant phasor is zero. The angular position of a principal maximum is $\theta = \left(\dfrac{\lambda}{2\pi d}\right)\phi_{\text{max}}$. The angular position of the adjacent minimum is $\theta_{\text{min}}^{\pm} = \left(\dfrac{\lambda}{2\pi d}\right)\phi_{\text{min}}^{\pm}$.

$\theta_{\text{min}}^{+} = \left(\dfrac{\lambda}{2\pi d}\right)\left(\phi_{\text{max}} + \dfrac{2\pi}{N}\right) = \theta + \left(\dfrac{\lambda}{2\pi d}\right)\left(\dfrac{2\pi}{N}\right) = \theta + \dfrac{\lambda}{Nd}$

$\theta_{\text{min}}^{-} = \left(\dfrac{\lambda}{2\pi d}\right)\left(\phi_{\text{max}} - \dfrac{2\pi}{N}\right) = \theta - \dfrac{\lambda}{Nd}$

The angular width of the principal maximum is $\theta_{\text{min}}^{+} - \theta_{\text{min}}^{-} = \dfrac{2\lambda}{Nd}$, as was to be shown.

EVALUATE: The angular width of the principal maximum decreases like $1/N$ as N increases.

36.61.   **IDENTIFY** and **SET UP:**   Draw the specified phasor diagrams. There is totally destructive interference between two slits when their phasors are in opposite directions.

EXECUTE: **(a)** For eight slits, the phasor diagrams must have eight vectors. The diagrams for each specified value of ϕ are sketched in Figure 36.61a. In each case the phasors all sum to zero.

(b) The additional phasor diagrams for $\phi = 3\pi/2$ and $3\pi/4$ are sketched in Figure 36.61b.

For $\phi = \dfrac{3\pi}{4}$, $\phi = \dfrac{5\pi}{4}$, and $\phi = \dfrac{7\pi}{4}$, totally destructive interference occurs between slits four apart. For $\phi = \dfrac{3\pi}{2}$, totally destructive interference occurs with every second slit.

EVALUATE: At a minimum the phasors for all slits sum to zero.

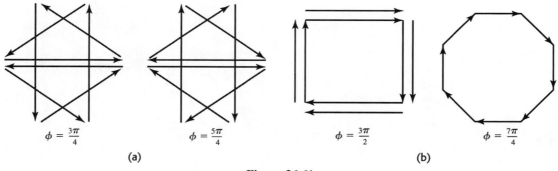

$\phi = \dfrac{3\pi}{4}$ $\phi = \dfrac{5\pi}{4}$ $\phi = \dfrac{3\pi}{2}$ $\phi = \dfrac{7\pi}{4}$

(a)   (b)

Figure 36.61

36.65.   **IDENTIFY** and **SET UP:**   The condition for an intensity maximum is $d\sin\theta = m\lambda$, $m = 0, \pm1, \pm2,\dots$ Third order means $m = 3$. The longest observable wavelength is the one that gives $\theta = 90°$ and hence $\theta = 1$.

EXECUTE: 6500 lines/cm so 6.50×10^{5} lines/m and $d = \dfrac{1}{6.50\times10^{5}}$ m $= 1.538\times10^{-6}$ m

$\lambda = \dfrac{d\sin\theta}{m} = \dfrac{(1.538\times10^{-6} \text{ m})(1)}{3} = 5.13\times10^{-7} \text{ m} = 513 \text{ nm}$

EVALUATE: The longest wavelength that can be obtained decreases as the order increases.

36.71.   **IDENTIFY** and **SET UP:**   Resolved by Rayleigh's criterion means the angular separation θ of the objects is given by $\theta = 1.22\lambda / D$. $\theta = y/s$, where $y = 75.0$ m is the distance between the two objects and s is their distance from the astronaut (her altitude).

EXECUTE: $\dfrac{y}{s} = 1.22\dfrac{\lambda}{D}$

$$s = \frac{yD}{1.22\lambda} = \frac{(75.0 \text{ m})(4.00 \times 10^{-3} \text{ m})}{1.22(500 \times 10^{-9} \text{ m})} = 4.92 \times 10^{5} \text{ m} = 492 \text{ km}$$

EVALUATE: In practice, this diffraction limit of resolution is not achieved. Defects of vision and distortion by the earth's atmosphere limit the resolution more than diffraction does.

RELATIVITY

37

37.1. **IDENTIFY** and **SET UP:** Consider the distance A to O' and B to O' as observed by an observer on the ground (Figure 37.1).

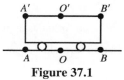

Figure 37.1

EXECUTE: Simultaneous to observer on train means light pulses from A' and B' arrive at O' at the same time. To observer at O light from A' has a longer distance to travel than light from B' so O will conclude that the pulse from $A(A')$ started before the pulse at $B(B')$. To observer at O bolt A appeared to strike first.

EVALUATE: Section 37.2 shows that if they are simultaneous to the observer on the ground then an observer on the train measures that the bolt at B' struck first.

37.5. **(a) IDENTIFY** and **SET UP:** $\Delta t_0 = 2.60 \times 10^{-8}$ s; $\Delta t = 4.20 \times 10^{-7}$ s. In the lab frame the pion is created and decays at different points, so this time is not the proper time.

EXECUTE: $\Delta t = \dfrac{\Delta t_0}{\sqrt{1 - u^2/c^2}}$ says $1 - \dfrac{u^2}{c^2} = \left(\dfrac{\Delta t_0}{\Delta t}\right)^2$

$$\frac{u}{c} = \sqrt{1 - \left(\frac{\Delta t_0}{\Delta t}\right)^2} = \sqrt{1 - \left(\frac{2.60 \times 10^{-8}\ \text{s}}{4.20 \times 10^{-7}\ \text{s}}\right)^2} = 0.998;\ u = 0.998c$$

EVALUATE: $u < c$, as it must be, but u/c is close to unity and the time dilation effects are large.

(b) IDENTIFY and **SET UP:** The speed in the laboratory frame is $u = 0.998c$; the time measured in this frame is Δt, so the distance as measured in this frame is $d = u\Delta t$

EXECUTE: $d = (0.998)(2.998 \times 10^8\ \text{m/s})(4.20 \times 10^{-7}\ \text{s}) = 126$ m

EVALUATE: The distance measured in the pion's frame will be different because the time measured in the pion's frame is different (shorter).

37.7. **IDENTIFY** and **SET UP:** A clock moving with respect to an observer appears to run more slowly than a clock at rest in the observer's frame. The clock in the spacecraft measures the proper time Δt_0. $\Delta t = 365$ days $= 8760$ hours.

EXECUTE: The clock on the moving spacecraft runs slow and shows the smaller elapsed time.

$\Delta t_0 = \Delta t \sqrt{1 - u^2/c^2} = (8760\ \text{h})\sqrt{1 - (4.80 \times 10^6/3.00 \times 10^8)^2} = 8758.88$ h. The difference in elapsed times is $8760\ \text{h} - 8758.88\ \text{h} = 1.12\ \text{h}$.

37.9. **IDENTIFY** and **SET UP:** $l = l_0\sqrt{1 - u^2/c^2}$. The length measured when the spacecraft is moving is $l = 74.0$ m; l_0 is the length measured in a frame at rest relative to the spacecraft.

EXECUTE: $l_0 = \dfrac{l}{\sqrt{1 - u^2/c^2}} = \dfrac{74.0\ \text{m}}{\sqrt{1 - (0.600c/c)^2}} = 92.5$ m.

EVALUATE: $l_0 > l$. The moving spacecraft appears to an observer on the planet to be shortened along the direction of motion.

37.11. **IDENTIFY** and **SET UP:** The 2.2 μs lifetime is Δt_0 and the observer on earth measures Δt. The atmosphere is moving relative to the muon so in its frame the height of the atmosphere is l and l_0 is 10 km.

EXECUTE: **(a)** The greatest speed the muon can have is c, so the greatest distance it can travel in 2.2×10^{-6} s is $d = vt = (3.00 \times 10^8\ \text{m/s})(2.2 \times 10^{-6}\ \text{s}) = 660$ m $= 0.66$ km.

(b) $\Delta t = \dfrac{\Delta t_0}{\sqrt{1-u^2/c^2}} = \dfrac{2.2\times10^{-6}\text{ s}}{\sqrt{1-(0.999)^2}} = 4.9\times10^{-5}\text{ s}$

$d = vt = (0.999)(3.00\times10^8\text{ m/s})(4.9\times10^{-5}\text{ s}) = 15\text{ km}$

In the frame of the earth the muon can travel 15 km in the atmosphere during its lifetime.

(c) $l = l_0\sqrt{1-u^2/c^2} = (10\text{ km})\sqrt{1-(0.999)^2} = 0.45\text{ km}$

In the frame of the muon the height of the atmosphere is less than the distance it moves during its lifetime.

37.13. (a) $l_0 = 3600\text{ m}$.

$$l = l_0\sqrt{1-\dfrac{u^2}{c^2}} = l_0(3600\text{ m})\sqrt{1-\dfrac{(4.00\times10^7\text{ m/s})^2}{(3.00\times10^8\text{ m/s})^2}} = (3600\text{ m})(0.991) = 3568\text{ m}.$$

(b) $\Delta t_0 = \dfrac{l_0}{u} = \dfrac{3600\text{ m}}{4.00\times10^7\text{ m/s}} = 9.00\times10^{-5}\text{ s}.$

(c) $\Delta t = \dfrac{l}{u} = \dfrac{3568\text{ m}}{4.00\times10^7\text{ m/s}} = 8.92\times10^{-5}\text{ s}.$

37.17. **IDENTIFY:** The relativistic velocity addition formulas apply since the speeds are close to that of light.

SET UP: The relativistic velocity addition formula is $v'_x = \dfrac{v_x - u}{1 - \dfrac{uv_x}{c^2}}.$

EXECUTE: (a) For the pursuit ship to catch the cruiser, the distance between them must be decreasing, so the velocity of the cruiser relative to the pursuit ship must be directed toward the pursuit ship.

(b) Let the unprimed frame be Tatooine and let the primed frame be the pursuit ship. We want the velocity v' of the cruiser knowing the velocity of the primed frame u and the velocity of the cruiser v in the unprimed frame (Tatooine).

$$v'_x = \dfrac{v_x - u}{1 - \dfrac{uv_x}{c^2}} = \dfrac{0.600c - 0.800c}{1 - (0.600)(0.800)} = -0.385c$$

The result implies that the cruiser is moving toward the pursuit ship at $0.385c$.

EVALUATE: The nonrelativistic formula would have given $-0.200c$, which is considerably different from the correct result.

37.19. **IDENTIFY** and **SET UP:** Reference frames S and S' are shown in Figure 37.19.

Frame S is at rest in the laboratory. Frame S' is attached to particle 1.

lab frame

Figure 37.19

u is the speed of S' relative to S; this is the speed of particle 1 as measured in the laboratory. Thus $u = +0.650c$. The speed of particle 2 in S' is $0.950c$. Also, since the two particles move in opposite directions, 2 moves in the $-x'$ direction and $v'_x = -0.950c$. We want to calculate v_x, the speed of particle 2 in frame S; use Eq.(37.23).

EXECUTE: $v_x = \dfrac{v'_x + u}{1 + uv'_x/c^2} = \dfrac{-0.950c + 0.650c}{1 + (0.950c)(-0.650c)/c^2} = \dfrac{-0.300c}{1 - 0.6175} = -0.784c.$ The speed of the second particle, as measured in the laboratory, is $0.784c$.

EVALUATE: The incorrect Galilean expression for the relative velocity gives that the speed of the second particle in the lab frame is $0.300c$. The correct relativistic calculation gives a result more than twice this.

37.23. **IDENTIFY** and **SET UP:** The reference frames are shown in Figure 37.23.

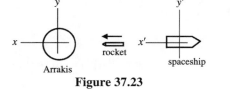

S = Arrakis frame
S' = spaceship frame
The object is the rocket.

Figure 37.23

u is the velocity of the spaceship relative to Arrakis.

$v_x = +0.360c$; $v'_x = +0.920c$

(In each frame the rocket is moving in the positive coordinate direction.)

Use the Lorentz velocity transformation equation, Eq.(37.22): $v'_x = \dfrac{v_x - u}{1 - uv_x/c^2}.$

EXECUTE: $v_x' = \dfrac{v_x - u}{1 - uv_x/c^2}$ so $v_x' - u\left(\dfrac{v_x v_x'}{c^2}\right) = v_x - u$ and $u\left(1 - \dfrac{v_x v_x'}{c^2}\right) = v_x - v_x'$

$$u = \frac{v_x - v_x'}{1 - v_x v_x'/c^2} = \frac{0.360c - 0.920c}{1 - (0.360c)(0.920c)/c^2} = \frac{0.560c}{0.6688} = -0.837c$$

The speed of the spacecraft relative to Arrakis is $0.837c = 2.51\times10^8$ m/s. The minus sign in our result for u means that the spacecraft is moving in the $-x$-direction, so it is moving away from Arrakis.

EVALUATE: The incorrect Galilean expression also says that the spacecraft is moving away from Arrakis, but with speed $0.920c - 0.360c = 0.560c$.

37.25.   **IDENTIFY** and **SET UP:**   Source and observer are approaching, so use Eq.(37.25): $f = \sqrt{\dfrac{c + u}{c - u}}\, f_0$. Solve for u, the speed of the light source relative to the observer.

(a) EXECUTE: $f^2 = \left(\dfrac{c + u}{c - u}\right) f_0^2$

$(c - u)f^2 = (c + u)f_0^2$ and $u = \dfrac{c(f^2 - f_0^2)}{f^2 + f_0^2} = c\left(\dfrac{(f/f_0)^2 - 1}{(f/f_0)^2 + 1}\right)$

$\lambda_0 = 675$ nm, $\lambda = 575$ nm

$$u = \left(\frac{(675 \text{ nm}/575 \text{ nm})^2 - 1}{(675 \text{ nm}/575 \text{ nm})^2 + 1}\right) c = 0.159c = (0.159)(2.998\times10^8 \text{ m/s}) = 4.77\times10^7 \text{ m/s; definitely speeding}$$

(b) 4.77×10^7 m/s $= (4.77\times10^7$ m/s$)(1$ km/1000 m$)(3600$ s/1 h$) = 1.72\times10^8$ km/h. Your fine would be 1.72×10^8 (172 million dollars).

EVALUATE: The source and observer are approaching, so $f > f_0$ and $\lambda < \lambda_0$. Our result gives $u < c$, as it must.

37.27.   **IDENTIFY** and **SET UP:**   If $\vec{F}$ is parallel to $\vec{v}$ then $\vec{F}$ changes the magnitude of $\vec{v}$ and not its direction.

$$F = \frac{dp}{dt} = \frac{d}{dt}\left(\frac{mv}{\sqrt{1 - v^2/c^2}}\right)$$

Use the chain rule to evaluate the derivative: $\dfrac{d}{dt} f(v(t)) = \dfrac{df}{dv}\dfrac{dv}{dt}$.

EXECUTE: **(a)** $F = \dfrac{m}{(1 - v^2/c^2)^{1/2}}\left(\dfrac{dv}{dt}\right) + \dfrac{mv}{(1 - v^2/c^2)^{3/2}}\left(-\dfrac{1}{2}\right)\left(-\dfrac{2v}{c^2}\right)\left(\dfrac{dv}{dt}\right)$

$F = \dfrac{dv}{dt}\dfrac{m}{(1 - v^2/c^2)^{3/2}}\left(1 - \dfrac{v^2}{c^2} + \dfrac{v^2}{c^2}\right) = \dfrac{dv}{dt}\dfrac{m}{(1 - v^2/c^2)^{3/2}}$

But $\dfrac{dv}{dt} = a$, so $a = (F/m)(1 - v^2/c^2)^{3/2}$.

EVALUATE: Our result agrees with Eq.(37.30).

(b) IDENTIFY and **SET UP:**   If $\vec{F}$ is perpendicular to $\vec{v}$ then $\vec{F}$ changes the direction of $\vec{v}$ and not its magnitude.

$$\vec{F} = \frac{d}{dt}\left(\frac{m\vec{v}}{\sqrt{1 - v^2/c^2}}\right).$$

$\vec{a} = d\vec{v}/dt$ but the magnitude of v in the denominator of Eq.(37.29) is constant.

EXECUTE: $F = \dfrac{ma}{\sqrt{1 - v^2/c^2}}$ and $a = (F/m)(1 - v^2/c^2)^{1/2}$.

EVALUATE: This result agrees with Eq.(37.33).

37.33.   **IDENTIFY** and **SET UP:**   Use Eqs.(37.38) and (37.39).

EXECUTE: (a) $E = mc^2 + K$, so $E = 4.00mc^2$ means $K = 3.00mc^2 = 4.50\times10^{-10}$ J

(b) $E^2 = (mc^2)^2 + (pc)^2$; $E = 4.00mc^2$, so $15.0(mc^2)^2 = (pc)^2$

$p = \sqrt{15}\, mc = 1.94\times10^{-18}$ kg·m/s

(c) $E = mc^2/\sqrt{1 - v^2/c^2}$

$E = 4.00mc^2$ gives $1 - v^2/c^2 = 1/16$ and $v = \sqrt{15/16}\, c = 0.968c$

EVALUATE: The speed is close to c since the kinetic energy is greater than the rest energy. Nonrelativistic expressions relating E, K, p and v will be very inaccurate.

37.35.   **IDENTIFY:**   Use $E = mc^2$ to relate the mass increase to the energy increase.
(a) SET UP: Your total energy E increases because your gravitational potential energy mgy increases.
EXECUTE: $\Delta E = mg\Delta y$

$\Delta E = (\Delta m)c^2$ so $\Delta m = \Delta E / c^2 = mg(\Delta y)/c^2$

$\Delta m / m = (g\Delta y)/c^2 = (9.80 \text{ m/s}^2)(30 \text{ m})/(2.998 \times 10^8 \text{ m/s})^2 = 3.3 \times 10^{-13}\%$

This increase is much, much too small to be noticed.
(b) SET UP: The energy increases because potential energy is stored in the compressed spring.
EXECUTE: $\Delta E = \Delta U = \frac{1}{2}kx^2 = \frac{1}{2}(2.00 \times 10^4 \text{ N/m})(0.060 \text{ m})^2 = 36.0 \text{ J}$

$\Delta m = (\Delta E)/c^2 = 4.0 \times 10^{-16} \text{ kg}$

Energy increases so mass increases. The mass increase is much, much too small to be noticed.
EVALUATE: In both cases the energy increase corresponds to a mass increase. But since c^2 is a very large number the mass increase is very small.

37.39.   **IDENTIFY and SET UP:**   The total energy is given in terms of the momentum by Eq.(37.39). In terms of the total energy E, the kinetic energy K is $K = E - mc^2$ (from Eq.37.38). The rest energy is mc^2.
EXECUTE: **(a)** $E = \sqrt{(mc^2)^2 + (pc)^2} =$

$\sqrt{[(6.64 \times 10^{-27})(2.998 \times 10^8)^2]^2 + [(2.10 \times 10^{-18})(2.998 \times 10^8)]^2}$ J

$E = 8.67 \times 10^{-10}$ J
(b) $mc^2 = (6.64 \times 10^{-27} \text{ kg})(2.998 \times 10^8 \text{ m/s})^2 = 5.97 \times 10^{-10}$ J

$K = E - mc^2 = 8.67 \times 10^{-10}$ J $- 5.97 \times 10^{-10}$ J $= 2.70 \times 10^{-10}$ J
(c) $\dfrac{K}{mc^2} = \dfrac{2.70 \times 10^{-10} \text{ J}}{5.97 \times 10^{-10} \text{ J}} = 0.452$

EVALUATE: The incorrect nonrelativistic expressions for K and p give $K = p^2/2m = 3.3 \times 10^{-10}$ J; the correct relativistic value is less than this.

37.43.   **IDENTIFY and SET UP:**   Use Eq.(23.12) and conservation of energy to relate the potential difference to the kinetic energy gained by the electron. Use Eq.(37.36) to calculate the kinetic energy from the speed.
EXECUTE: **(a)** $K = q\Delta V = e\Delta V$

$K = mc^2\left(\dfrac{1}{\sqrt{1 - v^2/c^2}} - 1\right) = 4.025mc^2 = 3.295 \times 10^{-13} \text{ J} = 2.06 \text{ MeV}$

$\Delta V = K/e = 2.06 \times 10^6$ V
(b) From part (a), $K = 3.30 \times 10^{-13}$ J $= 2.06$ MeV
EVALUATE: The speed is close to c and the kinetic energy is four times the rest mass.

37.45.   **IDENTIFY:**   The relativistic expression for the kinetic energy is $K = (\gamma - 1)mc^2$, where $\gamma = \dfrac{1}{\sqrt{1-x}}$ and $x = v^2/c^2$.

The Newtonian expression for the kinetic energy is $K_N = \dfrac{1}{2}mv^2$.

SET UP: Solve for v such that $K = \dfrac{3}{2}K_N$.

EXECUTE: $(\gamma - 1)mc^2 = \dfrac{3}{4}mv^2$. $\dfrac{1}{\sqrt{1-x}} - 1 = \dfrac{3}{4}x$. $\dfrac{1}{1-x} = \left(1 + \dfrac{3}{4}x\right)^2$. After a little algebra this becomes

$9x^2 + 15x - 8 = 0$. $x = \dfrac{1}{18}\left(-15 \pm \sqrt{(15)^2 + 4(9)(8)}\right)$. The positive root is $x = 0.425$. $x = v^2/c^2$, so

$v = \sqrt{x}\, c = 0.652c$.
EVALUATE: The fractional increase of the relativistic expression above the nonrelativistic one increases as v increases.

37.49.   **(a) IDENTIFY and SET UP:** $\Delta t_0 = 2.60 \times 10^{-8}$ s is the proper time, measured in the pion's frame. The time measured in the lab must satisfy $d = c\Delta t$, where $u \approx c$. Calculate Δt and then use Eq.(37.6) to calculate u.
EXECUTE: $\Delta t = \dfrac{d}{c} = \dfrac{1.20 \times 10^3 \text{ m}}{2.998 \times 10^8 \text{ m/s}} = 4.003 \times 10^{-6}$ s

$\Delta t = \dfrac{\Delta t_0}{\sqrt{1 - u^2/c^2}}$ so $(1 - u^2/c^2)^{1/2} = \dfrac{\Delta t_0}{\Delta t}$ and $(1 - u^2/c^2) = \left(\dfrac{\Delta t_0}{\Delta t}\right)^2$

Write $u = (1 - \Delta)c$ so that $(u/c)^2 = (1 - \Delta)^2 = 1 - 2\Delta + \Delta^2 \approx 1 - 2\Delta$ since Δ is small.

Using this in the above gives $1-(1-2\Delta) = \left(\dfrac{\Delta t_0}{\Delta t}\right)^2$

$$\Delta = \frac{1}{2}\left(\frac{\Delta t_0}{\Delta t}\right)^2 = \frac{1}{2}\left(\frac{2.60\times10^{-8}\text{ s}}{4.003\times10^{-6}\text{ s}}\right)^2 = 2.11\times10^{-5}$$

EVALUATE: An alternative calculation is to say that the length of the tube must contract relative to the moving pion so that the pion travels that length before decaying. The contracted length must be
$l = c\Delta t_0 = (2.998\times10^8\text{ m/s})(2.60\times10^{-8}\text{ s}) = 7.79$ m.

$$l = l_0\sqrt{1-u^2/c^2} \text{ so } 1-u^2/c^2 = \left(\frac{l}{l_0}\right)^2$$

Then $u = (1-\Delta)c$ gives $\Delta = \dfrac{1}{2}\left(\dfrac{l}{l_0}\right)^2 = \dfrac{1}{2}\left(\dfrac{7.79\text{ m}}{1.20\times10^3\text{ m}}\right)^2 = 2.11\times10^{-5}$, which checks.

(b) IDENTIFY and **SET UP:** $E = \gamma mc^2$ (Eq.(37.38)).

EXECUTE: $\gamma = \dfrac{1}{\sqrt{1-u^2/c^2}} = \dfrac{1}{\sqrt{2\Delta}} = \dfrac{1}{\sqrt{2(2.11\times10^{-5})}} = 154$

$E = 154(139.6\text{ MeV}) = 2.15\times10^4\text{ MeV} = 21.5$ GeV

EVALUATE: The total energy is 154 times the rest energy.

37.51.   **IDENTIFY** and **SET UP:**   There must be a length contraction such that the length a becomes the same as b; $l_0 = a$, $l = b$. l_0 is the distance measured by an observer at rest relative to the spacecraft. Use Eq.(37.16) and solve for u.

EXECUTE: $\dfrac{l}{l_0} = \sqrt{1-u^2/c^2}$ so $\dfrac{b}{a} = \sqrt{1-u^2/c^2}$;

$a = 1.40b$ gives $b/1.40b = \sqrt{1-u^2/c^2}$ and thus $1-u^2/c^2 = 1/(1.40)^2$

$u = \sqrt{1-1/(1.40)^2}\,c = 0.700c = 2.10\times10^8$ m/s

EVALUATE: A length on the spacecraft in the direction of the motion is shortened. A length perpendicular to the motion is unchanged.

37.53.   **(a)** $E = \gamma mc^2$ and $\gamma = 10 = \dfrac{1}{\sqrt{1-(v/c)^2}} \Rightarrow \dfrac{v}{c} = \sqrt{\dfrac{\gamma^2-1}{\gamma^2}} \Rightarrow \dfrac{v}{c} = \sqrt{\dfrac{99}{100}} = 0.995.$

(b) $(pc)^2 = m^2v^2\gamma^2c^2$, $E^2 = m^2c^4\left[\left(\dfrac{v}{c}\right)^2\gamma^2 + 1\right]$

$$\Rightarrow \frac{E^2-(pc)^2}{E^2} = \frac{1}{1+\gamma^2\left(\dfrac{v}{c}\right)^2} = \frac{1}{1+(10/(0.995))^2} = 0.01 = 1\%.$$

37.55.   **IDENTIFY:**   Since the speed is very close to the speed of light, we must use the relativistic formula for kinetic energy.

SET UP: The relativistic formula for kinetic energy is $K = mc^2\left(\dfrac{1}{\sqrt{1-v^2/c^2}} - 1\right)$ and the relativistic mass is

$m_{\text{rel}} = \dfrac{m}{\sqrt{1-v^2/c^2}}.$

EXECUTE: **(a)** $K = 7\times10^{12}\text{ eV} = 1.12\times10^{-6}$ J . Using this value in the relativistic kinetic energy formula and

substituting the mass of the proton for m, we get $K = mc^2\left(\dfrac{1}{\sqrt{1-v^2/c^2}} - 1\right)$

which gives $\dfrac{1}{\sqrt{1-v^2/c^2}} = 7.45\times10^3$ and $1-\dfrac{v^2}{c^2} = \dfrac{1}{(7.45\times10^3)^2}$. Solving for v gives

$1-\dfrac{v^2}{c^2} = \dfrac{(c+v)(c-v)}{c^2} = \dfrac{2(c-v)}{c}$, since $c + v \approx 2c$. Substituting $v = (1-\Delta)c$, we have.

$1-\dfrac{v^2}{c^2} = \dfrac{2(c-v)}{c} = \dfrac{2[c-(1-\Delta)c]}{c} = 2\Delta$. Solving for Δ gives $\Delta = \dfrac{1-v^2/c^2}{2} = \dfrac{\dfrac{1}{(7.45\times10^3)^2}}{2} = 9\times10^{-9}$, to one

significant digit.

(b) Using the relativistic mass formula and the result that $\dfrac{1}{\sqrt{1-v^2/c^2}}=7.45\times10^3$, we have

$$m_{\text{rel}}=\frac{m}{\sqrt{1-v^2/c^2}}=m\left(\frac{1}{\sqrt{1-v^2/c^2}}\right)=(7\times10^3)m\text{ , to one significant digit.}$$

EVALUATE: At such high speeds, the proton's mass is over 7000 times as great as its rest mass.

37.57.   **IDENTIFY** and **SET UP:**   In crown glass the speed of light is $v=\dfrac{c}{n}$. Calculate the kinetic energy of an electron that has this speed.

EXECUTE: $v=\dfrac{2.998\times10^8\text{ m/s}}{1.52}=1.972\times10^8\text{ m/s.}$

$K=mc^2(\gamma-1)$

$mc^2=(9.109\times10^{-31}\text{ kg})(2.998\times10^8\text{ m/s})^2=8.187\times10^{-14}\text{ J}(1\text{ eV}/1.602\times10^{-19}\text{ J})=0.5111\text{ MeV}$

$\gamma=\dfrac{1}{\sqrt{1-v^2/c^2}}=\dfrac{1}{\sqrt{1-((1.972\times10^8\text{ m/s})/(2.998\times10^8\text{ m/s}))^2}}=1.328$

$K=mc^2(\gamma-1)=(0.5111\text{ MeV})(1.328-1)=0.168\text{ MeV}$

EVALUATE: No object can travel faster than the speed of light in vacuum but there is nothing that prohibits an object from traveling faster than the speed of light in some material.

37.59.   **IDENTIFY** and **SET UP:**   Let S be the lab frame and S' be the frame of the proton that is moving in the $+x$ direction, so $u=+c/2$. The reference frames and moving particles are shown in Figure 37.59. The other proton moves in the $-x$ direction in the lab frame, so $v=-c/2$. A proton has rest mass $m_p=1.67\times10^{-27}\text{ kg}$ and rest energy $m_p c^2=938\text{ MeV}$.

EXECUTE: (a) $v'=\dfrac{v-u}{1-uv/c^2}=\dfrac{-c/2-c/2}{1-(c/2)(-c/2)/c^2}=-\dfrac{4c}{5}$

The speed of each proton relative to the other is $\dfrac{4}{5}c$.

(b) In nonrelativistic mechanics the speeds just add and the speed of each relative to the other is c.

(c) $K=\dfrac{mc^2}{\sqrt{1-v^2/c^2}}-mc^2$

(i) Relative to the lab frame each proton has speed $v=c/2$. The total kinetic energy of each proton is

$K=\dfrac{938\text{ MeV}}{\sqrt{1-\left(\dfrac{1}{2}\right)^2}}-(938\text{ MeV})=145\text{ MeV}$.

(ii) In its rest frame one proton has zero speed and zero kinetic energy and the other has speed $\dfrac{4}{5}c$. In this frame the kinetic energy of the moving proton is $K=\dfrac{938\text{ MeV}}{\sqrt{1-\left(\dfrac{4}{5}\right)^2}}-(938\text{ MeV})=625\text{ MeV}$

(d) (i) Each proton has speed $v=c/2$ and kinetic energy

$K=\dfrac{1}{2}mv^2=\left(\dfrac{1}{2}m\right)(c/2)^2=\dfrac{mc^2}{8}=\dfrac{938\text{ MeV}}{8}=117\text{ MeV}$

(ii) One proton has speed $v=0$ and the other has speed c. The kinetic energy of the moving proton is $K=\dfrac{1}{2}mc^2=\dfrac{938\text{ MeV}}{2}=469\text{ MeV}$

EVALUATE: The relativistic expression for K gives a larger value than the nonrelativistic expression. The kinetic energy of the system is different in different frames.

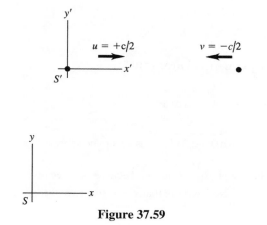

Figure 37.59

37.63. **IDENTIFY and SET UP:** Use Eq.(37.30), with $a = dv/dt$, to obtain an expression for dv/dt. Separate the variables v and t and integrate to obtain an expression for $v(t)$. In this expression, let $t \to \infty$.

EXECUTE: $a = \dfrac{dv}{dt} = \dfrac{F}{m}(1 - v^2/c^2)^{3/2}$. (One-dimensional motion is assumed, and all the F, v, and a refer to x-components.)

$$\frac{dv}{(1 - v^2/c^2)^{3/2}} = \left(\frac{F}{m}\right) dt$$

Integrate from $t = 0$, when $v = 0$, to time t, when the velocity is v.

$$\int_0^v \frac{dv}{(1 - v^2/c^2)^{3/2}} = \int_0^t \left(\frac{F}{m}\right) dt$$

Since F is constant, $\int_0^t \left(\dfrac{F}{m}\right) dt = \dfrac{Ft}{m}$. In the velocity integral make the change of variable $y = v/c$; then $dy = dv/c$.

$$\int_0^v \frac{dv}{(1 - v^2/c^2)^{3/2}} = c \int_0^{v/c} \frac{dy}{(1 - y^2)^{3/2}} = c \left[\frac{y}{(1 - y^2)^{1/2}}\right]_0^{v/c} = \frac{v}{\sqrt{1 - v^2/c^2}}$$

Thus $\dfrac{v}{\sqrt{1 - v^2/c^2}} = \dfrac{Ft}{m}$.

Solve this equation for v:

$$\frac{v^2}{1 - v^2/c^2} = \left(\frac{Ft}{m}\right)^2 \text{ and } v^2 = \left(\frac{Ft}{m}\right)^2 (1 - v^2/c^2)$$

$$v^2 \left(1 + \left(\frac{Ft}{mc}\right)^2\right) = \left(\frac{Ft}{m}\right)^2 \text{ so } v = \frac{(Ft/m)}{\sqrt{1 + (Ft/mc)^2}} = c \frac{Ft}{\sqrt{m^2 c^2 + F^2 t^2}}$$

As $t \to \infty$, $\dfrac{Ft}{\sqrt{m^2 c^2 + F^2 t^2}} \to \dfrac{Ft}{\sqrt{F^2 t^2}} \to 1$, so $v \to c$.

EVALUATE: Note that $\dfrac{Ft}{\sqrt{m^2 c^2 + F^2 t^2}}$ is always less than 1, so $v < c$ always and v approaches c only when $t \to \infty$.

37.65. **(a) IDENTIFY and SET UP:** Use the Lorentz coordinate transformation (Eq.37.21) for (x_1, t_1) and (x_2, t_2):

$$x_1' = \frac{x_1 - ut_1}{\sqrt{1 - u^2/c^2}}, \quad x_2' = \frac{x_2 - ut_2}{\sqrt{1 - u^2/c^2}}$$

$$t_1' = \frac{t_1 - ux_1/c^2}{\sqrt{1 - u^2/c^2}}, \quad t_2' = \frac{t_2 - ux_2/c^2}{\sqrt{1 - u^2/c^2}}$$

Same point in S' implies $x_1' = x_2'$. What then is $\Delta t' = t_2' - t_1'$?

EXECUTE: $x_1' = x_2'$ implies $x_1 - ut_1 = x_2 - ut_2$

$$u(t_2 - t_1) = x_2 - x_1 \text{ and } u = \frac{x_2 - x_1}{t_2 - t_1} = \frac{\Delta x}{\Delta t}$$

From the time transformation equations,

$$\Delta t' = t_2' - t_1' = \frac{1}{\sqrt{1-u^2/c^2}}(\Delta t - u\Delta x/c^2)$$

Using the result that $u = \frac{\Delta x}{\Delta t}$ gives

$$\Delta t' = \frac{1}{\sqrt{1-(\Delta x)^2/((\Delta t)^2 c^2)}}(\Delta t - (\Delta x)^2/((\Delta t)c^2))$$

$$\Delta t' = \frac{\Delta t}{\sqrt{(\Delta t)^2 - (\Delta x)^2/c^2}}(\Delta t - (\Delta x)^2/((\Delta t)c^2))$$

$$\Delta t' = \frac{(\Delta t)^2 - (\Delta x)^2/c^2}{\sqrt{(\Delta t)^2 - (\Delta x)^2/c^2}} = \sqrt{(\Delta t)^2 - (\Delta x/c)^2},\text{ as was to be shown.}$$

This equation doesn't have a physical solution (because of a negative square root) if $(\Delta x/c)^2 > (\Delta t)^2$ or $\Delta x \geq c\Delta t$.

(b) IDENTIFY and SET UP: Now require that $t_2' = t_1'$ (the two events are simultaneous in S') and use the Lorentz coordinate transformation equations.

EXECUTE: $t_2' = t_1'$ implies $t_1 - ux_1/c^2 = t_2 - ux_2/c^2$

$$t_2 - t_1 = \left(\frac{x_2 - x_1}{c^2}\right)u \text{ so } \Delta t = \left(\frac{\Delta x}{c^2}\right)u \text{ and } u = \frac{c^2\Delta t}{\Delta x}$$

From the Lorentz transformation equations,

$$\Delta x' = x_2' - x_1' = \left(\frac{1}{\sqrt{1-u^2/c^2}}\right)(\Delta x - u\Delta t).$$

Using the result that $u = c^2\Delta t/\Delta x$ gives

$$\Delta x' = \frac{1}{\sqrt{1-c^2(\Delta t)^2/(\Delta x)^2}}(\Delta x - c^2(\Delta t)^2/\Delta x)$$

$$\Delta x' = \frac{\Delta x}{\sqrt{(\Delta x)^2 - c^2(\Delta t)^2}}(\Delta x - c^2(\Delta t)^2/\Delta x)$$

$$\Delta x' = \frac{(\Delta x)^2 - c^2(\Delta t)^2}{\sqrt{(\Delta x)^2 - c^2(\Delta t)^2}} = \sqrt{(\Delta x)^2 - c^2(\Delta t)^2}$$

(c) IDENTIFY and SET UP: The result from part (b) is $\Delta x' = \sqrt{(\Delta x)^2 - c^2(\Delta t)^2}$

Solve for Δt: $(\Delta x')^2 = (\Delta x)^2 - c^2(\Delta t)^2$

EXECUTE: $\Delta t = \frac{\sqrt{(\Delta x)^2 - (\Delta x')^2}}{c} = \frac{\sqrt{(5.00\text{ m})^2 - (2.50\text{ m})^2}}{2.998\times10^8\text{ m/s}} = 1.44\times10^{-8}$ s

EVALUATE: This provides another illustration of the concept of simultaneity (Section 37.2): events observed to be simultaneous in one frame are not simultaneous in another frame that is moving relative to the first.

37.67.   **IDENTIFY and SET UP:**   An increase in wavelength corresponds to a decrease in frequency ($f = c/\lambda$), so the

atoms are moving away from the earth. Receding, so use Eq.(37.26): $f = \sqrt{\frac{c-u}{c+u}}f_0$

EXECUTE: Solve for u: $(f/f_0)^2(c+u) = c - u$ and $u = c\left(\frac{1-(f/f_0)^2}{1+(f/f_0)^2}\right)$

$f = c/\lambda, f_0 = c/\lambda_0$ so $f/f_0 = \lambda_0/\lambda$

$$u = c\left(\frac{1-(\lambda_0/\lambda)^2}{1+(\lambda_0/\lambda)^2}\right) = c\left(\frac{1-(656.3/953.4)^2}{1+(656.3/953.4)^2}\right) = 0.357c = 1.07\times10^8\text{ m/s}$$

EVALUATE: The relative speed is large, 36% of c. The cosmological implication of such observations will be discussed in Section 44.6.

37.71.   The crux of this problem is the question of simultaneity. To be "in the barn at one time" for the runner is different than for a stationary observer in the barn. The diagram in Figure 37.71a shows the rod fitting into the barn at time $t = 0$, according to the stationary observer. The diagram in Figure 37.71b is in the runner's frame of reference. The front of the rod enters the barn at time t_1 and leaves the back of the barn at time t_2. However, the back of the rod does not enter the front of the barn until the later time t_3.

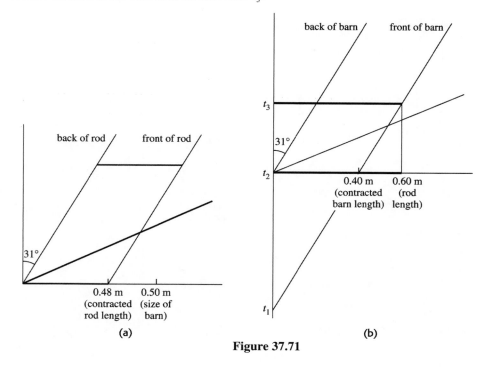

(a)

(b)

Figure 37.71

PHOTONS, ELECTRONS, AND ATOMS

38

38.5. **IDENTIFY** and **SET UP:** Eq.(38.2) relates the photon energy and wavelength. $c = f\lambda$ relates speed, frequency and wavelength for an electromagnetic wave.

EXECUTE: **(a)** $E = hf$ so $f = \dfrac{E}{h} = \dfrac{(2.45 \times 10^6 \text{ eV})(1.602 \times 10^{-19} \text{ J/1 eV})}{6.626 \times 10^{-34} \text{ J} \cdot \text{s}} = 5.92 \times 10^{20}$ Hz

(b) $c = f\lambda$ so $\lambda = \dfrac{c}{f} = \dfrac{2.998 \times 10^8 \text{ m/s}}{5.92 \times 10^{20} \text{ Hz}} = 5.06 \times 10^{-13}$ m

(c) EVALUATE: λ is comparable to a nuclear radius. Note that in doing the calculation the energy in MeV was converted to the SI unit of Joules.

38.7. **IDENTIFY** and **SET UP:** Eq.(38.3): $\dfrac{1}{2}mv_{max}^2 = hf - \phi = \dfrac{hc}{\lambda} - \phi$. Take the work function ϕ from Table 38.1. Solve for v_{max}. Note that we wrote f as c/λ.

EXECUTE: $\dfrac{1}{2}mv_{max}^2 = \dfrac{(6.626 \times 10^{-34} \text{ J} \cdot \text{s})(2.998 \times 10^8 \text{ m/s})}{235 \times 10^{-9} \text{ m}} - (5.1 \text{ eV})(1.602 \times 10^{-19} \text{ J/1 eV})$

$\dfrac{1}{2}mv_{max}^2 = 8.453 \times 10^{-19} \text{ J} - 8.170 \times 10^{-19} \text{ J} = 2.83 \times 10^{-20}$ J

$v_{max} = \sqrt{\dfrac{2(2.83 \times 10^{-20} \text{ J})}{9.109 \times 10^{-31} \text{ kg}}} = 2.49 \times 10^5$ m/s

EVALUATE: The work function in eV was converted to joules for use in Eq.(38.3). A photon with $\lambda = 235$ nm has energy greater then the work function for the surface.

38.9. **IDENTIFY** and **SET UP:** $c = f\lambda$. The source emits $(0.05)(75 \text{ J}) = 3.75$ J of energy as visible light each second. $E = hf$, with $h = 6.63 \times 10^{-34}$ J·s.

EXECUTE: **(a)** $f = \dfrac{c}{\lambda} = \dfrac{3.00 \times 10^8 \text{ m/s}}{600 \times 10^{-9} \text{ m}} = 5.00 \times 10^{14}$ Hz

(b) $E = hf = (6.63 \times 10^{-34} \text{ J} \cdot \text{s})(5.00 \times 10^{14} \text{ Hz}) = 3.32 \times 10^{-19}$ J. The number of photons emitted per second is

$\dfrac{3.75 \text{ J}}{3.32 \times 10^{-19} \text{ J/photon}} = 1.13 \times 10^{19}$ photons.

(c) No. The frequency of the light depends on the energy of each photon. The number of photons emitted per second is proportional to the power output of the source.

38.11. **IDENTIFY:** Protons have mass and photons are massless.

(a) SET UP: For a particle with mass, $K = p^2/2m$.

EXECUTE: $p_2 = 2p_1$ means $K_2 = 4K_1$.

(b) SET UP: For a photon, $E = pc$.

EXECUTE: $p_2 = 2p_1$ means $E_2 = 2E_1$.

EVALUATE: The relation between E and p is different for particles with mass and particles without mass.

38.13. **(a) IDENTIFY:** First use Eq.(38.4) to find the work function ϕ.

SET UP: $eV_0 = hf - \phi$ so $\phi = hf - eV_0 = \dfrac{hc}{\lambda} - eV_0$

EXECUTE: $\phi = \dfrac{(6.626 \times 10^{-34} \text{ J} \cdot \text{s})(2.998 \times 10^8 \text{ m/s})}{254 \times 10^{-9} \text{ m}} - (1.602 \times 10^{-19} \text{ C})(0.181 \text{ V})$

$\phi = 7.821 \times 10^{-19} \text{ J} - 2.900 \times 10^{-20} \text{ J} = 7.531 \times 10^{-19} \text{ J}(1 \text{ eV}/1.602 \times 10^{-19} \text{ J}) = 4.70$ eV

IDENTIFY and **SET UP:**   The threshold frequency f_{th} is the smallest frequency that still produces photoelectrons. It corresponds to $K_{max} = 0$ in Eq.(38.3), so $hf_{th} = \phi$.

EXECUTE: $f = \dfrac{c}{\lambda}$ says $\dfrac{hc}{\lambda_{th}} = \phi$

$\lambda_{th} = \dfrac{hc}{\phi} = \dfrac{(6.626 \times 10^{-34}\ \text{J} \cdot \text{s})(2.998 \times 10^{8}\ \text{m/s})}{7.531 \times 10^{-19}\ \text{J}} = 2.64 \times 10^{-7}\ \text{m} = 264\ \text{nm}$

(b) EVALUATE: As calculated in part (a), $\phi = 4.70$ eV. This is the value given in Table 38.1 for copper.

38.15.   **IDENTIFY** and **SET UP:**   Balmer's formula is $\dfrac{1}{\lambda} = R\left(\dfrac{1}{2^2} - \dfrac{1}{n^2}\right)$. For the H_γ spectral line $n = 5$. Once we have λ, calculate f from $f = c/\lambda$ and E from Eq.(38.2).

EXECUTE: **(a)** $\dfrac{1}{\lambda} = R\left(\dfrac{1}{2^2} - \dfrac{1}{5^2}\right) = R\left(\dfrac{25-4}{100}\right) = R\left(\dfrac{21}{100}\right)$.

Thus $\lambda = \dfrac{100}{21R} = \dfrac{100}{21(1.097 \times 10^7)}$ m $= 4.341 \times 10^{-7}$ m $= 434.1$ nm.

(b) $f = \dfrac{c}{\lambda} = \dfrac{2.998 \times 10^8\ \text{m/s}}{4.341 \times 10^{-7}\ \text{m}} = 6.906 \times 10^{14}$ Hz

(c) $E = hf = (6.626 \times 10^{-34}\ \text{J} \cdot \text{s})(6.906 \times 10^{14}\ \text{Hz}) = 4.576 \times 10^{-19}\ \text{J} = 2.856$ eV

EVALUATE: Section 38.3 shows that the longest wavelength in the Balmer series (H_α) is 656 nm and the shortest is 365 nm. Our result for H_γ falls within this range. The photon energies for hydrogen atom transitions are in the eV range, and our result is of this order.

38.19.   **IDENTIFY** and **SET UP:**   The wavelength of the photon is related to the transition energy $E_i - E_f$ of the atom by

$E_i - E_f = \dfrac{hc}{\lambda}$ where $hc = 1.240 \times 10^{-6}$ eV·m.

EXECUTE: **(a)** The minimum energy to ionize an atom is when the upper state in the transition has $E = 0$, so $E_1 = -17.50$ eV. For $n = 5 \rightarrow n = 1$, $\lambda = 73.86$ nm and $E_5 - E_1 = \dfrac{1.240 \times 10^{-6}\ \text{eV} \cdot \text{m}}{73.86 \times 10^{-9}\ \text{m}} = 16.79$ eV.

$E_5 = -17.50$ eV $+ 16.79$ eV $= -0.71$ eV. For $n = 4 \rightarrow n = 1$, $\lambda = 75.63$ nm and $E_4 = -1.10$ eV. For $n = 3 \rightarrow n = 1$, $\lambda = 79.76$ nm and $E_3 = -1.95$ eV. For $n = 2 \rightarrow n = 1$, $\lambda = 94.54$ nm and $E_2 = -4.38$ eV.

(b) $E_i - E_f = E_4 - E_2 = -1.10$ eV $- (-4.38$ eV$) = 3.28$ eV and $\lambda = \dfrac{hc}{E_i - E_f} = \dfrac{1.240 \times 10^{-6}\ \text{eV} \cdot \text{m}}{3.28\ \text{eV}} = 378$ nm

EVALUATE: The $n = 4 \rightarrow n = 2$ transition energy is smaller than the $n = 4 \rightarrow n = 1$ transition energy so the wavelength is longer. In fact, this wavelength is longer than for any transition that ends in the $n = 1$ state.

38.21.   **(a) IDENTIFY:**   If the particles are treated as point charges, $U = \dfrac{1}{4\pi\epsilon_0} \dfrac{q_1 q_2}{r}$.

SET UP: $q_1 = 2e$ (alpha particle); $q_2 = 82e$ (gold nucleus); r is given so we can solve for U.

EXECUTE: $U = (8.987 \times 10^9\ \text{N} \cdot \text{m}^2/\text{C}^2) \dfrac{(2)(82)(1.602 \times 10^{-19}\ \text{C})^2}{6.50 \times 10^{-14}\ \text{m}} = 5.82 \times 10^{-13}$ J

$U = 5.82 \times 10^{-13}$ J$(1$ eV$/1.602 \times 10^{-19}$ J$) = 3.63 \times 10^6$ eV $= 3.63$ MeV

(b) IDENTIFY: Apply conservation of energy: $K_1 + U_1 = K_2 + U_2$.

SET UP: Let point 1 be the initial position of the alpha particle and point 2 be where the alpha particle momentarily comes to rest. Alpha particle is initially far from the lead nucleus implies $r_1 \approx \infty$ and $U_1 = 0$. Alpha particle stops implies $K_2 = 0$.

EXECUTE: Conservation of energy thus says $K_1 = U_2 = 5.82 \times 10^{-13}$ J $= 3.63$ MeV.

(c) $K = \dfrac{1}{2}mv^2$ so $v = \sqrt{\dfrac{2K}{m}} = \sqrt{\dfrac{2(5.82 \times 10^{-13}\ \text{J})}{6.64 \times 10^{-27}\ \text{kg}}} = 1.32 \times 10^7$ m/s

EVALUATE: $v/c = 0.044$, so it is ok to use the nonrelativistic expression to relate K and v. When the alpha particle stops, all its initial kinetic energy has been converted to electrostatic potential energy.

38.23.   **IDENTIFY** and **SET UP:**   Use the energy to calculate n for this state. Then use the Bohr equation, Eq.(38.10), to calculate L.

EXECUTE: $E_n = -(13.6\text{ eV})/n^2$, so this state has $n = \sqrt{13.6/1.51} = 3$. In the Bohr model. $L = n\hbar$ so for this state $L = 3\hbar = 3.16 \times 10^{-34}$ kg·m²/s.

EVALUATE: We will find in Section 41.1 that the modern quantum mechanical description gives a different result.

38.25.   **IDENTIFY:**   The force between the electron and the nucleus in Be^{3+} is $F = \dfrac{1}{4\pi\epsilon_0}\dfrac{Ze^2}{r^2}$, where $Z = 4$ is the nuclear charge. All the equations for the hydrogen atom apply to Be^{3+} if we replace e^2 by Ze^2.

(a) SET UP: Modify Eq.(38.18).

EXECUTE: $E_n = -\dfrac{1}{\epsilon_0}\dfrac{me^4}{8n^2h^2}$ (hydrogen) becomes

$$E_n = -\frac{1}{\epsilon_0}\frac{m(Ze^2)^2}{8n^2h^2} = Z^2\left(-\frac{1}{\epsilon_0}\frac{me^4}{8n^2h^2}\right) = Z^2\left(-\frac{13.60\text{ eV}}{n^2}\right) \text{ (for } Be^{3+})$$

The ground-level energy of Be^{3+} is $E_1 = 16\left(-\dfrac{13.60\text{ eV}}{1^2}\right) = -218$ eV.

EVALUATE: The ground-level energy of Be^{3+} is $Z^2 = 16$ times the ground-level energy of H.

(b) SET UP: The ionization energy is the energy difference between the $n \to \infty$ level energy and the $n = 1$ level energy.

EXECUTE: The $n \to \infty$ level energy is zero, so the ionization energy of Be^{3+} is 218 eV.

EVALUATE: This is 16 times the ionization energy of hydrogen.

(c) SET UP: $\dfrac{1}{\lambda} = R\left(\dfrac{1}{n_1^2} - \dfrac{1}{n_2^2}\right)$ just as for hydrogen but now R has a different value.

EXECUTE: $R_H = \dfrac{me^4}{8\epsilon_0 h^3 c} = 1.097 \times 10^7$ m⁻¹ for hydrogen becomes

$R_{Be} = Z^2\dfrac{me^4}{8\epsilon_0 h^3 c} = 16(1.097 \times 10^7\text{ m}^{-1}) = 1.755 \times 10^8$ m⁻¹ for Be^{3+}.

For $n = 2$ to $n = 1$, $\dfrac{1}{\lambda} = R_{Be}\left(\dfrac{1}{1^2} - \dfrac{1}{2^2}\right) = 3R/4$.

$\lambda = 4/(3R) = 4/(3(1.755 \times 10^8\text{ m}^{-1})) = 7.60 \times 10^{-9}$ m = 7.60 nm.

EVALUATE: This wavelength is smaller by a factor of 16 compared to the wavelength for the corresponding transition in the hydrogen atom.

(d) SET UP: Modify Eq.(38.12): $r_n = \epsilon_0\dfrac{n^2h^2}{\pi me^2}$ (hydrogen).

EXECUTE: $r_n = \epsilon_0\dfrac{n^2h^2}{\pi m(Ze^2)}$ (Be^{3+}).

EVALUATE: For a given n the orbit radius for Be^{3+} is smaller by a factor of $Z = 4$ compared to the corresponding radius for hydrogen.

38.29.   **IDENTIFY and SET UP:**   The number of photons emitted each second is the total energy emitted divided by the energy of one photon. The energy of one photon is given by Eq.(38.2). $E = Pt$ gives the energy emitted by the laser in time t.

EXECUTE: In 1.00 s the energy emitted by the laser is $(7.50 \times 10^{-3}\text{ W})(1.00\text{ s}) = 7.50 \times 10^{-3}$ J.

The energy of each photon is $E = \dfrac{hc}{\lambda} = \dfrac{(6.626 \times 10^{-34}\text{ J·s})(2.998 \times 10^8\text{ m/s})}{10.6 \times 10^{-6}\text{ m}} = 1.874 \times 10^{-20}$ J.

Therefore $\dfrac{7.50 \times 10^{-3}\text{ J/s}}{1.874 \times 10^{-20}\text{ J/photon}} = 4.00 \times 10^{17}$ photons/s

EVALUATE: The number of photons emitted per second is extremely large.

38.31.   **IDENTIFY:**   Apply Eq.(38.21): $\dfrac{n_{5s}}{n_{3p}} = e^{-(E_{5s} - E_{3p})/kT}$

SET UP: From Fig.38.24a in the textbook, $E_{5s} = 20.66$ eV and $E_{3p} = 18.70$ eV

EXECUTE: $E_{5s} - E_{3p} = 20.66$ eV $- 18.70$ eV $= 1.96$ eV$(1.602 \times 10^{-19}$ J/1 eV$) = 3.140 \times 10^{-19}$ J

(a) $\dfrac{n_{5s}}{n_{3p}} = e^{-(3.140\times10^{-19}\text{ J})/[(1.38\times10^{-23}\text{ J/K})(300\text{ K})]} = e^{-75.79} = 1.2\times10^{-33}$

(b) $\dfrac{n_{5s}}{n_{3p}} = e^{-(3.140\times10^{-19}\text{ J})/[(1.38\times10^{-23}\text{ J/K})(600\text{ K})]} = e^{-37.90} = 3.5\times10^{-17}$

(c) $\dfrac{n_{5s}}{n_{3p}} = e^{-(3.140\times10^{-19}\text{ J})/[(1.38\times10^{-23}\text{ J/K})(1200\text{ K})]} = e^{-18.95} = 5.9\times10^{-9}$

(d) EVALUATE**:** At each of these temperatures the number of atoms in the $5s$ excited state, the initial state for the transition that emits 632.8 nm radiation, is quite small. The ratio increases as the temperature increases.

38.37. **I**DENTIFY**:** Apply Eq.(38.23): $\lambda' - \lambda = \dfrac{h}{mc}(1 - \cos\phi) = \lambda_{\text{C}}(1 - \cos\phi)$

SET **U**P**:** Solve for λ': $\lambda' = \lambda + \lambda_{\text{C}}(1 - \cos\phi)$

The largest λ' corresponds to $\phi = 180°$, so $\cos\phi = -1$.

EXECUTE**:** $\lambda' = \lambda + 2\lambda_{\text{C}} = 0.0665\times10^{-9}$ m $+ 2(2.426\times10^{-12}$ m$) = 7.135\times10^{-11}$ m $= 0.0714$ nm. This wavelength occurs at a scattering angle of $\phi = 180°$.

EVALUATE**:** The incident photon transfers some of its energy and momentum to the electron from which it scatters. Since the photon loses energy its wavelength increases, $\lambda' > \lambda$.

38.39. **I**DENTIFY **and S**ET **U**P**:** The shift in wavelength of the photon is $\lambda' - \lambda = \dfrac{h}{mc}(1 - \cos\phi)$ where λ' is the wavelength after the scattering and $\dfrac{h}{mc} = \lambda_{\text{c}} = 2.426\times10^{-12}$ m. The energy of a photon of wavelength λ is $E = \dfrac{hc}{\lambda} = \dfrac{1.24\times10^{-6}\text{ eV}\cdot\text{m}}{\lambda}$. Conservation of energy applies to the collision, so the energy lost by the photon equals the energy gained by the electron.

EXECUTE**:** **(a)** $\lambda' - \lambda = \lambda_{\text{c}}(1 - \cos\phi) = (2.426\times10^{-12}$ m$)(1 - \cos35.0°) = 4.39\times10^{-13}$ m $= 4.39\times10^{-4}$ nm

(b) $\lambda' = \lambda + 4.39\times10^{-4}$ nm $= 0.04250$ nm $+ 4.39\times10^{-4}$ nm $= 0.04294$ nm

(c) $E_\lambda = \dfrac{hc}{\lambda} = 2.918\times10^4$ eV and $E_{\lambda'} = \dfrac{hc}{\lambda'} = 2.888\times10^4$ eV so the photon loses 300 eV of energy.

(d) Energy conservation says the electron gains 300 eV of energy.

38.43. **(a)** $H = Ae\sigma T^4$; $A = \pi r^2 l$

$$T = \left(\dfrac{H}{Ae\sigma}\right)^{1/4} = \left(\dfrac{100\text{ W}}{2\pi(0.20\times10^{-3}\text{ m})(0.30\text{ m})(0.26)(5.671\times10^{-8}\text{ W/m}^2\cdot\text{K}^4)}\right)^{1/4}$$

$T = 2.06\times10^3$ K

(b) $\lambda_{\text{m}}T = 2.90\times10^{-3}$ m $\cdot$ K; $\lambda_{\text{m}} = 1410$ nm

Much of the emitted radiation is in the infrared.

38.45. **I**DENTIFY **and S**ET **U**P**:** The wavelength λ_{m} where the Planck distribution peaks is given by Eq.(38.30).

EXECUTE**:** $\lambda_{\text{m}} = \dfrac{2.90\times10^{-3}\text{ m}\cdot\text{K}}{2.728\text{ K}} = 1.06\times10^{-3}$ m $= 1.06$ mm.

EVALUATE**:** This wavelength is in the microwave portion of the electromagnetic spectrum. This radiation is often referred to as the "microwave background" (Section 44.7). Note that in Eq.(38.30), T must be in kelvins.

38.51. **I**DENTIFY **and S**ET **U**P**:** Use $c = f\lambda$ to relate frequency and wavelength and use $E = hf$ to relate photon energy and frequency.

EXECUTE**:** **(a)** One photon dissociates one AgBr molecule, so we need to find the energy required to dissociate a single molecule. The problem states that it requires 1.00×10^5 J to dissociate one mole of AgBr, and one mole contains Avogadro's number (6.02×10^{23}) of molecules, so the energy required to dissociate one AgBr is

$\dfrac{1.00\times10^5\text{ J/mol}}{6.02\times10^{23}\text{ molecules/mol}} = 1.66\times10^{-19}$ J/molecule.

The photon is to have this energy, so $E = 1.66\times10^{-19}$ J$(1\text{eV}/1.602\times10^{-19}$ J$) = 1.04$ eV.

(b) $E = \dfrac{hc}{\lambda}$ so $\lambda = \dfrac{hc}{E} = \dfrac{(6.626\times10^{-34}\text{ J}\cdot\text{s})(2.998\times10^8\text{ m/s})}{1.66\times10^{-19}\text{ J}} = 1.20\times10^{-6}$ m $= 1200$ nm

(c) $c = f\lambda$ so $f = \dfrac{c}{\lambda} = \dfrac{2.998\times10^8\text{ m/s}}{1.20\times10^{-6}\text{ m}} = 2.50\times10^{14}$ Hz

(d) $E = hf = (6.626 \times 10^{-34} \text{ J} \cdot \text{s})(100 \times 10^6 \text{ Hz}) = 6.63 \times 10^{-26}$ J

$E = 6.63 \times 10^{-26}$ J$(1 \text{ eV}/1.602 \times 10^{-19}$ J$) = 4.14 \times 10^{-7}$ eV

(e) EVALUATE: A photon with frequency $f = 100$ MHz has too little energy, by a large factor, to dissociate a AgBr molecule. The photons in the visible light from a firefly do individually have enough energy to dissociate AgBr. The huge number of 100 MHz photons can't compensate for the fact that individually they have too little energy.

38.55.   **(a) IDENTIFY:** Apply the photoelectric effect equation, Eq.(38.4).

SET UP: $eV_0 = hf - \phi = (hc/\lambda) - \phi$. Call the stopping potential V_{01} for λ_1 and V_{02} for λ_2. Thus $eV_{01} = (hc/\lambda_1) - \phi$ and $eV_{02} = (hc/\lambda_2) - \phi$. Note that the work function ϕ is a property of the material and is independent of the wavelength of the light.

EXECUTE: Subtracting one equation from the other gives $e(V_{02} - V_{01}) = hc\left(\dfrac{\lambda_1 - \lambda_2}{\lambda_1 \lambda_2}\right)$.

(b) $\Delta V_0 = \dfrac{(6.626 \times 10^{-34} \text{ J} \cdot \text{s})(2.998 \times 10^8 \text{ m/s})}{1.602 \times 10^{-19} \text{ C}}\left(\dfrac{295 \times 10^{-9} \text{ m} - 265 \times 10^{-9} \text{ m}}{(295 \times 10^{-9} \text{ m})(265 \times 10^{-9} \text{ m})}\right) = 0.476$ V.

EVALUATE: $e\Delta V_0$, which is 0.476 eV, is the increase in photon energy from 295 nm to 265 nm. The stopping potential increases when λ deceases because the photon energy increases when the wavelength decreases.

38.57.   **IDENTIFY and SET UP:** The energy added to mass m of the blood to heat it to $T_f = 100$ °C and to vaporize it is $Q = mc(T_f - T_i) + mL_v$, with $c = 4190$ J/kg·K and $L_v = 2.256 \times 10^6$ J/kg. The energy of one photon is $E = \dfrac{hc}{\lambda} = \dfrac{1.99 \times 10^{-25} \text{ J} \cdot \text{m}}{\lambda}$.

EXECUTE: **(a)** $Q = (2.0 \times 10^{-9} \text{ kg})(4190 \text{ J/kg} \cdot \text{K})(100°\text{C} - 33°\text{C}) + (2.0 \times 10^{-9} \text{ kg})(2.256 \times 10^6 \text{ J/kg}) = 5.07 \times 10^{-3}$ J
The pulse must deliver 5.07 mJ of energy.

(b) $P = \dfrac{\text{energy}}{t} = \dfrac{5.07 \times 10^{-3} \text{ J}}{450 \times 10^{-6} \text{ s}} = 11.3$ W

(c) One photon has energy $E = \dfrac{hc}{\lambda} = \dfrac{1.99 \times 10^{-25} \text{ J} \cdot \text{m}}{585 \times 10^{-9} \text{ m}} = 3.40 \times 10^{-19}$ J. The number N of photons per pulse is the energy per pulse divided by the energy of one photon: $N = \dfrac{5.07 \times 10^{-3} \text{ J}}{3.40 \times 10^{-19} \text{ J/photon}} = 1.49 \times 10^{16}$ photons

38.59.   **(a) IDENTIFY and SET UP:** Apply Eq.(38.20): $m_r = \dfrac{m_1 m_2}{m_1 + m_2} = \dfrac{207 m_e m_p}{207 m_e + m_p}$

EXECUTE: $m_r = \dfrac{207(9.109 \times 10^{-31} \text{ kg})(1.673 \times 10^{-27} \text{ kg})}{207(9.109 \times 10^{-31} \text{ kg}) + 1.673 \times 10^{-27} \text{ kg}} = 1.69 \times 10^{-28}$ kg

We have used m_e to denote the electron mass.

(b) IDENTIFY: In Eq.(38.18) replace $m = m_e$ by m_r: $E_n = -\dfrac{1}{\epsilon_0^2} \dfrac{m_r e^4}{8n^2 h^2}$.

SET UP: Write as $E_n = \left(\dfrac{m_r}{m_H}\right)\left(-\dfrac{1}{\epsilon_0^2}\dfrac{m_H e^4}{8n^2 h^2}\right)$, since we know that $\dfrac{1}{\epsilon_0^2}\dfrac{m_H e^4}{8h^2} = 13.60$ eV. Here m_H denotes the reduced mass for the hydrogen atom; $m_H = 0.99946(9.109 \times 10^{-31} \text{ kg}) = 9.104 \times 10^{-31}$ kg.

EXECUTE: $E_n = \left(\dfrac{m_r}{m_H}\right)\left(-\dfrac{13.60 \text{ eV}}{n^2}\right)$

$E_1 = \dfrac{1.69 \times 10^{-28} \text{ kg}}{9.104 \times 10^{-31} \text{ kg}}(-13.60 \text{ eV}) = 186(-13.60 \text{ eV}) = -2.53$ keV

(c) SET UP: From part (b), $E_n = \left(\dfrac{m_r}{m_H}\right)\left(-\dfrac{R_H ch}{n^2}\right)$, where $R_H = 1.097 \times 10^7$ m^{-1} is the Rydberg constant for the hydrogen atom. Use this result in $\dfrac{hc}{\lambda} = E_i - E_f$ to find an expression for $1/\lambda$. The initial level for the transition is the $n_i = 2$ level and the final level is the $n_f = 1$ level.

EXECUTE: $\dfrac{hc}{\lambda} = \dfrac{m_r}{m_H}\left(-\dfrac{R_H ch}{n_i^2} - \left(-\dfrac{R_H ch}{n_f^2}\right)\right)$

$$\frac{1}{\lambda} = \frac{m_r}{m_H} R_H \left(\frac{1}{n_f^2} - \frac{1}{n_i^2} \right)$$

$$\frac{1}{\lambda} = \frac{1.69 \times 10^{-28} \text{ kg}}{9.104 \times 10^{-31} \text{ kg}} (1.097 \times 10^7 \text{ m}^{-1}) \left(\frac{1}{1^2} - \frac{1}{2^2} \right) = 1.527 \times 10^9 \text{ m}^{-1}$$

$\lambda = 0.655$ nm

EVALUATE: From Example 38.6 the wavelength of the radiation emitted in this transition in hydrogen is 122 nm. The wavelength for muonium is $\frac{m_H}{m_r} = 5.39 \times 10^{-3}$ times this. The reduced mass for hydrogen is very close to the electron mass because the electron mass is much less then the proton mass: $m_p / m_e = 1836$. The muon mass is $207 m_e = 1.886 \times 10^{-28}$ kg. The proton is only about 10 times more massive than the muon, so the reduced mass is somewhat smaller than the muon mass. The muon-proton atom has much more strongly bound energy levels and much shorter wavelengths in its spectrum than for hydrogen.

38.61. **IDENTIFY** and **SET UP:** $\lambda' = \lambda + \frac{h}{mc}(1 - \cos\phi)$

$\phi = 180°$ so $\lambda' = \lambda + \frac{2h}{mc} = 0.09485$ m. Use Eq.(38.5) to calculate the momentum of the scattered photon. Apply conservation of energy to the collision to calculate the kinetic energy of the electron after the scattering. The energy of the photon is given by Eq.(38.2),

EXECUTE: (a) $p' = h / \lambda' = 6.99 \times 10^{-24}$ kg·m/s.

(b) $E = E' + E_e$; $hc / \lambda = hc / \lambda' + E_e$

$$E_e = hc \left(\frac{1}{\lambda} - \frac{1}{\lambda'} \right) = (hc) \frac{\lambda' - \lambda}{\lambda \lambda'} = 1.129 \times 10^{-16} \text{ J} = 705 \text{ eV}$$

EVALUATE: The energy of the incident photon is 13.8 keV, so only about 5% of its energy is transferred to the electron. This corresponds to a fractional shift in the photon's wavelength that is also 5%.

38.63. **IDENTIFY** and **SET UP:** The H_α line in the Balmer series corresponds to the $n = 3$ to $n = 2$ transition.

$$E_n = -\frac{13.6 \text{ eV}}{n^2}. \quad \frac{hc}{\lambda} = \Delta E.$$

EXECUTE: (a) The atom must be given an amount of energy $E_3 - E_1 = -(13.6 \text{ eV}) \left(\frac{1}{3^2} - \frac{1}{1^2} \right) = 12.1 \text{ eV}$.

(b) There are three possible transitions. $n = 3 \rightarrow n = 1$: $\Delta E = 12.1$ eV and $\lambda = \frac{hc}{\Delta E} = 103$ nm ;

$n = 3 \rightarrow n = 2$: $\Delta E = -(13.6 \text{ eV}) \left(\frac{1}{3^2} - \frac{1}{2^2} \right) = 1.89$ eV and $\lambda = 657$ nm ; $n = 2 \rightarrow n = 1$:

$\Delta E = -(13.6 \text{ eV}) \left(\frac{1}{2^2} - \frac{1}{1^2} \right) = 10.2$ eV and $\lambda = 122$ nm .

38.65. (a) **IDENTIFY** and **SET UP:** The photon energy is given to the electron in the atom. Some of this energy overcomes the binding energy of the atom and what is left appears as kinetic energy of the free electron. Apply $hf = E_f - E_i$, the energy given to the electron in the atom when a photon is absorbed.

EXECUTE: The energy of one photon is $\frac{hc}{\lambda} = \frac{(6.626 \times 10^{-34} \text{ J} \cdot \text{s})(2.998 \times 10^8 \text{ m/s})}{85.5 \times 10^{-9} \text{ m}}$

$\frac{hc}{\lambda} = 2.323 \times 10^{-18}$ J(1 eV/1.602×10^{-19} J) = 14.50 eV.

The final energy of the electron is $E_f = E_i + hf$. In the ground state of the hydrogen atom the energy of the electron is $E_i = -13.60$ eV. Thus $E_f = -13.60$ eV + 14.50 eV = 0.90 eV.

(b) **EVALUATE:** At thermal equilibrium a few atoms will be in the $n = 2$ excited levels, which have an energy of -13.6 eV/4 = -3.40 eV, 10.2 eV greater than the energy of the ground state. If an electron with $E = -3.40$ eV gains 14.5 eV from the absorbed photon, it will end up with 14.5 eV − 3.4 eV = 11.1 eV of kinetic energy.

38.69. **IDENTIFY:** The energy of the peak-intensity photons must be equal to the energy difference between the $n = 1$ and the $n = 4$ states. Wien's law allows us to calculate what the temperature of the blackbody must be for it to radiate with its peak intensity at this wavelength.

SET UP: In the Bohr model, the energy of an electron in shell n is $E_n = -\dfrac{13.6 \text{ eV}}{n^2}$, and Wien's displacement law

is $\lambda_m = \dfrac{2.90 \times 10^{-3} \text{ m} \cdot \text{K}}{T}$. The energy of a photon is $E = hf = hc/\lambda$.

EXECUTE: First find the energy (ΔE) that a photon would need to excite the atom. The ground state of the atom is $n = 1$ and the third excited state is $n = 4$. This energy is the *difference* between the two energy levels. Therefore

$$\Delta E = (-13.6 \text{ eV})\left(\frac{1}{4^2} - \frac{1}{1^2}\right) = 12.8 \text{ eV}.$$ Now find the wavelength of the photon having this amount of energy.

$hc/\lambda = 12.8$ eV and

$$\lambda = (4.136 \times 10^{-15} \text{ eV} \cdot \text{s})(3.00 \times 10^8 \text{ m/s})/(12.8 \text{ eV}) = 9.73 \times 10^{-8} \text{ m}$$

Now use Wien's law to find the temperature. $T = (0.00290 \text{ m} \cdot \text{K})/(9.73 \times 10^{-8} \text{ m}) = 2.98 \times 10^4 \text{ K}$.

EVALUATE: This temperature is well above ordinary room temperatures, which is why hydrogen atoms are not in excited states during everyday conditions.

38.71.   **IDENTIFY:**   Apply conservation of energy and conservation of linear momentum to the system of atom plus photon.

(a) SET UP: Let E_{tr} be the transition energy, E_{ph} be the energy of the photon with wavelength λ', and E_r be the kinetic energy of the recoiling atom. Conservation of energy gives $E_{ph} + E_r = E_{tr}$.

$$E_{ph} = \frac{hc}{\lambda'} \text{ so } \frac{hc}{\lambda'} = E_{tr} - E_r \text{ and } \lambda' = \frac{hc}{E_{tr} - E_r}.$$

EXECUTE: If the recoil energy is neglected then the photon wavelength is $\lambda = hc/E_{tr}$.

$$\Delta\lambda = \lambda' - \lambda = hc\left(\frac{1}{E_{tr} - E_r} - \frac{1}{E_{tr}}\right) = \left(\frac{hc}{E_{tr}}\right)\left(\frac{1}{1 - E_r/E_{tr}} - 1\right)$$

$$\frac{1}{1 - E_r/E_{tr}} = \left(1 - \frac{E_r}{E_{tr}}\right)^{-1} \approx 1 + \frac{E_r}{E_{tr}} \text{ since } \frac{E_r}{E_{tr}} \ll 1$$

(We have used the binomial theorem, Appendix B.)

Thus $\Delta\lambda = \dfrac{hc}{E_{tr}}\left(\dfrac{E_r}{E_{tr}}\right)$, or since $E_{tr} = hc/\lambda$, $\Delta\lambda = \left(\dfrac{E_r}{hc}\right)\lambda^2$.

SET UP: Use conservation of linear momentum to find E_r: Assuming that the atom is initially at rest, the momentum p_r of the recoiling atom must be equal in magnitude and opposite in direction to the momentum $p_{ph} = h/\lambda$ of the emitted photon: $h/\lambda = p_r$.

EXECUTE: $E_r = \dfrac{p_r^2}{2m}$, where m is the mass of the atom, so $E_r = \dfrac{h^2}{2m\lambda^2}$.

Use this result in the above equation: $\Delta\lambda = \left(\dfrac{E_r}{hc}\right)\lambda^2 = \left(\dfrac{h^2}{2m\lambda^2}\right)\left(\dfrac{\lambda^2}{hc}\right) = \dfrac{h}{2mc}$;

note that this result for $\Delta\lambda$ is independent of the atomic transition energy.

(b) For a hydrogen atom $m = m_p$ and $\Delta\lambda = \dfrac{h}{2m_p c} = \dfrac{6.626 \times 10^{-34} \text{ J} \cdot \text{s}}{2(1.673 \times 10^{-27} \text{ kg})(2.998 \times 10^8 \text{ m/s})} = 6.61 \times 10^{-16} \text{ m}$

EVALUATE: The correction is independent of n. The wavelengths of photons emitted in hydrogen atom transitions are on the order of $100 \text{ nm} = 10^{-7}$ m, so the recoil correction is exceedingly small.

38.73.   **IDENTIFY** and **SET UP:**   Find the average change in wavelength for one scattering and use that in $\Delta\lambda$ in Eq.(38.23) to calculate the average scattering angle ϕ.

EXECUTE: **(a)** The wavelength of a 1 MeV photon is

$$\lambda = \frac{hc}{E} = \frac{(4.136 \times 10^{-15} \text{ eV} \cdot \text{s})(2.998 \times 10^8 \text{ m/s})}{1 \times 10^6 \text{ eV}} = 1 \times 10^{-12} \text{ m}$$

The total change in wavelength therefore is $500 \times 10^{-9} \text{ m} - 1 \times 10^{-12} \text{ m} = 500 \times 10^{-9}$ m.

If this shift is produced in 10^{26} Compton scattering events, the wavelength shift in each scattering event is

$$\Delta\lambda = \frac{500 \times 10^{-9} \text{ m}}{1 \times 10^{26}} = 5 \times 10^{-33} \text{ m}.$$

(b) Use this $\Delta\lambda$ in $\Delta\lambda = \dfrac{h}{mc}(1-\cos\phi)$ and solve for ϕ. We anticipate that ϕ will be very small, since $\Delta\lambda$ is

much less than h/mc, so we can use $\cos\phi \approx 1 - \phi^2/2$.

$$\Delta\lambda = \frac{h}{mc}(1-(1-\phi^2/2)) = \frac{h}{2mc}\phi^2$$

$$\phi = \sqrt{\frac{2\Delta\lambda}{(h/mc)}} = \sqrt{\frac{2(5\times10^{-33}\ \text{m})}{2.426\times10^{-12}\ \text{m}}} = 6.4\times10^{-11}\ \text{rad} = (4\times10^{-9})^\circ$$

ϕ in radians is much less than 1 so the approximation we used is valid.

(c) IDENTIFY and **SET UP:**   We know the total transit time and the total number of scatterings, so we can calculate the average time between scatterings.

EXECUTE: The total time to travel from the core to the surface is $(10^6\ \text{y})(3.156\times10^7\ \text{s/y}) = 3.2\times10^{13}\ \text{s}$. There are

10^{26} scatterings during this time, so the average time between scatterings is $t = \dfrac{3.2\times10^{13}\ \text{s}}{10^{26}} = 3.2\times10^{-13}\ \text{s}$.

The distance light travels in this time is $d = ct = (3.0\times10^8\ \text{m/s})(3.2\times10^{-13}\ \text{s}) = 0.1\ \text{mm}$

EVALUATE: The photons are on the average scattered through a very small angle in each scattering event. The average distance a photon travels between scatterings is very small.

38.75.   **(a) IDENTIFY** and **SET UP:**   Conservation of energy applied to the collision gives $E_\lambda = E_{\lambda'} + E_e$, where E_e is the kinetic energy of the electron after the collision and E_λ and $E_{\lambda'}$ are the energies of the photon before and after the collision. The energy of a photon is related to its wavelength according to Eq.(38.2).

EXECUTE: $E_e = hc\left(\dfrac{1}{\lambda} - \dfrac{1}{\lambda'}\right) = hc\left(\dfrac{\lambda'-\lambda}{\lambda\lambda'}\right)$

$$E_e = (6.626\times10^{-34}\ \text{J}\cdot\text{s})(2.998\times10^8\ \text{m/s})\left(\frac{0.0032\times10^{-9}\ \text{m}}{(0.1100\times10^{-9}\ \text{m})(0.1132\times10^{-9}\ \text{m})}\right)$$

$E_e = 5.105\times10^{-17}\ \text{J} = 319\ \text{eV}$

$E_e = \dfrac{1}{2}mv^2$ so $v = \sqrt{\dfrac{2E_e}{m}} = \sqrt{\dfrac{2(5.105\times10^{-17}\ \text{J})}{9.109\times10^{-31}\ \text{kg}}} = 1.06\times10^7\ \text{m/s}$

(b) The wavelength λ of a photon with energy E_e is given by $E_e = hc/\lambda$ so

$$\lambda = \frac{hc}{E_e} = \frac{(6.626\times10^{-34}\ \text{J}\cdot\text{s})(2.998\times10^8\ \text{m/s})}{5.105\times10^{-17}\ \text{J}} = 3.89\ \text{nm}$$

EVALUATE: Only a small portion of the incident photon's energy is transferred to the struck electron; this is why the wavelength calculated in part (b) is much larger than the wavelength of the incident photon in the Compton scattering.

THE WAVE NATURE OF PARTICLES

39.5. **IDENTIFY** and **SET UP:** The de Broglie wavelength is $\lambda = \dfrac{h}{p} = \dfrac{h}{mv}$. In the Bohr model, $mv r_n = n(h/2\pi)$,

so $mv = nh/(2\pi r_n)$. Combine these two expressions and obtain an equation for λ in terms of n. Then

$\lambda = h\left(\dfrac{2\pi r_n}{nh}\right) = \dfrac{2\pi r_n}{n}$.

EXECUTE: **(a)** For $n=1$, $\lambda = 2\pi r_1$ with $r_1 = a_0 = 0.529 \times 10^{-10}$ m, so $\lambda = 2\pi(0.529 \times 10^{-10}$ m$) = 3.32 \times 10^{-10}$ m

$\lambda = 2\pi r_1$; the de Broglie wavelength equals the circumference of the orbit.

(b) For $n=4$, $\lambda = 2\pi r_4/4$.

$r_n = n^2 a_0$ so $r_4 = 16 a_0$.

$\lambda = 2\pi(16 a_0)/4 = 4(2\pi a_0) = 4(3.32 \times 10^{-10}$ m$) = 1.33 \times 10^{-9}$ m

$\lambda = 2\pi r_4/4$; the de Broglie wavelength is $\dfrac{1}{n} = \dfrac{1}{4}$ times the circumference of the orbit.

EVALUATE: As n increases the momentum of the electron increases and its de Broglie wavelength decreases. For any n, the circumference of the orbits equals an integer number of de Broglie wavelengths.

39.9. **IDENTIFY** and **SET UP:** A photon has zero mass and its energy and wavelength are related by Eq.(38.2). An electron has mass. Its energy is related to its momentum by $E = p^2/2m$ and its wavelength is related to its momentum by Eq.(39.1).

EXECUTE: **(a)** photon

$E = \dfrac{hc}{\lambda}$ so $\lambda = \dfrac{hc}{E} = \dfrac{(6.626 \times 10^{-34}\ \text{J} \cdot \text{s})(2.998 \times 10^8\ \text{m/s})}{(20.0\ \text{eV})(1.602 \times 10^{-19}\ \text{J/eV})} = 62.0$ nm

electron

$E = p^2/(2m)$ so $p = \sqrt{2mE} = \sqrt{2(9.109 \times 10^{-31}\ \text{kg})(20.0\ \text{eV})(1.602 \times 10^{-19}\ \text{J/eV})} = 2.416 \times 10^{-24}\ \text{kg} \cdot \text{m/s}$

$\lambda = h/p = 0.274$ nm

(b) photon $E = hc/\lambda = 7.946 \times 10^{-19}$ J $= 4.96$ eV

electron $\lambda = h/p$ so $p = h/\lambda = 2.650 \times 10^{-27}\ \text{kg} \cdot \text{m/s}$

$E = p^2/(2m) = 3.856 \times 10^{-24}$ J $= 2.41 \times 10^{-5}$ eV

(c) **EVALUATE:** You should use a probe of wavelength approximately 250 nm. An electron with $\lambda = 250$ nm has much less energy than a photon with $\lambda = 250$ nm, so is less likely to damage the molecule. Note that $\lambda = h/p$ applies to all particles, those with mass and those with zero mass. $E = hf = hc/\lambda$ applies only to photons and $E = p^2/2m$ applies only to particles with mass.

39.11. **IDENTIFY** and **SET UP:** Use Eq.(39.1).

EXECUTE: $\lambda = \dfrac{h}{p} = \dfrac{h}{mv} = \dfrac{6.626 \times 10^{-34}\ \text{J} \cdot \text{s}}{(5.00 \times 10^{-3}\ \text{kg})(340\ \text{m/s})} = 3.90 \times 10^{-34}$ m

EVALUATE: This wavelength is extremely short; the bullet will not exhibit wavelike properties.

39.13. **(a)** $\lambda = 0.10$ nm. $p = mv = h/\lambda$ so $v = h/(m\lambda) = 7.3 \times 10^6$ m/s.

(b) $E = \dfrac{1}{2}mv^2 = 150$ eV

(c) $E = hc/\lambda = 12$ KeV

(d) The electron is a better probe because for the same λ it has less energy and is less damaging to the structure being probed.

39.15. For $m = 1$, $\lambda = d\sin\theta = \dfrac{h}{\sqrt{2mE}}$.

$$E = \frac{h^2}{2md^2\sin^2\theta} = \frac{(6.63\times10^{-34}\ \text{J}\cdot\text{s})^2}{2(1.675\times10^{-27}\ \text{kg})\,(9.10\times10^{-11}\ \text{m})^2\sin^2(28.6°)} = 6.91\times10^{-20}\ \text{J} = 0.432\ \text{eV}.$$

39.17. The condition for a maximum is $d\sin\theta = m\lambda$. $\lambda = \dfrac{h}{p} = \dfrac{h}{Mv}$, so $\theta = \arcsin\left(\dfrac{mh}{dMv}\right)$. (Careful! Here, m is the order of the maximum, whereas M is the incoming particle mass.)

(a) $m = 1 \Rightarrow \theta_1 = \arcsin\left(\dfrac{h}{dMv}\right)$

$$= \arcsin\left(\frac{6.63\times10^{-34}\ \text{J}\cdot\text{s}}{(1.60\times10^{-6}\ \text{m})\,(9.11\times10^{-31}\ \text{kg})\,(1.26\times10^{4}\ \text{m/s})}\right) = 2.07°.$$

$$m = 2 \Rightarrow \theta_2 = \arcsin\left(\frac{(2)\,(6.63\times10^{-34}\ \text{J}\cdot\text{s})}{(1.60\times10^{-6}\ \text{m})\,(9.11\times10^{-31}\ \text{kg})\,(1.26\times10^{4}\ \text{m/s})}\right) = 4.14°.$$

(b) For small angles (in radians!) $y \cong D\theta$, so $y_1 \approx (50.0\ \text{cm})\,(2.07°)\left(\dfrac{\pi\ \text{radians}}{180°}\right) = 1.81\ \text{cm}$,

$$y_2 \approx (50.0\ \text{cm})\,(4.14°)\left(\frac{\pi\ \text{radians}}{180°}\right) = 3.61\ \text{cm} \ \text{and} \ y_2 - y_1 = 3.61\ \text{cm} - 1.81\ \text{cm} = 1.81\ \text{cm}.$$

39.19. **(a) IDENTIFY** and **SET UP:** Use $\Delta x\Delta p_x \geq h/2\pi$ to calculate Δx and obtain Δv_x from this.

EXECUTE: $\Delta p_x \geq \dfrac{h}{2\pi\Delta x} = \dfrac{6.626\times10^{-34}\ \text{J}\cdot\text{s}}{2\pi(1.00\times10^{-6}\ \text{m})} = 1.055\times10^{-28}\ \text{kg}\cdot\text{m/s}$

$\Delta v_x = \dfrac{\Delta p_x}{m} = \dfrac{1.055\times10^{-28}\ \text{kg}\cdot\text{m/s}}{1200\ \text{kg}} = 8.79\times10^{-32}\ \text{m/s}$

(b) EVALUATE: Even for this very small Δx the minimum Δv_x required by the Heisenberg uncertainty principle is very small. The uncertainty principle does not impose any practical limit on the simultaneous measurements of the positions and velocities of ordinary objects.

39.25. $\Delta E\Delta t = \dfrac{h}{2\pi}$. $\Delta E = \dfrac{h}{2\pi\Delta t} = \dfrac{(6.63\times10^{-34}\ \text{J}\cdot\text{s})}{2\pi(7.6\times10^{-21}\ \text{s})} = 1.39\times10^{-14}\ \text{J} = 8.69\times10^{4}\ \text{eV} = 0.0869\ \text{MeV}.$

$$\frac{\Delta E}{E} = \frac{0.0869\ \text{MeV}/c^2}{3097\ \text{MeV}/c^2} = 2.81\times10^{-5}.$$

39.27. **IDENTIFY** and **SET UP:** For a photon $E_{ph} = \dfrac{hc}{\lambda} = \dfrac{1.99\times10^{-25}\ \text{J}\cdot\text{m}}{\lambda}$. For an electron $E_e = \dfrac{p^2}{2m} = \dfrac{1}{2m}\left(\dfrac{h}{\lambda}\right)^2 = \dfrac{h^2}{2m\lambda^2}$.

EXECUTE: **(a)** <u>photon</u> $E_{ph} = \dfrac{1.99\times10^{-25}\ \text{J}\cdot\text{m}}{10.0\times10^{-9}\ \text{m}} = 1.99\times10^{-17}\ \text{J}$

<u>electron</u> $E_e = \dfrac{(6.63\times10^{-34}\ \text{J}\cdot\text{s})^2}{2(9.11\times10^{-31}\ \text{kg})(10.0\times10^{-9}\ \text{m})^2} = 2.41\times10^{-21}\ \text{J}$

$\dfrac{E_{ph}}{E_e} = \dfrac{1.99\times10^{-17}\ \text{J}}{2.41\times10^{-21}\ \text{J}} = 8.26\times10^{3}$

(b) The electron has much less energy so would be less damaging.

EVALUATE: For a particle with mass, such as an electron, $E \sim \lambda^{-2}$. For a massless photon $E \sim \lambda^{-1}$.

39.29. **IDENTIFY** and **SET UP:** $\psi(x) = A\sin kx$. The position probability density is given by $|\psi(x)|^2 = A^2\sin^2 kx$.

EXECUTE: **(a)** The probability is highest where $\sin kx = 1$ so $kx = 2\pi x/\lambda = n\pi/2$, $n = 1,\ 3,\ 5,\ldots$
$x = n\lambda/4$, $n = 1,\ 3,\ 5,\ldots$ so $x = \lambda/4,\ 3\lambda/4,\ 5\lambda/4,\ldots$

(b) The probability of finding the particle is zero where $|\psi|^2 = 0$, which occurs where $\sin kx = 0$ and
$kx = 2\pi x/\lambda = n\pi$, $n = 0,\ 1,\ 2,\ldots$
$x = n\lambda/2$, $n = 0,\ 1,\ 2,\ldots$ so $x = 0,\ \lambda/2,\ \lambda,\ 3\lambda/2,\ldots$

EVALUATE: The situation is analogous to a standing wave, with the probability analogous to the square of the amplitude of the standing wave.

39.35.   **IDENTIFY:**   To describe a real situation, a wave function must be normalizable.

SET UP: $|\psi|^2\, dV$ is the probability that the particle is found in volume dV. Since the particle must be *somewhere*, ψ must have the property that $\int |\psi|^2\, dV = 1$ when the integral is taken over all space.

EXECUTE: **(a)** For normalization of the one-dimensional wave function, we have

$$1 = \int_{-\infty}^{\infty} |\psi|^2\, dx = \int_{-\infty}^{0} \left(Ae^{bx}\right)^2 dx + \int_{0}^{\infty} \left(Ae^{-bx}\right)^2 dx = \int_{-\infty}^{0} A^2 e^{2bx}\, dx + \int_{0}^{\infty} A^2 e^{-2bx}\, dx.$$

$$1 = A^2 \left\{ \left.\frac{e^{2bx}}{2b}\right|_{-\infty}^{0} + \left.\frac{e^{-2bx}}{-2b}\right|_{0}^{\infty} \right\} = \frac{A^2}{b}, \text{ which gives } A = \sqrt{b} = \sqrt{2.00 \text{ m}^{-1}} = 1.41 \text{ m}^{-1/2}$$

(b) The graph of the wavefunction versus x is given in Figure 39.35.

(c) (i) $P = \int_{-0.500 \text{ m}}^{+5.00 \text{ m}} |\psi|^2\, dx = 2 \int_{0}^{+5.00 \text{ m}} A^2 e^{-2bx}\, dx$, where we have used the fact that the wave function is an even function of x. Evaluating the integral gives

$$P = \frac{-A^2}{b}\left(e^{-2b(0.500 \text{ m})} - 1\right) = \frac{-(2.00 \text{ m}^{-1})}{2.00 \text{ m}^{-1}}\left(e^{-2.00} - 1\right) = 0.865$$

There is a little more than an 86% probability that the particle will be found within 50 cm of the origin.

(ii) $P = \int_{-\infty}^{0} \left(Ae^{bx}\right)^2 dx = \int_{-\infty}^{0} A^2 e^{2bx}\, dx = \frac{A^2}{2b} = \frac{2.00 \text{ m}^{-1}}{2(2.00 \text{ m}^{-1})} = \frac{1}{2} = 0.500$

There is a 50-50 chance that the particle will be found to the left of the origin, which agrees with the fact that the wave function is symmetric about the y-axis.

(iii) $P = \int_{0.500 \text{ m}}^{1.00 \text{ m}} A^2 e^{-2bx}\, dx$

$$= \frac{A^2}{-2b}\left(e^{-2(2.00 \text{ m}^{-1})(1.00 \text{ m})} - e^{-2(2.00 \text{ m}^{-1})(0.500 \text{ m})}\right) = -\frac{1}{2}\left(e^{-4} - e^{-2}\right) = 0.0585$$

EVALUATE: There is little chance of finding the particle in regions where the wave function is small.

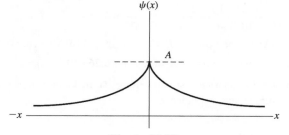

Figure 39.35

39.37. $-\dfrac{\hbar^2}{2m}\dfrac{d^2\psi}{dx^2} + U\psi = BE_1\psi_1 + CE_2\psi_2$. If ψ were a solution with energy E, then $BE_1\psi_1 + CE_2\psi_2 = BE\psi_1 + CE\psi_2$ or $B(E_1 - E)\psi_1 = C(E - E_2)\psi_2$. This would mean that ψ_1 is a constant multiple of ψ_2, and ψ_1 and ψ_2 would be wave functions with the same energy. However, $E_1 \neq E_2$, so this is not possible, and ψ cannot be a solution to Eq. (39.18).

39.41.   **IDENTIFY:**   The electrons behave like waves and produce a double-slit interference pattern after passing through the slits.

SET UP: The first angle at which destructive interference occurs is given by $d \sin \theta = \lambda/2$. The de Broglie wavelength of each of the electrons is $\lambda = h/mv$.

EXECUTE: **(a)** First find the wavelength of the electrons. For the first dark fringe, we have $d \sin \theta = \lambda/2$, which gives $(1.25 \text{ nm})(\sin 18.0°) = \lambda/2$, and $\lambda = 0.7725$ nm. Now solve the de Broglie wavelength equation for the speed of the electron:

$$v = \frac{h}{m\lambda} = \frac{6.626 \times 10^{-34} \text{ J} \cdot \text{s}}{(9.11 \times 10^{-31} \text{ kg})(0.7725 \times 10^{-9} \text{ m})} = 9.42 \times 10^5 \text{ m/s}$$

which is about 0.3% the speed of light, so they are *nonrelativistic*.

(b) Energy conservation gives $eV = \frac{1}{2} mv^2$ and

$$V = mv^2/2e = (9.11 \times 10^{-31} \text{ kg})(9.42 \times 10^5 \text{ m})^2 / [2(1.60 \times 10^{-19} \text{ C})] = 2.52 \text{ V}$$

EVALUATE: The hole must be much smaller than the wavelength of visible light for the electrons to show diffraction.

39.43.   **IDENTIFY:**   Both the electrons and photons behave like waves and exhibit single-slit diffraction after passing through their respective slits.

SET UP: The energy of the photon is $E = hc/\lambda$ and the de Broglie wavelength of the electron is $\lambda = h/mv = h/p$. Destructive interference for a single slit first occurs when $a \sin \theta = \lambda$.

EXECUTE: **(a)** For the photon: $\lambda = hc/E$ and $a\sin\theta = \lambda$. Since the a and θ are the same for the photons and electrons, they must both have the same wavelength. Equating these two expressions for λ gives $a\sin\theta = hc/E$.

For the electron, $\lambda = h/p = \dfrac{h}{\sqrt{2mK}}$ and $a\sin\theta = \lambda$. Equating these two expressions for λ gives $a\sin\theta = \dfrac{h}{\sqrt{2mK}}$.

Equating the two expressions for $a\sin\theta$ gives $hc/E = \dfrac{h}{\sqrt{2mK}}$, which gives $E = c\sqrt{2mK} = (4.05\times10^{-7}\ \text{J}^{1/2})\sqrt{K}$

(b) $\dfrac{E}{K} = \dfrac{c\sqrt{2mK}}{K} = \sqrt{\dfrac{2mc^2}{K}}$. Since $v \ll c$, $mc^2 > K$, so the square root is >1. Therefore $E/K > 1$, meaning that the photon has more energy than the electron.

EVALUATE: As we have seen in Problem 39.10, when a photon and a particle have the same wavelength, the photon has more energy than the particle.

39.47.   **(a)** Recall $\lambda = \dfrac{h}{p} = \dfrac{h}{\sqrt{2mE}} = \dfrac{h}{\sqrt{2mq\Delta V}}$. So for an electron:

$\lambda = \dfrac{6.63\times10^{-34}\ \text{J}\cdot\text{s}}{\sqrt{2(9.11\times10^{-31}\ \text{kg})(1.60\times10^{-19}\ \text{C})(125\ \text{V})}} \Rightarrow \lambda = 1.10\times10^{-10}\ \text{m}.$

(b) For an alpha particle: $\lambda = \dfrac{6.63\times10^{-34}\ \text{J}\cdot\text{s}}{\sqrt{2(6.64\times10^{-27}\ \text{kg})2(1.60\times10^{-19}\ \text{C})(125\ \text{V})}} = 9.10\times10^{-13}\ \text{m}.$

39.49.   **IDENTIFY** and **SET UP:**   Combining the two equations in the hint gives $PC = \sqrt{K(K+2mc^2)}$ and $\lambda = \dfrac{hc}{\sqrt{K(K+2mc^2)}}$.

EXECUTE: **(a)** With $K = 3mc^2$ this becomes $\lambda = \dfrac{hc}{\sqrt{3mc^2(3mc^2+2mc^2)}} = \dfrac{h}{\sqrt{15}mc}$.

(b) (i) $K = 3mc^2 = 3(9.109\times10^{-31}\ \text{kg})(2.998\times10^8\ \text{m/s})^2 = 2.456\times10^{-13}\ \text{J} = 1.53\ \text{MeV}$

$\lambda = \dfrac{h}{\sqrt{15}mc} = \dfrac{6.626\times10^{-34}\ \text{J}\cdot\text{s}}{\sqrt{15}(9.109\times10^{-31}\ \text{kg})(2.998\times10^8\ \text{m/s})} = 6.26\times10^{-13}\ \text{m}$

(ii) K is proportional to m, so for a proton $K = (m_{\text{p}}/m_{\text{e}})(1.53\ \text{MeV}) = 1836(1.53\ \text{MeV}) = 2810\ \text{MeV}$

λ is proportional to $1/m$, so for a proton $\lambda = (m_{\text{e}}/m_{\text{p}})(6.26\times10^{-13}\ \text{m}) = (1/1836)(6.626\times10^{-13}\ \text{m}) = 3.41\times10^{-16}\ \text{m}$

EVALUATE: The proton has a larger rest mass energy so its kinetic energy is larger when $K = 3mc^2$. The proton also has larger momentum so has a smaller λ.

39.51.   **(a)** **IDENTIFY** and **SET UP:** $\Delta x\Delta p_x \geq h/2\pi$

Estimate Δx as $\Delta x \approx 5.0\times10^{-15}\ \text{m}$.

EXECUTE: Then the minimum allowed Δp_x is $\Delta p_x \approx \dfrac{h}{2\pi\Delta x} = \dfrac{6.626\times10^{-34}\ \text{J}\cdot\text{s}}{2\pi(5.0\times10^{-15}\ \text{m})} = 2.1\times10^{-20}\ \text{kg}\cdot\text{m/s}$

(b) **IDENTIFY** and **SET UP:**   Assume $p \approx 2.1\times10^{-20}\ \text{kg}\cdot\text{m/s}$. Use Eq.(37.39) to calculate E, and then $K = E - mc^2$.

EXECUTE: $E = \sqrt{(mc^2)^2 + (pc)^2}$

$mc^2 = (9.109\times10^{-31}\ \text{kg})(2.998\times10^8\ \text{m/s})^2 = 8.187\times10^{-14}\ \text{J}$

$pc = (2.1\times10^{-20}\ \text{kg}\cdot\text{m/s})(2.998\times10^8\ \text{m/s}) = 6.296\times10^{-12}\ \text{J}$

$E = \sqrt{(8.187\times10^{-14}\ \text{J})^2 + (6.296\times10^{-12}\ \text{J})^2} = 6.297\times10^{-12}\ \text{J}$

$K = E - mc^2 = 6.297\times10^{-12}\ \text{J} - 8.187\times10^{-14}\ \text{J} = 6.215\times10^{-12}\ \text{J}(1\ \text{eV}/1.602\times10^{-19}\ \text{J}) = 39\ \text{MeV}$

(c) **IDENTIFY** and **SET UP:**   The Coulomb potential energy for a pair of point charges is given by Eq.(23.9). The proton has charge $+e$ and the electron has charge $-e$.

EXECUTE: $U = -\dfrac{ke^2}{r} = -\dfrac{(8.988\times10^9\ \text{N}\cdot\text{m}^2/\text{C}^2)(1.602\times10^{-19}\ \text{C})^2}{5.0\times10^{-15}\ \text{m}} = -4.6\times10^{-14}\ \text{J} = -0.29\ \text{MeV}$

EVALUATE: The kinetic energy of the electron required by the uncertainty principle would be much larger than the magnitude of the negative Coulomb potential energy. The total energy of the electron would be large and positive and the electron could not be bound within the nucleus.

39.53.   **IDENTIFY** and **SET UP:** $\Delta E\Delta t \geq \dfrac{h}{2\pi}$. Take the minimum uncertainty product, so $\Delta E = \dfrac{h}{2\pi\Delta t}$, with

$\Delta t = 8.4\times10^{-17}\ \text{s}.\quad m = 264m_{\text{e}}.\quad \Delta m = \dfrac{\Delta E}{c^2}.$

EXECUTE: $\Delta E = \dfrac{6.63\times10^{-34}\ J\cdot s}{2\pi(8.4\times10^{-17}\ s)} = 1.26\times10^{-18}\ J$. $\Delta m = \dfrac{1.26\times10^{-18}\ J}{(3.00\times10^{8}\ m/s)^2} = 1.4\times10^{-35}\ kg$.

$\dfrac{\Delta m}{m} = \dfrac{1.4\times10^{-35}\ kg}{(264)(9.11\times10^{-31}\ kg)} = 5.8\times10^{-8}$

39.55. **IDENTIFY** and **SET UP:** Use Eq.(39.1) to relate your wavelength and speed.

EXECUTE: (a) $\lambda = \dfrac{h}{mv}$, so $v = \dfrac{h}{m\lambda} = \dfrac{6.626\times10^{-34}\ J\cdot s}{(60.0\ kg)(1.0\ m)} = 1.1\times10^{-35}\ m/s$

(b) $t = \dfrac{distance}{velocity} = \dfrac{0.80\ m}{1.1\times10^{-35}\ m/s} = 7.3\times10^{34}\ s(1\ y/3.156\times10^{7}\ s) = 2.3\times10^{27}\ y$

Since you walk through doorways much more quickly than this, you will not experience diffraction effects.

EVALUATE: A 1 kg object moving at 1 m/s has a de Broglie wavelength $\lambda = 6.6\times10^{-34}\ m$, which is exceedingly small. An object like you has a very, very small λ at ordinary speeds and does not exhibit wavelike properties.

39.59. (a) The maxima occur when $2d\sin\theta = m\lambda$ as described in Section 38.7.

(b) $\lambda = \dfrac{h}{p} = \dfrac{h}{\sqrt{2mE}}$. $\lambda = \dfrac{(6.63\times10^{-34}\ J\cdot s)}{\sqrt{2(9.11\times10^{-37}\ kg)(71.0\ eV)(1.60\times10^{-19}\ J/eV)}} = 1.46\times10^{-10}\ m = 0.146\ nm$.

$\theta = \sin^{-1}\left(\dfrac{m\lambda}{2d}\right)$ (Note: This m is the order of the maximum, not the mass.)

$$\Rightarrow \sin^{-1}\left(\dfrac{(1)(1.46\times10^{-10}\ m)}{2(9.10\times10^{-11}\ m)}\right) = 53.3°.$$

(c) The work function of the metal acts like an attractive potential increasing the kinetic energy of incoming electrons by $e\phi$. An increase in kinetic energy is an increase in momentum that leads to a smaller wavelength. A smaller wavelength gives a smaller angle θ (see part (b)).

39.61. (a) **IDENTIFY** and **SET UP:** $U = A|x|$. Eq.(7.17) relates force and potential. The slope of the function $A|x|$ is not continuous at $x = 0$ so we must consider the regions $x > 0$ and $x < 0$ separately.

EXECUTE: For $x > 0$, $|x| = x$ so $U = Ax$ and $F = -\dfrac{d(Ax)}{dx} = -A$. For $x < 0$, $|x| = -x$ so $U = -Ax$ and

$F = -\dfrac{d(-Ax)}{dx} = +A$. We can write this result as $F = -A|x|/x$, valid for all x except for $x = 0$.

(b) **IDENTIFY** and **SET UP:** Use the uncertainty principle, expressed as $\Delta p\Delta x \approx h$, and as in Problem 39.50 estimate Δp by p and Δx by x. Use this to write the energy E of the particle as a function of x. Find the value of x that gives the minimum E and then find the minimum E.

EXECUTE: $E = K + U = \dfrac{p^2}{2m} + A|x|$

$px \approx h$, so $p \approx h/x$

Then $E \approx \dfrac{h^2}{2mx^2} + A|x|$.

For $x > 0, E = \dfrac{h^2}{2mx^2} + Ax$.

To find the value of x that gives minimum E set $\dfrac{dE}{dx} = 0$.

$0 = \dfrac{-2h^2}{2mx^3} + A$

$x^3 = \dfrac{h^2}{mA}$ and $x = \left(\dfrac{h^2}{mA}\right)^{1/3}$

With this x the minimum E is

$$E = \frac{h^2}{2m}\left(\frac{mA}{h^2}\right)^{2/3} + A\left(\frac{h^2}{mA}\right)^{1/3} = \frac{1}{2}h^{2/3}m^{-1/3}A^{2/3} + h^{2/3}m^{-1/3}A^{2/3}$$

$$E = \frac{3}{2}\left(\frac{h^2A^2}{m}\right)^{1/3}$$

EVALUATE: The potential well is shaped like a V. The larger A is the steeper the slope of U and the smaller the region to which the particle is confined and the greater is its energy. Note that for the x that minimizes E, $2K = U$.

39.65.   **(a) IDENTIFY** and **SET UP:**   Let the y-direction be from the thrower to the catcher, and let the x-direction be horizontal and perpendicular to the y-direction. A cube with volume $V = 125 \text{ cm}^3 = 0.125 \times 10^{-3} \text{ m}^3$ has side length $l = V^{1/3} = (0.125 \times 10^{-3} \text{ m}^3)^{1/3} = 0.050 \text{ m}$. Thus estimate Δx as $\Delta x \approx 0.050 \text{ m}$. Use the uncertainty principle to estimate Δp_x.

EXECUTE: $\Delta x \Delta p_x \geq h/2\pi$ then gives $\Delta p_x \approx \dfrac{h}{2\pi\Delta x} = \dfrac{0.0663 \text{ J}\cdot\text{s}}{2\pi(0.050 \text{ m})} = 0.21 \text{ kg}\cdot\text{m/s}$

(The value of h in this other universe has been used.)

(b) IDENTIFY and **SET UP:** $\Delta x = (\Delta v_x)t$ is the uncertainty in the x-coordinate of the ball when it reaches the catcher, where t is the time it takes the ball to reach the second student. Obtain Δv_x from Δp_x.

EXECUTE: The uncertainty in the ball's horizontal velocity is $\Delta v_x = \dfrac{\Delta p_x}{m} = \dfrac{0.21 \text{ kg}\cdot\text{m/s}}{0.25 \text{ kg}} = 0.84 \text{ m/s}$

The time it takes the ball to travel to the second student is $t = \dfrac{12 \text{ m}}{6.0 \text{ m/s}} = 2.0 \text{ s}$. The uncertainty in the x-coordinate of the ball when it reaches the second student that is introduced by Δv_x is $\Delta x = (\Delta v_x)t = (0.84 \text{ m/s})(2.0 \text{ s}) = 1.7 \text{ m}$. The ball could miss the second student by about 1.7 m.

EVALUATE: A game of catch would be very different in this universe. We don't notice the effects of the uncertainty principle in everyday life because h is so small.

39.67.   **(a) IDENTIFY** and **SET UP:**   The probability is $P = |\psi|^2 dV$ with $dV = 4\pi r^2 dr$

EXECUTE: $|\psi|^2 = A^2 e^{-2\alpha r^2}$ so $P = 4\pi A^2 r^2 e^{-2\alpha r^2}\, dr$

(b) IDENTIFY and **SET UP:** P is maximum where $\dfrac{dP}{dr} = 0$

EXECUTE: $\dfrac{d}{dr}(r^2 e^{-2\alpha r^2}) = 0$

$2re^{-2\alpha r^2} - 4\alpha r^3 e^{-2\alpha r^2} = 0$ and this reduces to $2r - 4\alpha r^3 = 0$

$r = 0$ is a solution of the equation but corresponds to a minimum not a maximum. Seek r not equal to 0 so divide by r and get $2 - 4\alpha r^2 = 0$

This gives $r = \dfrac{1}{\sqrt{2\alpha}}$ (We took the positive square root since r must be positive.)

EVALUATE: This is different from the value of r, $r = 0$, where $|\psi|^2$ is a maximum. At $r = 0$, $|\psi|^2$ has a maximum but the volume element $dV = 4\pi r^2 dr$ is zero here so P does not have a maximum at $r = 0$.

39.69.   **(a)** $\psi(x) = \displaystyle\int_0^\infty B(k)\cos kx\, dk = \int_0^{k_0}\left(\frac{1}{k_0}\right)\cos kx\, dk = \frac{\sin kx}{k_0 x}\bigg|_0^{k_0} = \frac{\sin k_0 x}{k_0 x}$

(b) $\psi(x)$ has a maximum value at the origin $x = 0$. $\psi(x_0) = 0$ when $k_0 x_0 = \pi$ so $x_0 = \dfrac{\pi}{k_0}$. Thus the width of this function $w_x = 2x_0 = \dfrac{2\pi}{k_0}$. If $k_0 = \dfrac{2\pi}{L}, w_x = L$. $B(k)$ versus k is graphed in Figure 39.69a. The graph of $\psi(x)$ versus x is in Figure 39.69b.

(c) If $k_0 = \dfrac{\pi}{L}, w_x = 2L$.

(d) $w_p w_x = \left(\dfrac{hw_k}{2\pi}\right)\left(\dfrac{2\pi}{k_0}\right) = \dfrac{hw_k}{k_0} = \dfrac{hk_0}{k_0} = h$. The uncertainty principle states that $w_p w_x \geq \dfrac{h}{2\pi}$. For us, no matter what

k_0 is, $w_p w_x = h$, which is greater than $\dfrac{h}{2\pi}$.

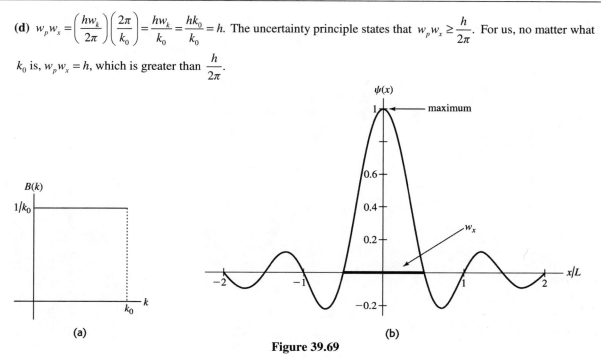

(a)

(b)

Figure 39.69

QUANTUM MECHANICS

40

40.1. **IDENTIFY** and **SET UP:** The energy levels for a particle in a box are given by $E_n = \dfrac{n^2 h^2}{8mL^2}$.

EXECUTE: (a) The lowest level is for $n = 1$, and $E_1 = \dfrac{(1)(6.626 \times 10^{-34}\text{ J·s})^2}{8(0.20\text{ kg})(1.5\text{ m})^2} = 1.2 \times 10^{-67}$ J.

(b) $E = \dfrac{1}{2}mv^2$ so $v = \sqrt{\dfrac{2E}{m}} = \sqrt{\dfrac{2(1.2 \times 10^{-67}\text{ J})}{0.20\text{ kg}}} = 1.1 \times 10^{-33}$ m/s. If the ball has this speed the time it would take it

to travel from one side of the table to the other is $t = \dfrac{1.5\text{ m}}{1.1 \times 10^{-33}\text{ m/s}} = 1.4 \times 10^{33}$ s.

(c) $E_1 = \dfrac{h^2}{8mL^2}$, $E_2 = 4E_1$, so $\Delta E = E_2 - E_1 = 3E_1 = 3(1.2 \times 10^{-67}\text{ J}) = 3.6 \times 10^{-67}$ J

(d) **EVALUATE:** No, quantum mechanical effects are not important for the game of billiards. The discrete, quantized nature of the energy levels is completely unobservable.

40.5. **IDENTIFY** and **SET UP:** Eq.(40.9) gives the energy levels. Use this to obtain an expression for $E_2 - E_1$ and use the value given for this energy difference to solve for L.

EXECUTE: Ground state energy is $E_1 = \dfrac{h^2}{8mL^2}$; first excited state energy is $E_2 = \dfrac{4h^2}{8mL^2}$. The energy separation

between these two levels is $\Delta E = E_2 - E_1 = \dfrac{3h^2}{8mL^2}$. This gives $L = h\sqrt{\dfrac{3}{8m\Delta E}} =$

$L = 6.626 \times 10^{-34}\text{ J·s}\sqrt{\dfrac{3}{8(9.109 \times 10^{-31}\text{ kg})(3.0\text{ eV})(1.602 \times 10^{-19}\text{ J/1 eV})}} = 6.1 \times 10^{-10}$ m $= 0.61$ nm.

EVALUATE: This energy difference is typical for an atom and L is comparable to the size of an atom.

40.7. **IDENTIFY** and **SET UP:** For the $n = 2$ first excited state the normalized wave function is given by Eq.(40.13).

$\psi_2(x) = \sqrt{\dfrac{2}{L}}\sin\left(\dfrac{2\pi x}{L}\right)$. $|\psi_2(x)|^2 dx = \dfrac{2}{L}\sin^2\left(\dfrac{2\pi x}{L}\right)dx$. Examine $|\psi_2(x)|^2\,dx$ and find where it is zero and where it is

maximum.

EXECUTE: (a) $|\psi_2|^2\,dx = 0$ implies $\sin\left(\dfrac{2\pi x}{L}\right) = 0$

$\dfrac{2\pi x}{L} = m\pi$, $m = 0, 1, 2, \ldots$; $x = m(L/2)$

For $m = 0$, $x = 0$; for $m = 1$, $x = L/2$; for $m = 2$, $x = L$

The probability of finding the particle is zero at $x = 0$, $L/2$, and L.

(b) $|\psi_2|^2\,dx$ is maximum when $\sin\left(\dfrac{2\pi x}{L}\right) = \pm 1$

$\dfrac{2\pi x}{L} = m(\pi/2)$, $m = 1, 3, 5, \ldots$; $x = m(L/4)$

For $m = 1$, $x = L/4$; for $m = 3$, $x = 3L/4$

The probability of finding the particle is largest at $x = L/4$ and $3L/4$.

(c) **EVALUATE:** The answers to part (a) correspond to the zeros of $|\psi|^2$ shown in Fig.40.5 in the textbook and the

answers to part (b) correspond to the two values of x where $|\psi|^2$ in the figure is maximum.

40.9.   **(a) IDENTIFY** and **SET UP:** $\psi = A\cos kx$. Calculate $d\psi^2/dx^2$ and substitute into Eq.(40.3) to see if this equation is satisfied.

EXECUTE: Eq.(40.3): $-\dfrac{h^2}{8\pi^2 m}\dfrac{d^2\psi}{dx^2} = E\psi$

$\dfrac{d\psi}{dx} = A(-k\sin kx) = -Ak\sin kx$

$\dfrac{d^2\psi}{dx^2} = -Ak(k\cos kx) = -Ak^2\cos kx$

Thus Eq.(40.3) requires $-\dfrac{h^2}{8\pi^2 m}(-Ak^2\cos kx) = E(A\cos kx)$.

This says $-\dfrac{h^2 k^2}{8\pi^2 m} = E; \ \ k = \dfrac{\sqrt{2mE}}{(h/2\pi)} = \dfrac{\sqrt{2mE}}{\hbar}$

$\psi = A\cos kx$ is a solution to Eq.(40.3) if $k = \dfrac{\sqrt{2mE}}{\hbar}$.

(b) EVALUATE: The wave function for a particle in a box with rigid walls at $x=0$ and $x=L$ must satisfy the boundary conditions $\psi = 0$ at $x=0$ and $\psi = 0$ at $x=L$. $\psi(0) = A\cos 0 = A$, since $\cos 0 = 1$. Thus ψ is not 0 at $x=0$ and this wave function isn't acceptable because it doesn't satisfy the required boundary condition, even though it is a solution to the Schrödinger equation.

40.11.   Recall $\lambda = \dfrac{h}{p} = \dfrac{h}{\sqrt{2mE}}$.

(a) $E_1 = \dfrac{h^2}{8mL^2} \Rightarrow \lambda_1 = \dfrac{h}{\sqrt{2mh^2/8mL^2}} = 2L = 2(3.0\times10^{-10} \text{ m}) = 6.0\times10^{-10}$ m. The wavelength is twice the width of the box. $p_1 = \dfrac{h}{\lambda_1} = \dfrac{(6.63\times10^{-34} \text{ J}\cdot\text{s})}{6.0\times10^{-10} \text{ m}} = 1.1\times10^{-24}$ kg·m/s

(b) $E_2 = \dfrac{4h^2}{8mL^2} \Rightarrow \lambda_2 = L = 3.0\times10^{-10}$ m. The wavelength is the same as the width of the box.

$p_2 = \dfrac{h}{\lambda_2} = 2p_1 = 2.2\times10^{-24}$ kg·m/s.

(c) $E_3 = \dfrac{9h^2}{8mL^2} \Rightarrow \lambda_3 = \dfrac{2}{3}L = 2.0\times10^{-10}$ m. The wavelength is two-thirds the width of the box.

$p_3 = 3p_1 = 3.3\times10^{-24}$ kg·m/s.

40.15. $E_1 = 0.625E_\infty = 0.625\dfrac{\pi^2\hbar^2}{2mL^2}; \ \ E_1 = 2.00 \text{ eV} = 3.20\times10^{-19}$ J

$L = \pi\hbar\left(\dfrac{0.625}{2(9.109\times10^{-31} \text{ kg})(3.20\times10^{-19} \text{ J})}\right)^{1/2} = 3.43\times10^{-10}$ m

40.19.   **IDENTIFY:**   Find the transition energy ΔE and set it equal to the energy of the absorbed photon. Use $E = hc/\lambda$ to find the wavelength of the photon.

SET UP: $U_0 = 6E_\infty$, as in Fig.40.8 in the textbook, so $E_1 = 0.625E_\infty$ and $E_3 = 5.09E_\infty$ with $E_\infty = \dfrac{\pi^2\hbar^2}{2mL^2}$. In this problem the particle bound in the well is a proton, so $m = 1.673\times10^{-27}$ kg.

EXECUTE: $E_\infty = \dfrac{\pi^2\hbar^2}{2mL^2} = \dfrac{\pi^2(1.055\times10^{-34} \text{ J}\cdot\text{s})^2}{2(1.673\times10^{-27} \text{ kg})(4.0\times10^{-15} \text{ m})^2} = 2.052\times10^{-12}$ J. The transition energy is

$\Delta E = E_3 - E_1 = (5.09 - 0.625)E_\infty = 4.465E_\infty$. $\Delta E = 4.465(2.052\times10^{-12} \text{ J}) = 9.162\times10^{-12}$ J

The wavelength of the photon that is absorbed is related to the transition energy by $\Delta E = hc/\lambda$, so

$\lambda = \dfrac{hc}{\Delta E} = \dfrac{(6.626\times10^{-34} \text{ J}\cdot\text{s})(2.998\times10^8 \text{ m/s})}{9.162\times10^{-12} \text{ J}} = 2.2\times10^{-14}$ m = 22 fm.

EVALUATE: The wavelength of the photon is comparable to the size of the box.

40.23.   **IDENTIFY** and **SET UP:**   Use Eq.(39.1), where $K = p^2/2m$ and $E = K + U$.

   EXECUTE: $\lambda = h/p = h/\sqrt{2mK}$, so $\lambda\sqrt{K}$ is constant

   $\lambda_1\sqrt{K_1} = \lambda_2\sqrt{K_2}$; λ_1 and K_1 are for $x > L$ where $K_1 = 2U_0$ and λ_2 and K_2 are for $0 < x < L$ where

   $K_2 = E - U_0 = U_0$

   $\dfrac{\lambda_1}{\lambda_2} = \sqrt{\dfrac{K_2}{K_1}} = \sqrt{\dfrac{U_0}{2U_0}} = \dfrac{1}{\sqrt{2}}$

   EVALUATE: When the particle is passing over the barrier its kinetic energy is less and its wavelength is larger.

40.25.   **IDENTIFY** and **SET UP:**   The probability is $T = Ae^{-2\kappa L}$, with $A = 16\dfrac{E}{U_0}\left(1 - \dfrac{E}{U_0}\right)$ and $\kappa = \dfrac{\sqrt{2m(U_0 - E)}}{\hbar}$.

   $E = 32$ eV, $U_0 = 41$ eV, $L = 0.25 \times 10^{-9}$ m. Calculate T.

   EXECUTE: **(a)** $A = 16\dfrac{E}{U_0}\left(1 - \dfrac{E}{U_0}\right) = 16\dfrac{32}{41}\left(1 - \dfrac{32}{41}\right) = 2.741$.

   $\kappa = \dfrac{\sqrt{2m(U_0 - E)}}{\hbar}$

   $\kappa = \dfrac{\sqrt{2(9.109 \times 10^{-31}\ \text{kg})(41\ \text{eV} - 32\ \text{eV})(1.602 \times 10^{-19}\ \text{J/eV})}}{1.055 \times 10^{-34}\ \text{J} \cdot \text{s}} = 1.536 \times 10^{10}\ \text{m}^{-1}$

   $T = Ae^{-2\kappa L} = (2.741)e^{-2(1.536 \times 10^{10}\ \text{m}^{-1})(0.25 \times 10^{-9}\ \text{m})} = 2.741e^{-7.68} = 0.0013$

   (b) The only change in the mass m, which appears in κ.

   $\kappa = \dfrac{\sqrt{2m(U_0 - E)}}{\hbar}$

   $\kappa = \dfrac{\sqrt{2(1.673 \times 10^{-27}\ \text{kg})(41\ \text{eV} - 32\ \text{eV})(1.602 \times 10^{-19}\ \text{J/eV})}}{1.055 \times 10^{-34}\ \text{J} \cdot \text{s}} = 6.584 \times 10^{11}\ \text{m}^{-1}$

   Then $T = Ae^{-2\kappa L} = (2.741)e^{-2(6.584 \times 10^{11}\ \text{m}^{-1})(0.25 \times 10^{-9}\ \text{m})} = 2.741e^{-392.2} = 10^{-143}$

   EVALUATE: The more massive proton has a much smaller probability of tunneling than the electron does.

40.27.   **IDENTIFY** and **SET UP:**   The energy levels are given by Eq.(40.26), where $\omega = \sqrt{\dfrac{k'}{m}}$.

   EXECUTE: $\omega = \sqrt{\dfrac{k'}{m}} = \sqrt{\dfrac{110\ \text{N/m}}{0.250\ \text{kg}}} = 21.0$ rad/s

   The ground state energy is given by Eq.(40.26):

   $E_0 = \dfrac{1}{2}\hbar\omega = \dfrac{1}{2}(1.055 \times 10^{-34}\ \text{J} \cdot \text{s})(21.0\ \text{rad/s}) = 1.11 \times 10^{-33}\ \text{J}(1\ \text{eV}/1.602 \times 10^{-19}\ \text{J}) = 6.93 \times 10^{-15}$ eV

   $E_n = \left(n + \dfrac{1}{2}\right)\hbar\omega,\ E_{(n+1)} = \left(n + 1 + \dfrac{1}{2}\right)\hbar\omega$

   The energy separation between these adjacent levels is

   $\Delta E = E_{n+1} - E_n = \hbar\omega = 2E_0 = 2(1.11 \times 10^{-33}\ \text{J}) = 2.22 \times 10^{-33}\ \text{J} = 1.39 \times 10^{-14}$ eV

   EVALUATE: These energies are extremely small; quantum effects are not important for this oscillator.

40.31.   **IDENTIFY** and **SET UP:**   Use the energies given in Eq.(40.26) to solve for the amplitude A and maximum speed v_{max} of the oscillator. Use these to estimate Δx and Δp_x and compute the uncertainty product $\Delta x \Delta p_x$.

   EXECUTE: The total energy of a Newtonian oscillator is given by $E = \frac{1}{2}k'A^2$ where k' is the force constant and A is the amplitude of the oscillator. Set this equal to the energy $E = (n + \frac{1}{2})\hbar\omega$ of an excited level that has quantum

   number n, where $\omega = \sqrt{\dfrac{k'}{m}}$, and solve for A: $\frac{1}{2}k'A^2 = (n + \frac{1}{2})\hbar\omega$

   $A = \sqrt{\dfrac{(2n+1)\hbar\omega}{k'}}$

   The total energy of the Newtonian oscillator can also be written as $E = \frac{1}{2}mv_{\text{max}}^2$. Set this equal to $E = (n + \frac{1}{2})\hbar\omega$ and solve for v_{max}: $\frac{1}{2}mv_{\text{max}}^2 = (n + \frac{1}{2})\hbar\omega$

   $v_{\text{max}} = \sqrt{\dfrac{(2n+1)\hbar\omega}{m}}$

Thus the maximum linear momentum of the oscillator is $p_{max} = mv_{max} = \sqrt{(2n+1)\hbar m\omega}$. Assume that A represents the uncertainty Δx in position and that p_{max} is the corresponding uncertainty Δp_x in momentum. Then the uncertainty product is $\Delta x \Delta p_x = \sqrt{\dfrac{(2n+1)\hbar\omega}{k'}}\sqrt{(2n+1)\hbar m\omega} = (2n+1)\hbar\omega\sqrt{\dfrac{m}{k'}} = (2n+1)\hbar\omega\left(\dfrac{1}{\omega}\right) = (2n+1)\hbar$.

EVALUATE: For $n=1$ this gives $\Delta x \Delta p_x = 3\hbar$, in agreement with the result derived in Section 40.4. The uncertainty product $\Delta x \Delta p_x$ increases with n.

40.33. **IDENTIFY:** We model the atomic vibration in the crystal as a harmonic oscillator.

SET UP: The energy levels of a harmonic oscillator are given by $E_n = \left(n+\dfrac{1}{2}\right)\hbar\sqrt{\dfrac{k'}{m}} = \left(n+\dfrac{1}{2}\right)\hbar\omega$.

EXECUTE: (a) The ground state energy of a simple harmonic oscillator is

$$E_0 = \frac{1}{2}\hbar\omega = \frac{1}{2}\hbar\sqrt{\frac{k'}{m}} = \frac{(1.055\times10^{-34}\text{ J}\cdot\text{s})}{2}\sqrt{\frac{12.2\text{ N/m}}{3.82\times10^{-26}\text{ kg}}} = 9.43\times10^{-22}\text{ J} = 5.89\times10^{-3}\text{ eV}$$

(b) $E_4 - E_3 = \hbar\omega = 2E_0 = 0.0118$ eV, so $\lambda = \dfrac{hc}{E} = \dfrac{(6.63\times10^{-34}\text{ J}\cdot\text{s})(3.00\times10^8\text{ m/s})}{1.88\times10^{-21}\text{ J}} = 106\ \mu m$

(c) $E_{n+1} - E_n = \hbar\omega = 2E_0 = 0.0118$ eV

EVALUATE: These energy differences are much smaller than those due to electron transitions in the hydrogen atom.

40.35. **IDENTIFY:** Let I refer to the region $x<0$ and let II refer to the region $x>0$, so $\psi_I(x) = Ae^{ik_1 x} + Be^{-ik_1 x}$ and $\psi_{II}(x) = Ce^{ik_2 x}$. Set $\psi_I(0) = \psi_{II}(0)$ and $\dfrac{d\psi_I}{dx} = \dfrac{d\psi_{II}}{dx}$ at $x=0$.

SET UP: $\dfrac{d}{dx}(e^{ikx}) = ike^{ikx}$.

EXECUTE: $\psi_I(0) = \psi_{II}(0)$ gives $A + B = C$. $\dfrac{d\psi_I}{dx} = \dfrac{d\psi_{II}}{dx}$ at $x=0$ gives $ik_1 A - ik_1 B = ik_2 C$. Solving this pair of equations for B and C gives $B = \left(\dfrac{k_1 - k_2}{k_1 + k_2}\right)A$ and $C = \left(\dfrac{2k_2}{k_1 + k_2}\right)A$.

EVALUATE: The probability of reflection is $R = \dfrac{B^2}{A^2} = \dfrac{(k_1 - k_2)^2}{(k_1 + k_2)^2}$. The probability of transmission is

$T = \dfrac{C^2}{A^2} = \dfrac{4k_1^2}{(k_1 + k_2)^2}$. Note that $R + T = 1$.

40.37. **IDENTIFY and SET UP:** The energy levels are given by Eq.(40.9): $E_n = \dfrac{n^2 h^2}{8mL^2}$. Calculate ΔE for the transition and set $\Delta E = hc/\lambda$, the energy of the photon.

EXECUTE: (a) Ground level, $n=1$, $E_1 = \dfrac{h^2}{8mL^2}$

First excited level, $n=2$, $E_2 = \dfrac{4h^2}{8mL^2}$

The transition energy is $\Delta E = E_2 - E_1 = \dfrac{3h^2}{8mL^2}$. Set the transition energy equal to the energy hc/λ of the emitted photon. This gives $\dfrac{hc}{\lambda} = \dfrac{3h^2}{8mL^2}$.

$\lambda = \dfrac{8mcL^2}{3h} = \dfrac{8(9.109\times10^{-31}\text{ kg})(2.998\times10^8\text{ m/s})(4.18\times10^{-9}\text{ m})^2}{3(6.626\times10^{-34}\text{ J}\cdot\text{s})}$

$\lambda = 1.92\times10^{-5}$ m $= 19.2\ \mu m$.

(b) Second excited level has $n=3$ and $E_3 = \dfrac{9h^2}{8mL^2}$. The transition energy is $\Delta E = E_3 - E_2 = \dfrac{9h^2}{8mL^2} - \dfrac{4h^2}{8mL^2} = \dfrac{5h^2}{8mL^2}$.

$\dfrac{hc}{\lambda} = \dfrac{5h^2}{8mL^2}$ so $\lambda = \dfrac{8mcL^2}{5h} = \dfrac{3}{5}(19.2\ \mu m) = 11.5\ \mu m$.

EVALUATE: The energy spacing between adjacent levels increases with n, and this corresponds to a shorter wavelength and more energetic photon in part (b) than in part (a).

40.39.   **IDENTIFY:**   The probability of the particle being between x_1 and x_2 is $\int_{x_1}^{x_2}|\psi|^2 dx$, where ψ is the normalized wave function for the particle.

(a) SET UP: The normalized wave function for the ground state is $\psi_1 = \sqrt{\dfrac{2}{L}}\sin\left(\dfrac{\pi x}{L}\right)$.

EXECUTE: The probability P of the particle being between $x = L/4$ and $x = 3L/4$ is

$P = \int_{L/4}^{3L/4}|\psi_1|^2 dx = \dfrac{2}{L}\int_{L/4}^{3L/4}\sin^2\left(\dfrac{\pi x}{L}\right)dx$. Let $y = \pi x/L$; $dx = (L/\pi)\,dy$ and the integration limits become $\pi/4$ and $3\pi/4$.

$P = \dfrac{2}{L}\left(\dfrac{L}{\pi}\right)\int_{\pi/4}^{3\pi/4}\sin^2 y\,dy = \dfrac{2}{\pi}\left[\dfrac{1}{2}y - \dfrac{1}{4}\sin 2y\right]_{\pi/4}^{3\pi/4}$

$P = \dfrac{2}{\pi}\left[\dfrac{3\pi}{8} - \dfrac{\pi}{8} - \dfrac{1}{4}\sin\left(\dfrac{3\pi}{2}\right) + \dfrac{1}{4}\sin\left(\dfrac{\pi}{2}\right)\right]$

$P = \dfrac{2}{\pi}\left(\dfrac{\pi}{4} - \dfrac{1}{4}(-1) + \dfrac{1}{4}(1)\right) = \dfrac{1}{2} + \dfrac{1}{\pi} = 0.818$. (Note: The integral formula $\int\sin^2 y\,dy = \dfrac{1}{2}y - \dfrac{1}{4}\sin 2y$ was used.)

(b) SET UP: The normalized wave function for the first excited state is $\psi_2 = \sqrt{\dfrac{2}{L}}\sin\left(\dfrac{2\pi x}{L}\right)$

EXECUTE: $P = \int_{L/4}^{3L/4}|\psi_2|^2 dx = \dfrac{2}{L}\int_{L/4}^{3L/4}\sin^2\left(\dfrac{2\pi x}{L}\right)dx$. Let $y = 2\pi x/L$; $dx = (L/2\pi)\,dy$ and the integration limits become $\pi/2$ and $3\pi/2$.

$P = \dfrac{2}{L}\left(\dfrac{L}{2\pi}\right)\int_{\pi/2}^{3\pi/2}\sin^2 y\,dy = \dfrac{1}{\pi}\left[\dfrac{1}{2}y - \dfrac{1}{4}\sin 2y\right]_{\pi/2}^{3\pi/2} = \dfrac{1}{\pi}\left(\dfrac{3\pi}{4} - \dfrac{\pi}{4}\right) = 0.500$

(c) EVALUATE: These results are consistent with Fig.40.4b in the textbook. That figure shows that $|\psi|^2$ is more concentrated near the center of the box for the ground state than for the first excited state; this is consistent with the answer to part (a) being larger than the answer to part (b). Also, this figure shows that for the first excited state half the area under $|\psi|^2$ curve lies between $L/4$ and $3L/4$, consistent with our answer to part (b).

40.41.   **IDENTIFY and SET UP:**   The normalized wave function for the $n = 2$ first excited level is $\psi_2 = \sqrt{\dfrac{2}{L}}\sin\left(\dfrac{2\pi x}{L}\right)$.

$P = |\psi(x)|^2 dx$ is the probability that the particle will be found in the interval x to $x + dx$.

EXECUTE: **(a)** $x = L/4$

$\psi(x) = \sqrt{\dfrac{2}{L}}\sin\left(\left(\dfrac{2\pi}{L}\right)\left(\dfrac{L}{4}\right)\right) = \sqrt{\dfrac{2}{L}}\sin\left(\dfrac{\pi}{2}\right) = \sqrt{\dfrac{2}{L}}$.

$P = (2/L)dx$

(b) $x = L/2$

$\psi(x) = \sqrt{\dfrac{2}{L}}\sin\left(\left(\dfrac{2\pi}{L}\right)\left(\dfrac{L}{2}\right)\right) = \sqrt{\dfrac{2}{L}}\sin(\pi) = 0$

$P = 0$

(c) $x = 3L/4$

$\psi(x) = \sqrt{\dfrac{2}{L}}\sin\left(\left(\dfrac{2\pi}{L}\right)\left(\dfrac{3L}{4}\right)\right) = \sqrt{\dfrac{2}{L}}\sin\left(\dfrac{3\pi}{2}\right) = -\sqrt{\dfrac{2}{L}}$.

$P = (2/L)dx$

EVALUATE: Our results are consistent with the $n = 2$ part of Fig.40.5 in the textbook. $|\psi|^2$ is zero at the center of the box and is symmetric about this point.

40.47.   **IDENTIFY and SET UP:**   When κL is large, then $e^{\kappa L}$ is large and $e^{-\kappa L}$ is small. When κL is small, $\sinh \kappa L \to \kappa L$. Consider both κL large and κL small limits.

EXECUTE: **(a)** $T = \left[1 + \dfrac{(U_0 \sinh \kappa L)^2}{4E(U_0 - E)}\right]^{-1}$

$\sinh \kappa L = \dfrac{e^{\kappa L} - e^{-\kappa L}}{2}$

For $\kappa L \gg 1$, $\sinh \kappa L \to \dfrac{e^{\kappa L}}{2}$ and $T \to \left[1 + \dfrac{U_0^2 e^{2\kappa L}}{16E(U_0 - E)}\right]^{-1} = \dfrac{16E(U_0 - E)}{16E(U_0 - E) + U_0^2 e^{2\kappa L}}$

For $\kappa L \gg 1$, $16E(U_0 - E) + U_0^2 e^{2\kappa L} \to U_0^2 e^{2\kappa L}$

$T \to \dfrac{16E(U_0 - E)}{U_0^2 e^{2\kappa L}} = 16\left(\dfrac{E}{U_0}\right)\left(1 - \dfrac{E}{U_0}\right)e^{-2\kappa L}$, which is Eq.(40.21).

(b) $\kappa L = \dfrac{L\sqrt{2m(U_0 - E)}}{\hbar}$. So $\kappa L \gg 1$ when L is large (barrier is wide) or $U_0 - E$ is large. (E is small compared to U_0.)

(c) $\kappa = \dfrac{\sqrt{2m(U_0 - E)}}{\hbar}$; κ becomes small as E approaches U_0. For κ small, $\sinh \kappa L \to \kappa L$ and

$T \to \left[1 + \dfrac{U_0^2 \kappa^2 L^2}{4E(U_0 - E)}\right]^{-1} = \left[1 + \dfrac{U_0^2 2m(U_0 - E)L^2}{\hbar^2 \, 4E(U_0 - E)}\right]^{-1}$ (using the definition of κ)

Thus $T \to \left[1 + \dfrac{2U_0^2 L^2 m}{4E\hbar^2}\right]^{-1}$

$U_0 \to E$ so $\dfrac{U_0^2}{E} \to E$ and $T \to \left[1 + \dfrac{2EL^2 m}{4\hbar^2}\right]^{-1}$

But $k^2 = \dfrac{2mE}{\hbar^2}$, so $T \to \left[1 + \left(\dfrac{kL}{2}\right)^2\right]^{-1}$, as was to be shown.

EVALUATE: When κL is large Eq.(40.20) applies and T is small. When $E \to U_0$, T does not approach unity.

40.49. **IDENTIFY** and **SET UP:** Calculate the angular frequency ω of the pendulum and apply Eq.(40.26) for the energy levels.

EXECUTE: $\omega = \dfrac{2\pi}{T} = \dfrac{2\pi}{0.500 \text{ s}} = 4\pi \text{ s}^{-1}$

The ground-state energy is $E_0 = \dfrac{1}{2}\hbar\omega = \dfrac{1}{2}(1.055 \times 10^{-34} \text{ J} \cdot \text{s})(4\pi \text{ s}^{-1}) = 6.63 \times 10^{-34}$ J.

$E_0 = 6.63 \times 10^{-34}$ J$(1 \text{ eV}/1.602 \times 10^{-19} \text{ J}) = 4.14 \times 10^{-15}$ eV

$E_n = \left(n + \dfrac{1}{2}\right)\hbar\omega$

$E_{n+1} = \left(n + 1 + \dfrac{1}{2}\right)\hbar\omega$

The energy difference between the adjacent energy levels is
$\Delta E = E_{n+1} - E_n = \hbar\omega = 2E_0 = 1.33 \times 10^{-33}$ J $= 8.30 \times 10^{-15}$ eV

EVALUATE: These energies are much too small to detect. Quantum effects are not important for ordinary size objects.

40.51. **IDENTIFY:** If the given wave function is a solution to the Schrödinger equation, we will get an identity when we substitute that wave function into the Schrödinger equation.

SET UP: The given wave function is $\psi_0(x) = A_0 e^{-\alpha^2 x^2/2}$ and the Schrödinger equation is

$-\dfrac{\hbar}{2m}\dfrac{d^2\psi(x)}{dx^2} + \dfrac{k'x^2}{2}\psi(x) = E \, \psi(x)$.

EXECUTE: **(a)** Start by taking the derivatives: $\psi_0(x) = A_0 e^{-\alpha^2 x^2/2}$. $\dfrac{d\psi_0(x)}{dx} = -\alpha^2 x A_0 e^{-\alpha^2 x^2/2}$.

$\dfrac{d^2\psi_0(x)}{dx^2} = -A_0 \alpha^2 e^{-\alpha^2 x^2/2} + (\alpha^2)^2 x^2 A_0 e^{-\alpha^2 x^2/2}$. $\dfrac{d^2\psi_0(x)}{dx^2} = [-\alpha^2 + (\alpha^2)^2 x^2]\,\psi_0(x)$.

$-\dfrac{\hbar}{2m}\dfrac{d^2\psi_0(x)}{dx^2} = -\dfrac{\hbar^2}{2m}[-\alpha^2 + (\alpha^2)^2 x^2]\,\psi_0(x)$. Equation (40.22) is $-\dfrac{\hbar}{2m}\dfrac{d^2\psi(x)}{dx^2} + \dfrac{k'x^2}{2}\psi(x) = E \, \psi(x)$. Substituting

the above result into that equation gives $-\dfrac{\hbar^2}{2m}[-\alpha^2 + (\alpha^2)^2 x^2]\,\psi_0(x) + \dfrac{k'x^2}{2}\psi_0(x) = E \, \psi_0(x)$. Since $\alpha^2 = \dfrac{m\omega}{\hbar}$ and

$\omega = \sqrt{\dfrac{k'}{m}}$, the coefficient of x^2 is $-\dfrac{\hbar^2}{2m}(\alpha^2)^2 + \dfrac{k'}{2} = -\dfrac{\hbar^2}{2m}\left(\dfrac{m\omega}{\hbar}\right)^2 + \dfrac{m\omega^2}{2} = 0$.

(b) $A_0 = \left(\dfrac{m\omega}{\hbar\pi}\right)^{1/4}$

(c) The classical turning points are at $A = \pm\sqrt{\dfrac{\hbar}{\omega m}}$. The probability density function $|\psi|^2$ is

$$|\psi_0(x)|^2 = A_0^2 e^{-\alpha^2 x^2} = \left(\frac{m\omega}{\hbar\pi}\right)^{1/2} e^{-m\omega x^2/\hbar}.\text{ At } x = 0,\ |\psi_0|^2 = \left(\frac{m\omega}{\hbar\pi}\right)^{1/2}.$$

$$\frac{d|\psi_0(x)|^2}{dx} = \left(\frac{m\omega}{\hbar\pi}\right)^{1/2}(-\alpha^2 2x)e^{-\alpha^2 x^2} = -2\frac{m\omega}{\hbar}\left(\frac{m\omega}{\hbar\pi}\right)^{1/2} xe^{-\alpha^2 x^2}.\text{ At } x = 0,\ \frac{d|\psi_0(x)|^2}{dx} = 0.$$

$$\frac{d^2|\psi_0(x)|^2}{dx^2} = -2\frac{m\omega}{\hbar}\left(\frac{m\omega}{\hbar\pi}\right)^{1/2}[1 - 2\alpha^2 x^2]e^{-\alpha^2 x^2}.\text{ At } x = 0,\ \frac{d^2|\psi_0(x)|^2}{dx^2} < 0. \text{ Therefore, at } x = 0,\text{ the first derivative is}$$

zero and the second derivative is negative. Therefore, the probability density function has a maximum at $x = 0$.

EVALUATE: $\psi_0(x) = A_0 e^{-\alpha^2 x^2/2}$ is a solution to equation (40.22) if $-\dfrac{\hbar^2}{2m}(-\alpha^2)\psi_0(x) = E\ \psi_0(x)$ or

$E = \dfrac{\hbar^2\alpha^2}{2m} = \dfrac{\hbar\omega}{2}$. $E_0 = \dfrac{\hbar\omega}{2}$ corresponds to $n = 0$ in Equation (40.26).

40.53. **IDENTIFY** and **SET UP:** Evaluate $\partial^2\psi/\partial x^2$, $\partial^2\psi/\partial y^2$, and $\partial^2\psi/\partial z^2$ for the proposed ψ and put Eq.(40.29). Use that ψ_{n_x}, ψ_{n_y}, and ψ_{n_z} are each solutions to Eq.(40.22).

EXECUTE: **(a)** $-\dfrac{\hbar^2}{2m}\left(\dfrac{\partial^2\psi}{\partial x^2} + \dfrac{\partial^2\psi}{\partial y^2} + \dfrac{\partial^2\psi}{\partial z^2}\right) + U\psi = E\psi$

ψ_{n_x}, ψ_{n_y}, ψ_{n_z} are each solutions of Eq.(40.22), so $-\dfrac{\hbar^2}{2m}\dfrac{d^2\psi_{n_x}}{dx^2} + \dfrac{1}{2}k'x^2\psi_{n_x} = E_{n_x}\psi_{n_x}$

$-\dfrac{\hbar^2}{2m}\dfrac{d^2\psi_{n_y}}{dy^2} + \dfrac{1}{2}k'y^2\psi_{n_y} = E_{n_y}\psi_{n_y}$

$-\dfrac{\hbar^2}{2m}\dfrac{d^2\psi_{n_z}}{dz^2} + \dfrac{1}{2}k'z^2\psi_{n_z} = E_{n_z}\psi_{n_z}$

$\psi = \psi_{n_x}(x)\psi_{n_y}(y)\psi_{n_z}(z),\ U = \dfrac{1}{2}k'x^2 + \dfrac{1}{2}k'y^2 + \dfrac{1}{2}k'z^2$

$\dfrac{\partial^2\psi}{\partial x^2} = \left(\dfrac{d^2\psi_{n_x}}{dx^2}\right)\psi_{n_y}\psi_{n_z},\ \dfrac{\partial^2\psi}{\partial y^2} = \left(\dfrac{d^2\psi_{n_y}}{dy^2}\right)\psi_{n_x}\psi_{n_z},\ \dfrac{\partial^2\psi}{\partial z^2} = \left(\dfrac{d^2\psi_{n_z}}{dz^2}\right)\psi_{n_x}\psi_{n_y}.$

So $-\dfrac{\hbar^2}{2m}\left(\dfrac{\partial^2\psi}{\partial x^2} + \dfrac{\partial^2\psi}{\partial y^2} + \dfrac{\partial^2\psi}{\partial z^2}\right) + U\psi = \left(-\dfrac{\hbar^2}{2m}\dfrac{d^2\psi_{n_x}}{dx^2} + \dfrac{1}{2}k'x^2\psi_{n_x}\right)\psi_{n_y}\psi_{n_z}$

$+ \left(-\dfrac{\hbar^2}{2m}\dfrac{d^2\psi_{n_y}}{dy^2} + \dfrac{1}{2}k'y^2\psi_{n_y}\right)\psi_{n_x}\psi_{n_z} + \left(-\dfrac{\hbar^2}{2m}\dfrac{d^2\psi_{n_z}}{dz^2} + \dfrac{1}{2}k'z^2\psi_{n_z}\right)\psi_{n_x}\psi_{n_y}$

$-\dfrac{\hbar^2}{2m}\left(\dfrac{\partial^2\psi}{\partial x^2} + \dfrac{\partial^2\psi}{\partial y^2} + \dfrac{\partial^2\psi}{\partial z^2}\right) + U\psi = (E_{n_x} + E_{n_y} + E_{n_z})\psi$

Therefore, we have shown that this ψ is a solution to Eq.(40.29), with energy

$E_{n_x n_y n_z} = E_{n_x} + E_{n_y} + E_{n_z} = \left(n_x + n_y + n_z + \dfrac{3}{2}\right)\hbar\omega$

(b) and **(c)** The ground state has $n_x = n_y = n_z = 0$, so the energy is $E_{000} = \dfrac{3}{2}\hbar\omega$. There is only one set of n_x, n_y and n_z that give this energy.

First-excited state: $n_x = 1,\ n_y = n_z = 0$ or $n_y = 1,\ n_x = n_z = 0$ or $n_z = 1,\ n_x = n_y = 0$ and $E_{100} = E_{010} = E_{001} = \dfrac{5}{2}\hbar\omega$

There are three different sets of n_x, n_y, n_z quantum numbers that give this energy, so there are three different quantum states that have this same energy.

EVALUATE: For the three-dimensional isotropic harmonic oscillator, the wave function is a product of one-dimensional harmonic oscillator wavefunctions for each dimension. The energy is a sum of energies for three one-dimensional oscillators. All the excited states are degenerate, with more than one state having the same energy.

ATOMIC STRUCTURE

41.1. **IDENTIFY** and **SET UP:** $L = \sqrt{l(l+1)}\hbar$. $L_z = m_l\hbar$. $l = 0,\ 1,\ 2,...,\ n-1$. $m_l = 0,\ \pm1,\ \pm2,...,\ \pm l$. $\cos\theta = L_z/L$.

EXECUTE: **(a)** $\underline{l=0}$: $L = 0$, $L_z = 0$. $\underline{l=1}$: $L = \sqrt{2}\hbar$, $L_z = \hbar, 0, -\hbar$. $\underline{l=2}$: $L = \sqrt{6}\hbar$, $L_z = 2\hbar, \hbar, 0, -\hbar, -2\hbar$.

(b) In each case $\cos\theta = L_z/L$. $L = 0$: θ not defined. $L = \sqrt{2}\hbar$: $45.0°$, $90.0°$, $135.0°$. $L = \sqrt{6}\hbar$: $35.3°$, $65.9°$, $90.0°$, $114.1°$, $144.7°$.

EVALUATE: There is no state where $\vec{L}$ is totally aligned along the z axis.

41.3. **IDENTIFY** and **SET UP:** The magnitude of the orbital angular momentum L is related to the quantum number l by Eq.(41.4): $L = \sqrt{l(l+1)}\hbar$, $l = 0,\ 1,\ 2,...$

EXECUTE: $l(l+1) = \left(\dfrac{L}{\hbar}\right)^2 = \left(\dfrac{4.716\times10^{-34}\ \text{kg}\cdot\text{m}^2/\text{s}}{1.055\times10^{-34}\ \text{J}\cdot\text{s}}\right)^2 = 20$

And then $l(l+1) = 20$ gives that $l = 4$.

EVALUATE: l must be integer.

41.5. **IDENTIFY** and **SET UP:** The angular momentum L is related to the quantum number l by Eq.(41.4), $L = \sqrt{l(l+1)}\hbar$. The maximum l, l_{max}, for a given n is $l_{\text{max}} = n-1$.

EXECUTE: For $n = 2$, $l_{\text{max}} = 1$ and $L = \sqrt{2}\hbar = 1.414\hbar$.

For $n = 20$, $l_{\text{max}} = 19$ and $L = \sqrt{(19)(20)}\hbar = 19.49\hbar$.

For $n = 200$, $l_{\text{max}} = 199$ and $L = \sqrt{(199)(200)}\hbar = 199.5\hbar$.

EVALUATE: As n increases, the maximum L gets closer to the value $n\hbar$ postulated in the Bohr model.

41.11. **IDENTIFY** and **SET UP:** Eq.(41.8) gives $a = \dfrac{4\pi\epsilon_0\hbar^2}{m_r e^2} = \dfrac{\epsilon_0 h^2}{\pi m_r e^2}$.

EXECUTE: **(a)** $m_r = m$

$a = \dfrac{\epsilon_0 h^2}{\pi m_r e^2} = \dfrac{(8.854\times10^{-12}\ \text{C}^2/\text{N}\cdot\text{m}^2)(6.626\times10^{-34}\ \text{J}\cdot\text{s})^2}{\pi(9.109\times10^{-31}\ \text{kg})(1.602\times10^{-19}\ \text{C})^2} = 0.5293\times10^{-10}\ \text{m}$

(b) $m_r = m/2$

$a = 2\left(\dfrac{\epsilon_0 h^2}{\pi m_r e^2}\right) = 1.059\times10^{-10}\ \text{m}$

(c) $m_r = 185.8m$

$a = \dfrac{1}{185.8}\left(\dfrac{\epsilon_0 h^2}{\pi m_r e^2}\right) = 2.849\times10^{-13}\ \text{m}$

EVALUATE: a is the radius for the $n = 1$ level in the Bohr model. When the reduced mass m_r increases, a decreases. For positronium and muonium the reduced mass effect is large.

41.15. **IDENTIFY** and **SET UP:** The interaction energy between an external magnetic field and the orbital angular momentum of the atom is given by Eq.(41.18). The energy depends on m_l with the most negative m_l value having the lowest energy.

EXECUTE: **(a)** For the $5g$ level, $l = 4$ and there are $2l+1 = 9$ different m_l states. The $5g$ level is split into 9 levels by the magnetic field.

(b) Each m_l level is shifted in energy an amount given by $U = m_l \mu_B B$. Adjacent levels differ in m_l by one, so $\Delta U = \mu_B B$.

$$\mu_B = \frac{e\hbar}{2m} = \frac{(1.602\times10^{-19}\text{ C})(1.055\times10^{-34}\text{ J}\cdot\text{s})}{2(9.109\times10^{-31}\text{ kg})} = 9.277\times10^{-24}\text{ A}\cdot\text{m}^2$$

$$\Delta U = \mu_B B = (9.277\times10^{-24}\text{ A/m}^2)(0.600\text{ T}) = 5.566\times10^{-24}\text{ J}(1\text{ eV}/1.602\times10^{-19}\text{ J}) = 3.47\times10^{-5}\text{ eV}$$

(c) The level of highest energy is for the largest m_l, which is $m_l = l = 4$; $U_4 = 4\mu_B B$. The level of lowest energy is for the smallest m_l, which is $m_l = -l = -4$; $U_{-4} = -4\mu_B B$. The separation between these two levels is

$$U_4 - U_{-4} = 8\mu_B B = 8(3.47\times10^{-5}\text{ eV}) = 2.78\times10^{-4}\text{ eV}.$$

EVALUATE: The energy separations are proportional to the magnetic field. The energy of the $n = 5$ level in the absence of the external magnetic field is $(-13.6\text{ eV})/5^2 = -0.544\text{ eV}$, so the interaction energy with the magnetic field is much less than the binding energy of the state.

41.17. $3p \Rightarrow n = 3, l = 1, \Delta U = \mu_B B \Rightarrow B = \dfrac{U}{\mu_B} = \dfrac{(2.71\times10^{-5}\text{ eV})}{(5.79\times10^{-5}\text{ eV/T})} = 0.468\text{ T}$

(b) Three: $m_l = 0, \pm 1$.

41.19. **IDENTIFY** and **SET UP:** The interaction energy is $U = -\vec{\mu}\cdot\vec{B}$, with μ_z given by Eq.(41.22).

EXECUTE: $U = -\vec{\mu}\cdot\vec{B} = +\mu_z B$, since the magnetic field is in the negative z-direction.

$\mu_z = -(2.00232)\left(\dfrac{e}{2m}\right)S_z$, so $U = -(2.00232)\left(\dfrac{e}{2m}\right)S_z B$

$S_z = m_s\hbar$, so $U = -2.00232\left(\dfrac{e\hbar}{2m}\right)m_s B$

$\dfrac{e\hbar}{2m} = \mu_B = 5.788\times10^{-5}\text{ eV/T}$

$U = -2.00232\mu_B m_s B$

The $m_s = +\dfrac{1}{2}$ level has lower energy.

$$\Delta U = U\left(m_s = -\frac{1}{2}\right) - U\left(m_s = +\frac{1}{2}\right) = -2.00232\mu_B B\left(-\frac{1}{2} - \left(+\frac{1}{2}\right)\right) = +2.00232\mu_B B$$

$$\Delta U = +2.00232(5.788\times10^{-5}\text{ eV/T})(1.45\text{ T}) = 1.68\times10^{-4}\text{ eV}$$

EVALUATE: The interaction energy with the electron spin is the same order of magnitude as the interaction energy with the orbital angular momentum for states with $m_l \neq 0$. But a $1s$ state has $l = 0$ and $m_l = 0$, so there is no orbital magnetic interaction.

41.21. **IDENTIFY** and **SET UP:** j can have the values $l + 1/2$ and $l - 1/2$.

EXECUTE: If j takes the values $7/2$ and $9/2$ it must be that $l - 1/2 = 7/2$ and $l = 8/2 = 4$. The letter that labels this l is g.

EVALUATE: l must be an integer.

41.23. **IDENTIFY** and **SET UP:** For a classical particle $L = I\omega$. For a uniform sphere with mass m and radius R, $I = \dfrac{2}{5}mR^2$, so $L = \left(\dfrac{2}{5}mR^2\right)\omega$. Solve for ω and then use $v = r\omega$ to solve for v.

EXECUTE: **(a)** $L = \sqrt{\dfrac{3}{4}}\hbar$ so $\dfrac{2}{5}mR^2\omega = \sqrt{\dfrac{3}{4}}\hbar$

$$\omega = \frac{5\sqrt{3/4}\hbar}{2mR^2} = \frac{5\sqrt{3/4}(1.055\times10^{-34}\text{ J}\cdot\text{s})}{2(9.109\times10^{-31}\text{ kg})(1.0\times10^{-17}\text{ m})^2} = 2.5\times10^{30}\text{ rad/s}$$

(b) $v = r\omega = (1.0\times10^{-17}\text{ m})(2.5\times10^{30}\text{ rad/s}) = 2.5\times10^{13}\text{ m/s}$.

EVALUATE: This is much greater than the speed of light c, so the model cannot be valid.

41.27. **IDENTIFY** and **SET UP:** The energy of an atomic level is given in terms of n and Z_{eff} by Eq.(41.27),

$$E_n = -\left(\frac{Z_{\text{eff}}^2}{n^2}\right)(13.6\text{ eV}).$$ The ionization energy for a level with energy $-E_n$ is $+E_n$.

EXECUTE: $n = 5$ and $Z_{\text{eff}} = 2.771$ gives $E_5 = -\dfrac{(2.771)^2}{5^2}(13.6\text{ eV}) = -4.18\text{ eV}$

The ionization energy is 4.18 eV.

EVALUATE: The energy of an atomic state is proportional to Z_{eff}^2.

41.29.   **IDENTIFY** and **SET UP:**   Use the exclusion principle to determine the ground-state electron configuration, as in Table 41.3. Estimate the energy by estimating Z_{eff}, taking into account the electron screening of the nucleus.

EXECUTE: **(a)** $Z = 7$ for nitrogen so a nitrogen atom has 7 electrons. N^{2+} has 5 electrons: $1s^2 2s^2 2p$.

(b) $Z_{eff} = 7 - 4 = 3$ for the $2p$ level.

$$E_n = -\left(\frac{Z_{eff}^2}{n^2}\right)(13.6 \text{ eV}) = -\frac{3^2}{2^2}(13.6 \text{ eV}) = -30.6 \text{ eV}$$

(c) $Z = 15$ for phosphorus so a phosphorus atom has 15 electrons.
P^{2+} has 13 electrons: $1s^2 2s^2 2p^6 3s^2 3p$

(d) $Z_{eff} = 15 - 12 = 3$ for the $3p$ level.

$$E_n = -\left(\frac{Z_{eff}^2}{n^2}\right)(13.6 \text{ eV}) = -\frac{3^2}{3^2}(13.6 \text{ eV}) = -13.6 \text{ eV}$$

EVALUATE: In these ions there is one electron outside filled subshells, so it is a reasonable approximation to assume full screening by these inner-subshell electrons.

41.31.   **IDENTIFY** and **SET UP:**   Estimate Z_{eff} by considering electron screening and use Eq.(41.27) to calculate the energy. Z_{eff} is calculated as in Example 41.8.

EXECUTE: **(a)** The element Be has nuclear charge $Z = 4$. The ion Be^+ has 3 electrons. The outermost electron sees the nuclear charge screened by the other two electrons so $Z_{eff} = 4 - 2 = 2$.

$$E_n = -\left(\frac{Z_{eff}^2}{n^2}\right)(13.6 \text{ eV}) \text{ so } E_2 = -\frac{2^2}{2^2}(13.6 \text{ eV}) = -13.6 \text{ eV}$$

(b) The outermost electron in Ca^+ sees a $Z_{eff} = 2$. $E_4 = -\frac{2^2}{4^2}(13.6 \text{ eV}) = -3.4 \text{ eV}$

EVALUATE: For the electron in the highest l-state it is reasonable to assume full screening by the other electrons, as in Example 41.8. The highest l-states of Be^+, Mg^+, Ca^+, etc. all have a $Z_{eff} = 2$. But the energies are different because for each ion the outermost sublevel has a different n quantum number.

41.33.   **(a)** $Z = 20$: $f = (2.48 \times 10^{15} \text{ Hz})(20 - 1)^2 = 8.95 \times 10^{17} \text{ Hz}$.

$$E = hf = (4.14 \times 10^{-15} \text{ eV} \cdot \text{s})(8.95 \times 10^{17} \text{ Hz}) = 3.71 \text{ keV}. \quad \lambda = \frac{c}{f} = \frac{3.00 \times 10^8 \text{ m/s}}{8.95 \times 10^{17} \text{ Hz}} = 3.35 \times 10^{-10} \text{ m}.$$

(b) $Z = 27$: $f = 1.68 \times 10^{18} \text{ Hz}$. $E = 6.96 \text{ keV}$. $\lambda = 1.79 \times 10^{-10} \text{ m}$.

(c) $Z = 48$: $f = 5.48 \times 10^{18} \text{ Hz}$, $E = 22.7 \text{ keV}$, $\lambda = 5.47 \times 10^{-11} \text{ m}$.

41.39.   **(a) IDENTIFY** and **SET UP:**   The energy is given by Eq.(38.18), which is identical to Eq.(41.3). The potential energy is given by Eq.(23.9), with $q = +Ze$ and $q_0 = -e$.

EXECUTE: $E_{1s} = -\frac{1}{(4\pi\epsilon_0)^2} \frac{me^4}{2\hbar^2}$; $U(r) = -\frac{1}{4\pi\epsilon_0} \frac{e^2}{r}$

$E_{1s} = U(r)$ gives $-\frac{1}{(4\pi\epsilon_0)^2} \frac{me^4}{2\hbar^2} = -\frac{1}{4\pi\epsilon_0} \frac{e^2}{r}$

$r = \frac{(4\pi\epsilon_0)2\hbar^2}{me^2} = 2a$

EVALUATE: The turning point is twice the Bohr radius.

(b) IDENTIFY and **SET UP:**   For the $1s$ state the probability that the electron is in the classically forbidden region is $P(r > 2a) = \int_{2a}^{\infty} |\psi_{1s}|^2 dV = 4\pi \int_{2a}^{\infty} |\psi_{1s}|^2 r^2 dr$. The normalized wave function of the $1s$ state of hydrogen is given in Example 41.3: $\psi_{1s}(r) = \frac{1}{\sqrt{\pi a^3}} e^{-r/a}$. Evaluate the integral; the integrand is the same as in Example 41.3.

EXECUTE: $P(r > 2a) = 4\pi \left(\frac{1}{\pi a^3}\right) \int_{2a}^{\infty} r^2 e^{-2r/a} dr$

Use the integral formula $\int r^2 e^{-\alpha r} dr = -e^{-\alpha r}\left(\frac{r^2}{\alpha} + \frac{2r}{\alpha^2} + \frac{2}{\alpha^3}\right)$, with $\alpha = 2/a$.

$$P(r > 2a) = -\frac{4}{a^3}\left[e^{-2r/a}\left(\frac{ar^2}{2} + \frac{a^2 r}{2} + \frac{a^3}{4}\right)\right]_{2a}^{\infty} = +\frac{4}{a^3}e^{-4}(2a^3 + a^3 + a^3/4)$$

$P(r > 2a) = 4e^{-4}(13/4) = 13e^{-4} = 0.238$.

EVALUATE: These is a 23.8% probability of the electron being found in the classically forbidden region, where classically its kinetic energy would be negative.

41.41. $\psi_{2s}(r) = \dfrac{1}{\sqrt{32\pi a^3}}\left(2 - \dfrac{r}{a}\right)e^{-r/2a}$

(a) **IDENTIFY and SET UP:**   Let $I = \int_0^\infty |\psi_{2s}|^2\, dV = 4\pi \int_0^\infty |\psi_{2s}|^2 r^2 dr$. If ψ_{2s} is normalized then we will find that

$I = 1$.

EXECUTE: $I = 4\pi\left(\dfrac{1}{32\pi a^3}\right)\int_0^\infty \left(2 - \dfrac{r}{a}\right)^2 e^{-r/a} r^2 dr = \dfrac{1}{8a^3}\int_0^\infty \left(4r^2 - \dfrac{4r^3}{a} + \dfrac{r^4}{a^2}\right)e^{-r/a} dr$

Use the integral formula $\int_0^\infty x^n e^{-\alpha x}dx = \dfrac{n!}{\alpha^{n+1}}$, with $\alpha = 1/a$

$I = \dfrac{1}{8a^3}\left(4(2!)(a^3) - \dfrac{4}{a}(3!)(a)^4 + \dfrac{1}{a^2}(4!)(a)^5\right) = \dfrac{1}{8}(8 - 24 + 24) = 1$; this ψ_{2s} is normalized.

(b) **SET UP:**   For a spherically symmetric state such as the $2s$, the probability that the electron will be found at $r < 4a$ is $P(r < 4a) = \int_0^{4a} |\psi_{2s}|^2 dV = 4\pi \int_0^{4a} |\psi_{2s}|^2 r^2 dr$.

EXECUTE: $P(r < 4a) = \dfrac{1}{8a^3}\int_0^{4a}\left(4r^2 - \dfrac{4r^3}{a} + \dfrac{r^4}{a^2}\right)e^{-r/a} dr$

Let $P(r < 4a) = \dfrac{1}{8a^3}(I_1 + I_2 + I_3)$.

$I_1 = 4\int_0^{4a} r^2 e^{-r/a} dr$

Use the integral formula $\int r^2 e^{-\alpha r} dr = -e^{-\alpha r}\left(\dfrac{r^2}{\alpha} + \dfrac{2r}{\alpha^2} + \dfrac{2}{\alpha^3}\right)$ with $\alpha = 1/a$.

$I_1 = -4[e^{-r/a}(r^2 a + 2ra^2 + 2a^3)]_0^{4a} = (-104e^{-4} + 8)a^3$.

$I_2 = -\dfrac{4}{a}\int_0^{4a} r^3 e^{-r/a} dr$

Use the integral formula $\int r^3 e^{-\alpha r} dr = -e^{-\alpha r}\left(\dfrac{r^3}{\alpha} + \dfrac{3r^2}{\alpha^2} + \dfrac{6r}{a^3} + \dfrac{6}{\alpha^4}\right)$ with $\alpha = 1/a$.

$I_2 = \dfrac{4}{a}[e^{-r/a}(r^3 a + 3r^2 a^2 + 6ra^3 + 6a^4)]_0^{4a} = (568e^{-4} - 24)a^3$.

$I_3 = \dfrac{1}{a^2}\int_0^{4a} r^4 e^{-r/a} dr$

Use the integral formula $\int r^4 e^{-\alpha r} dr = -e^{-\alpha r}\left(\dfrac{r^4}{\alpha} + \dfrac{4r^3}{\alpha^2} + \dfrac{12r^2}{\alpha^3} + \dfrac{24r}{a^4} + \dfrac{24}{a^5}\right)$ with $\alpha = 1/a$.

$I_3 = -\dfrac{1}{a^2}[e^{-r/a}(r^4 a + 4r^3 a^2 + 12r^2 a^3 + 24ra^4 + 24a^5)]_0^{4a} = (-824e^{-4} + 24)a^3$.

Thus $P(r < 4a) = \dfrac{1}{8a^3}(I_1 + I_2 + I_3) = \dfrac{1}{8a^3}a^3([8 - 24 + 24] + e^{-4}[-104 + 568 - 824])$

$P(r < 4a) = \dfrac{1}{8}(8 - 360e^{-4}) = 1 - 45e^{-4} = 0.176$.

EVALUATE: There is an 82.4% probability that the electron will be found at $r > 4a$. In the Bohr model the electron is for certain at $r = 4a$; this is a poor description of the radial probability distribution for this state.

41.43.   **IDENTIFY:**   Use Figure 41.2 in the textbook to relate θ_L to L_z and L: $\cos\theta_L = \dfrac{L_z}{L}$ so $\theta_L = \arccos\left(\dfrac{L_z}{L}\right)$

(a) **SET UP:**   The smallest angle $(\theta_L)_{\min}$ is for the state with the largest L and the largest L_z. This is the state with $l = n - 1$ and $m_l = l = n - 1$.

EXECUTE: $L_z = m_l \hbar = (n-1)\hbar$

$L = \sqrt{l(l+1)}\hbar = \sqrt{(n-1)n}\hbar$

$(\theta_L)_{\min} = \arccos\left(\dfrac{(n-1)\hbar}{\sqrt{(n-1)n}\hbar}\right) = \arccos\left(\dfrac{(n-1)}{\sqrt{(n-1)n}}\right) = \arccos\left(\sqrt{\dfrac{n-1}{n}}\right) = \arccos(\sqrt{1 - 1/n})$.

EVALUATE: Note that $(\theta_L)_{min}$ approaches $0°$ as $n \to \infty$.

(b) SET UP: The largest angle $(\theta_L)_{max}$ is for $l = n-1$ and $m_l = -l = -(n-1)$.

EXECUTE: A similar calculation to part (a) yields $(\theta_L)_{max} = \arccos(-\sqrt{1-1/n})$

EVALUATE: Note that $(\theta_L)_{max}$ approaches $180°$ as $n \to \infty$.

41.49.   **IDENTIFY:** The presence of an external magnetic field shifts the energy levels up or down, depending upon the value of m_l.

SET UP: The energy difference due to the magnetic field is $\Delta E = \mu_B B$ and the energy of a photon is $E = hc/\lambda$.

EXECUTE: For the p state, $m_l = 0$ or ± 1, and for the s state $m_l = 0$. Between any two adjacent lines, $\Delta E = \mu_B B$. Since the change in the wavelength $(\Delta\lambda)$ is very small, the energy change (ΔE) is also very small, so we can use

differentials. $E = hc/\lambda$. $|dE| = \dfrac{hc}{\lambda^2} d\lambda$ and $\Delta E = \dfrac{hc\Delta\lambda}{\lambda^2}$. Since $\Delta E = \mu_B B$, we get $\mu_B B = \dfrac{hc\Delta\lambda}{\lambda^2}$ and $B = \dfrac{hc\Delta\lambda}{\mu_B\lambda^2}$.

$B = (4.136 \times 10^{-15} \text{ eV} \cdot \text{s})(3.00 \times 10^8 \text{ m/s})(0.0462 \text{ nm})/(5.788 \times 10^{-5} \text{ eV/T})(575.050 \text{ nm})^2 = 3.00 \text{ T}$

EVALUATE: Even a strong magnetic field produces small changes in the energy levels, and hence in the wavelengths of the emitted light.

41.51.   **IDENTIFY:**   The ratio according to the Boltzmann distribution is given by Eq.(38.21): $\dfrac{n_1}{n_0} = e^{-(E_1-E_0)/kT}$, where 1 is

the higher energy state and 0 is the lower energy state.

SET UP: The interaction energy with the magnetic field is $U = -\mu_z B = 2.00232 \left(\dfrac{e\hbar}{2m}\right) m_s B$ (Example 41.5.). The

energy of the $m_s = +\dfrac{1}{2}$ level is increased and the energy of the $m_s = -\dfrac{1}{2}$ level is decreased.

$\dfrac{n_{1/2}}{n_{-1/2}} = e^{-(U_{1/2}-U_{-1/2})/kT}$

EXECUTE: $U_{1/2} - U_{-1/2} = 2.00232\left(\dfrac{e\hbar}{2m}\right)B\left(\dfrac{1}{2} - \left(-\dfrac{1}{2}\right)\right) = 2.00232\left(\dfrac{e\hbar}{2m}\right)B = 2.00232\mu_B B$

$\dfrac{n_{1/2}}{n_{-1/2}} = e^{-(2.00232)\mu_B B/kT}$

(a) $B = 5.00 \times 10^{-5} \text{ T}$

$\dfrac{n_{1/2}}{n_{-1/2}} = e^{-2.00232(9.274\times10^{-24} \text{ A/m}^2)(5.00\times10^{-5} \text{ T})/([1.381\times10^{-23} \text{ J/K}][300 \text{ K}])}$

$\dfrac{n_{1/2}}{n_{-1/2}} = e^{-2.24\times10^{-7}} = 0.99999978 = 1 - 2.2\times10^{-7}$

(b) $B = 5.00 \times 10^{-5} \text{ T}$, $\dfrac{n_{1/2}}{n_{-1/2}} = e^{-2.24\times10^{-3}} = 0.9978$

(c) $B = 5.00 \times 10^{-5} \text{ T}$, $\dfrac{n_{1/2}}{n_{-1/2}} = e^{-2.24\times10^{-2}} = 0.978$

EVALUATE: For small fields the energy separation between the two spin states is much less than kT for $T = 300 \text{ K}$ and the states are equally populated. For $B = 5.00 \text{ T}$ the energy spacing is large enough for there to be a small excess of atoms in the lower state.

41.53.   **IDENTIFY and SET UP:** m_s can take on 4 different values: $m_s = -\dfrac{3}{2}, -\dfrac{1}{2}, +\dfrac{1}{2}, +\dfrac{3}{2}$. Each nlm_l state can have 4

electrons, each with one of the four different m_s values. Apply the exclusion principle to determine the electron configurations.

EXECUTE: **(a)** For a filled $n = 1$ shell, the electron configuration would be $1s^4$; four electrons and $Z = 4$. For a filled $n = 2$ shell, the electron configuration would be $1s^4 2s^4 2p^{12}$; twenty electrons and $Z = 20$.

(b) Sodium has $Z = 11$; 11 electrons. The ground-state electron configuration would be $1s^4 2s^4 2p^3$.

EVALUATE: The chemical properties of each element would be very different.

41.55.   **(a) IDENTIFY and SET UP:**   The energy of the photon equals the transition energy of the atom: $\Delta E = hc/\lambda$. The energies of the states are given by Eq.(41.3).

EXECUTE: $E_n = -\dfrac{13.60 \text{ eV}}{n^2}$ so $E_2 = -\dfrac{13.60 \text{ eV}}{4}$ and $E_1 = -\dfrac{13.60 \text{ eV}}{1}$

$\Delta E = E_2 - E_1 = 13.60 \text{ eV}\left(-\dfrac{1}{4}+1\right) = \dfrac{3}{4}(13.60 \text{ eV}) = 10.20 \text{ eV} = (10.20 \text{ eV})(1.602\times10^{-19} \text{ J/eV}) = 1.634\times10^{-18} \text{ J}$

$\lambda = \dfrac{hc}{\Delta E} = \dfrac{(6.626\times10^{-34} \text{ J}\cdot\text{s})(2.998\times10^8 \text{ m/s})}{1.634\times10^{-18} \text{ J}} = 1.22\times10^{-7} \text{ m} = 122 \text{ nm}$

(b) IDENTIFY and **SET UP:** Calculate the change in ΔE due to the orbital magnetic interaction energy, Eq.(41.17), and relate this to the shift $\Delta\lambda$ in the photon wavelength.

EXECUTE: The shift of a level due to the energy of interaction with the magnetic field in the z-direction is $U = m_l \mu_{\text{B}} B$. The ground state has $m_l = 0$ so is unaffected by the magnetic field. The $n = 2$ initial state has $m_l = -1$ so its energy is shifted downward an amount $U = m_l \mu_{\text{B}} B = (-1)(9.274\times10^{-24} \text{ A/m}^2)(2.20 \text{ T}) =$

$(-2.040\times10^{-23} \text{ J})(1 \text{ eV}/1.602\times10^{-19} \text{ J}) = 1.273\times10^{-4} \text{ eV}$

Note that the shift in energy due to the magnetic field is a very small fraction of the 10.2 eV transition energy. Problem 39.56c shows that in this situation $|\Delta\lambda/\lambda| = |\Delta E/E|$. This gives

$|\Delta\lambda| = \lambda|\Delta E/E| = 122 \text{ nm}\left(\dfrac{1.273\times10^{-4} \text{ eV}}{10.2 \text{ eV}}\right) = 1.52\times10^{-3} \text{ nm} = 1.52 \text{ pm}.$

EVALUATE: The upper level in the transition is lowered in energy so the transition energy is decreased. A smaller ΔE means a larger λ; the magnetic field increases the wavelength. The fractional shift in wavelength, $\Delta\lambda/\lambda$ is small, only 1.2×10^{-5}.

41.57. IDENTIFY: Estimate the atomic transition energy and use Eq.(38.6) to relate this to the photon wavelength.
(a) SET UP: vanadium, $Z = 23$
minimum wavelength; corresponds to largest transition energy
EXECUTE: The highest occupied shell is the N shell ($n = 4$). The highest energy transition is $N \rightarrow K$, with transition energy $\Delta E = E_N - E_K$. Since the shell energies scale like $1/n^2$ neglect E_N relative to E_K, so $\Delta E = E_K = (Z-1)^2(13.6 \text{ eV}) = (23-1)^2(13.6 \text{ eV}) = 6.582\times10^3 \text{ eV} = 1.055\times10^{-15} \text{ J}$. The energy of the emitted photon equals this transition energy, so the photon's wavelength is given by $\Delta E = hc/\lambda$ so $\lambda = hc/\Delta E$.

$\lambda = \dfrac{(6.626\times10^{-34} \text{ J}\cdot\text{s})(2.998\times10^8 \text{ m/s})}{1.055\times10^{-15} \text{ J}} = 1.88\times10^{-10} \text{ m} = 0.188 \text{ nm}.$

SET UP: maximum wavelength; corresponds to smallest transition energy, so for the K_α transition
EXECUTE: The frequency of the photon emitted in this transition is given by Moseley's law (Eq.41.29):
$f = (2.48\times10^{15} \text{ Hz})(Z-1)^2 = (2.48\times10^{15} \text{ Hz})(23-1)^2 = 1.200\times10^{18} \text{ Hz}$

$\lambda = \dfrac{c}{f} = \dfrac{2.998\times10^8 \text{ m/s}}{1.200\times10^{18} \text{ Hz}} = 2.50\times10^{-10} \text{ m} = 0.250 \text{ nm}$

(b) rhenium, $Z = 45$
Apply the analysis of part (a), just with this different value of Z.
minimum wavelength
$\Delta E = E_K = (Z-1)^2(13.6 \text{ eV}) = (45-1)^2(13.6 \text{ eV}) = 2.633\times10^4 \text{ eV} = 4.218\times10^{-15} \text{ J}.$

$\lambda = hc/\Delta E = \dfrac{(6.626\times10^{-34} \text{ J}\cdot\text{s})(2.998\times10^8 \text{ m/s})}{4.218\times10^{-15} \text{ J}} = 4.71\times10^{-11} \text{ m} = 0.0471 \text{ nm}.$

maximum wavelength
$f = (2.48\times10^{15} \text{ Hz})(Z-1)^2 = (2.48\times10^{15} \text{ Hz})(45-1)^2 = 4.801\times10^{18} \text{ Hz}$

$\lambda = \dfrac{c}{f} = \dfrac{2.998\times10^8 \text{ m/s}}{4.801\times10^{18} \text{ Hz}} = 6.24\times10^{-11} \text{ m} = 0.0624 \text{ nm}$

EVALUATE: Our calculated wavelengths have values corresponding to x rays. The transition energies increase when Z increases and the photon wavelengths decrease.

MOLECULES AND CONDENSED MATTER

42.1. **(a)** $K = \frac{3}{2}kT \Rightarrow T = \frac{2K}{3k} = \frac{2(7.9 \times 10^{-4} \text{ eV})(1.60 \times 10^{-19} \text{ J/eV})}{3(1.38 \times 10^{-23} \text{ J/K})} = 6.1 \text{ K}$

(b) $T = \frac{2(4.48 \text{ eV})(1.60 \times 10^{-19} \text{ J/eV})}{3(1.38 \times 10^{-23} \text{ J/K})} = 34,600 \text{ K}.$

(c) The thermal energy associated with room temperature (300 K) is much greater than the bond energy of He_2 (calculated in part (a)), so the typical collision at room temperature will be more than enough to break up He_2. However, the thermal energy at 300 K is much less than the bond energy of H_2, so we would expect it to remain intact at room temperature.

42.3. **IDENTIFY:** The energy given to the photon comes from a transition between rotational states.

SET UP: The rotational energy of a molecule is $E = l(l+1)\frac{\hbar^2}{2I}$ and the energy of the photon is $E = hc/\lambda$.

EXECUTE: Use the energy formula, the energy difference between the $l = 3$ and $l = 1$ rotational levels of the molecule is $\Delta E = \frac{\hbar^2}{2I}[3(3+1) - 1(1+1)] = \frac{5\hbar^2}{I}$. Since $\Delta E = hc/\lambda$, we get $hc/\lambda = 5\hbar^2/I$. Solving for I gives

$$I = \frac{5\hbar\lambda}{2\pi c} = \frac{5(1.055 \times 10^{-34} \text{ J·s})(1.780 \text{ nm})}{2\pi(3.00 \times 10^8 \text{ m/s})} = 4.981 \times 10^{-52} \text{ kg·m}^2.$$

Using $I = m_r r_0^2$, we can solve for r_0: $r_0 = \sqrt{\frac{I(m_N + m_H)}{m_N m_H}} = \sqrt{\frac{(4.981 \times 10^{-52} \text{ kg·m}^2)(2.33 \times 10^{-26} \text{ kg} + 1.67 \times 10^{-27} \text{ kg})}{(2.33 \times 10^{-26} \text{ kg})(1.67 \times 10^{-27} \text{ kg})}}$

$r_0 = 5.65 \times 10^{-13} \text{ m}$

EVALUATE: This separation is much smaller than the diameter of a typical atom and is not very realistic. But we are treating a *hypothetical* NH molecule.

42.7. **IDENTIFY** and **SET UP:** Set $K = E_1$ from Example 42.2. Use $K = \frac{1}{2}I\omega^2$ to solve for ω and $v = r\omega$ to solve for v.

EXECUTE: **(a)** From Example 42.2, $E_1 = 0.479 \text{ meV} = 7.674 \times 10^{-23} \text{ J}$ and $I = 1.449 \times 10^{-46} \text{ kg·m}^2$

$K = \frac{1}{2}I\omega^2$ and $K = E$ gives $\omega = \sqrt{2E_1/I} = 1.03 \times 10^{12} \text{ rad/s}$

(b) $v_1 = r_1\omega_1 = (0.0644 \times 10^{-9} \text{ m})(1.03 \times 10^{12} \text{ rad/s}) = 66.3 \text{ m/s}$ (carbon)

$v_2 = r_2\omega_2 = (0.0484 \times 10^{-9} \text{ m})(1.03 \times 10^{12} \text{ rad/s}) = 49.8 \text{ m/s}$ (oxygen)

(c) $T = 2\pi/\omega = 6.10 \times 10^{-12} \text{ s}$

EVALUATE: From the information in Example 42.3 we can calculate the vibrational period to be $T = 2\pi/\omega = 2\pi\sqrt{m_r/k'} = 1.5 \times 10^{-14} \text{ s}$. The rotational motion is over an order of magnitude slower than the vibrational motion.

42.9. **IDENTIFY** and **SET UP:** The energy of a rotational level with quantum number l is $E_l = l(l+1)\hbar^2/2I$ (Eq.(42.3)). $I = m_r r^2$, with the reduced mass m_r given by Eq.(42.4). Calculate I and ΔE and then use $\Delta E = hc/\lambda$ to find λ.

EXECUTE: **(a)** $m_r = \frac{m_1 m_2}{m_1 + m_2} = \frac{m_{Li} m_H}{m_{Li} + m_H} = \frac{(1.17 \times 10^{-26} \text{ kg})(1.67 \times 10^{-27} \text{ kg})}{1.17 \times 10^{-26} \text{ kg} + 1.67 \times 10^{-27} \text{ kg}} = 1.461 \times 10^{-27} \text{ kg}$

$I = m_r r^2 = (1.461 \times 10^{-27} \text{ kg})(0.159 \times 10^{-9} \text{ m})^2 = 3.694 \times 10^{-47} \text{ kg·m}^2$

$l = 3: \ E = 3(4)\left(\frac{\hbar^2}{2I}\right) = 6\left(\frac{\hbar^2}{I}\right)$

$$l = 4: \quad E = 4(5)\left(\frac{\hbar^2}{2I}\right) = 10\left(\frac{\hbar^2}{I}\right)$$

$$\Delta E = E_4 - E_3 = 4\left(\frac{\hbar^2}{I}\right) = 4\left(\frac{(1.055\times10^{-34}\ \text{J}\cdot\text{s})^2}{3.694\times10^{-47}\ \text{kg}\cdot\text{m}^2}\right) = 1.20\times10^{-21}\ \text{J} = 7.49\times10^{-3}\ \text{eV}$$

(b) $\Delta E = hc/\lambda$ so $\lambda = \dfrac{hc}{\Delta E} = \dfrac{(4.136\times10^{-15}\ \text{eV})(2.998\times10^8\ \text{m/s})}{7.49\times10^{-3}\ \text{eV}} = 166\,\mu\text{m}$

EVALUATE: LiH has a smaller reduced mass than CO and λ is somewhat smaller here than the λ calculated for CO in Example 42.2

42.13.   **(a) IDENTIFY** and **SET UP:**   Use $\omega = \sqrt{k'/m_r}$ and $\omega = 2\pi f$ to calculate k'. The atomic masses are used in Eq.(42.4) to calculate m_r.

EXECUTE: $f = \dfrac{\omega}{2\pi} = \dfrac{1}{2\pi}\sqrt{\dfrac{k'}{m_r}}$, so $k' = m_r(2\pi f)^2$

$$m_r = \frac{m_1 m_2}{m_1 + m_2} = \frac{m_H m_F}{m_H + m_F} = \frac{(1.67\times10^{-27}\ \text{kg})(3.15\times10^{-26}\ \text{kg})}{1.67\times10^{-27}\ \text{kg} + 3.15\times10^{-26}\ \text{kg}} = 1.586\times10^{-27}\ \text{kg}$$

$k' = m_r(2\pi f)^2 = (1.586\times10^{-27}\ \text{kg})(2\pi[1.24\times10^{14}\ \text{Hz}])^2 = 963\ \text{N/m}$

(b) IDENTIFY and **SET UP:**   The energy levels are given by Eq.(42.7). $E_n = (n+\frac{1}{2})\hbar\omega = (n+\frac{1}{2})hf$, since $\hbar\omega = (h/2\pi)\omega$ and $(\omega/2\pi) = f$. The energy spacing between adjacent levels is $\Delta E = E_{n+1} - E_n = (n+1+\frac{1}{2}-n-\frac{1}{2})hf = hf$, independent of n.

EXECUTE: $\Delta E = hf = (6.626\times10^{-34}\ \text{J}\cdot\text{s})(1.24\times10^{14}\ \text{Hz}) = 8.22\times10^{-20}\ \text{J} = 0.513\ \text{eV}$

(c) IDENTIFY and **SET UP:**   The photon energy equals the transition energy so $\Delta E = hc/\lambda$.

EXECUTE: $hf = hc/\lambda$ so $\lambda = \dfrac{c}{f} = \dfrac{2.998\times10^8\ \text{m/s}}{1.24\times10^{14}\ \text{Hz}} = 2.42\times10^{-6}\ \text{m} = 2.42\ \mu\text{m}$

EVALUATE: This photon is infrared, which is typical for vibrational transitions.

42.15.   **IDENTIFY** and **SET UP:**   Find the volume occupied by each atom. The density is the average mass of Na and Cl divided by this volume.
EXECUTE: Each atom occupies a cube with side length 0.282 nm. Therefore, the volume occupied by each atom is $V = (0.282\times10^{-9}\ \text{m})^3 = 2.24\times10^{-29}\ \text{m}^3$. In NaCl there are equal numbers of Na and Cl atoms, so the average mass of the atoms in the crystal is $m = \frac{1}{2}(m_{Na} + m_{Cl}) = \frac{1}{2}(3.82\times10^{-26}\ \text{kg} + 5.89\times10^{-26}\ \text{kg}) = 4.855\times10^{-26}\ \text{kg}$

The density then is $\rho = \dfrac{m}{V} = \dfrac{4.855\times10^{-26}\ \text{kg}}{2.24\times10^{-29}\ \text{m}^3} = 2.17\times10^3\ \text{kg/m}^3$.

EVALUATE: The density of water is $1.00\times10^3\ \text{kg/m}^3$, so our result is reasonable.

42.19. $\Delta E = \dfrac{hc}{\lambda} = \dfrac{(6.63\times10^{-34}\ \text{J}\cdot\text{s})(3.00\times10^8\ \text{m/s})}{9.31\times10^{-13}\ \text{m}} = 2.14\times10^{-13}\ \text{J} = 1.34\times10^6\ \text{eV}$. So the number of electrons that can be

excited to the conduction band is $n = \dfrac{1.34\times10^6\ \text{eV}}{1.12\ \text{eV}} = 1.20\times10^6$ electrons

42.23.   **(a) IDENTIFY** and **SET UP:**   The three-dimensional Schrödinger equation is $-\dfrac{\hbar^2}{2m}\left(\dfrac{\partial^2\psi}{\partial x^2} + \dfrac{\partial^2\psi}{\partial y^2} + \dfrac{\partial^2\psi}{\partial z^2}\right) + U\psi = E\psi$

(Eq.40.29). For free electrons, $U = 0$. Evaluate $\partial^2\psi/\partial x^2$, $\partial^2\psi/\partial y^2$, and $\partial^2\psi/\partial z^2$ for ψ as given by Eq.(42.10). Put the results into Eq.(40.20) and see if the equation is satisfied.

EXECUTE: $\dfrac{\partial\psi}{\partial x} = \dfrac{n_x\pi}{L}A\cos\left(\dfrac{n_x\pi x}{L}\right)\sin\left(\dfrac{n_y\pi y}{L}\right)\sin\left(\dfrac{n_z\pi z}{L}\right)$

$\dfrac{\partial^2\psi}{\partial x^2} = -\left(\dfrac{n_x\pi}{L}\right)^2 A\sin\left(\dfrac{n_x\pi x}{L}\right)\sin\left(\dfrac{n_y\pi y}{L}\right)\sin\left(\dfrac{n_z\pi z}{L}\right) = -\left(\dfrac{n_x\pi}{L}\right)^2\psi$

Similarly $\dfrac{\partial^2\psi}{\partial y^2} = -\left(\dfrac{n_y\pi}{L}\right)^2\psi$ and $\dfrac{\partial^2\psi}{\partial z^2} = -\left(\dfrac{n_z\pi}{L}\right)^2\psi$.

Therefore, $-\dfrac{\hbar^2}{2m}\left(\dfrac{\partial^2\psi}{\partial x^2}+\dfrac{\partial^2\psi}{\partial y^2}+\dfrac{\partial^2\psi}{\partial z^2}\right)=\dfrac{\hbar^2}{2m}\left(\dfrac{\pi^2}{L^2}\right)(n_x^2+n_y^2+n_z^2)\psi=\dfrac{(n_x^2+n_y^2+n_z^2)\pi^2\hbar^2}{2mL^2}\psi$

This equals $E\psi$, with $E=\dfrac{(n_x^2+n_y^2+n_z^2)\pi^2\hbar^2}{2mL^2}$, which is Eq.(42.11).

EVALUATE: ψ given by Eq.(42.10) is a solution to Eq.(40.29), with E as given by Eq.(42.11).

(b) IDENTIFY and SET UP: Find the set of quantum numbers n_x, n_y, and n_z that give the lowest three values of E. The degeneracy is the number of sets n_x, n_y, n_z and m_s that give the same E.

EXECUTE: Ground level: lowest E so $n_x=n_y=n_z=1$ and $E=\dfrac{3\pi^2\hbar^2}{2mL^2}$. No other combination of n_x, n_y, and n_z gives this same E, so the only degeneracy is the degeneracy of two due to spin.

First excited level: next lower E so one n equals 2 and the others equal 1. $E=(2^2+1^2+1^2)\dfrac{\pi^2\hbar^2}{2mL^2}=\dfrac{6\pi^2\hbar^2}{2mL^2}$

There are three different sets of n_x, n_y, n_z values that give this E:

$n_x=2,\ n_y=1,\ n_z=1;\ n_x=1,\ n_y=2,\ n_z=1;\ n_x=1,\ n_y=1;\ n_z=2$

This gives a degeneracy of 3 so the total degeneracy, with the factor of 2 from spin, is 6.
Second excited level: next lower E so two of n_x, n_y, n_z equal 2 and the other equals 1.

$E=(2^2+2^2+1^2)\dfrac{\pi^2\hbar^2}{2mL^2}=\dfrac{9\pi^2\hbar^2}{2mL^2}$

There are different sets of n_x, n_y, n_z values that give this E:

$n_x=2,\ n_y=2,\ n_z=1;\ n_x=2,\ n_y=1,\ n_z=2;\ n_x=1,\ n_y=2,\ n_z=2.$

Thus, as for the first excited level, the total degeneracy, including spin, is 6.

EVALUATE: The wavefunction for the 3-dimensional box is a product of the wavefunctions for a 1-dimensional box in the x, y, and z coordinates and the energy is the sum of energies for three 1-dimensional boxes. All levels except for the ground level have a degeneracy greater than two. Compare to the 3-dimensional isotropic harmonic oscillator treated in Problem 40.53.

42.25. **(a) IDENTIFY and SET UP:** The electron contribution to the molar heat capacity at constant volume of a metal is

$C_V=\left(\dfrac{\pi^2 KT}{2E_F}\right)R.$

EXECUTE: $C_V=\dfrac{\pi^2(1.381\times10^{-23}\text{ J/K})(300\text{ K})}{2(5.48\text{ eV})(1.602\times10^{-19}\text{ J/eV})}R=0.0233R.$

(b) EVALUATE: The electron contribution found in part (a) is $0.0233R=0.194$ J/mol·K. This is $0.194/25.3=7.67\times10^{-3}=0.767\%$ of the total C_V.

(c) Only a small fraction of C_V is due to the electrons. Most of C_V is due to the vibrational motion of the ions.

42.27. **IDENTIFY:** The probability is given by the Fermi-Dirac distribution.

SET UP: The Fermi-Dirac distribution is $f(E)=\dfrac{1}{e^{(E-E_F)/kT}+1}$.

EXECUTE: We calculate the value of $f(E)$, where $E=8.520$ eV, $E_F=8.500$ eV, $k=1.38\times10^{-23}$ J/K $=8.625\times10^{-5}$ eV/K, and $T=20°\text{C}=293$ K. The result is $f(E)=0.312=31.2\%$.

EVALUATE: Since the energy is close to the Fermi energy, the probability is quite high that the state is occupied by an electron.

42.29. **IDENTIFY:** Use Eq.(42.17), $f(E)=\dfrac{1}{e^{(E-E_F)/kT}+1}$. Solve for $E-E_F$.

SET UP: $e^{(E-E_F)/kT}=\dfrac{1}{f(E)}-1$

The problem states that $f(E)=4.4\times10^{-4}$ for E at the bottom of the conduction band.

EXECUTE: $e^{(E-E_F)/kT}=\dfrac{1}{4.4\times10^{-4}}-1=2.272\times10^3.$

$E-E_F=kT\ln(2.272\times10^3)=(1.3807\times10^{-23}\text{ J/T})(300\text{ K})\ln(2.272\times10^3)=3.201\times10^{-20}\text{ J}=0.20\text{ eV}$

$E_F=E-0.20$ eV; the Fermi level is 0.20 eV below the bottom of the conduction band.

EVALUATE: The energy gap between the Fermi level and bottom of the conduction band is large compared to kT at $T=300$ K and as a result $f(E)$ is small.

42.31. **IDENTIFY and SET UP:** The voltage-current relation is given by Eq.(42.23): $I = I_s(e^{eV/kT} - 1)$. Use the current for $V = +15.0$ mV to solve for the constant I_s.

EXECUTE: **(a)** Find I_s: $V = +15.0 \times 10^{-3}$ V gives $I = 9.25 \times 10^{-3}$ A

$$\frac{eV}{kT} = \frac{(1.602 \times 10^{-19} \text{ C})(15.0 \times 10^{-3} \text{ V})}{(1.381 \times 10^{-23} \text{ J/K})(300 \text{ K})} = 0.5800$$

$$I_s = \frac{I}{e^{eV/kT} - 1} = \frac{9.25 \times 10^{-3} \text{ A}}{e^{0.5800} - 1} = 1.177 \times 10^{-2} = 11.77 \text{ mA}$$

Then can calculate I for $V = 10.0$ mV: $\dfrac{eV}{kT} = \dfrac{(1.602 \times 10^{-19} \text{ C})(10.0 \times 10^{-3} \text{ V})}{(1.381 \times 10^{-23} \text{ J/K})(300 \text{ K})} = 0.3867$

$$I = I_s(e^{eV/kT} - 1) = (11.77 \text{ mA})(e^{0.3867} - 1) = 5.56 \text{ mA}$$

(b) $\dfrac{eV}{kT}$ has the same magnitude as in part (a) but not V is negative so $\dfrac{eV}{kT}$ is negative.

$V = -15.0$ mV: $\dfrac{eV}{kT} = -0.5800$ and $I = I_s(e^{eV/kT} - 1) = (11.77 \text{ mA})(e^{-0.5800} - 1) = -5.18 \text{ mA}$

$V = -10.0$ mV: $\dfrac{eV}{kT} = -0.3867$ and $I = I_s(e^{eV/kT} - 1) = (11.77 \text{ mA})(e^{-0.3867} - 1) = -3.77 \text{ mA}$

EVALUATE: There is a directional asymmetry in the current, with a forward-bias voltage producing more current than a reverse-bias voltage of the same magnitude, but the voltage is small enough for the asymmetry not be pronounced. Compare to Example 42.11, where more extreme voltages are considered.

42.35. **IDENTIFY and SET UP:** Eq.(21.14) gives the electric dipole moment as $p = qd$, where the dipole consists of charges $\pm q$ separated by distance d.

EXECUTE: **(a)** Point charges $+e$ and $-e$ separated by distance d, so

$p = ed = (1.602 \times 10^{-19} \text{ C})(0.24 \times 10^{-9} \text{ m}) = 3.8 \times 10^{-29} \text{ C} \cdot \text{m}$

(b) $p = qd$ so $q = \dfrac{p}{d} = \dfrac{3.0 \times 10^{-29} \text{ C} \cdot \text{m}}{0.24 \times 10^{-9} \text{ m}} = 1.3 \times 10^{-19} \text{ C}$

(c) $\dfrac{q}{e} = \dfrac{1.3 \times 10^{-19} \text{ C}}{1.602 \times 10^{-19} \text{ C}} = 0.81$

(d) $q = \dfrac{p}{d} = \dfrac{1.5 \times 10^{-30} \text{ C} \cdot \text{m}}{0.16 \times 10^{-9} \text{ m}} = 9.37 \times 10^{-21} \text{ C}$

$\dfrac{q}{e} = \dfrac{9.37 \times 10^{-21} \text{ C}}{1.602 \times 10^{-19} \text{ C}} = 0.058$

EVALUATE: The fractional ionic character for the bond in HI is much less than the fractional ionic character for the bond in NaCl. The bond in HI is mostly covalent and not very ionic.

42.37. **(a) IDENTIFY:** $E(\text{Na}) + E(\text{Cl}) = E(\text{Na}^+) + E(\text{Cl}^-) + U(r)$. Solving for $U(r)$ gives

$U(r) = -[E(\text{Na}^+) - E(\text{Na})] + [E(\text{Cl}) - E(\text{Cl}^-)]$.

SET UP: $[E(\text{Na}^+) - E(\text{Na})]$ is the ionization energy of Na, the energy required to remove one electron, and is equal to 5.1 eV. $[E(\text{Cl}) - E(\text{Cl}^-)]$ is the electron affinity of Cl, the magnitude of the decrease in energy when an electron is attached to a neutral Cl atom, and is equal to 3.6 eV.

EXECUTE: $U = -5.1$ eV $+ 3.6$ eV $= -1.5$ eV $= -2.4 \times 10^{-19}$ J, and $-\dfrac{1}{4\pi\epsilon_0}\dfrac{e^2}{r} = -2.4 \times 10^{-19}$ J

$r = \left(\dfrac{1}{4\pi\epsilon_0}\right)\dfrac{e^2}{2.4 \times 10^{-19} \text{ J}} = (8.988 \times 10^9 \text{ N} \cdot \text{m}^2/\text{C}^2)\dfrac{(1.602 \times 10^{-19} \text{ C})^2}{2.4 \times 10^{-19} \text{ J}}$

$r = 9.6 \times 10^{-10}$ m $= 0.96$ nm

(b) ionization energy of K $= 4.3$ eV; electron affinity of Br $= 3.5$ eV

Thus $U = -4.3$ eV $+ 3.5$ eV $= -0.8$ eV $= -1.28 \times 10^{-19}$ J, and $-\dfrac{1}{4\pi\epsilon_0}\dfrac{e^2}{r} = -1.28 \times 10^{-19}$ J

$r = \left(\dfrac{1}{4\pi\epsilon_0}\right)\dfrac{e^2}{1.28 \times 10^{-19} \text{ J}} = (8.988 \times 10^9 \text{ N} \cdot \text{m}^2/\text{C}^2)\dfrac{(1.602 \times 10^{-19} \text{ C})^2}{1.28 \times 10^{-19} \text{ J}}$

$r = 1.8 \times 10^{-9}$ m $= 1.8$ nm

EVALUATE: K has a smaller ionization energy than Na and the electron affinities of Cl and Br are very similar, so it takes less energy to make $K^+ + Br^-$ from $K + Br$ than to make $Na^+ + Cl^-$ from $Na + Cl$. Thus, the stabilization distance is larger for KBr than for NaCl.

42.39.   (a) IDENTIFY: The rotational energies of a molecule depend on its moment of inertia, which in turn depends on the separation between the atoms in the molecule.

SET UP: Problem 42.38 gives $I = 2.71 \times 10^{-47}$ kg·m². $I = m_r r^2$. Calculate m_r and solve for r.

EXECUTE: $m_r = \dfrac{m_H m_{Cl}}{m_H + m_{Cl}} = \dfrac{(1.67 \times 10^{-27} \text{ kg})(5.81 \times 10^{-26} \text{ kg})}{1.67 \times 10^{-27} \text{ kg} + 5.81 \times 10^{-26} \text{ kg}} = 1.623 \times 10^{-27}$ kg

$r = \sqrt{\dfrac{I}{m_r}} = \sqrt{\dfrac{2.71 \times 10^{-47} \text{ kg·m}^2}{1.623 \times 10^{-27} \text{ kg}}} = 1.29 \times 10^{-10}$ m $= 0.129$ nm

EVALUATE: This is a typical atomic separation for a diatomic molecule; see Example 42.2 for the corresponding distance for CO.

(b) IDENTIFY: Each transition is from the level l to the level $l-1$. The rotational energies are given by Eq.(42.3). The transition energy is related to the photon wavelength by $\Delta E = hc / \lambda$.

SET UP: $E_l = l(l+1)\hbar^2 / 2I$, so $\Delta E = E_l - E_{l-1} = [l(l+1) - l(l-1)]\left(\dfrac{\hbar^2}{2I}\right) = l\left(\dfrac{\hbar^2}{I}\right)$.

EXECUTE: $l\left(\dfrac{\hbar^2}{I}\right) = \dfrac{hc}{\lambda}$

$l = \dfrac{2\pi c I}{\hbar \lambda} = \dfrac{2\pi(2.998 \times 10^8 \text{ m/s})(2.71 \times 10^{-47} \text{ kg·m}^2)}{(1.055 \times 10^{-34} \text{ J·s})\lambda} = \dfrac{4.843 \times 10^{-4} \text{ m}}{\lambda}$

For $\lambda = 60.4 \ \mu$m, $l = \dfrac{4.843 \times 10^{-4} \text{ m}}{60.4 \times 10^{-6} \text{ m}} = 8$.

For $\lambda = 69.0 \ \mu$m, $l = \dfrac{4.843 \times 10^{-4} \text{ m}}{69.0 \times 10^{-6} \text{ m}} = 7$.

For $\lambda = 80.4 \ \mu$m, $l = \dfrac{4.843 \times 10^{-4} \text{ m}}{80.4 \times 10^{-6} \text{ m}} = 6$.

For $\lambda = 96.4 \ \mu$m, $l = \dfrac{4.843 \times 10^{-4} \text{ m}}{96.4 \times 10^{-6} \text{ m}} = 5$.

For $\lambda = 120.4 \ \mu$m, $l = \dfrac{4.843 \times 10^{-4} \text{ m}}{120.4 \times 10^{-6} \text{ m}} = 4$.

EVALUATE: In each case l is an integer, as it must be.

(c) IDENTIFY and SET UP: Longest λ implies smallest ΔE, and this is for the transition from $l = 1$ to $l = 0$.

EXECUTE: $\Delta E = l\left(\dfrac{\hbar^2}{I}\right) = (1)\dfrac{(1.055 \times 10^{-34} \text{ J·s})^2}{2.71 \times 10^{-47} \text{ kg·m}^2} = 4.099 \times 10^{-22}$ J

$\lambda = \dfrac{hc}{\Delta E} = \dfrac{(6.626 \times 10^{-34} \text{ J·s})(2.998 \times 10^8 \text{ m/s})}{4.099 \times 10^{-22} \text{ J}} = 4.85 \times 10^{-4}$ m $= 485 \ \mu$m.

EVALUATE: This is longer than any wavelengths in part (b).

(d) IDENTIFY: What changes is m_r, the reduced mass of the molecule.

SET UP: The transition energy is $\Delta E = l\left(\dfrac{\hbar^2}{I}\right)$ and $\Delta E = \dfrac{hc}{\lambda}$, so $\lambda = \dfrac{2\pi c I}{l\hbar}$ (part (b)). $I = m_r r^2$, so λ is directly proportional to m_r. $\dfrac{\lambda(HCl)}{m_r(HCl)} = \dfrac{\lambda(DCl)}{m_r(DCl)}$ so $\lambda(DCl) = \lambda(HCl)\dfrac{m_r(DCl)}{m_r(HCl)}$

EXECUTE: The mass of a deuterium atom is approximately twice the mass of a hydrogen atom, so $m_D = 3.34 \times 10^{-27}$ kg.

$m_r(DCl) = \dfrac{m_D m_{Cl}}{m_D + m_{Cl}} = \dfrac{(3.34 \times 10^{-27} \text{ kg})(5.81 \times 10^{-27} \text{ kg})}{3.34 \times 10^{-27} \text{ kg} + 5.81 \times 10^{-26} \text{ kg}} = 3.158 \times 10^{-27}$ kg

$\lambda(DCl) = \lambda(HCl)\left(\dfrac{3.158 \times 10^{-27} \text{ kg}}{1.623 \times 10^{-27} \text{ kg}}\right) = (1.946)\lambda(HCl)$

$l = 8 \rightarrow l = 7; \lambda = (60.4 \ \mu\text{m})(1.946) = 118 \ \mu\text{m}$

$l = 7 \rightarrow l = 6; \lambda = (69.0 \ \mu\text{m})(1.946) = 134 \ \mu\text{m}$

$l = 6 \rightarrow l = 5; \lambda = (80.4 \ \mu\text{m})(1.946) = 156 \ \mu\text{m}$

$l = 5 \rightarrow l = 4; \lambda = (96.4 \ \mu\text{m})(1.946) = 188 \ \mu\text{m}$

$l = 4 \rightarrow l = 3; \lambda = (120.4 \ \mu\text{m})(1.946) = 234 \ \mu\text{m}$

EVALUATE: The moment of inertia increases when H is replaced by D, so the transition energies decrease and the wavelengths increase. The larger the rotational inertia the smaller the rotational energy for a given l (Eq.42.3).

42.43.   **IDENTIFY** and **SET UP:** $E_l = l(l+1)\hbar^2 / 2I$, so E_l and the transition energy ΔE depend on I. Different isotopic molecules have different I.

EXECUTE: **(a)** Calculate I for Na^{35}Cl: $m_r = \dfrac{m_{\text{Na}}m_{\text{Cl}}}{m_{\text{Na}} + m_{\text{Cl}}} = \dfrac{(3.8176 \times 10^{-26} \ \text{kg})(5.8068 \times 10^{-26} \ \text{kg})}{3.8176 \times 10^{-26} \ \text{kg} + 5.8068 \times 10^{-26} \ \text{kg}} = 2.303 \times 10^{-26} \ \text{kg}$

$I = m_r r^2 = (2.303 \times 10^{-26} \ \text{kg})(0.2361 \times 10^{-9} \ \text{m})^2 = 1.284 \times 10^{-45} \ \text{kg} \cdot \text{m}^2$

$\underline{l = 2 \rightarrow l = 1 \text{ transition}}$

$\Delta E = E_2 - E_1 = (6 - 2)\left(\dfrac{\hbar^2}{2I}\right) = \dfrac{2\hbar^2}{I} = \dfrac{2(1.055 \times 10^{-34} \ \text{J} \cdot \text{s})^2}{1.284 \times 10^{-45} \ \text{kg} \cdot \text{m}^2} = 1.734 \times 10^{-23} \ \text{J}$

$\Delta E = \dfrac{hc}{\lambda}$ so $\lambda = \dfrac{hc}{\Delta E} = \dfrac{(6.626 \times 10^{-34} \ \text{J} \cdot \text{s})(2.998 \times 10^8 \ \text{m/s})}{1.734 \times 10^{-23} \ \text{J}} = 1.146 \times 10^{-2} \ \text{m} = 1.146 \ \text{cm}$

$\underline{l = 1 \rightarrow l = 0 \text{ transition}}$

$\Delta E = E_1 - E_0 = (2 - 0)\left(\dfrac{\hbar^2}{2I}\right) = \dfrac{\hbar^2}{I} = \dfrac{1}{2}(1.734 \times 10^{-23} \ \text{J}) = 8.67 \times 10^{-24} \ \text{J}$

$\lambda = \dfrac{hc}{\Delta E} = \dfrac{(6.626 \times 10^{-34} \ \text{J} \cdot \text{s})(2.998 \times 10^8 \ \text{m/s})}{8.67 \times 10^{-24} \ \text{J}} = 2.291 \ \text{cm}$

(b) Calculate I for Na^{37}Cl: $m_r = \dfrac{m_{\text{Na}}m_{\text{Cl}}}{m_{\text{Na}} + m_{\text{Cl}}} = \dfrac{(3.8176 \times 10^{-26} \ \text{kg})(6.1384 \times 10^{-26} \ \text{kg})}{3.8176 \times 10^{-26} \ \text{kg} + 6.1384 \times 10^{-26} \ \text{kg}} = 2.354 \times 10^{-26} \ \text{kg}$

$I = m_r r^2 = (2.354 \times 10^{-26} \ \text{kg})(0.2361 \times 10^{-9} \ \text{m})^2 = 1.312 \times 10^{-45} \ \text{kg} \cdot \text{m}^2$

$\underline{l = 2 \rightarrow l = 1 \text{ transition}}$

$\Delta E = \dfrac{2\hbar^2}{I} = \dfrac{2(1.055 \times 10^{-34} \ \text{J} \cdot \text{s})^2}{1.312 \times 10^{-45} \ \text{kg} \cdot \text{m}^2} = 1.697 \times 10^{-23} \ \text{J}$

$\lambda = \dfrac{hc}{\Delta E} = \dfrac{(6.626 \times 10^{-34} \ \text{J} \cdot \text{s})(2.998 \times 10^8 \ \text{m/s})}{1.697 \times 10^{-23} \ \text{J}} = 1.171 \times 10^{-2} \ \text{m} = 1.171 \ \text{cm}$

$\underline{l = 1 \rightarrow l = 0 \text{ transition}}$

$\Delta E = \dfrac{\hbar^2}{I} = \dfrac{1}{2}(1.697 \times 10^{-23} \ \text{J}) = 8.485 \times 10^{-24} \ \text{J}$

$\lambda = \dfrac{hc}{\Delta E} = \dfrac{(6.626 \times 10^{-34} \ \text{J} \cdot \text{s})(2.998 \times 10^8 \ \text{m/s})}{8.485 \times 10^{-24} \ \text{J}} = 2.341 \ \text{cm}$

The differences in the wavelengths for the two isotopes are:

$l = 2 \rightarrow l = 1$ transition: $1.171 \ \text{cm} - 1.146 \ \text{cm} = 0.025 \ \text{cm}$

$l = 1 \rightarrow l = 0$ transition: $2.341 \ \text{cm} - 2.291 \ \text{cm} = 0.050 \ \text{cm}$

EVALUATE: Replacing ^{35}Cl by ^{37}Cl increases I, decreases ΔE and increases λ. The effect on λ is small but measurable.

42.47.   **IDENTIFY** and **SET UP:**   Use Eq.(42.6) to calculate I. The energy levels are given by Eq.(42.9). The transition energy ΔE is related to the photon wavelength by $\Delta E = hc / \lambda$.

EXECUTE: **(a)** $m_r = \dfrac{m_{\text{H}}m_{\text{I}}}{m_{\text{H}} + m_{\text{I}}} = \dfrac{(1.67 \times 10^{-27} \ \text{kg})(2.11 \times 10^{-25} \ \text{kg})}{1.67 \times 10^{-27} \ \text{kg} + 2.11 \times 10^{-25} \ \text{kg}} = 1.657 \times 10^{-27} \ \text{kg}$

$I = m_r r^2 = (1.657 \times 10^{-27} \ \text{kg})(0.160 \times 10^{-9} \ \text{m})^2 = 4.24 \times 10^{-47} \ \text{kg} \cdot \text{m}^2$

(b) The energy levels are $E_{nl} = l(l+1)\left(\dfrac{\hbar^2}{2I}\right) + (n + \tfrac{1}{2})\hbar\sqrt{\dfrac{k'}{m_r}}$ (Eq.(42.9))

$\sqrt{\dfrac{k'}{m}} = \omega = 2\pi f$ so $E_{nl} = l(l+1)\left(\dfrac{\hbar^2}{2I}\right) + (n + \tfrac{1}{2})hf$

(i) transition $n=1 \rightarrow n=0$, $l=1 \rightarrow l=0$

$$\Delta E = (2-0)\left(\frac{\hbar^2}{2I}\right) + (1+\tfrac{1}{2}-\tfrac{1}{2})hf = \frac{\hbar^2}{I} + hf$$

$$\Delta E = \frac{hc}{\lambda} \text{ so } \lambda = \frac{hc}{\Delta E} = \frac{hc}{(\hbar^2/I)+hf} = \frac{c}{(\hbar/2\pi I)+f}$$

$$\frac{\hbar}{2\pi I} = \frac{1.055\times10^{-34} \text{ J}\cdot\text{s}}{2\pi(4.24\times10^{-47} \text{ kg}\cdot\text{m}^2)} = 3.960\times10^{11} \text{ Hz}$$

$$\lambda = \frac{c}{(\hbar/2\pi I)+f} = \frac{2.998\times10^8 \text{ m/s}}{3.960\times10^{11} \text{ Hz}+6.93\times10^{13} \text{ Hz}} = 4.30 \ \mu\text{m}$$

(ii) transition $n=1 \rightarrow n=0$, $l=2 \rightarrow l=1$

$$\Delta E = (6-2)\left(\frac{\hbar^2}{2I}\right) + hf = \frac{2\hbar^2}{I} + hf$$

$$\lambda = \frac{c}{2(\hbar/2\pi I)+f} = \frac{2.998\times10^8 \text{ m/s}}{2(3.960\times10^{11} \text{ Hz})+6.93\times10^{13} \text{ Hz}} = 4.28 \ \mu\text{m}$$

(iii) transition $n=2 \rightarrow n=1$, $l=2 \rightarrow l=3$

$$\Delta E = (6-12)\left(\frac{\hbar^2}{2I}\right) + hf = -\frac{3\hbar^2}{I} + hf$$

$$\lambda = \frac{c}{-3(\hbar/2\pi I)+f} = \frac{2.998\times10^8 \text{ m/s}}{-3(3.960\times10^{11} \text{ Hz})+6.93\times10^{13} \text{ Hz}} = 4.40 \ \mu\text{m}$$

EVALUATE: The vibrational energy change for the $n=1 \rightarrow n=0$ transition is the same as for the $n=2 \rightarrow n=1$ transition. The rotational energies are much smaller than the vibrational energies, so the wavelengths for all three transitions don't differ much.

41.51.   **IDENTIFY** and **SET UP:**   Use the description of the bcc lattice in Fig.42.12c in the textbook to calculate the number of atoms per unit cell and then the number of atoms per unit volume.

EXECUTE: **(a)** Each unit cell has one atom at its center and 8 atoms at its corners that are each shared by 8 other unit cells. So there are $1+8/8=2$ atoms per unit cell.

$$\frac{n}{V} = \frac{2}{(0.35\times10^{-9} \text{ m})^3} = 4.66\times10^{28} \text{ atoms/m}^3$$

(b) $E_{F0} = \dfrac{3^{2/3}\pi^{4/3}\hbar^2}{2m}\left(\dfrac{N}{V}\right)^{2/3}$

In this equation N/V is the number of free electrons per m^3. But the problem says to assume one free electron per atom, so this is the same as n/V calculated in part (a).

$m=9.109\times10^{-31}$ kg (the electron mass), so $E_{F0}=7.563\times10^{-19}$ J $=4.7$ eV

EVALUATE: Our result for metallic lithium is similar to that calculated for copper in Example 42.8.

42.53.   **(a) IDENTIFY** and **SET UP:** $p = -\dfrac{dE_{tot}}{dV}$. Relate E_{tot} to E_{F0} and evaluate the derivative.

EXECUTE: $E_{tot} = NE_{av} = \dfrac{3N}{5}E_{F0} = \dfrac{3}{5}\left(\dfrac{3^{2/3}\pi^{4/3}\hbar^2}{2m}\right)N^{5/3}V^{-2/3}$

$\dfrac{dE_{tot}}{dV} = \dfrac{3}{5}\left(\dfrac{3^{2/3}\pi^{4/3}\hbar^2}{2m}\right)N^{5/3}\left(-\dfrac{2}{3}V^{-5/3}\right)$ so $p = \left(\dfrac{3^{2/3}\pi^{4/3}\hbar^2}{5m}\right)\left(\dfrac{N}{V}\right)^{5/3}$, as was to be shown.

(b) $N/V = 8.45\times10^{28}$ m^{-3}

$$p = \left(\frac{3^{2/3}\pi^{4/3}(1.055\times10^{-34} \text{ J}\cdot\text{s})^2}{5(9.109\times10^{-31} \text{ kg})}\right)(8.45\times10^{28} \text{ m}^{-3})^{5/3} = 3.81\times10^{10} \text{ Pa} = 3.76\times10^5 \text{ atm.}$$

(c) EVALUATE: Normal atmospheric pressure is about 10^5 Pa, so these pressures are extremely large. The electrons are held in the metal by the attractive force exerted on them by the copper ions.

NUCLEAR PHYSICS

43.3. **IDENTIFY:** Calculate the spin magnetic energy shift for each spin state of the $1s$ level. Calculate the energy splitting between these states and relate this to the frequency of the photons.

SET UP: When the spin component is parallel to the field the interaction energy is $U = -\mu_z B$. When the spin component is antiparallel to the field the interaction energy is $U = +\mu_z B$. The transition energy for a transition between these two states is $\Delta E = 2\mu_z B$, where $\mu_z = 2.7928\mu_n$. The transition energy is related to the photon frequency by $\Delta E = hf$, so $2\mu_z B = hf$.

EXECUTE: $B = \dfrac{hf}{2\mu_z} = \dfrac{(6.626\times 10^{-34}\,\text{J}\cdot\text{s})(22.7\times 10^6\,\text{Hz})}{2(2.7928)(5.051\times 10^{-27}\,\text{J/T})} = 0.533\,\text{T}$

EVALUATE: This magnetic field is easily achievable. Photons of this frequency have wavelength $\lambda = c/f = 13.2$ m. These are radio waves.

43.5. **IDENTIFY:** Calculate the spin magnetic energy shift for each spin component. Calculate the energy splitting between these states and relate this to the frequency of the photons.

(a) SET UP: From Example 43.2, when the z-component of $\vec{S}$ (and $\vec{\mu}$) is parallel to $\vec{B}$, $U = -|\mu_z|B = -2.7928\mu_n B$. When the z-component of $\vec{S}$ (and $\vec{\mu}$) is antiparallel to $\vec{B}$, $U = -|\mu_z|B = +2.7928\mu_n B$. The state with the proton spin component parallel to the field lies lower in energy. The energy difference between these two states is $\Delta E = 2(2.7928\mu_n B)$.

EXECUTE: $\Delta E = hf$ so $f = \dfrac{\Delta E}{h} = \dfrac{2(2.7928\mu_n)B}{h} = \dfrac{2(2.7928)(5.051\times 10^{-27}\,\text{J/T})(1.65\,\text{T})}{6.626\times 10^{-34}\,\text{J}\cdot\text{s}}$

$f = 7.03\times 10^7\,\text{Hz} = 7.03\,\text{MHz}$

And then $\lambda = \dfrac{c}{f} = \dfrac{2.998\times 10^8\,\text{m/s}}{7.03\times 10^7\,\text{Hz}} = 4.26\,\text{m}$

EVALUATE: From Figure 32.4 in the textbook, these are radio waves.

(b) SET UP: From Eqs. (27.27) and (41.22) and Fig.41.14 in the textbook, the state with the z-component of $\vec{\mu}$ parallel to $\vec{B}$ has lower energy. But, since the charge of the electron is negative, this is the state with the electron spin component antiparallel to $\vec{B}$. That is, for the $m_s = -\frac{1}{2}$ state lies lower in energy.

EXECUTE: For the $m_s = +\frac{1}{2}$ state, $U = +(2.00232)\left(\dfrac{e}{2m}\right)\left(+\dfrac{\hbar}{2}\right)B = +\frac{1}{2}(2.00232)\left(\dfrac{e\hbar}{2m}\right)B = +\frac{1}{2}(2.00232)\mu_B B$.

For the $m_s = -\frac{1}{2}$ state, $U = -\frac{1}{2}(2.00232)\mu_B B$. The energy difference between these two states is $\Delta E = (2.00232)\mu_B B$.

$\Delta E = hf$ so $f = \dfrac{\Delta E}{h} = \dfrac{2.00232\mu_B B}{h} = \dfrac{(2.00232)(9.274\times 10^{-24}\,\text{J/T})(1.65\,\text{T})}{6.626\times 10^{-34}\,\text{J}\cdot\text{s}} = 4.62\times 10^{10}\,\text{Hz} = 46.2\,\text{GHz}$. And

$\lambda = \dfrac{c}{f} = \dfrac{2.998\times 10^8\,\text{m/s}}{4.62\times 10^{10}\,\text{Hz}} = 6.49\times 10^{-3}\,\text{m} = 6.49\,\text{mm}$.

EVALUATE: From Figure 32.4 in the textbook, these are microwaves. The interaction energy with the magnetic field is inversely proportional to the mass of the particle, so it is less for the proton than for the electron. The smaller transition energy for the proton produces a larger wavelength.

43.7. **IDENTIFY and SET UP:** The text calculates that the binding energy of the deuteron is 2.224 MeV. A photon that breaks the deuteron up into a proton and a neutron must have at least this much energy.

$E = \dfrac{hc}{\lambda}$ so $\lambda = \dfrac{hc}{E}$

EXECUTE: $\lambda = \dfrac{(4.136\times10^{-15} \text{ eV}\cdot\text{s})(2.998\times10^{8} \text{ m/s})}{2.224\times10^{6} \text{ eV}} = 5.575\times10^{-13} \text{ m} = 0.5575 \text{ pm}.$

EVALUATE: This photon has gamma-ray wavelength.

43.11.   **(a) IDENTIFY:**   Find the energy equivalent of the mass defect.

SET UP: A $^{11}_{5}\text{B}$ atom has 5 protons, $11-5=6$ neutrons, and 5 electrons. The mass defect therefore is

$\Delta M = 5m_{\text{p}} + 6m_{\text{n}} + 5m_{\text{e}} - M(^{11}_{5}\text{B}).$

EXECUTE: $\Delta M = 5(1.0072765 \text{ u}) + 6(1.0086649 \text{ u}) + 5(0.0005485799 \text{ u}) - 11.009305 \text{ u} = 0.08181 \text{ u}.$ The energy equivalent is $E_{\text{B}} = (0.08181 \text{ u})(931.5 \text{ MeV/u}) = 76.21 \text{ MeV}.$

(b) IDENTIFY and SET UP: Eq.(43.11): $E_{\text{B}} = C_1 A - C_2 A^{2/3} - C_3 Z(Z-1)/A^{1/3} - C_4 (A-2Z)^2/A$

The fifth term is zero since Z is odd but N is even. $A=11$ and $Z=5$.

EXECUTE: $E_{\text{B}} = (15.75 \text{ MeV})(11) - (17.80 \text{ MeV})(11)^{2/3} - (0.7100 \text{ MeV})5(4)/11^{1/3} - (23.69 \text{ MeV})(11-10)^2/11.$

$E_{\text{B}} = +173.25 \text{ MeV} - 88.04 \text{ MeV} - 6.38 \text{ MeV} - 2.15 \text{ MeV} = 76.68 \text{ MeV}$

The percentage difference between the calculated and measured E_{B} is $\dfrac{76.68 \text{ MeV} - 76.21 \text{ MeV}}{76.21 \text{ MeV}} = 0.6\%$

EVALUATE: Eq.(43.11) has a greater percentage accuracy for ^{62}Ni. The semi-empirical mass formula is more accurate for heavier nuclei.

43.13.   **IDENTIFY**   In each case determine how the decay changes A and Z of the nucleus. The β^{+} and β^{-} particles have charge but their nucleon number is $A=0$.

(a) SET UP: α-decay: Z increases by 2, $A = N + Z$ decreases by 4 (an α particle is a $^{4}_{2}\text{He}$ nucleus)

EXECUTE: $^{239}_{94}\text{Pu} \rightarrow {}^{4}_{2}\text{He} + {}^{235}_{92}\text{U}$

(b) SET UP: β^{-} decay: Z increases by 1, $A = N + Z$ remains the same (a β^{-} particle is an electron, $^{0}_{-1}\text{e}$)

EXECUTE: $^{24}_{11}\text{Na} \rightarrow {}^{0}_{-1}\text{e} + {}^{24}_{12}\text{Mg}$

(c) SET UP β^{+} decay: Z decreases by 1, $A = N + Z$ remains the same (a β^{+} particle is a positron, $^{0}_{+1}\text{e}$)

EXECUTE: $^{15}_{8}\text{O} \rightarrow {}^{0}_{+1}\text{e} + {}^{15}_{7}\text{N}$

EVALUATE: In each case the total charge and total number of nucleons for the decay products equals the charge and number of nucleons for the parent nucleus; these two quantities are conserved in the decay.

43.17.   If β^{-} decay of ^{14}C is possible, then we are considering the decay $^{14}_{6}\text{C} \rightarrow {}^{14}_{7}\text{N} + \beta^{-}$.

$\Delta m = M(^{14}_{6}\text{C}) - M(^{14}_{7}\text{N}) - m_{\text{e}}$

$\Delta m = (14.003242 \text{ u} - 6(0.000549 \text{ u})) - (14.003074 \text{ u} - 7(0.000549 \text{ u})) - 0.0005491 \text{ u}$

$\Delta m = +1.68\times10^{-4} \text{ u}.$ So $E = (1.68\times10^{-4} \text{ u})(931.5 \text{ MeV/u}) = 0.156 \text{ MeV} = 156 \text{ keV}$

43.19.   **(a)** As in the example, $(0.000898 \text{ u})(931.5 \text{ MeV/u}) = 0.836 \text{ MeV}.$

(b) $0.836 \text{ MeV} - 0.122 \text{ MeV} - 0.014 \text{ MeV} = 0.700 \text{ MeV}.$

43.21.   **IDENTIFY and SET UP:** $T_{1/2} = \dfrac{\ln 2}{\lambda}$ The mass of a single nucleus is $124m_{\text{p}} = 2.07\times10^{-25} \text{ kg}$.

$|\Delta N/\Delta t| = 0.350 \text{ Ci} = 1.30\times10^{10} \text{ Bq}$; $|\Delta N/\Delta t| = \lambda N$

EXECUTE: $N = \dfrac{6.13\times10^{-3} \text{ kg}}{2.07\times10^{-25} \text{ kg}} = 2.96\times10^{22}$; $\lambda = \dfrac{\Delta N/\Delta t}{N} = \dfrac{1.30\times10^{10} \text{ Bq}}{2.96\times10^{22}} = 4.39\times10^{-13} \text{ s}^{-1}$

$T_{1/2} = \dfrac{\ln 2}{\lambda} = 1.58\times10^{12} \text{ s} = 5.01\times10^{4} \text{ yr}$

43.23.   **IDENTIFY and SET UP:**   As discussed in Section 43.4, the activity $A = |dN/dt|$ obeys the same decay equation as Eq. (43.17): $A = A_0 e^{-\lambda t}$. For ^{14}C, $T_{1/2} = 5730$ y and $\lambda = \ln 2/T_{1/2}$ so $A = A_0 e^{-(\ln 2)t/T_{1/2}}$; Calculate A at each t; $A_0 = 180.0$ decays/min.

EXECUTE: **(a)** $t = 1000$ y, $A = 159$ decays/min

(b) $t = 50,000$ y, $A = 0.43$ decays/min

EVALUATE: The time in part (b) is 8.73 half-lives, so the decay rate has decreased by a factor or $(\frac{1}{2})^{8.73}$.

43.29.   **IDENTIFY and SET UP:**   Calculate the number N of ^{14}C atoms in the sample and then use Eq. (43.17) to find the decay constant λ. Eq. (43.18) then gives $T_{1/2}$.

EXECUTE: Find the total number of carbon atoms in the sample.

$n = m/M$;

$N_{tot} = nN_A = mN_A/M = (12.0 \times 10^{-3} \text{ kg})(6.022 \times 10^{23} \text{ atoms/mol})/(12.011 \times 10^{-3} \text{ kg/mol})$

$N_{tot} = 6.016 \times 10^{23}$ atoms, so $(1.3 \times 10^{-12})(6.016 \times 10^{23}) = 7.82 \times 10^{11}$ carbon-14 atoms

$\Delta N / \Delta t = -180$ decays/min $= -3.00$ decays/s

$\Delta N / \Delta t = -\lambda N; \quad \lambda = \dfrac{-\Delta N / \Delta t}{N} = 3.836 \times 10^{-12} \text{ s}^{-1}$

$T_{1/2} = (\ln 2)/\lambda = 1.807 \times 10^{11} \text{ s} = 5730 \text{ y}$

EVALUATE: The value we calculated agrees with the value given in Section 43.4.

43.33.   **IDENTIFY** and **SET UP:**   Find λ from the half-life and the number N of nuclei from the mass of one nucleus and the mass of the sample. Then use Eq.(43.16) to calculate $|dN/dt|$, the number of decays per second.

EXECUTE: (a) $|dN/dt| = \lambda N$

$\lambda = \dfrac{0.693}{T_{1/2}} = \dfrac{0.693}{(1.28 \times 10^9 \text{ y})(3.156 \times 10^7 \text{ s/1 y})} = 1.715 \times 10^{-17} \text{ s}^{-1}$

The mass of ^{40}K atom is approximately 40 u, so the number of ^{40}K nuclei in the sample is

$N = \dfrac{1.63 \times 10^{-9} \text{ kg}}{40 \text{ u}} = \dfrac{1.63 \times 10^{-9} \text{ kg}}{40(1.66054 \times 10^{-27} \text{ kg})} = 2.454 \times 10^{16}$.

Then $|dN/dt| = \lambda N = (1.715 \times 10^{-17} \text{ s}^{-1})(2.454 \times 10^{16}) = 0.421$ decays/s

(b) $|dN/dt| = (0.421 \text{ decays/s})(1 \text{ Ci}/(3.70 \times 10^{10} \text{ decays/s})) = 1.14 \times 10^{-11} \text{ Ci}$

EVALUATE: The very small sample still contains a very large number of nuclei. But the half life is very large, so the decay rate is small.

43.35.   1 rad $= 10^{-2}$ Gy, so 1 Gy = 100 rad and the dose was 500 rad.

rem = (rad)(RBE) = (500 rad)(4.0) = 2000 rem.   1 Gy $= 1$ J/kg, so 5.0 J/kg .

43.39.   (a) We need to know how many decays per second occur.

$$\lambda = \dfrac{0.693}{T_{1/2}} = \dfrac{0.693}{(12.3 \text{ y}) (3.156 \times 10^7 \text{ s/y})} = 1.79 \times 10^{-9} \text{s}^{-1}.$$

The number of tritium atoms is $N_0 = \dfrac{1}{\lambda}\left|\dfrac{dN}{dt}\right| = \dfrac{(0.35 \text{ Ci}) (3.70 \times 10^{10} \text{ Bq/Ci})}{1.79 \times 10^{-9} \text{ s}^{-1}} = 7.2540 \times 10^{18}$ nuclei .

The number of remaining nuclei after one week is

$N = N_0 e^{-\lambda t} = (7.25 \times 10^{18}) e^{-(1.79 \times 10^{-9} \text{ s}^{-1})(7)(24)(3600\text{s})} = 7.2462 \times 10^{18}$ nuclei. $\Delta N = N_0 - N = 7.8 \times 10^{15}$ decays. So the

energy absorbed is $E_{total} = \Delta N \, E_\gamma = (7.8 \times 10^{15}) (5000 \text{ eV}) (1.60 \times 10^{-19} \text{ J/eV}) = 6.24$ J.   The absorbed dose is

$\dfrac{(6.24 \text{ J})}{(50 \text{ kg})} = 0.125$ J/kg $= 12.5$ rad. Since RBE = 1, then the equivalent dose is 12.5 rem.

(b) In the decay, antinetrinos are also emitted. These are not absorbed by the body, and so some of the energy of the decay is lost (about 12 keV).

43.41.   (a) **IDENTIFY** and **SET UP:**   Determine X by balancing the charge and nucleon number on the two sides of the reaction equation.

EXECUTE: X must have $A = 2 + 14 - 10 = 6$ and $Z = 1 + 7 - 5 = 3$. Thus X is $^{6}_{3}$Li and the reaction is

$^{2}_{1}\text{H} + ^{14}_{7}\text{N} \rightarrow ^{6}_{3}\text{Li} + ^{10}_{5}\text{B}$

(b) **IDENTIFY** and **SET UP:**   Calculate the mass decrease and find its energy equivalent.

EXECUTE: The neutral atoms on each side of the reaction equation have a total of 8 electrons, so the electron masses cancel when neutral atom masses are used. The neutral atom masses are found in Table 43.2.

mass of $^{2}_{1}\text{H} + ^{14}_{7}\text{N}$ is 2.014102 u $+ 14.003074$ u $= 16.017176$ u

mass of $^{6}_{3}\text{Li} + ^{10}_{5}\text{B}$ is 6.015121 u $+ 10.012937$ u $= 16.028058$ u

The mass increases, so energy is absorbed by the reaction. The Q value is

(16.017176 u $- 16.028058$ u)(931.5 MeV/u) $= -10.14$ MeV

(c) **IDENTIFY** and **SET UP:**   The available energy in the collision, the kinetic energy K_{cm} in the center of mass reference frame, is related to the kinetic energy K of the bombarding particle by Eq. (43.24).

EXECUTE: The kinetic energy that must be available to cause the reaction is 10.14 MeV. Thus $K_{cm} = 10.14$ MeV. The mass M of the stationary target $\binom{14}{7}N$ is $M = 14$ u. The mass m of the colliding particle $\binom{2}{1}H$ is 2 u. Then by Eq. (43.24) the minimum kinetic energy K that the ^2_1H must have is

$$K = \left(\frac{M+m}{M}\right) K_{cm} = \left(\frac{14 \text{ u} + 2 \text{ u}}{14 \text{ u}}\right)(10.14 \text{ MeV}) = 11.59 \text{ MeV}$$

EVALUATE: The projectile $\binom{2}{1}H$ is much lighter than the target $\binom{14}{7}N$ so K is not much larger than K_{cm}. The K we have calculated is what is required to allow the mass increase. We would also need to check to see if at this energy the projectile can overcome the Coulomb repulsion to get sufficiently close to the target nucleus for the reaction to occur.

43.43. **IDENTIFY** and **SET UP:** Determine X by balancing the charge and the nucleon number on the two sides of the reaction equation.

EXECUTE: X must have $A = +2 + 9 - 4 = 7$ and $Z = +1 + 4 - 2 = 3$. Thus X is ^7_3Li and the reaction is

$$^2_1H + ^9_4Be = ^7_3Li + ^4_2He$$

(b) IDENTIFY and **SET UP:** Calculate the mass decrease and find its energy equivalent.

EXECUTE: If we use the neutral atom masses then there are the same number of electrons (five) in the reactants as in the products. Their masses cancel, so we get the same mass defect whether we use nuclear masses or neutral atom masses. The neutral atoms masses are given in Table 43.2.

$^2_1H + ^9_4Be$ has mass 2.014102 u + 9.012182 u = 11.26284 u

$^7_3Li + ^4_2He$ has mass 7.016003 u + 4.002603 u = 11.018606 u

The mass decrease is 11.026284 u − 11.018606 u = 0.007678 u.

This corresponds to an energy release of 0.007678 u(931.5 MeV/1 u) = 7.152 MeV.

(c) IDENTIFY and **SET UP:** Estimate the threshold energy by calculating the Coulomb potential energy when the ^2_1H and ^9_4Be nuclei just touch. Obtain the nuclear radii from Eq. (43.1).

EXECUTE: The radius R_{Be} of the ^9_4Be nucleus is $R_{Be} = (1.2 \times 10^{-15} \text{ m})(9)^{1/3} = 2.5 \times 10^{-15}$ m.

The radius R_H of the ^2_1H nucleus is $R_H = (1.2 \times 10^{-15} \text{ m})(2)^{1/3} = 1.5 \times 10^{-15}$ m.

The nuclei touch when their center-to-center separation is

$R = R_{Be} + R_H = 4.0 \times 10^{-15}$ m.

The Coulomb potential energy of the two reactant nuclei at this separation is

$$U = \frac{1}{4\pi\epsilon_0}\frac{q_1 q_2}{r} = \frac{1}{4\pi\epsilon_0}\frac{e(4e)}{r}$$

$$U = (8.988 \times 10^9 \text{ N} \cdot \text{m}^2/\text{C}^2)\frac{4(1.602 \times 10^{-19} \text{ C})^2}{(4.0 \times 10^{-15} \text{ m})(1.602 \times 10^{-19} \text{ J/eV})} = 1.4 \text{ MeV}$$

This is an estimate of the threshold energy for this reaction.

EVALUATE: The reaction releases energy but the total initial kinetic energy of the reactants must be 1.4 MeV in order for the reacting nuclei to get close enough to each other for the reaction to occur. The nuclear force is strong but is very short-range.

43.49. Nuclei: $^A_Z X^{Z+} \rightarrow ^{A-4}_{Z-2}Y^{(Z-2)+} + ^4_2He^{2+}$. Add the mass of Z electrons to each side and we find:

$\Delta m = M(^A_Z X) - M(^{A-4}_{Z-2}Y) - M(^4_2He)$, where now we have the mass of the neutral atoms. So as long as the mass of the original neutral atom is greater than the sum of the neutral products masses, the decay can happen.

43.51. $^A_Z X^{Z+} \rightarrow ^A_{Z-1}Y^{(Z-1)+} + \beta^+$. Adding $(Z-1)$ electrons to both sides yields $^A_Z X^+ \rightarrow ^A_{Z-1}Y + \beta^+$. So in terms of masses:

$\Delta m = M\left(^A_Z X^+\right) - M\left(^A_{Z-1}Y\right) - m_e = \left(M\left(^A_Z X\right) - m_e\right) - M\left(^A_{Z-1}Y\right) - m_e = M\left(^A_Z X\right) - M\left(^A_{Z-1}Y\right) - 2m_e$. So the decay will occur as long as the original neutral mass is greater than the sum of the neutral product mass and two electron masses.

43.55. **(a) IDENTIFY** and **SET UP:** The heavier nucleus will decay into the lighter one.

EXECUTE: $^{25}_{13}Al$ will decay into $^{25}_{12}Mg$.

(b) IDENTIFY and **SET UP:** Determine the emitted particle by balancing A and Z in the decay reaction.

EXECUTE: This gives $^{25}_{13}Al \rightarrow ^{25}_{12}Mg + ^0_{+1}e$. The emitted particle must have charge $+e$ and its nucleon number must be zero. Therefore, it is a β^+ particle, a positron.

(c) IDENTIFY and **SET UP:** Calculate the energy defect ΔM for the reaction and find the energy equivalent of ΔM. Use the nuclear masses for $^{25}_{13}Al$ and $^{25}_{12}Mg$, to avoid confusion in including the correct number of electrons if neutral atom masses are used.

EXECUTE: The nuclear mass for $^{25}_{13}Al$ is $M_{nuc}\left(^{25}_{13}Al\right) = 24.990429$ u − 13(0.000548580 u) = 24.983297 u.

The nuclear mass for $_{12}^{25}$Mg is $M_{\text{nuc}}\left(_{12}^{25}\text{Mg}\right)=24.985837\ \text{u}-12(0.000548580\ \text{u})=24.979254\ \text{u}$.

The mass defect for the reaction is

$\Delta M=M_{\text{nuc}}\left(_{13}^{25}\text{Al}\right)-M_{\text{nuc}}\left(_{12}^{25}\text{Mg}\right)-M\left(_{+1}^{0}\text{e}\right)=24.983297\ \text{u}-24.979254\ \text{u}-0.00054858\ \text{u}=0.003494\ \text{u}$

$Q=(\Delta M)c^2=0.003494\ \text{u}(931.5\ \text{MeV/1 u})=3.255\ \text{MeV}$

EVALUATE: The mass decreases in the decay and energy is released. Note: $_{13}^{25}$Al can also decay into $_{12}^{25}$Mg by the electron capture.

$_{13}^{25}\text{Al}+_{-1}^{0}\text{e}\rightarrow_{12}^{25}\text{Mg}$

The $_{-1}^{0}$ electron in the reaction is an orbital electron in the neutral $_{13}^{25}$Al atom. The mass defect can be calculated using the nuclear masses:

$\Delta M=M_{\text{nuc}}\left(_{13}^{25}\text{Al}\right)+M\left(_{-1}^{0}\text{e}\right)-M_{\text{nuc}}\left(_{12}^{25}\text{Mg}\right)=24.983287\ \text{u}+0.00054858\ \text{u}-24.979254\ \text{u}=0.004592\ \text{u}$.

$Q=(\Delta M)^2\ c^2=(0.004592\ \text{u})(931.5\ \text{MeV/1 u})=4.277\ \text{MeV}$

The mass decreases in the decay and energy is released.

43.57.   **IDENTIFY and SET UP:**   The amount of kinetic energy released is the energy equivalent of the mass change in the decay. $m_e=0.0005486\ \text{u}$ and the atomic mass of $_7^{14}$N is $14.003074\ \text{u}$. The energy equivalent of 1 u is 931.5 MeV. ^{14}C has a half-life of $T_{1/2}=5730\ \text{yr}=1.81\times10^{11}\ \text{s}$. The RBE for an electron is 1.0.

EXECUTE: **(a)** $_6^{14}\text{C}\rightarrow\text{e}^-+_7^{14}\text{N}+\bar{\nu}_e$

(b) The mass decrease is $\Delta M=m\left(_6^{14}\text{C}\right)-\left[m_e+m\left(_7^{14}\text{N}\right)\right]$. Use nuclear masses, to avoid difficulty in accounting for atomic electrons. The nuclear mass of $_6^{14}$C is $14.003242\ \text{u}-6m_e=13.999950\ \text{u}$.

The nuclear mass of $_7^{14}$N is $14.003074\ \text{u}-7m_e=13.999234\ \text{u}$.

$\Delta M=13.999950\ \text{u}-13.999234\ \text{u}-0.000549\ \text{u}=1.67\times10^{-4}\ \text{u}$. The energy equivalent of ΔM is 0.156 MeV.

(c) The mass of carbon is $(0.18)(75\ \text{kg})=13.5\ \text{kg}$. From Example 43.9, the activity due to 1 g of carbon in a living organism is 0.255 Bq. The number of decay/s due to 13.5 kg of carbon is $(13.5\times10^3)(0.255\ \text{Bq/g})=3.4\times10^3\ \text{decays/s}$.

(d) Each decay releases 0.156 MeV so $3.4\times10^3\ \text{decays/s}$ releases $530\ \text{MeV/s}=8.5\times10^{-11}\ \text{J/s}$.

(e) The total energy absorbed in 1 yr is $(8.5\times10^{-11}\ \text{J/s})(3.156\times10^7\ \text{s})=2.7\times10^{-3}\ \text{J}$. The absorbed dose is

$\dfrac{2.7\times10^{-3}\ \text{J}}{75\ \text{kg}}=3.6\times10^{-5}\ \text{J/kg}=36\ \mu\text{Gy}=3.6\ \text{mrad}$. With $\text{RBE}=1.0$, the equivalent dose is $36\ \mu\text{Sv}=3.6\ \text{mrem}$.

43.59.   **IDENTIFY and SET UP:**   Find the energy equivalent of the mass decrease. Part of the released energy appears as the emitted photon and the rest as kinetic energy of the electron.

EXECUTE: $_{79}^{198}\text{Au}\rightarrow_{80}^{198}\text{Hg}+_{-1}^{0}\text{e}$

The mass change is $197.968225\ \text{u}-197.966752\ \text{u}=1.473\times10^{-3}\ \text{u}$

(The neutral atom masses include 79 electrons before the decay and 80 electrons after the decay. This one additional electron in the products accounts correctly for the electron emitted by the nucleus.) The total energy released in the decay is $(1.473\times10^{-3}\ \text{u})(931.5\ \text{MeV/u})=1.372\ \text{MeV}$. This energy is divided between the energy of the emitted photon and the kinetic energy of the β^- particle. Thus the β^- particle has kinetic energy equal to $1.372\ \text{MeV}-0.412\ \text{MeV}=0.960\ \text{MeV}$.

EVALUATE: The emitted electron is much lighter than the $_{80}^{198}$Hg nucleus, so the electron has almost all the final kinetic energy. The final kinetic energy of the ^{198}Hg nucleus is very small.

43.61.   **IDENTIFY and SET UP:**   The decay is energetically possible if the total mass decreases. Determine the nucleus produced by the decay by balancing A and Z on both sides of the equation. $_7^{13}\text{N}\rightarrow_{+1}^{0}\text{e}+_6^{13}\text{C}$. To avoid confusion in including the correct number of electrons with neutral atom masses, use nuclear masses, obtained by subtracting the mass of the atomic electrons from the neutral atom masses.

EXECUTE: The nuclear mass for $_7^{13}$N is $M_{\text{nuc}}\left(_7^{13}\text{N}\right)=13.005739\ \text{u}-7(0.00054858\ \text{u})=13.001899\ \text{u}$.

The nuclear mass for $_6^{13}$C is $M_{\text{nuc}}\left(_6^{13}\text{C}\right)=13.003355\ \text{u}-6(0.00054858\ \text{u})=13.000064\ \text{u}$.

The mass defect for the reaction is

$\Delta M=M_{\text{nuc}}\left(_7^{13}\text{N}\right)-M_{\text{nuc}}\left(_6^{13}\text{C}\right)-M\left(_{+1}^{0}\text{e}\right)$. $\Delta M=13.001899\ \text{u}-13.000064\ \text{u}-0.00054858\ \text{u}=0.001286\ \text{u}$.

EVALUATE: The mass decreases in the decay, so energy is released. This decay is energetically possible.

43.65.   IDENTIFY and SET UP: One-half of the sample decays in a time of $T_{1/2}$.

EXECUTE: (a) $\dfrac{10\times10^9 \text{ yr}}{200,000 \text{ yr}} = 5.0\times10^4$

(b) $(\frac{1}{2})^{5.0\times10^4}$. This exponent is too large for most hand-held calculators. But $(\frac{1}{2}) = 10^{-0.301}$ so

$(\frac{1}{2})^{5.0\times10^4} = (10^{-0.301})^{5.0\times10^4} = 10^{-15,000}$

43.67.   IDENTIFY: Use Eq. (43.17) to relate the initial number of radioactive nuclei, N_0, to the number , N, left after time t.

SET UP: We have to be careful; after ^{87}Rb has undergone radioactive decay it is no longer a rubidium atom. Let N_{85} be the number of ^{85}Rb atoms; this number doesn't change. Let N_0 be the number of ^{87}Rb atoms on earth when the solar system was formed. Let N be the present number of ^{87}Rb atoms.

EXECUTE: The present measurements say that $0.2783 = N/(N+N_{85})$.

$(N+N_{85})(0.2783) = N$, so $N = 0.3856 N_{85}$. The percentage we are asked to calculate is $N_0/(N_0+N_{85})$.

N and N_0 are related by $N = N_0 e^{-\lambda t}$ so $N_0 = e^{+\lambda t} N$.

Thus $\dfrac{N_0}{N_0+N_{85}} = \dfrac{Ne^{\lambda t}}{Ne^{\lambda t}+N_{85}} = \dfrac{(0.3855 e^{\lambda t})N_{85}}{(0.3856 e^{\lambda t})N_{85}+N_{85}} = \dfrac{0.3856 e^{\lambda t}}{0.3856 e^{\lambda t}+1}$.

$t = 4.6\times10^9$ y; $\lambda = \dfrac{0.693}{T_{1/2}} = \dfrac{0.693}{4.75\times10^{10} \text{ y}} = 1.459\times10^{-11}$ y^{-1}

$e^{\lambda t} = e^{(1.459\times10^{-11} \text{ y}^{-1})(4.6\times10^9 \text{ y})} = e^{0.16711} = 1.0694$

Thus $\dfrac{N_0}{N_0+N_{85}} = \dfrac{(0.3856)(1.0694)}{(0.3856)(1.0694)+1} = 29.2\%$.

EVALUATE: The half-life for ^{87}Rb is a factor of 10 larger than the age of the solar system, so only a small fraction of the ^{87}Rb nuclei initially present have decayed; the percentage of rubidium atoms that are radioactive is only a bit less now than it was when the solar system was formed.

43.69.   IDENTIFY and SET UP: Find the energy emitted and the energy absorbed each second. Convert the absorbed energy to absorbed dose and to equivalent dose.

EXECUTE: (a)   First find the number of decays each second:

2.6×10^{-4} Ci$\left(\dfrac{3.70\times10^{10} \text{ decays/s}}{1 \text{ Ci}}\right) = 9.6\times10^6$ decays/s

The average energy per decay is 1.25 MeV, and one-half of this energy is deposited in the tumor. The energy delivered to the tumor per second then is

$\frac{1}{2}(9.6\times10^6$ decays/s$)(1.25\times10^6$ eV/decay$)(1.602\times10^{-19}$ J/eV$) = 9.6\times10^{-7}$ J/s.

(b)   The absorbed dose is the energy absorbed divided by the mass of the tissue:

$\dfrac{9.6\times10^{-7} \text{ J/s}}{0.500 \text{ kg}} = (1.9\times10^{-6} \text{ J/kg}\cdot\text{s})(1 \text{ rad}/(0.01 \text{ J/kg})) = 1.9\times10^{-4}$ rad/s

(c)   equivalent dose (REM) = RBE × absorbed dose (rad)

In one second the equivalent dose is $0.70(1.9\times10^{-4} \text{ rad}) = 1.3\times10^{-4}$ rem.

(d) $(200 \text{ rem}/1.3\times10^{-4} \text{ rem/s}) = 1/5\times10^6$ s$(1 \text{ h}/3600 \text{ s}) = 420$ h $= 17$ days.

EVALUATE: The activity of the source is small so that absorbed energy per second is small and it takes several days for an equivalent dose of 200 rem to be absorbed by the tumor. A 200 rem dose equals 2.00 Sv and this is large enough to damage the tissue of the tumor.

43.71.   IDENTIFY and SET UP: The number of radioactive nuclei left after time t is given by $N = N_0 e^{-\lambda t}$. The problem says $N/N_0 = 0.21$; solve for t.

EXECUTE: $0.21 = e^{-\lambda t}$ so $\ln(0.21) = -\lambda t$ and $t = -\ln(0.21)/\lambda$

Example 43.9 gives $\lambda = 1.209\times10^{-4}$ y^{-1} for ^{14}C. Thus $t = \dfrac{-\ln(0.21)}{1.209\times10^{-4} \text{ y}} = 1.3\times10^4$ y.

EVALUATE: The half-life of ^{14}C is 5730 y, so our calculated t is more than two half-lives, so the fraction remaining is less than $(\frac{1}{2})^2 = \frac{1}{4}$.

43.73.   (a) IDENTIFY and SET UP: Use Eq.(43.1) to calculate the radius R of a $^2_1$H nucleus. Calculate the Coulomb potential energy (Eq.23.9) of the two nuclei when they just touch.

EXECUTE: The radius of $_1^2\text{H}$ is $R = (1.2\times10^{-15}\text{ m})(2)^{1/3} = 1.51\times10^{-15}$ m. The barrier energy is the Coulomb

potential energy of two $_1^2\text{H}$ nuclei with their centers separated by twice this distance:

$$U = \frac{1}{4\pi\epsilon_0}\frac{e^2}{r} = (8.988\times10^9\text{ N}\cdot\text{m}^2/\text{C}^2)\frac{(1.602\times10^{-19}\text{ C})^2}{2(1.51\times10^{-15}\text{ m})} = 7.64\times10^{-14}\text{ J} = 0.48\text{ MeV}$$

(b) IDENTIFY and **SET UP:**   Find the energy equivalent of the mass decrease.

EXECUTE: $_1^2\text{H} + {}_1^2\text{H} \rightarrow {}_2^3\text{He} + {}_0^1\text{n}$

If we use neutral atom masses there are two electrons on each side of the reaction equation, so their masses cancel. The neutral atom masses are given in Table 43.2.

$_1^2\text{H} + {}_1^2\text{H}$ has mass $2(2.014102\text{ u}) = 4.028204$ u

$_2^3\text{He} + {}_0^1\text{n}$ has mass $3.016029\text{ u} + 1.008665\text{ u} = 4.024694$ u

The mass decrease is $4.028204\text{ u} - 4.024694\text{ u} = 3.510\times10^{-3}$ u. This corresponds to a liberated energy of $(3.510\times10^{-3}\text{ u})(931.5\text{ MeV/u}) = 3.270\text{ MeV}$, or $(3.270\times10^6\text{ eV})(1.602\times10^{-19}\text{ J/eV}) = 5.239\times10^{-13}$ J.

(c) IDENTIFY and **SET UP:**   We know the energy released when two $_1^2\text{H}$ nuclei fuse. Find the number of reactions

obtained with one mole of $_1^2\text{H}$.

EXECUTE: Each reaction takes two $_1^2\text{H}$ nuclei. Each mole of D_2 has 6.022×10^{23} molecules, so 6.022×10^{23} pairs

of atoms. The energy liberated when one mole of deuterium undergoes fusion is $(6.022\times10^{23})(5.239\times10^{-13}\text{ J}) =$

3.155×10^{11} J/mol.

EVALUATE: The energy liberated per mole is more than a million times larger than from chemical combustion of one mole of hydrogen gas.

PARTICLE PHYSICS AND COSMOLOGY

44

44.1. **(a) IDENTIFY** and **SET UP:** Use Eq.(37.36) to calculate the kinetic energy K.

EXECUTE: $K = mc^2 \left(\dfrac{1}{\sqrt{1 - v^2/c^2}} - 1 \right) = 0.1547mc^2$

$m = 9.109 \times 10^{-31}$ kg, so $K = 1.27 \times 10^{-14}$ J

(b) IDENTIFY and **SET UP:** The total energy of the particles equals the sum of the energies of the two photons. Linear momentum must also be conserved.

EXECUTE: The total energy of each electron or positron is $E = K + mc^2 = 1.1547mc^2 = 9.46 \times 10^{-13}$ J. The total energy of the electron and positron is converted into the total energy of the two photons. The initial momentum of the system in the lab frame is zero (since the equal-mass particles have equal speeds in opposite directions), so the final momentum must also be zero. The photons must have equal wavelengths and must be traveling in opposite directions. Equal λ means equal energy, so each photon has energy 9.46×10^{-14} J.

(c) IDENTIFY and **SET UP:** Use Eq. (38.2) to relate the photon energy to the photon wavelength.

EXECUTE: $E = hc/\lambda$ so $\lambda = hc/E = hc/(9.46 \times 10^{-14} \text{ J}) = 2.10$ pm

EVALUATE: The wavelength calculated in Example 44.1 is 2.43 pm. When the particles also have kinetic energy, the energy of each photon is greater, so its wavelength is less.

44.3. **IDENTIFY** and **SET UP:** By momentum conservation the two photons must have equal and opposite momenta. Then $E = pc$ says the photons must have equal energies. Their total energy must equal the rest mass energy $E = mc^2$ of the pion. Once we have found the photon energy we can use $E = hf$ to calculate the photon frequency and use $\lambda = c/f$ to calculate the wavelength.

EXECUTE: The mass of the pion is $270m_e$, so the rest energy of the pion is $270(0.511 \text{ MeV}) = 138$ MeV. Each photon has half this energy, or 69 MeV. $E = hf$ so $f = \dfrac{E}{h} = \dfrac{(69 \times 10^6 \text{ eV})(1.602 \times 10^{-19} \text{ J/eV})}{6.626 \times 10^{-34} \text{ J} \cdot \text{s}} = 1.7 \times 10^{22}$ Hz

$\lambda = \dfrac{c}{f} = \dfrac{2.998 \times 10^8 \text{ m/s}}{1.7 \times 10^{22} \text{ Hz}} = 1.8 \times 10^{-14} \text{ m} = 18$ fm.

EVALUATE: These photons are in the gamma ray part of the electromagnetic spectrum.

44.5. **(a)** $\Delta m = m_{\pi+} - m_{\mu+} = 270 \, m_e - 207 \, m_e = 63 \, m_e \Rightarrow E = 63(0.511 \text{ MeV}) = 32$ MeV.

(b) A positive muon has less mass than a positive pion, so if the decay from muon to pion was to happen, you could always find a frame where energy was not conserved. This cannot occur.

44.11. **(a) IDENTIFY** and **SET UP:** Eq. (44.7) says $\omega = |q|B/m$ so $B = m\omega/|q|$. And since $\omega = 2\pi f$, this becomes $B = 2\pi m f /|q|$.

EXECUTE: A deuteron is a deuterium nucleus $\left({}^2_1\text{H} \right)$. Its charge is $q = +e$. Its mass is the mass of the neutral ${}^2_1\text{H}$ atom (Table 43.2) minus the mass of the one atomic electron:

$m = 2.014102 \text{ u} - 0.0005486 \text{ u} = 2.013553 \text{ u} (1.66054 \times 10^{-27} \text{ kg/1 u}) = 3.344 \times 10^{-27}$ kg

$B = \dfrac{2\pi m f}{|q|} = \dfrac{2\pi(3.344 \times 10^{-27} \text{ kg})(9.00 \times 10^6 \text{ Hz})}{1.602 \times 10^{-19} \text{ C}} = 1.18$ T

(b) Eq.(44.8): $K = \dfrac{q^2 B^2 R^2}{2m} = \dfrac{[(1.602 \times 10^{-19}\ \text{C})(1.18\ \text{T})(0.320\ \text{m})]^2}{2(3.344 \times 10^{-27}\ \text{kg})}$.

$K = 5.471 \times 10^{-13}\ \text{J} = (5.471 \times 10^{-13}\ \text{J})(1\ \text{eV}/1.602 \times 10^{-19}\ \text{J}) = 3.42\ \text{MeV}$

$K = \frac{1}{2} m v^2$ so $v = \sqrt{\dfrac{2K}{m}} = \sqrt{\dfrac{2(5.471 \times 10^{-13}\ \text{J})}{3.344 \times 10^{-27}\ \text{kg}}} = 1.81 \times 10^7\ \text{m/s}$

EVALUATE: $v/c = 0.06$, so it is ok to use the nonrelativistic expression for kinetic energy.

44.13.   **(a) IDENTIFY and SET UP:**   The masses of the target and projectile particles are equal, so Eq. (44.10) can be used.

$E_a^2 = 2mc^2(E_m + mc^2)$. E_a is specified; solve for the energy E_m of the beam particles.

EXECUTE: $E_m = \dfrac{E_a^2}{2mc^2} - mc^2$

The mass for the alpha particle can be calculated by subtracting two electron masses from the $_2^4\text{He}$ atomic mass:

$m = m_\alpha = 4.002603\ \text{u} - 2(0.0005486\ \text{u}) = 4.001506\ \text{u}$

Then $mc^2 = (4.001506\ \text{u})(931.5\ \text{MeV/u}) = 3.727\ \text{GeV}$.

$E_m = \dfrac{E_a^2}{2mc^2} - mc^2 = \dfrac{(16.0\ \text{GeV})^2}{2(3.727\ \text{GeV})} - 3.727\ \text{GeV} = 30.6\ \text{GeV}$.

(b) Each beam must have $\frac{1}{2} E_a = 8.0\ \text{GeV}$.

EVALUATE: For a stationary target the beam energy is nearly twice the available energy. In a colliding beam experiment all the energy is available and each beam needs to have just half the required available energy.

44.15.   **(a) IDENTIFY and SET UP:**   For a proton beam on a stationary proton target and since E_a is much larger than the proton rest energy we can use Eq.(44.11): $E_a^2 = 2mc^2 E_m$.

EXECUTE: $E_m = \dfrac{E_a^2}{2mc^2} = \dfrac{(77.4\ \text{GeV})^2}{2(0.938\ \text{GeV})} = 3200\ \text{GeV}$

(b) IDENTIFY and SET UP: For colliding beams the total momentum is zero and the available energy E_a is the total energy for the two colliding particles.

EXECUTE: For proton-proton collisions the colliding beams each have the same energy, so the total energy of each beam is $\frac{1}{2} E_a = 38.7\ \text{GeV}$.

EVALUATE: For a stationary target less than 3% of the beam energy is available for conversion into mass. The beam energy for a colliding beam experiment is a factor of $(1/83)$ times smaller than the required energy for a stationary target experiment.

44.19.   **IDENTIFY and SET UP:**   Find the energy equivalent of the mass decrease.

EXECUTE: The mass decrease is $m(\Sigma^+) - m(\text{p}) - m(\pi^0)$ and the energy released is

$mc^2(\Sigma^+) - mc^2(\text{p}) - mc^2(\pi^0) = 1189\ \text{MeV} - 938.3\ \text{MeV} - 135.0\ \text{MeV} = 116\ \text{MeV}$. (The mc^2 values for each particle were taken from Table 44.3.)

EVALUATE: The mass of the decay products is less than the mass of the original particle, so the decay is energetically allowed and energy is released.

44.23.   **IDENTIFY and SET UP:**   Compare the sum of the strangeness quantum numbers for the particles on each side of the decay equation. The strangeness quantum numbers for each particle are given Table 44.3.

EXECUTE: **(a)** $\text{K}^+ \rightarrow \mu^+ + \nu_\mu$; $S_{\text{K}+} = +1$, $S_{\mu+} = 0$, $S_{\nu_\mu} = 0$

$S = 1$ initially; $S = 0$ for the products; S is <u>not conserved</u>

(b) $\text{n} + \text{K}^+ \rightarrow \text{p} + \pi^0$; $S_\text{n} = 0$, $S_{\text{K}+} = +1$, $S_\text{p} = 0$, $S_{\pi^0} = 0$

$S = 1$ initially; $S = 0$ for the products; S is <u>not conserved</u>

(c) $\text{K}^+ + \text{K}^- \rightarrow \pi^0 + \pi^0$; $S_{\text{K}+} = +1$; $S_{\text{K}-} = -1$; $S_{\pi^0} = 0$

$S = +1 - 1 = 0$ initially; $S = 0$ for the products; S is <u>conserved</u>

(d) $\text{p} + \text{K}^- \rightarrow \Lambda^0 + \pi^0$; $S_\text{p} = 0$, $S_{\text{K}-} = -1$, $S_{\Lambda^0} = -1$, $S_{\pi^0} = 0$.

$S = -1$ initially; $S = -1$ for the products; S is <u>conserved</u>

EVALUATE: Strangeness is not a conserved quantity in weak interactions and strangeness non-conserving reactions or decays can occur.

44.27.   **IDENTIFY and SET UP:**   Each value for the combination is the sum of the values for each quark. Use Table 44.4.

EXECUTE: **(a)** *uds*

$Q = \frac{2}{3} e - \frac{1}{3} e - \frac{1}{3} e = 0$

$B = \frac{1}{3} + \frac{1}{3} + \frac{1}{3} = 1$

$S = 0 + 0 - 1 = -1$

$C = 0 + 0 + 0 = 0$

(b) $c\bar{u}$

The values for $\bar{u}$ are the negative for those for u.

$Q = \frac{2}{3}e - \frac{2}{3}e = 0$

$B = \frac{1}{3} - \frac{1}{3} = 0$

$S = 0 + 0 = 0$

$C = +1 + 0 = +1$

(c) ddd

$Q = -\frac{1}{3}e - \frac{1}{3}e - \frac{1}{3}e = -e$

$B = \frac{1}{3} + \frac{1}{3} + \frac{1}{3} = +1$

$S = 0 + 0 + 0 = 0$

$C = 0 + 0 + 0 = 0$

(d) $d\bar{c}$

$Q = -\frac{1}{3}e - \frac{2}{3}e = -e$

$B = \frac{1}{3} - \frac{1}{3} = 0$

$S = 0 + 0 = 0$

$C = 0 - 1 = -1$

EVALUATE: The charge, baryon number, strangeness and charm quantum numbers of a particle are determined by the particle's quark composition.

44.29. **(a)** The antiparticle must consist of the antiquarks so $\bar{n} = \bar{u}\bar{d}\bar{d}$.

(b) So $n = udd$ is not its own antiparticle.

(c) $\psi = c\bar{c}$ so $\bar{\psi} = \bar{c}c = \psi$ so the ψ is its own antiparticle.

44.33. **(a)** **IDENTIFY** and **SET UP:** Use Eq.(44.14) to calculate v.

EXECUTE: $v = \left[\frac{(\lambda_0/\lambda_s)^2 - 1}{(\lambda_0/\lambda_s)^2 + 1}\right]c = \left[\frac{(658.5 \text{ nm}/590 \text{ nm})^2 - 1}{(658.5 \text{ nm}/590 \text{ nm})^2 + 1}\right]c = 0.1094c$

$v = (0.1094)(2.998 \times 10^8 \text{ m/s}) = 3.28 \times 10^7 \text{ m/s}$

(b) **IDENTIFY** and **SET UP:** Use Eq.(44.15) to calculate r.

EXECUTE: $r = \frac{v}{H_0} = \frac{3.28 \times 10^4 \text{ km/s}}{(71 \text{ (km/s)/Mpc})(1 \text{ Mpc}/3.26 \text{ Mly})} = 1510 \text{ Mly}$

EVALUATE: The red shift $\lambda_0/\lambda_s - 1$ for this galaxy is 0.116. It is therefore about twice as far from earth as the galaxy in Examples 44.9 and 44.10, that had a red shift of 0.053.

44.35. **(a)** **IDENTIFY** and **SET UP:** Hubble's law is Eq.(44.15), with $H_0 = 71$ (km/s)/(Mpc). 1 Mpc = 3.26 Mly.

EXECUTE: $r = 5210$ Mly so $v = H_0 r = ((71 \text{ km/s})/\text{Mpc})(1 \text{ Mpc}/3.26 \text{ Mly})(5210 \text{ Mly}) = 1.1 \times 10^5 \text{ km/s}$

(b) **IDENTIFY** and **SET UP:** Use v from part (a) in Eq. (44.13).

EXECUTE: $\frac{\lambda_0}{\lambda_s} = \sqrt{\frac{c+v}{c-v}} = \sqrt{\frac{1+v/c}{1-v/c}}$

$\frac{v}{c} = \frac{1.1 \times 10^8 \text{ m/s}}{2.9980 \times 10^8 \text{ m/s}} = 0.367$ so $\frac{\lambda_0}{\lambda_s} = \sqrt{\frac{1+0.367}{1-0.367}} = 1.5$

EVALUATE: The galaxy in Examples 44.9 and 44.10 is 710 Mly away so has a smaller recession speed and redshift than the galaxy in this problem.

44.37. **IDENTIFY** and **SET UP:** Find the energy equivalent of the mass decrease.

EXECUTE: **(a)** $p + {}^2_1H \rightarrow {}^3_2He$ or can write as ${}^1_1H + {}^2_1H \rightarrow {}^3_2He$

If neutral atom masses are used then the masses of the two atomic electrons on each side of the reaction will cancel.

Taking the atomic masses from Table 43.2, the mass decrease is $m({}^1_1H) + m({}^2_1H) - m({}^3_2He) = 1.007825$ u + 2.014102 u $-$ 3.016029 u = 0.005898 u. The energy released is the energy equivalent of this mass decrease: $(0.005898 \text{ u})(931.5 \text{ MeV/u}) = 5.494$ MeV

(b) ${}^1_0n + {}^3_2He \rightarrow {}^4_2He$

If neutral helium masses are used then the masses of the two atomic electrons on each side of the reaction equation will cancel. The mass decrease is $m\left(_0^1\text{n}\right)+m\left(_2^3\text{He}\right)-m\left(_2^4\text{He}\right)=1.008665\text{ u}+3.016029\text{ u}-4.002603\text{ u}=$
0.022091 u. The energy released is the energy equivalent of this mass decrease:
(0.022091 u)(931.15 MeV/u) = 20.58 MeV

EVALUATE: These are important nucleosynthesis reactions, discussed in Section 44.7.

44.41.   **IDENTIFY** and **SET UP:**   The Wien displacement law (Eq.38.30) sys $\lambda_m T$ equals a constant. Use this to relate $\lambda_{m,1}$ at T_1 to $\lambda_{m,2}$ at T_2.

EXECUTE: $\lambda_{m,1}T_1 = \lambda_{m,2}T_2$

$$\lambda_{m,1} = \lambda_{m,2}\left(\frac{T_2}{T_1}\right) = 1.062\times10^{-3}\text{ m}\left(\frac{2.728\text{ K}}{3000\text{ K}}\right) = 966\text{ nm}$$

EVALUATE: The peak wavelength was much less when the temperature was much higher.

44.43.   **IDENTIFY** and **SET UP:**   For colliding beams the available energy is twice the beam energy. For a fixed-target experiment only a portion of the beam energy is available energy (Eqs.44.9 and 44.10).
EXECUTE: **(a)** $E_a = 2(7.0\text{ TeV}) = 14.0\text{ TeV}$

(b) Need $E_a = 14.0\text{ TeV} = 14.0\times10^6$ MeV. Since the target and projectile particles are both protons Eq. (44.10) can be used: $E_a^2 = 2mc^2(E_m + mc^2)$

$$E_m = \frac{E_a^2}{2mc^2} - mc^2 = \frac{(14.0\times10^6\text{ MeV})^2}{2(938.3\text{ MeV})} - 938.3\text{ MeV} = 1.0\times10^{11}\text{ MeV} = 1.0\times10^5\text{ TeV.}$$

EVALUATE: This shows the great advantage of colliding beams at relativistic energies.

44.47.   **IDENTIFY:**   With a stationary target, only part of the initial kinetic energy of the moving proton is available. Momentum conservation tells us that there must be nonzero momentum after the collision, which means that there must also be left over kinetic energy. Therefore not all of the initial energy is available.

SET UP: The available energy is given by $E_a^2 = 2mc^2\left(E_m + mc^2\right)$ for two particles of equal mass when one is initially stationary. The *minimum* available energy must be equal to the rest mass energies of the products, which in this case is two protons, a K^+ and a K^-. The available energy must be at least the sum of the final rest masses.
EXECUTE: The minimum amount of available energy must be
$E_a = 2m_p + m_{K^+} + m_{K^-} = 2(938.3\text{ MeV}) + 493.7\text{ MeV} + 493.7\text{ MeV} = 2864\text{ MeV} = 2.864\text{ GeV}$

Solving the available energy formula for E_m gives $E_a^2 = 2mc^2\left(E_m + mc^2\right)$ and

$$E_m = \frac{E_a^2}{2mc^2} - mc^2 = \frac{(2864\text{ MeV})^2}{2(938.3\text{ MeV})} - 938.3\text{ MeV} = 3432.6\text{ MeV}$$

Recalling that E_m is the *total* energy of the proton, including its rest mass energy (RME), we have
$$K = E_m - RME = 3432.6\text{ MeV} - 938.3\text{ MeV} = 2494\text{ MeV} = 2.494\text{ GeV}$$
Therefore the threshold kinetic energy is $K = 2494\text{ MeV} = 2.494\text{ GeV}$.
EVALUATE: Considerably less energy would be needed if the experiment were done using colliding beams of protons.

44.49.   **IDENTIFY** and **SET UP:**   Apply conservation of linear momentum to the collision. A photon has momentum $p = h/\lambda$, in the direction it is traveling. The energy of a photon is $E = pc = \dfrac{hc}{\lambda}$. All the mass of the electron and positron is converted to the total energy of the two photons, according to $E = mc^2$. The mass of an electron and of a positron is $m_e = 9.11\times10^{-31}$ kg

EXECUTE: **(a)** In the lab frame the initial momentum of the system is zero, since the electron and positron have equal speeds in opposite directions. According to momentum conservation, the final momentum of the system must also be zero. A photon has momentum, so the momentum of a single photon is not zero.
(b) For the two photons to have zero total momentum they must have the same magnitude of momentum and move in opposite directions. Since $E = pc$, equal p means equal E.

(c) $2E_{ph} = 2m_e c^2$ so $E_{ph} = m_e c^2$

$$E_{ph} = \frac{hc}{\lambda} \text{ so } \frac{hc}{\lambda} = m_e c^2 \text{ and } \lambda = \frac{h}{m_e c} = \frac{6.63\times10^{-34}\text{ J}\cdot\text{s}}{(9.11\times10^{-31}\text{ kg})(3.00\times10^8\text{ m/s})} = 2.43\text{ pm}$$

These are gamma ray photons.
EVALUATE: The total charge of the electron/positron system is zero and the photons have no charge, so charge is conserved in the particle-antiparticle annihilation.

44.51.   (a) The baryon number is 0, the charge is $+e$, the strangeness is 1, all lepton numbers are zero, and the particle is K^+.

(b) The baryon number is 0, the charge is $-e$, the strangeness is 0, all lepton numbers are zero, and the particle is π^-.

(c) The baryon numbers is –1, the charge is 0, the strangeness is zero, all lepton numbers are 0, and the particle is an antineutron.

(d) The baryon number is 0, the charge is $+e$, the strangeness is 0, the muonic lepton number is –1, all other lepton numbers are 0, and the particle is μ^+.

44.55.   (a) The number of protons in a kilogram is

$$(1.00 \text{ kg})\left(\frac{6.023\times10^{23} \text{ molecules/mol}}{18.0\times10^{-3} \text{ kg/mol}}\right)(2 \text{ protons/molecule}) = 6.7\times10^{25}.$$

Note that only the protons in the hydrogen atoms are considered as possible sources of proton decay. The energy per decay is $m_p c^2 = 938.3 \text{ MeV} = 1.503\times10^{-10}$ J, and so the energy deposited in a year, per kilogram, is

$$(6.7\times10^{25})\left(\frac{\ln(2)}{1.0\times10^{18} \text{ y}}\right)(1 \text{ y})(1.50\times10^{-10} \text{ J}) = 7.0\times10^{-3} \text{ Gy} = 0.70 \text{ rad}$$

(b) For an RBE of unity, the equivalent dose is (1) (0.70 rad) = 0.70 rem.

44.57.   (a) For this model, $\dfrac{dR}{dt} = HR$, so $\dfrac{dR/dt}{R} = \dfrac{HR}{R} = H$, presumed to be the same for all points on the surface.

(b) For constant θ, $\dfrac{dr}{dt} = \dfrac{dR}{dt}\theta = HR\theta = Hr$.

(c) See part (a), $H_0 = \dfrac{dR/dt}{R}$.

(d) The equation $\dfrac{dR}{dt} = H_0 R$ is a differential equation, the solution to which, for constant H_0, is $R(t) = R_0 e^{H_0 t}$, where R_0 is the value of R at $t = 0$. This equation may be solved by separation of variables, as $\dfrac{dR/dt}{R} = \dfrac{d}{dt}\ln(R) = H_0$ and integrating both sides with respect to time.

(e) A constant H_0 would mean a constant critical density, which is inconsistent with uniform expansion.